Victory and Occupation

HISTORY OF U. S. MARINE CORPS

OPERATIONS IN WORLD WAR II

VOLUME V

by

BENIS M. FRANK

HENRY I. SHAW, JR.

Historical Branch, G-3 Division, Headquarters, U. S. Marine Corps

1968

Part 1

I

Other Volumes in this Series

I

LtCol Frank O. Hough, Maj Verle E. Ludwig, and Henry I. Shaw, Jr.

Pearl Harbor to Guadalcanal

II

Henry I. Shaw, Jr. and Maj Douglas T. Kane

Isolation of Rabaul

III

Henry I. Shaw, Jr., Bernard C. Nalty, and Edwin T. Turnbladh

Central Pacific Drive

Library of Congress Catalog Card No. 58–60002

PCN 19000262800

For sale by the Superintendent of Documents, U.S. Government Printing Office
Washington, D.C. 20402 - Price $11.75

Foreword

This book represents the final work in the five-volume history of Marine Corps operations in World War II. The story of the Okinawa campaign, told earlier in a separate monograph, has been reevaluated and rewritten to detail events in proper proportion to each other and in a correct perspective to the war as a whole. New material, particularly from Japanese sources and from the recorded interviews conducted with senior Marine Corps officers who participated in the Marine Corps Oral History Program, has been included to provide fresh insight into the Marine Corps' contribution to the final victory of the Pacific War.

These pages cover Marine Corps activities in the Okinawa invasion and the occupations of Japan and North China as well as the little-known story of Marine prisoners of war. The book relates the Corps' postwar demobilization and reorganization programs as well. By 1945, amphibious warfare doctrine and techniques had become highly developed. While new and improved weapons were employed in the Okinawa campaign, the landing operation itself realistically demonstrated the soundness of fundamental amphibious doctrine developed over the years by the Navy and the Marine Corps. Again, as at Guadalcanal, the battle for Okinawa clearly reemphasized the fact that basic Marine Corps tactics and techniques were sound. An outgrowth of the lessons learned at Okinawa was the establishment of a balanced air-ground amphibious force in readiness which has become the hallmark of the present-day Marine Corps. Many of the senior officers and commanders at Okinawa were prewar teachers and planners who had participated in the early operations of the war in the Pacific. The successful application at Okinawa of the knowledge, expertise, and experiences of these individuals against a fanatic foe fighting a last-ditch battle to protect his homeland was a vital factor in the final victory over Japan.

The assault and capture of Okinawa represents the most ambitious joint Army-Navy-Marine Corps operation in the history of the Pacific War. Statistically, in comparison to previous assaults in this war zone, the numbers of men, ships, and planes as well as the tons of munitions and supplies employed in this campaign stagger the imagination. But, had the enemy not capitulated in face of the American victories in the western Pacific and as a result of the atom bomb attacks on Hiroshima and Nagasaki, the personnel and logistics figures reflecting the requirements for the planned assault on Japan would have been even more overwhelming. Fortunately for both sides, the war ended before more blood was shed.

After participating in several Central Pacific landings, I returned to the United States and was assigned to Headquarters Marine Corps. From this vantage point, I observed the conduct of Marine Corps operations in the late stages of the war, when ground, sea, and air forces drove relentlessly towards the heart of the Japanese Empire. I also viewed with great pride the outstanding performance of duty of Marine occupation troops in Japan and North China. Here, small units and individual Marines proved themselves and the validity of Marine Corps training and discipline under conditions that were often trying. The fund of command experience acquired by junior officers and noncommissioned officers in a variety of circumstances has since been drawn on constantly in peace and war.

Similarly, the discipline and training of Marines captured at the outbreak of the war and after was tried and found not wanting in face of trials that beggar the imagination. In their own way, against the ever-present threat of death, these men continued fighting the enemy by various means, including sabotage and escape. The heroism of such Marines equalled and at times surpassed the records of the men who were engaged in the march across the Pacific. The record of our Marine POWs in World War II is something we can all be proud of.

Like other active duty Marines at the end of the war, I, too, experienced the period of transition when the Corps reverted to a peacetime role in the defense of this nation. Responsive to its combat experiences in World War II, the Marine Corps made many tactical and organizational changes, as this book shows. Unchanged, however, was our highly prized esprit de corps, which, even as this is written, is being as jealously guarded as when our Corps was first formed.

When the roll of America's battle honors is read, the names of the World War II campaigns in which Marines fought—Wake Island to Okinawa—will strike a familiar ring to all who cherish liberty and freedom. I am proud of my association with the men who won these honors and to have shared their hardships and their victories.

Wallace M. Greene, Jr.

WALLACE M. GREENE, JR.
GENERAL, U.S. MARINE CORPS
COMMANDANT OF THE MARINE CORPS

Reviewed and approved
29 November 1967

Preface

After the amphibious assault of Guadalcanal, which marked the opening of the American offensive in the Pacific, the steadily accelerating tempo of successful operations against the enemy inexorably led to an Allied victory in the war. Highlighting American operations was the Navy-Marine Corps team's extensive reliance on the employment of amphibious warfare techniques developed in the years before the war and improved upon under combat conditions.

The Okinawa landing has been accurately depicted as representing the culmination of amphibious development in the Pacific War and as the most audacious and complex military effort undertaken by amphibious forces of the Pacific Fleet. This operation also marked the last major ground action of the war against Japan, and the touchstone to the decisive Allied victory here was the massive interservice effort which, as much as anything else, hastened enemy capitulation.

Victory at Okinawa and the subsequent end of the war did not signal any letdown in the number and types of missions facing the Marine Corps, for at the same time that the postwar demobilization program drastically reduced their strength, Fleet Marine Force units were assigned to occupation duty in Japan and North China and to re-establishing the Pacific garrisons. This book treats these and such other hitherto-unpublished matters as the tragic story of those Marines who became prisoners of war. Appearing here also for the first time is a full treatment of the development and organization of the Marine infantry division and the many changes it experienced during the course of the war. In addition, this book presents an overview of the salient facts concerning Marine Corps campaigns in the Pacific War first discussed in the previously published volumes of this series.

Our purpose in publishing this operational history in durable form is to make the Marine Corps record permanently available for study by military personnel and the general public as well as by serious students of military history. We have made a conscious effort to be objective in our treatment of the actions of Marines and of the men of other services who fought at their side. We have tried to write with understanding about our former enemies and in this effort have received invaluable help from the Japanese themselves. Few people so militant and unyielding in war have, in peace, been as dispassionate and analytical about their actions. We owe a special debt of gratitude to Mr. Susumu Nishiura, Chief of the War History Office, Defense Agency of Japan, and to the many researchers and historians of his office that reviewed our draft manuscripts.

This five-volume series was planned and outlined by Mr. Henry I. Shaw, Jr., Chief Historian, while Mr. Benis M. Frank was responsible for Volume V itself. Mr. Shaw wrote the story of Marines in North China and his earlier research and writing provided the basis for that part of the book concerning Marines in the occupation of Japan. Mr. Frank wrote the rest of this book, revising and editing it for publication. In his research on the Okinawa operation, Mr. Frank frequently consulted the material assembled for the monograph *Okinawa: Victory in the Pacific* by Major Charles S. Nichols, Jr., and Mr. Shaw. Mr. Frank also prepared all the appendices. Successive Heads of the Historical Branch—Major John H. Johnstone, Colonel Thomas G. Roe, Colonel Joseph F. Wagner, Jr., Lieutenant Colonel Richard J. Schening, and Colonel Frank C. Caldwell—made the final critical review of portions of the manuscript. The book was completed under the direction of Colonel Caldwell, current Head of the Branch.

A number of leading participants in the actions described have commented on the preliminary drafts of pertinent portions of the book. Their valuable assistance is gratefully acknowledged. Several senior officers, in particular General Alexander A. Vandegrift, General Lemuel C. Shepherd, Jr., General Gerald C. Thomas, Lieutenant General Keller E. Rockey, Lieutenant General Louis E. Woods, Lieutenant General Pedro A. del Valle, Lieutenant General Francis P. Mulcahy, Major General DeWitt Peck, Major General William A. Worton, Major General Ford O. Rogers, Major General Wilburt S. Brown, and Rear Admiral Charles J. Moore made valuable additions to their written comments during personal interviews. A number of these interviews were conducted by Mr. Frank in his capacity as Head of the Oral History Unit, Historical Branch, which administers the Marine Corps Oral History Program.

Special thanks are due to the historical agencies of the other services for their critical readings of draft chapters of this book. Outstanding among the many official historians who measurably assisted the authors were: the late Dr. John Miller, Jr., Deputy Chief Historian, and Dr. Stetson Conn, Chief Historian, Office of the Chief of Military History, Department of the Army; Dr. Dean C. Allard, Head, Operational Archives Branch, Naval History Division, Department of the Navy; and Dr. Robert F. Futrell, Historian, Historical Studies Branch, U. S. Air Force Historical Division, Aerospace Studies Institute, Maxwell Air Force Base.

Chief Warrant Officer Jo E. Kennedy, and his predecessors as Historical Branch Administrative Officer, Second Lieutentant Gerald S. Duncan and First Lieutenants John J. Hainsworth and D'Arcy E. Grisier, ably handled the many exacting duties involved in processing the volume from first drafts through final printed form. A number of the early preliminary typescripts were prepared by Mrs. Miriam R. Smallwood, Mrs. Joyce E. Bonnett, and Miss Alexandria Jozwick, while the remainder were done by Miss Kay P. Sue, who expertly handled the painstaking task of typing the final manuscript for the printer. Miss Sue also did much of the meticulous work demanded in preparing the index.

The maps were drafted by Sergeant Thomas L. Russell. Unless otherwise noted, official Department of Defense photographs have been used throughout the text.

H. NICKERSON, JR.
MAJOR GENERAL, U.S. MARINE CORPS
ASSISTANT CHIEF OF STAFF, G–3

Contents

PART V NORTH CHINA MARINES

PART VI CONCLUSION

APPENDICES

ILLUSTRATIONS

ILLUSTRATIONS—Continued

ILLUSTRATIONS—Continued PAGE

MAPS

MAPS—Continued PAGE

PART I

Prologue to the End

Strategic Background

In a report submitted to Secretary of the Navy James V. Forrestal on 12 March 1945, the Commander in Chief, United States Fleet (CominCh), Fleet Admiral Ernest J. King, stated that:

> The amphibious operations of the spring, summer and autumn of 1944 carried our forces such great distances across the Pacific that in February 1945 they were enabled to begin the assault upon the inner defenses of the Japanese Empire itself.[1]

Recognizing all that had been accomplished to the date of his report, Admiral King at the same time cautioned against complacency and warned of "a long, tough and laborious road ahead."[2]

Among the many factors leading to the favorable Allied posture in the Pacific at the beginning of 1945 was the strategic concept for the prosecution of the Pacific War adopted at the Cairo Conference (SEXTANT) in December 1943. In essence, President Roosevelt and Prime Minister Churchill agreed upon a grand plan that dictated the thrust of two concurrent and mutually supporting series of operations across the Pacific towards the heart of the Japanese Empire. These drives along separate approach axes would establish bases from which a massive effort could be launched against the Formosa-Luzon-China coastal areas in the spring of 1945.

One drive, to be mounted by Allied forces under General Douglas MacArthur, Commander in Chief, Southwest Pacific Area (CinCSWPA),[3] was to move along the northern coast of

[1] *The War Reports of General of the Army George C. Marshall, General of the Army H. H. Arnold, and Fleet Admiral Ernest J. King* (Philadelphia: J. B. Lippincott Company, 1947), p. 613, hereafter *War Reports.*

[2] *Ibid.*, p. 649.

[3] On 3 March 1942, the Combined Chiefs of Staff approved for the Western Pacific a dividing line that separated assigned spheres of command in that area. Burma and all Southeast Asia west of a north-south line between Java and Sumatra were added to General Sir Archibald V. Wavell's India Command responsibility, and the British Chiefs of Staff were charged with the strategic direction of this theater. The whole Pacific east of the new line was assigned to American Joint Chiefs of Staff control. The JCS then divided the Pacific into two strategic regions; the one in which the Navy would have paramount interests was the Pacific Ocean Areas, and the other in which the Army would be dominant was the Southwest Pacific Area. On 18 March 1942, MacArthur was designated CinCSWPA; on 3 April, Admiral Chester W. Nimitz, Commander in Chief, Pacific Fleet, was confirmed as the commander of the Pacific Ocean Areas. See LtCol Frank O. Hough, Maj Verle E. Ludwig, and Henry I. Shaw, Jr., *Pearl Harbor to Guadalcanal—History of U. S. Marine Corps Operations in World War II*, v. 1 (Washington: HistBr, G-3 Div, HQMC, 1958) pp. 86–87, hereafter Hough, Ludwig, and Shaw, *Pearl Harbor to Guadalcanal*, and Map I, Map Section of that volume, for additional information concerning these two American commands.

310-224 O - 69 - 2

New Guinea and thence to the Philippines; in the second, forces of Admiral Chester W. Nimitz, Commander in Chief, Pacific Ocean Area (CinCPOA), would push through the Central Pacific to the core of Japanese island defenses guarding the heart of the Empire. During this two-pronged advance, the major components of the Pacific Fleet, under Nimitz as Commander in Chief, Pacific Fleet (CinCPac), would support, as assigned, specific amphibious operations within both strategic command areas, and at the same time contain the Japanese fleet.

Almost immediately after the two heads of state had approved at SEXTANT the revised plan for the defeat of Japan, the Joint Chiefs of Staff (JCS), forwarded the directive to MacArthur and Nimitz, whose staffs and commands took steps to implement it. January 1944 opened with a landing at Saidor in New Guinea. At the end of the month, Central Pacific forces landed in the Marshalls and spent February thrusting deeply into the island group to collapse those outposts of the imperial defenses.[4] By the end of March, the Bismarck Archipelago barrier had been permanently breached and airfields and harbors seized in the Admiralties. MacArthur's forces began the drive up the New Guinea coast in April, with landings at Aitape and Hollandia. With the naval attack on and immobilization of Truk, the capture of Guam, Saipan, and Tinian in the Marianas during the summer of 1944, and the defeat of the Japanese fleet in the First Battle of the Philippine Sea, the Central Pacific drive cut the inner ring of Japanese island defenses in several places and consolidated footholds from which the drive westward was to continue. After the amphibious assaults on Peleliu and Angaur and the unopposed capture of Ulithi for use as a fleet anchorage and an advance base, Admiral Nimitz' forces stood poised on the threshold of the Japanese defenses ringing the Home Islands.[5]

By the end of July, Admiral William F. Halsey's South Pacific troops had advanced up the Solomons, and MacArthur's forces along hundreds of miles of the northern coast of New Guinea, in a series of leapfrogging operations. Thousands of Japanese soldiers on Bougainville, New Britain, New Ireland, and New Guinea itself were neutralized and isolated, and beyond hope of being effectively employed elsewhere. In September, MacArthur's forces occupied Morotai, southeast of the Philippines, before the planned landing on Mindanao.

In the course of naval covering strikes prior to the landings on Morotai and in the Western Carolines, Admiral Halsey's

---

[4] For the story of the Marshalls landing and the Central Pacific drive, see Henry I. Shaw, Jr., Bernard C. Nalty, and Edwin T. Turnbladh, *The Central Pacific Drive—History of U. S. Marine Corps Operations in World War II*, v. III (Washington: HistBr, G-3 Div, HQMC, 1966), hereafter Shaw, Nalty, and Turnbladh, *The Central Pacific Drive.*

[5] For the Peleliu operation and the story of Marine aviation in the Central Pacific, see George W. Garand and Truman R. Strobridge, "Operations in the Western Pacific—History of U. S. Marine Corps Operations in World War II," v. IV, to be published in 1969, hereafter Garand and Strobridge, "Western Pacific Operations."

fast carrier forces had discovered surprisingly weak enemy resistance in the central Philippines. In a follow-up to this discovery, the line of advance through the Southwest Pacific was reoriented northwards. Fully aware of "the necessity of being alert for symptoms of enemy weakness and of being ready to exploit them," Halsey recommended an early return of American troops to the Philippines in the Leyte-Samar area and cancellation of certain operations scheduled elsewhere.[6] His recommendation was approved by the JCS.

In the overall planning for the defeat of Japan, the strategists anticipated that the final phase of the Pacific War would involve a massive assault against the industrial heartland of the Empire by means of amphibious landings on the southern coast of Honshu in the area bounded by Shimonoseki in the south and the Kanto Plain near Tokyo in the north. Successful Allied operations in 1944 had brought ultimate victory into sight, and submarine blockade and air bombardment both had the Japanese viewing ultimate defeat, but some American commanders doubted the wisdom of using the Formosa-Luzon-China area as a springboard from which to launch the attack against Japan in 1945. More importantly, they believed that valuable time was being wasted and that a decision had to be made. In view of the SEXTANT Plan, and the advanced state of the operations against Japan, JCS planners were confronted with the problem of whether American forces should: (1) move on to Luzon and the rest of the Philippines, (2) invade only Luzon in the Philippines and also strike at Formosa and the China coast, or (3) attack the Philippines, Formosa, and the China coast. Arising out of the third option was an additional thorny problem—which area to attack first.

While Admiral King and some planners in Washington considered the possibility of entirely bypassing the Philippines, this concept was apparently only a minor aspect of the major effort by many officers to have Luzon, in particular, bypassed. The alternative to this was the seizure of Formosa. On the other hand, ample evidence exists to indicate that those who sought the Formosa objective did not intend this to be an exclusive operation, for they believed that the invasion of Luzon could proceed simultaneously with the Formosa operation or take place at a later date.[7]

Determined to return to the Philippines, MacArthur doubted the necessity of the Marianas campaign but generally approved the Palaus landings since they would directly support his impending operations. Admiral King took just the opposite view; he concluded that the occupation of the Marianas was essential and that the necessity of recapturing all of the Philippine Islands was questionable. Furthermore, he was firmly convinced that the main American effort should be bent in mounting a drive across the Central Pacific to Formosa

[6] FAdm William F. Halsey, USN, and LCdr J. Bryan, III, USNR, *Admiral Halsey's Story* (New York: Whittlesey House, McGraw-Hill Book Company, Inc., 1947), p. 199, hereafter Halsey and Bryan, *Halsey's Story*.

[7] Dir, Naval Hist, ltr to Hd, HistBr, G–3 Div, HQMC, dtd 4Nov65, hereafter *Dir, Naval Hist ltr I*.

and then on to the China coast.[8] Although various subordinate commanders in the Pacific Ocean Area held conflicting views regarding what course should be taken for the final phases of the war, the SEXTANT decision made it imperative that their staffs spend most of 1944 in planning for Operation CAUSEWAY, the invasion of Formosa, projected for the spring of 1945.

On the basis of a JCS directive issued on 12 March 1944,[9] the prevailing conflict was partially allayed. Admiral Nimitz was directed to land on 15 June in the Southern Marianas and on 15 September in the Palaus. General MacArthur was instructed to seize Hollandia in April and make plans for a landing on 15 November on Mindanao. Contained in the JCS order was a statement of long-range objectives that required Nimitz as CinCPOA to prepare the plans for an assault early in 1945 on Formosa, and assigned CinCSWPA the responsibility of planning for the recapture of Luzon "should such operations prove necessary prior to the move on Formosa." [10]

In view of the March JCS directive, which outlined the general concept of CAUSEWAY, Nimitz reconsidered and revised the troop list for the operation many times, and finally designated the task force commanders. Admiral Raymond A. Spruance, Commander, Fifth Fleet and Central Pacific Task Forces was to be in overall charge. Vice Admiral Richmond K. Turner was to command the expeditionary forces, and Lieutenant General Simon B. Buckner, Jr., USA, was to command the expeditionary troops and the Tenth Army.[11]

Further discussion regarding what the nature of Pacific strategy was to be following the Marianas operation continued after the JCS had directed the preparation of plans for CAUSEWAY. This topic was the subject of one of the periodic conferences which Admirals King and Nimitz and their key deputies held throughout the war, either at Pearl Harbor or San Francisco. At one such meeting on 6 May 1944, Vice Admiral Charles M. Cooke, Jr., King's chief of staff, pointed out that, although the JCS directive envisioned a landing on Formosa in February 1945, the best time for this operation—in view of other considerations—would probably be November-December 1944.[12] Cooke also noted that once Japan had been cut off from the mainland, her islands could be bombed and perhaps Kyushu even invaded.

During 1944, the Joint War Plans Committee (JWPC) had also considered

[8] FAdm Ernest H. King and Cdr Walter M. Whitehill, *Fleet Admiral King: A Naval Record* (New York: W. W. Norton & Co., Inc., 1952), p. 537, hereafter King and Whitehill, *King's Naval Record*.

[9] JCS 713/4, dtd 12Mar44, cited in Maj Charles S. Nichols, Jr., and Henry I. Shaw, Jr., *Okinawa: Victory in the Pacific* (Washington: HistBr, G-3 Div, HQMC, 1955), p. 12, hereafter Nichols and Shaw, *Okinawa Victory*.

[10] *Ibid.*

[11] CinCPOA msg of 10Aug44 to addees, cited in History of USAFMidPac and Predecessor Commands During World War II, 7Dec41–2Sep45, Hist of G-5 Sec, n.d., p. 174, (OCMH), hereafter *USAFMidPac G-5 Hist*.

[12] Minutes, 2d meeting, CominCh–CinCPac Pacific Conference, 6May44, p. 14 (OAB, NHD), hereafter *Minutes CominCh–CinCPac Conference* with date.

what the nature of future Pacific strategy should be. In early June, it issued a comprehensive study which far exceeded in scope and perspective the previous strategic positions taken by the Joint Chiefs and the Combined Chiefs of Staff (CCS), and outlined a series of campaigns that would lead to an assault on the Tokyo Plain by the end of 1945.[13] In this study, the JWPC pointed out that in view of the present and anticipated rate of advance of increasingly stronger American forces in the Pacific, it would appear that the "Inner Zone Defense of Japan" would be reached by spring 1945. The study concluded that the overall strategy approved at SEXTANT was inadequate, *i.e.*, future operations as planned extended only to the perimeter of the Formosa-Luzon line.

Instead, the JWPC recommended a new schedule or strategic concept for ending the war in the Pacific. The committee suggested that three phases precede the invasion of Japan: (1) During the period 1 April to 30 June 1945, American forces would seize positions in the Bonins and the Ryukyus from which they would launch an invasion against the central China coast in the Hangchow Bay area; (2) They would spend the time from 30 June to 30 September in consolidating and initially exploiting the China beachhead; and (3) The forces would land in Southern Kyushu 1 October and on the Tokyo Plain on Honshu on 31 December. This planning paper was passed to the Joint Staff Planners, who approved and forwarded it to the Joint Chiefs of Staff, who likewise favored the revised concept. On 11 July, the CCS received the study along with a JCS recommendation that the SEXTANT timetable for operations in the Pacific be changed to reflect the suggested JWPC schedule.

At one of the meetings held during the CominCh–CinCPac conference in the period 13–22 July 1944, Admiral King informed the conferees of the JCS action regarding the JWPC study. He also indicated that he believed Luzon could not be invaded before Formosa or Japan without the Americans first investigating what Saipan and Guam could offer in the way of fleet anchorages and base facilities for the support of the Luzon invasion forces.[14] Vice Admiral John H. Tower, Commander, Air Forces, Pacific Fleet, stated that neither the areas in American possession at that time or prospectively available would permit the establishment of naval and supply bases which would be adequate for the support of the future operations contemplated in the JWPC study.[15]

Along these lines, it was suggested that the feasibility and advisability of

[13] JPS 476, dtd 4Jun44, Subj: Operations Against Japan, Subsequent to Formosa, cited in Ray S. Cline, *Washington Command Post: The Operations Division—United States Army in World War II* (Washington: OCMH, DA, 1951), p. 337, hereafter Cline, *Washington Command Post.*

[14] *Minutes, CominCh–CinCPac Conference*, 13–22Jul44, p. 10. It should be noted that Saipan, invaded on 15 June, was not secured until 9 July, when the general mop-up began. The invasion of Guam, which had been delayed until 21 July, had caused a backup in the supply pipeline and it was contemplated that this situation could adversely affect subsequent operations.

[15] *Ibid.*, p. 13.

taking San Pedro Bay in Leyte Gulf as a fleet anchorage be investigated. Although considerable discussion of this recommendation resulted, no firm decision was made at this time.

Regarding the invasion of the Bonins, Admiral Tower stated that, because the United States plans for the establishment of VLR (very long range) bomber bases in the Marianas were close to being realized, steps to enhance their effectiveness should be taken at the earliest practicable date. This meant the seizure and development of positions in the Bonins, where fighter and bomber aircraft stationed on fields developed there could supplement and support the planned air raids on Japan. On the other hand, Admiral Tower added that a study of the prospective employment of fleet and assault forces did not indicate the Bonins could be taken until 1945, unless the timing of then currently planned operations could be drastically revised. Because these were of greater importance in the overall scheme for the defeat of Japan, the occupations of Iwo Jima and Chichi Jima would have to be deferred.[16]

Most of the others present at this meeting generally agreed with Admiral Tower's conclusions. Rear Admiral Forrest P. Sherman, Nimitz' chief of staff and head of his War Plans Division, pointed out that for the invasions of Leyte and Formosa—the two major operations of a decisive nature scheduled following the completion of the landings in the Palaus—American forces had been tailored down considerably. Sherman emphasized that if more ships and troops became available, they should be employed to supplement those already assigned to the landings on Leyte and Formosa. In no case, should they be diverted for such "minor operations" as the occupation of Iwo Jima and Chichi Jima.[17]

Admiral King agreed with this line of thinking, and added that it would be unwise to assault the Bonins until American forces were ready to invade Japan following the Formosa operation. Nonetheless, at this time, he directed CinCPac to prepare plans for the invasion of the Bonins.

As for determining those objectives that were to follow the capture of the Southern Marianas and the Pal us and were to be mounted before the ıvasion of Japan, even President R osevelt's visit to Pearl Harbor late in July to confer with MacArthur and Nimitz failed to resolve the impasse. When the conference began, Admiral Nimitz, the first to speak, presented the Navy position.

Contrary to general belief, no real controversy arose between Nimitz and MacArthur regarding the conduct of future operations against Japan. Nimitz made this quite clear in a letter to Admiral King, summarizing in a few words the discussions at the Pearl Harbor meetings. Nimitz told CominCh that:

> . . . our conferences with the Commander in Chief of the Army and Navy [President Roosevelt] and the Commander in Chief, Southwest Pacific Area [MacArthur] were quite satisfactory. The general trend of the discussion, like our own, was along the line of seeing

[16] *Ibid.*, p. 14.

[17] *Ibid.*

> MacArthur into the central Philippines, thereafter going direct to the Formosa Strait, and leaving the SWPA forces to work into Luzon under the cover of the Formosa operation. It was made clear that the time has not yet arrived for firm decisions on moves subsequent to Leyte.[18]

It is possible, however, that Nimitz and some of his staff had some doubts on the feasibility of the Formosa operation and the concept underlying the Formosa-first policy. Partial evidence for this is found in the fact that CinCPac staff members had prepared plans to seize Okinawa as a substitute for Formosa "well before such an operation gained serious consideration among high-level planners in Washington." [19]

After listening to the views of both MacArthur and Nimitz, the President returned to Washington without rendering a decision on the courses of action to be followed after the landings on Leyte. Nor does it appear that a firm decision for post-Leyte operations was expected. Although Nimitz may have entertained other opinions concerning future strategy, he was still operating under a JCS directive relative to the Formosa operation. On 23 August 1944, the CinCPac joint staff study of CAUSEWAY was published. In this document Admiral Nimitz indicated that he intended to invade Formosa after SWPA forces had established positions in the south and central Philippines. Following the successful operations on Formosa, the Ryukyus and the Bonins or the China coast were to be invaded as a prelude to the assault on Japan itself. A Luzon operation, as such, was not mentioned in this plan.

The dispute remained unresolved until 9 September, when, at the Quebec Conference (OCTAGON), the Combined Chiefs of Staff formally adopted and incorporated the JWPC concept within the SEXTANT schedule for the defeat of Japan, and in effect revised it. For planning purposes, the CCS then approved a new schedule of operations, which ended the campaigns of 1945 with a landing on Kyushu in October and on the Tokyo Plain in December.[20]

The Combined Chiefs also agreed that, if the Formosa operation materialized, it would be preceded by invasions of the Bonins in April, the Ryukyus in May, and the China coast in the period March to June 1945. On 15 September 1944, the JCS further clarified impending Pacific operations by cancelling the scheduled invasions of Mindanao and Yap and

[18] Adm Chester W. Nimitz ltr to Adm Ernest J. King, dtd 31Jul44 (OAB, NHD). This letter also indicates that King favored the establishment of positions in the southern or central Philippines.

[19] Robert R. Smith, *Triumph in the Philippines—United States Army in World War II—The War in the Pacific* (Washington: OCMH, DA, 1963), p. 9. For a thorough discussion of the Luzon-Formosa controversy, see *Ibid.*, chap 1, "The Debate Over Luzon," and Samuel Eliot Morison, *The Liberation of the Philippines—Luzon, Mindanao, the Visayas: 1944–1945—History of United States Naval Operations in World War II*, v. XIII (Boston: Little, Brown and Company, 1959), chap 1, "Planning for Luzon, October–December 1944."

[20] CCS 417/8, dtd 9Sep44, title: Opn for the Defeat of Japan; CCS 417/9, OCTAGON, dtd 11Sep44, title: Over-All Objective in War Against Japan; Min 173d Meeting CCS, 13Sep44, all cited in Cline, *Washington Command Post*, p. 339.

ADMIRAL NIMITZ *briefs General MacArthur, President Roosevelt, and Admiral Leahy at the July 1944 Pearl Harbor conference. (USA SC207297)*

COMBINED CHIEFS OF STAFF *meet with President Roosevelt and Prime Minister Churchill at the OCTAGON Conference in Quebec, September 1944. (USA SC194469)*

setting 20 October as the date for the invasion of Leyte.[21]

On 11 September, Lieutenant General Millard F. Harmon, Commanding General, Army Air Forces (AAF), Pacific Ocean Areas, proposed the abandonment of the Formosa operation in favor of amphibious landings in the Bonins and the Ryukyus. Harmon recommended the capture of Iwo Jima by 1 January 1945 and Luzon by 1 June 1945. Further, he suggested that POA troops seize Okinawa and Amami O Shima after MacArthur's forces recaptured Luzon; Kyushu was to be invaded in September 1945. Harmon also stated that he believed that the seizure and use of Luzon was an important consideration in the overall strategy of the Pacific War and that the launching of a major operation against Formosa would dilute some of the force being applied against the Japanese in other action areas.[22] In order to husband resources and to accelerate the march toward Japan, Harmon believed that the capture of Luzon for its airfields was imperative. Air operations launched from Luzon could neutralize Formosa and effectively cut Japanese communications to South China and Malaya.[23]

Less than a week later, after a review of the plans contemplated for the CAUSEWAY operation, Admiral Nimitz set forth his thoughts in a letter circulated to his senior commanders. In a key section of this letter, CinCPac recommended to consider the possibility:

> . . . of a re-orientation of a strategy in the Pacific which will provide for an advance northward with eventual assaults on the Empire itself, rather than intermediate action along the China Coast, thus indicating the probability of occupation of Iwo Jima and Okinawa with target dates as early as practicable after CAUSEWAY.[24]

He also directed Admirals Spruance and Turner and General Buckner to recommend suitable physical targets in the Formosa-Amoy-Pescadores areas for Operation CAUSEWAY. Criteria for the selections were the number of naval and air bases that would have to be established and the type and total of major troop units required. On 26 September, General Buckner submitted what he considered to be the primary objection to the entire projected operation; he said that the shortage of available supporting and service troops in the

[21] The invasion of Mindanao was restored to the plans for the recapture of the Philippines, and MacArthur's X Corps landed on the beaches of Illana Bay on 17 April 1945.

[22] General George C. Marshall, Chief of Staff of the Army, had favored the Formosa-first plan, and like Admiral King "had expressed the opinion that Japan itself, rather than Luzon, should be considered the substitute for Formosa." Smith, *op. cit.*, p. 9. In September, General Marshall revised his opinion and believed, that, in view of existing facts, the choice for the next operation would have to be Luzon. It seemed more logical to him to secure Luzon—which MacArthur promised to take in six weeks—than to concentrate on Formosa, which would take longer to capture. Marshall reasoned that if all of the resources that were to be poured into Formosa were diverted to Luzon, Admiral Nimitz could get ready to attack the Bonins and Ryukyus all the sooner, and the timetable for the invasion of Japan could be advanced.

[23] CGAAFPOA ltr to CinCPOA, dtd 11Sep44 (no file or serial number), cited in Nichols and Shaw, *Okinawa Victory*, p. 15.

[24] CinCPOA ltr to ComFifthFlt, ComGenTen, and ComPhibsPac, Ser 000113, dtd 16Sep44 (OAB, NHD).

POA for CAUSEWAY made it infeasible.[25] In an afterthought, on 4 October Buckner wrote Nimitz that the need for invading Formosa would be diminished greatly if plans for the invasion of Luzon came to fruition.[26]

The minutes of the CominCh-CinCPac conference held in San Francisco from 29 September to 1 October 1944, indicate that by this time, Admiral King had given serious thought to bypassing Formosa. He told Nimitz that, at his proposal, the Joint Logistics Committee (JLC) had made a survey of the resources available for the Formosa operation, and that the report of this committee was very discouraging. At the time of its survey, the JLC found that resources were not available for CAUSEWAY, and would not be available unless Germany capitulated a long time before it was expected to do so.

CinCPac then told King of General Buckner's requirements for additional men and equipment and that he, Nimitz, was in no position to dispute these figures. Nimitz then submitted a memorandum recommending changes for future Pacific operations based on the non-availability of necessary resources and the favorable results of recent carrier operations. Admiral Nimitz recommended that CinCPac forces support the SWPA invasion of Luzon with a target date of 20 December 1944, and the invasions of Iwo Jima on 20 January 1945 and of Okinawa on 1 March by POA Forces.[27]

CinCPac stated that the proposal for the SWPA forces to work up through the Philippines from Leyte by shore-to-shore operations had been discussed with President Roosevelt and General MacArthur at the Pearl Harbor conference in July. Because MacArthur had stated that he could not undertake these operations and in view of the insufficient resources for Formosa, Nimitz believed that the best way to keep pressure on the Japanese was for him to support the Lingayen Gulf operation proposed by MacArthur and to take the Bonins and the Ryukyus with POA forces.[28]

Admiral Sherman then told King that Nimitz expected to take Iwo Jima with two divisions and then to send in large numbers of construction personnel to build up the airfields rapidly. Following that, assuming that enemy air power on Formosa had been neutralized by carrier strikes assisted by shore-based air from Luzon, it was expected that Okinawa would be invaded on 1 March. King asked Nimitz why he was going to seize the Bonins if Okinawa was to be taken.

---

[25] CG, Tenth Army ltr to CinCPOA, dtd 26Sep44, Subj: Feasibility of CAUSEWAY, cited in *USAFMidPac G-5 Hist*, p. 177. Concerning the Marine troop requirements for Formosa, Lieutenant General Alexander A. Vandegrift, Commandant of the Marine Corps, informed Admiral King that many of the service forces General Buckner had said were needed to support the Marine component of the Tenth Army were, in fact, already organic to the Fleet Marine Force or else were neither suited nor required for Marine Corps amphibious operations. ComInCh-CNO Memo to JCS, dtd 4Sep44, Subj: Employment of Marine Divisions in "Formosa" Operations (OAB, NHD).

[26] CG, Tenth Army ltr to CinCPOA, dtd 4Oct44, Subj: CAUSEWAY Objectives, cited in *USAFMidPac G-5 Hist*, p. 179.

[27] *Minutes, ComInCh–CinCPac Conference*, 29Sep–1Oct44, p. 4.

[28] *Ibid.*, p. 5.

Nimitz replied that fighters based in the Bonins could give protection to the B-29s raiding Japan, and that the AAF wanted this added protection.

King returned to Washington and on 2 October proposed to the JCS a course of action for the Pacific. He stated that in view of the lack of necessary resources in the POA for CAUSEWAY, and because of the inability of the War Department to make up the deficit before the end of the war in Europe, he believed that operations should be mounted against Luzon, Iwo Jima, and the Ryukyus in succession. He also added that CAUSEWAY might be feasible at a later date if conditions in the Pacific and Europe warranted.[29] Concurring with King's proposal, the next day the JCS ordered MacArthur to invade Luzon on 20 December 1944 and Admiral Nimitz to land Marines on Iwo Jima on 20 January 1945. Following these operations, Operation ICEBERG was to be launched on 1 March 1945.[30] This date was flexible, however, since it would be affected by the: (1) Capture of Iwo Jima in time for the prompt release of fire support units and close air support squadrons required at Okinawa; (2) Prompt release of supporting naval forces and assault shipping from the Luzon operation; and (3) Attainment of undisputed control of the sea and air in the target area in preliminary strikes against the Ryukyus, Formosa, and Japan.[31]

With all attention and efforts now focused on the new objectives, the Formosan venture was reserved as a strategic goal for possible future reconsideration.[32] Although the basic command concept and troop list organization that had been set up for CAUSEWAY were retained for employment in ICEBERG,[33] there was much to be done between the time that the JCS ordered the capture of Okinawa and the actual date of the invasion.

[29] CominCh-CNO Memo to JCS dtd 2Oct44, Subj: Future Operations in the Pacific (OAB, NHD).

[30] JCS 713/19, dtd 3Oct44, cited in Nichols and Shaw, *Okinawa Victory*, p. 17.

[31] CinCPOA Joint Staff Study—ICEBERG, dtd 25Oct44, p. 1 (Okinawa AreaOp File, HistBr, HQMC), hereafter *ICEBERG Study*.

[32] JCS 713/19, dtd 3Oct44, cited in Nichols and Shaw, *op. cit.*, p. 17.

[33] Tenth Army AR, Ryukyus, 26Mar–30Jun45, dtd 3Sep45, chap 3, p. 3 (Okinawa Area Op File, HistBr, HQMC), hereafter *Tenth Army AR*.

The Japanese Situation[1]

As early as the spring of 1944, the high commands of the Japanese Army and Navy in the *Imperial General Headquarters* (IGHQ)[2] had, with some accuracy, predicted the trend of American strategy in Pacific War. The Japanese foresaw that the turning point of the conflict would begin developing in March or April in the Marianas. Further, the military chiefs were concerned with what Japanese strategy should be at this critical time.

Faced with an impending accelerated American drive in the Central Pacific, *IGHQ* issued an Army-Navy agreement for Japanese operations in that area. The Navy was given primary responsibility for denying the Allies bases from which further operations could be launched against other islands and finally Japan itself. By the spring of 1944, defenses in the Carolines, Marianas, and Volcano Islands were to be completed. Japanese Army units were to reinforce the island defenses and would operate under overall naval control in conducting ground operations.[3] A broader aspect of Japanese strategy was the decision to try to entrap and defeat decisively a major portion of U. S. naval forces. As island defenses were being strengthened, the Japanese Navy committed the bulk of its aerial strength—about 1,000 aircraft of which only 650 were operational—[4] to the Marianas and part of the remainder to the Carolines. Meanwhile, surface forces were to remain alert and ready to steam into combat when the time to strike arose.

Most *IGHQ* officers and government officials alike were supremely confident of winning the war and directed every

[1] Unless otherwise indicated, the material in this chapter is derived from: Foreign Histories Div, G–3, Hq, U. S. Army Japan, Japanese Monograph No. 45, rev. ed. 1959, History of Imperial General Headquarters, Army Section (OCMH), hereafter *IGHQ Hist;* Robert J. C. Butow, *Japan's Decision to Surrender* (Stanford: Stanford University Press, 1954), hereafter Butow, *Japan's Decision;* Takushiro Hattori, *Dai Toa Senso Zenshi* [The Complete History of The Greater East Asia War], MS trans, 4 vols (Tokyo: Matsu Publishing Co., 1955), v. IV (OCMH, DA), hereafter *War History;* Saburo Hayashi and Alvin Coox, *Kōgun* (Quantico: Marine Corps Association, 1959), hereafter Hayashi and Coox, *Kōgun;* Toshikasu Kase, *Journey to the Missouri*, David N. Rowe, ed. (New Haven: Yale University Press, 1950), hereafter Kase, *Journey to the Missouri;* USSBS, *Japan's Struggle to End the War* (Washington: July 1946), hereafter USSBS, *Japan's Struggle.*

[2] *IGHQ* was only a term used to denote the co-equal status and existence of the Tokyo-based headquarters of the *Army General Staff* and the *Navy General Staff* and their subordinate general and special staff sections. This duality of command typified the Japanese military system not only at the highest level, but in the lower echelons also.

[3] *IGHQ Hist*, p. 154.

[4] Hayashi and Coox, *Kōgun*, p. 106.

effort to ensure an ultimate Japanese victory. Not so certain that Japan was going to be the victor was an opposition group composed of former ministers, cabinet members, and elder statesmen *(Jushin)* who had opposed the war in the pre-Pearl Harbor period. Also in this group were some other influential Japanese leaders who, while not holding positions of power, had given mere lip service to their nation's involvement in a conflict. Rounding out the opposition were other formerly powerful men, who had "retired" in the early years of the war. The original doubts of the opposition gave it a basis for believing as early as the spring of 1944 that Japan was faced with inexorable defeat. These beliefs were buttressed by a demonstration of the American determination to fight aggressively and an ability to mount successful operations in the Pacific even before a second front had been opened in Europe. Alone, these two factors gave portents of disaster to those Japanese who were able to interpret them.[5]

Between September 1943 and February 1944, Rear Admiral Sokichi Takagi, chief of the Naval Ministry's research section, prepared a study of Japanese lessons learned in the fighting to that date. He maintained that it was impossible to continue the war and that it was manifestly impossible for Japan to win. He thus corroborated an estimate made by top Japanese naval officers before 1941. At that time, they concluded that unless the war was won before the end of 1943, Japan was doomed, for it did not have the resources to continue the war after that time.

Takagi's study and his conclusions were based on an analysis of fleet, air, and merchant shipping losses as of the last of 1943. He pointed out the serious difficulty Japan was facing in importing essential materials, high-level confusion regarding war aims and the direction of the war and the growing feeling among some political and military leaders that General Hideki Tojo, Prime Minister since 1941, should be removed from office.

Takagi stated also that both the possibility of American bombing raids on Japan and the inability of the Japanese to obtain essential raw and finished products dictated that the nation should seek a compromise peace immediately. In March he presented his findings orally to two influential naval officers, Admiral Mitsumasa Yonai, a former prime minister, and Vice Admiral Seibi Inouye, who employed the facts of the study to induce other members of the opposition to take firm steps to help change the course that Japan was travelling.[6]

Less than two months after the invasion of Normandy in June 1944, Japanese leaders began receiving reports of the massive numbers of men and amount of materiel that the Allies were able to land unopposed each day on the French coast. As a Japanese foreign ministry official later wrote:

> That was more than enough to dishearten us, the defenses of our home islands were far more vulnerable than the European invasion coast. Our amazement

[5] USSBS, *Japan's Struggle*, p. 2.

[6] *Ibid.*, p. 3.

> was boundless when we saw the American forces land on Saipan only ten days after D day in Europe. The Allies could execute simultaneous full-scale offensives in both European and Asiatic theaters![7]

By all accounts, Japanese and other, what really tipped the scales in favor of an eventual Allied victory in the Pacific, and more immediately caused the fall of the Tojo government, were the landings at Saipan and Japanese losses in the First Battle of the Philippine Sea. Only 1,350 miles from Tokyo, Saipan constituted one of the most vital points in the Japanese outer defense system. Toshikasu Kase, the foreign ministry official quoted above, wrote that the island:

> . . . was so strongly defended that it was considered impregnable. More than once I was told by the officers of the General Staff that Saipan was absolutely invincible. Our Supreme Command, however, made a strategic miscalculation. Anticipating an early attack on Palau Island, they transferred there the main fleet and the land-based air forces in order to deal a smashing blow to the hostile navy. The result was that Saipan, lacking both naval and air protection, proved surprisingly vulnerable.[8]

An even greater disaster befell the Japanese in the Battle of the Philippine Sea, 19–20 June 1944. This two-day conflict began when carrier-based aircraft of the Japanese *First Mobile Fleet* attacked Admiral Spruance's Fifth Fleet while it covered the Saipan operation. On the first day, two U. S. battleships, two carriers, and a heavy cruiser were damaged; the Japanese lost over 300 aircraft and two carriers. Pilots from Vice Admiral Marc A. Mitscher's fast carrier task force struck back violently the next day, sinking another enemy carrier and downing many Japanese planes. According to American estimates, their opponents suffered staggering losses in the two days: 426 carrier planes and 31 float planes. In addition, the Americans claimed that approximately 50 Guam-based aircraft had been destroyed.[9]

Japanese sources confirm the loss of carriers and state that four others of the nine committed in the fight were damaged. Enemy records show that of the 360 carrier-based aircraft sent to attack the American fleet, only 25 survived. "Although no battleships or cruisers were sunk, . . . the loss of aircraft carriers proved an almost fatal blow to the Japanese navy. With the loss of the decisive aerial and naval battles, the Marianas were lost." [10] Despite this thorough defeat, most Japanese were told that it was a glorious victory for them; "it was customary for GH [*IGHQ*] to make false announcements of victory in utter disregard of facts, and for the elated and complacent public to believe in them." [11]

Although the Japanese government did not announce its losses in the Battle of the Philippine Sea—or that it had even lost the battle—news of the fall of Saipan was made public. Upon learning this in July, an opposition group consisting mainly of *Jushin* determined to

[7] Kase, *Journey to the Missouri*, p. 90.

[8] *Ibid.*, p. 73.

[9] Samuel Eliot Morison, *New Guinea and the Marianas—History of United States Naval Operations in World War II*, v. VIII (Boston: Little, Brown and Company, 1960), pp. 319–321.

[10] *IGHQ Hist*, p. 172.

[11] Kase, *Journey to the Missouri*, p. 74.

overthrow the Tojo regime, and forced the Prime Minister to resign from office on 18 July 1944.

The problems facing Japan were hardly resolved with the appointment and installation of General Kuniaki Koiso as premier. The Japanese Army was still a political power, capable of dictating the rise, fall, and course of government, and Tojo and his followers remained uncontrite in their adherence to a chauvinistic program of Japanese conquest and supremacy. Although the Home Islands had not yet experienced the devastation and chaos to be brought by the vast Allied air raids, after the fall of Saipan a number of critical domestic problems affecting the war effort faced the Japanese government. The output of a number of essential items fell below peak requirements, and severe shipping losses reduced the amount of raw and finished material reaching Japanese shores to a point far short of needs.

On the home front, despite its unhappiness with Tojo and his handling of the war, the Japanese public was confident in ultimate victory. Those leaders in government opposed to the war, opposition leaders behind the scenes, and some of the war hawks, too, began to have greater misgivings as they learned of previous defeats and potential disasters. As this knowledge spread, the military factions slowly lost face and became discredited, but not until the last months of the war did they lose power.

Nonetheless, confident of their ability to guide Japan to what they considered would be a just victory, the military leaders made adjustment after adjustment in strategy and troop dispositions in one area after another as the Allied threat to the Home Islands intensified and accelerated. On the other hand, it is possible to understand their reluctance to view the situation realistically. From their earliest days, Japanese citizens were taught to believe that the one alternative to victory was death and that surrender was so disgraceful as to be unthinkable. And the high command planned, therefore, to continue the war, even on Japanese soil if necessary, but to fight to the finish in any case.

Even lower ranking Japanese Army and Navy officers, many of them products of a prewar conscript system, who very often came from peasant families, held the same beliefs as their seniors regarding honor and obedience and the disgrace of surrendering. The code of the *samurai* had been all-pervasive for many years and had influenced the attitude and outlook of nearly every facet of Japanese society.

IGHQ took steps for the defense of the homeland as early as the beginning of 1944, when it perceived the course that the war was taking and judged what future American strategy was to be. Japanese strategists believed that Allied forces would attack Japan proper from the direction of the Marianas and through the Philippines. The Tokyo headquarters prepared for this eventuality by setting up a defense line along the sea front connecting the Philippines, Formosa, the Ryukyu Islands, the Japanese homeland, and the Kurile Islands, and strengthened the garrisons on each. According to this plan, the Japanese would concentrate their full strength to destroy the Allied threat at whatever point it developed.

A schedule of four prepared reactions, called the *Sho-Go* operations, was drawn up. For the defense of Formosa and the Nansei group, *Sho-Go* No. 2, *IGHQ* placed the *Thirty-second Army* under the command of the *Formosa Army* in July 1944, and added two divisions to the order of battle of the former. In the 10 months between the landing on Saipan and the invasion of Okinawa, Japanese strength was built up in the Ryukyus from an estimated 10,000 to approximately 155,000 air, ground, and naval troops.[12]

For the defense of the Philippines, the high command had planned *Sho-Go* No. 1. Based on a decision of the *Imperial War Council* on 19 August 1944, Japan staked her national destiny on the outcome of the impending battle of Leyte.[13] It was here that the Army and Navy had to destroy the Americans. The critical losses sustained by the *Combined Fleet* in the four-day battle for Leyte Gulf, 23–26 October 1944, three days after the invasion of Leyte, and the inability of ground forces to contain the invaders, created a grave threat to Japanese hegemony in the Western Pacific and even more so to the safety of Japan proper. Allied task forces dominated the waters surrounding Japan proper and the East and South China Seas as well. An additional liability resulting from American successes was the concomitant loss of airdromes from which land-based planes could pummel Japan unmercifully.

Seeing that no good purpose would be served by prolonging the Leyte operation, *IGHQ* decided to withdraw Japanese forces from the island and to conduct delaying tactics elsewhere in the Philippines. The Luzon landing in January 1945 made it apparent that there was no further way of holding off the Americans. From November 1944 on, American air attacks on Tokyo, Nagoya, Osaka, and northern Kyushu increased in intensity, destroying great sections of these areas and seriously impeding the war effort.

In January, the overall *IGHQ* estimate of the situation concluded that although Japan and Germany had suffered many reverses, the Axis had exacted a heavy penalty of their enemy. In viewing the Japanese cause in the same way that the viewed the Emperor and sacred homeland—through an emotional and reverent haze—[14] the senior commanders concluded that "the final victory will be for those who will stand up against increasing hardship and will fight to the last with a firm belief in ultimate victory."[15] While it acknowledged that the defeat of Germany would mean the unleashing of tremendously powerful forces against Japan, *IGHQ* believed that one of the major American problems would be in the area of manpower mobilization. The Japanese commanders hoped that, tiring of the war, the American people would favor its end.[16]

---

[12] MID, WD, *Disposition and Movement of Japanese Ground Forces, 1941–1945*, dtd 10Dec45 (OAB, NHD).

[13] Hattori, *War History*, p. 1.

[14] USSBS, *Japan's Struggle*, p. 2.

[15] *IGHQ Hist*, p. 236.

[16] *Ibid.*

Believing that the United States wanted to terminate the war quickly, *IGHQ* speculated that American forces would take the shortest possible route leading to Japan. This estimate foresaw that after the landings in the Philippines, the Allies would move to Formosa, Okinawa, and Iwo Jima. Based on the fact that most Japanese supply lines to the south had been well interdicted early in 1945, and an interpretation of radio intelligence reports, it seemed very likely that Iwo Jima was to be attacked in the very near future.[17] *IGHQ* also speculated that American forces would land on mainland China in southern Kwantung and Hongkong.[18]

In the face of the impending invasion and to strengthen homeland defenses further, *IGHQ* planned a large-scale mobilization of all segments of the population. In October 1944, when the government invoked general mobilization, there were 6,390,000 reservists available for call-up. Of these, 4,690,000 were ready for immediate assignment to active duty. There was a problem, however, of achieving a proper balance in the armed forces, since a shortage of trained technical personnel existed. Moreover, of the approximately 87 percent of the Japanese adult population already employed in the vital food and munitions industries, 47 percent were reservists and not available unless the war effort was to be damaged.[19]

[17] *Ibid.*, p. 239.

[18] Chief, War History Office, Defense Agency of Japan, ltr to Hd, HistBr, G–3 Div, HQMC, dtd 4Feb66, hereafter *War History Office Comments.*

[19] *Ibid.*

Further, at this late stage in the war, all branches of science were mobilized in the faint hope that they could develop surprise attack weapons. Unfortunately for this program, students at Army schools and serving officers were not very well trained in scientific and technological subjects; because of the nature of their duties and the weapons which the Navy employed, naval officers were in a little better position. The Army, however, was and always had been the dominant military authority in Japan, and as in the past, determined how the country would fight a war. Nevertheless, as the Japanese war situation deteriorated, military leaders optimistically sought the development of miraculously effective weapons.

> Nonetheless, it became abundantly clear that the low scientific level of the nation could not possibly yield elaborate weapons. . . . The Army's attitude toward technology incurred many kinds of great criticism from private sources at the time, the major points being the following:
>
> The Army keeps matters tightly secret. The Army has a great predilection for bamboo-spear tactics, and has little understanding of technology. . . .[20]

Despite the many imposing obstacles looming ahead, *IGHQ* prepared to execute a protracted war in the Japanese islands. The command headquarters made itself the supreme authority for the operation of the war and took steps to see that the governmental structure would be revised so that the Prime Minister would have comparable authority over political matters. In addition, the entire nation was to be mobilized and all citizens capable of bearing

[20] Hayashi and Coox, *Kōgun*, pp. 118-119.

weapons were to be armed. Key industries as well as the communications and transportation facilities were to be reorganized and operated by the state along rigidly controlled lines.[21]

As Japan was already a corporate state, *i.e.*, a nation in which the government controlled every facet of industry and all other productive areas, and since military control of the state had been a fact of life from the time that Japan embarked upon her course of conquest in the early 1930s, there was little new in this revised policy, except for one phrase, ". . . in a military manner." This was a naked declaration of military ascendancy and control over all governmental functions. Even Tojo had given lip service to civilian primacy in nonmilitary matters. But opposition elements were not yet strong enough to take over the reins of government and to begin steps to sue for a negotiated peace; the militarists were still in power, and they continued preparations for a last-ditch fight.

IGHQ remained convinced the America was wearying of the war and that, even if this were not so, Japanese ground strength of some 4,747,000 men in uniform—a million and a half of whom were based in Japan—[22] was enough to prevent the Americans from reaching Japanese shores. If invaders did attempt to come ashore, homeland defense forces would drive them back into the sea. At the beginning of 1945 a large proportion of Japanese troop strength overall was tied down in China and Manchuria, however, and a smaller portion was isolated in the Central and Southwest Pacific, where replacements, reinforcements, and replenishment could not be sent. Nor could these units be withdrawn to Japan or elsewhere, so complete was the Allied encirclement. For all practical purposes, the units in the Pacific were lost to Japan and out of the war for good.

In late 1944 and early 1945, American bombings, fast carrier task force raids, and especially the submarine blockade had increased in intensity and reduced the Japanese north-south maritime shipments to a mere trickle, so that the economic structure of that country was slowly forced to a halt. Undoubtedly, the single most effective agent in this action was the blockade imposed by the ships of the U. S. Pacific submarine fleet. American submarines torpedoed or destroyed by gunfire 60 percent of the 2,117 Japanese merchant vessels, totalling 7,913,858 tons, sunk by American forces during the war. In addition, U. S. underseas forces accounted for 201 of the 686 enemy warships sunk in World War II.[23]

On 13 January 1945, *IGHQ* was startled to learn that an entire convoy of nine tankers and its escort squadron had been sunk off Qui Nhon, a town on the east coast of French Indo-China. In face of this crowning blow and to

[21] Hattori, *War History*, p. 5.

[22] OCMH, DA Estimate of Japanese Strength and Disposition of Forces, dtd Oct55, File No. 320.2, Geographic V–Japan (OCMH). This collation was derived from authoritative American intelligence sources published in 1945.

[23] JANAC, *Japanese Naval and Merchant Ship Losses* (Feb47), pp. vi, vii, cited in Samuel Eliot Morrison, *Leyte, June 1944–January 1945—History of United States Naval Operations in* World War II, v. XII (Boston: Little, Brown and Company, 1958), p. 413.

evade American planes and submarines, the Japanese devised a new system employing small convoys guarded by dispersed escorts instead of the larger convoys and concentrated escorts sent out previously. Even this method failed when air and naval bases on Luzon became operational and American attacks from the island quickened in pace with the submarine attacks.

At the beginning of March 1945, *IGHQ* stopped sending convoys to the south; northbound convoys carrying essential war material continued the attempt to reach Japanese ports, however. Some 70 to 80 percent of the ships never made it. Later in the month, Tokyo ordered shipping halted altogether.[24] The noose around Japan was drawing tighter and tighter.

Since February, *IGHQ* had received a mounting influx of reports of increasingly larger numbers of American convoys operating in the vicinity of the Marianas and Ulithi. On 12 February, Tokyo was alerted to the movement of a sizable task force heavily protected by carriers and headed towards Iwo Jima. On the 16th, *IGHQ* was certain that the Bonins were the American target. When the actual invasion of Iwo began three days later, there was litle doubt that Okinawa would be next.[25]

In March, the Army and Navy concluded yet another agreement concerning joint defense operations, this one establishing responsibilities for containing Allied advances into the East China Sea. According to the plan, when American task forces approached this area, Army and Navy air elements would mount massive attacks against the convoys. Included in the Japanese aerial formations were to be special aircraft flown by pilots trained in suicide tactics.

At the end of the month, Japanese air strength available for the defense of Okinawa was as follows:

8th Air Division (Army), Formosa; 120 fighters, 60 bombers, 10 reconnaissance planes, 250 special attack planes. The targets of the latter were American transports.

Sixth Air Army (Army), Japan; 90 fighters, 90 bombers, 45 reconnaissance aircraft, and 300 special attack planes were assigned to attack American transports; 60 fighters, 30 bombers, 20 reconnaissance aircraft, and 100 special attack planes were assigned to strike task force carriers, and Ryukuyan airfields when captured by the Americans. An additional 400 fighters and 45 reconnaissance planes were assigned to fly combat air patrols.

First Air Fleet (Navy), Formosa; 40 fighters, 40 bombers, 5 reconnaissance planes.

Third Air Fleet (Navy), Japan; 40 fighters, 30 bombers, and 20 reconnaissance planes.

Fifth Air Fleet (Navy), Japan; 200 fighters, 310 bombers, and 10 reconnaissance aircraft.

Tenth Air Fleet (Navy), Japan; 700 combat planes, 1,300 training planes. This fleet was a reserve force, and its aircraft were to be employed as special attack planes. According to the Army-Navy agreement, the Navy planes were to attack the U. S. task forces and the escort shipping guarding them. To en-

[24] Hattori, *War History*, p. 6.

[25] *IGHQ Hist*, p. 256.

large the number of special attack units, both the Army and the Navy were to indoctrinate their pilots "in the spirit of suicide attacks."[26]

From early January until the middle of March, American carrier-based pilots had battered Formosa and Okinawa in an aerial onslaught that showed no signs of letting up. It seemed inevitable to the *Imperial General Headquarters* that the U. S. move following Iwo would be against the Ryukyus. Late in March, the Tokyo command received word that American forces had steamed out of anchorages at Ulithi and in the Marianas. During the same period, fast carrier task force aircraft pummelled Okinawa with from 500 to 700 sorties daily. The prologue to the grand climax was reached on 26 March when the Kerama Retto was invaded; Okinawa's time was not far off.

Despite the clear indication that Okinawa was the major U. S. target, Japanese air strength had dwindled to the point where it was in no condition to contest the landing. The *Fifth Air Fleet*, with a major assignment in the defense of the Ryukyus, had been soundly crushed in February when American fast carriers visited Kyushu. The other major air commands slated for important roles in protecting Okinawa either were not yet deployed in positions from which they could fly out to hold back the impending invasion or, having been severely punished in earlier American attacks, were unable to strike back.

Japanese naval strength was hardly in better condition. The fleet was in woefully sad shape and unbalanced. The high toll in the loss of its carriers, destroyers, and aircraft had left it in a pitiable condition, while the overall shortage of fuel would have immobilized it in any case. By March 1945, it "was nothing but a partially paralyzed surviving unit." [27]

Following the news of the fall of Iwo Jima, the *Thirty-second Army* on Okinawa stood wary—listening, waiting, and watching for an invasion force to appear over the horizon. Its expectations were soon to be fulfilled.

[26] *IGHQ Hist*, pp. 274–275, 277–278.

[27] Hattori, *War History*, p. 26.

Marine Corps Order of Battle[1]

In July 1940, the Marine Corps had 28,000 men in uniform. The January 1945 strength figure of the Corps reflected some 421,605 Marines, men and women; before the end of the war, this number was to become even larger. Broken down, the Marine Corps January strength figure represented: FMF ground forces, 212,165; aviation, 125,162; sea-going Marines, 9,430; foreign and domestic naval and shore activities, 54,483; Women's Reserve, 18,365. In addition to the above and not included in the overall total were 16,017 doctors, hospital corpsmen, and other naval personnel assigned to the Marine Corps.[2]

In the beginning of 1945, no major Marine ground force as such was engaged in a major operation against the enemy. The 2d Division on Saipan and the 3d on Guam, where veterans and new replacements alike participated in on-the-job training, however, were mopping up survivors of the major Japanese defense garrisons which were defeated when those islands were officially declared secured in 1944.

The senior Marine commander in the Pacific at this time was Lieutenant General Holland M. Smith, Commanding General, Fleet Marine Force, Pacific

[1] Unless otherwise indicated, the material in this chapter is derived from: CMC AnRpt to SecNav for the Fiscal Year Ending 30Jun45, hereafter *CMC Rpt 1945* (CMC Rpts for other years will be cited *CMC Rpt* with year); DivAvn, HQMC, OpD, Jan45 (AdminFile, HistBr, HQMC), hereafter *DivAvn OpD* with date; G–1 Sec, Div P&P, HQMC, OpDs, 7Dec41-31Dec44 and Jan45 (AdminFile HistBr, HQMC), hereafter *G–1 OpD* with date; CO, 1st ProvHistPlat, FMF, MS, "Historical Outline of the Development of Fleet Marine Force, Pacific, 1941–1950 (Preliminary)," 1951 (HistBr, HQMC), hereafter *FMFPac Hist;* Kenneth W. Condit, Gerald Diamond, and Edwin T. Turnbladh, *Marine Corps Ground Training in World War II* (Washington: HistBr, G–3 Div, HQMC, 1956), hereafter Condit, Diamond, and Turnbladh, *Marine Corps Training;* George McMillan, *The Old Breed: A History of the First Marine Division in World War II* (Washington: Infantry Journal Press, 1949), hereafter McMillan, *The Old Breed;* Richard W. Johnston, *Follow Me! The Story of the Second Marine Division in World War II* (New York: Random House, 1948), hereafter Johnston, *2d MarDiv Hist;* 1stLts Robert A. Aurthur and Kenneth Cohlmia. *The Third Marine Division* (Washington: Infantry Journal Press, 1948), hereafter Aurthur and Cohlmia, *3d MarDiv Hist;* Carl W. Proehl, ed., *The Fourth Marine Division in World War II* (Washington: Infantry Journal Press, 1946), hereafter Proehl, *4th MarDiv Hist;* Howard M. Conner, *The Spearhead: The World War II History of the 5th Marine Division* (Washington: Infantry Journal Press, 1950), hereafter Conner, *5th MarDiv Hist;* Bevan G. Cass, ed., *History of the Sixth Marine Division* (Washington: Infantry Journal Press, 1948) hereafter Cass, *6th MarDiv Hist;* Robert Sherrod, *History of Marine Corps Aviation in World War II* (Washington: Combat Forces Press, 1952), hereafter Sherrod, *Marine Air Hist.*

[2] *G–1 OpDs,* 7Dec41-31Dec44 and Jan45; FMF Status Rpts, Ground and Air for Jan45, prepared by G–3 Sec, Div P&P, HQMC (Pers and Loc File, HistBr, HQMC), hereafter *FMF Grd* (or *Air*) *Status Rpt* with month. See App F for a complete location and strength breakdown of the FMF as of 30Apr45.

(FMFPac).[3] His was a type command which involved administrative control only over FMFPac units.[4] The majority of Marine combat troops were located in the Central Pacific under Admiral Spruance's control. By January 1945, six Marine divisions had been activated, grouped three each in two corps. Headquarters of Major General Roy S. Geiger's III Amphibious Corps (IIIAC) was on Guadalcanal, where Major General Lemuel C. Shepherd, Jr., based his 6th Marine Division. On Pavuvu in the Russells, approximately 65 miles northwest of Guadalcanal, was the 1st Marine Division commanded by Major General Pedro A. del Valle. Also in IIIAC was Major General Thomas E. Watson's 2d Marine Division on Saipan. The other major Marine ground command under FMFPac, V Amphibious Corps (VAC), was headed by Major General Harry Schmidt, whose headquarters was located on Maui, Hawaiian Islands. Also on Maui in VAC was Major General Clifton B. Cates' [5] 4th Marine Division. On the island of Hawaii was Major General Keller E. Rockey and his 5th Marine Division, and on Guam was the 3d Marine Division of Major General Graves B. Erskine.

All six divisions had long been in receipt of orders for their next combat assignments and were actively engaged in preparing for them. Authorized strength for a Marine division at this time was 856 officers and 16,069 enlisted Marines. The 1st, 2d, and 3d Divisions were short a few officers, while the 3d Division was short 271 enlisted men and the 6th Division, 513. Replacement drafts had been assigned to the divisions, however, and were in transit to join them either before or just after the impending landings. In January 1945, 11 replacement drafts, containing 14,331 Marines and naval corpsmen and doctors, were either en route to or in the process of joining the six Marine divisions at the staging areas.[6]

In addition to the combat divisions, there were other FMF organizations spread throughout the Pacific undertaking assigned support, garrison, or defense missions. A total of 74,474 Marines and naval personnel was involved in the operations of these units.[7] Two provisional field service commands, one at Guam and the other at Guadalcanal, and seven field depots and four service and supply battalions based in close proximity to the Marine divisions provided major supply support in the Pacific. Also available from FMFPac for support of and attachment to the two corps for upcoming operations were a variety of other units. These included 11 antiaircraft artillery battalions, 6 155mm gun and 6 155mm howitzer battalions, 3 armored amphibian battalions, 9 amphibian tractor battalions, and 6 amphibian truck companies. In January, most of these organizations were a part of the III and V

[3] For the history of FMFPac, see Garand and Strobridge, "Western Pacific Operations."

[4] Generally, a type commander has an administrative mission only and has no tactical responsibilities. For the Iwo Jima operation, General Smith was assigned an additional tactical responsibility as Commander, Expeditionary Troops.

[5] Generals Cates and Shepherd later became the 19th and 20th Commandants of the Marine Corps, respectively.

[6] *FMF Grd Status Rpt*, Jan 45.

[7] *Ibid.*

Amphibious Corps, although a few of the antiaircraft artillery battalions were still fulfilling island defense missions. Within FMFPac also were such other types of organizations as defense battalions, Joint Assault Signal Companies (JASCOs), provisional rocket detachments, war dog platoons, motor transport battalions, corps evacuation hospitals, bomb disposal companies, and separate engineer battalions; not all of them would be employed in future operations.

The highest level Marine aviation echelon in the Pacific, Aircraft, Fleet Marine Force, Pacific (AirFMFPac), was commanded by Major General Francis P. Mulcahy, whose headquarters was at Ewa, on the island of Oahu. A type command like FMFPac, AirFMFPac organized, administered, supplied, and deployed all Marine aviation squadrons in the Pacific, but controlled none. The four Marine aircraft wings (MAWs) in AirFMFPac were based on islands at widespread points in the Pacific. Included in the four wings was a total of 16 Marine aircraft groups (MAGs) holding 70 tactical squadrons broken down as follows: 28 fighter (VMFs); 14 scout bomber (VMSBs); 7 bomber (VMBs); 7 transport (VMRs); 5 night fighter (VMF(N)s); 5 observation (VMOs); and 3 torpedo bomber (VMTBs).

Major General Ralph J. Mitchell's 1st MAW headquarters was on Bougainville. He had under his command six MAGs; three of the groups and part of a fourth were assigned to CinCSWPA for the Philippines campaign; and one squadron of a fifth group was en route to join MacArthur. MAG–25, the latter group, and the portions of the other two not committed in the Philippines, were based on Emirau, Green Island, Manus, and Cape Torokina at Bougainville.[8] (See Map XI, Map Section.) A transport group, MAG–25 also had another designation, SCAT (South Pacific Combat Air Transport Command), and as such, with a few AAF transport squadrons assigned from time to time flew many extensive air supply missions all over the Pacific from the time it was activated, 24 November 1942, until the end of the war. When available for such an assignment, the group also evacuated casualties from captured islands which had strips capable of sustaining the operations of transport-type aircraft.

Headquarters of the 2d MAW was at Ewa. Major General James T. Moore had only one MAG within his command at this time, and it was based on Espiritu Santo, in the New Hebrides. The reason that the wing was so depleted in January was that in mid-1944 most of Moore's squadrons had been transferred to the 4th MAW, and at the end of the year his command became the nucleus for Tactical Air Force, Tenth Army, and assigned as a task unit for the invasion of Okinawa.

The 3d Marine Aircraft Wing was commanded for the first seven days in January 1945 by Colonel Ford O. Rogers, who was relieved on the 8th by Colonel Byron F. Johnson. The wing had its headquarters command and a group at Ewa, and a MAG based on

[8] Green Island is approximately 80 miles north of the tip of Bougainville, and Manus is 280 miles southeast of Emirau and 80 miles north of the northeast coast of New Guinea.

Midway. Four fighter squadrons of the Ewa-based group were on carriers and supported the Lingayen landings in January and later flew strikes against Luzon, Formosa, and Indochina. *Essex* carried two of the VMFs, and *Wasp* the other two. The primary mission of 3d MAW was to train AirFMFPac tactical squadrons and pilots in night-fighting, air-warning, and radar-bombing techniques.

Major General Louis E. Woods' 4th MAW was based on Majuro in the Marshalls, and with seven groups located on islands all over the Central Pacific, his was the largest of the four AirFMFPac wings. From airfields on such widely separated islands as those in the Marshalls, Marianas, and Palaus, Woods' squadrons took off day after day to neutralize bypassed Japanese defenses. Although boring and seemingly prosaic in nature, important benefits derived from the operations of the 4th Wing squadrons in neutralizing the Marshalls. "The 4th MAW's perfection of the napalm fire jelly formula was a big contribution to the rest of the Pacific. Also important was the development of the fighter bomber, that trusty weapon so sorely needed when more planes had to be had to save the fleet from the *Kamikaze*." [9]

There were two other major aviation commands in the Marine Corps; these were based on each coast of the United States and held similar missions. At San Diego, Marine Fleet Air, West Coast, administered, operated, trained, and equipped all Marine aviation organizations on the west coast; it also channelled personnel and materiel to AirFMFPac for further deployment in the Pacific. The east coast training command was the 9th MAW, with headquarters at the Marine Corps Air Station, Cherry Point, N. C. Within these two commands were 14 groups—6 on the west coast and 8 on the east. There were in turn a total of 100 tactical and 29 training squadrons of various types in the groups. On the west coast were 22 carrier squadrons, 16 fighter (VMF(CVS)) and 6 torpedo bombing (VMTB(CVS)), that were slated to go aboard new *Commencement Bay*-class escort carriers as soon as the latter had completed shake-down trials.

At the beginning of 1945, Marine Corps aviation had already passed through a period of expansion and was entering an era of consolidation with respect to its ultimate objectives in the war effort. Three months earlier, in October 1944, the Chief of Naval Operations had approved a plan to man four of the new class of escort carriers with Marine squadrons. Accordingly, that same month, the Marine Corps redesignated two groups already in existence on the west coast as Marine Air Support Groups (MASGs). By January 1945, the VMF(CVS)s and VMTB-(CVS)s were ready to begin a period of intense training from the decks of the carriers and at their former home bases, the Marine Corps Air Stations at El Centro and Mojave, in California. To each MASG was attached a Marine Carrier Group (MCVG), composed of a fighter and a torpedo bombing squadron; one of six Carrier Aircraft Service Detachments was to complete the

---

[9] Sherrod, *Marine Air Hist*, p. 246.

Marine complement in the carriers after the detachments had been organized on the west coast in February.

To enable Marine Corps aviation "to give support to the Fleet Marine Force in landing operations" more effectively in the future, the Navy approved a major revision of the Corps aviation structure at the beginning of 1945. Marine Corps planners envisioned a program encompassing balanced land-based and carrier-based aviation forces, which would be soundly supported by shore activities. Not all of the Marine carrier program went into effect before the end of the war, but its progress was such that the validity of basing Marine squadrons in the flattops was proven.

Fleet Marine Force ground training facilities in 1945 had advanced far beyond those in existence at the beginning of the war. In addition to base command housekeeping and training units, Camp Lejeune had 10 battalions undergoing infantry training in January 1945 and Camp Pendleton had 12. Besides this number, Camp Pendleton housed four replacement drafts, a total of some 5,000 Marines, who were awaiting shipment to the Pacific. With the record number of six Marine divisions in the field, the constant demand for replacements and the heavy burden imposed upon the training command continued incessantly.

As IIIAC and VAC completed their training phases and began combat loading for the trip to the target areas, the anticipated demands for replacement of expected casualties on Iwo Jima and Okinawa were already being met by the organization and training of new replacement drafts. As each Marine destined for assignment to a combat organization in the Pacific left the United States, he was aware that his was the same path taken by fellow Marines who had fought at such now-famous places as Guadalcanal, Bougainville, and Tarawa. His sole consolation, if one was needed, was the knowledge that, although he had not participated in the beginning of the fight, he might possibly be there to help end it.

PART II

Okinawa

The Target and the Enemy

BACKGROUND[1]

Once the Joint Chiefs of Staff decided on Okinawa as a future target, intensive planning and preparations were begun for the assault on this once obscure island. Large amounts of information of varying importance poured into the intelligence centers concerned with the impending operation, and were added to files already bulging with a store of knowledge of the Ryukyus group. Okinawa soon became the focus of attention of the CinCPac-CinCPOA headquarters and staff members who, in compliance with the JCS directive to Admiral Nimitz ". . . . to occupy one or more positions in the Nansei Shoto,"[2] filled in the details of an outline plan. A flurry of disciplined activity immediately engulfed the commands and staffs of the expeditionary forces assigned to the assault as they began their operational studies for ICEBERG, the code-name given to the approaching invasion.

The strategic importance of Okinawa was its location, and all other considerations stemmed from this. The Japanese viewed it as an integral link in a chain of islands, the Ryukyus or the Nansei Shoto, which formed an effective barrier to an Allied advance from the east or southeast towards the Chinese mainland, Korea, or the western coast of Japan. This group of islands was ideally situated to aid in the protection of the Japanese maritime lines of supply and communication to imperial conquests in southeast Asia. The island chain also provided the Japanese Navy with the only two substantial fleet anchorages south of the Home Islands[3] between

[1] Unless otherwise indicated, the material in this chapter is derived from: HQUSAFPOA G–2 Study of Okinawa Gunto, n.d., hereafter *HQUSAFPOA Study;* MIS, WD, Survey of the Nansei Shoto, dtd 15Feb43, hereafter *WD Survey; War Reports;* USSBS (Pac), NavAnalysisDiv, *The Campaigns of the Pacific War* (Washington, 1946), hereafter USSBS, *Campaigns;* Roy E. Appleman, *et. al., Okinawa: The Last Battle—U. S. Army in World War II—The War in the Pacific* (Washington: HistDiv, DA, 1948), hereafter, Appleman, *et. al., Okinawa Battle;* Wesley Frank Craven and James Lea Cate, eds., *The Pacific: Matterhorn to Nagasaki—The Army Air Forces in World War II,* v. 5 (Chicago: The University of Chicago Press, 1953), hereafter Craven and Cate, *Matterhorn to Nagasaki;* Jeter A. Isely and Philip A. Crowl, *The U. S. Marines and Amphibious War* (Princeton: Princeton University Press, 1951), hereafter Isely and Crowl, *Marines and Amphibious War;* Nichols and Shaw, *Okinawa Victory.* Documents not otherwise identified in this part are located in the following files of the HistBr, HQMC: Unit Historical Reports; Okinawa Area Operations; Publications; Aviation; and Okinawa Monograph and Comments. Because Appleman, *et. al., Okinawa Battle,* and Nichols and Shaw, *Okinawa Victory* have served as guidelines for pt II, hereafter they will be cited in direct reference only.

[2] *ICEBERG Study,* p. 1.

[3] The Home Islands were generally considered to consist of the four principal islands and the hundreds of smaller islands immediately adjacent to Hokkaido, Honshu, Shikoku, and Kyushu, which formed Japan.

Kyushu and Formosa, and numerous operating bases for aircraft of all types as well. (See Map 1.)

From the Allied point of view, the conquest of Okinawa would be most lucrative. As the largest island in the Ryukyus, it offered excellent locations for military and naval facilities. There was sufficient land area on the island on which to train and stage assault troops for subsequent operations against the heart of the Empire. Kyushu was only 350 nautical miles away, Formosa 330 miles distant, and Shanghai, 450. Two other major purposes of the impending invasion were to secure and develop airbase sites from which Allied aircraft could operate to gain air superiority over Japan. It was expected that by taking Okinawa, while at the same time subjecting the Home Islands to blockade and bombardment, Japanese military forces and their will to resist would be severely weakened.

OKINAWA: HISTORY, LAND, AND PEOPLE [4]

Before Commodore Matthew C. Perry, USN, visited Okinawa in 1853–54, few Americans had ever heard of the island. This state of ignorance did not change much in nearly a century, but American preinvasion studies in 1944 soon shed some light on this all-but-unknown area.

The course of Okinawa history—from the Chinese invasions about 600 A. D. until Japanese annexation in 1879—was dominated by an amalgamation of Chinese and Japanese cultural and political determinants. For many years, the Chinese influence reigned supreme. After the first Chinese-Okinawan contacts had been made, they warred against each other until the island peoples were subdued. Shortly after 1368, when the Ming Dynasty came to power, China demanded payment of tribute from Satsudo, the King of Okinawa. The payment was given along with his pledge of fealty as a Chinese subject.

In the midst of incessant Okinawan dynastic squabbles, Chinese control remained loose and intermittent until 1609, when the Japanese overran the island, devastating all that stood in their way. The king of Okinawa then reigning was taken prisoner, and a Japanese local government was established temporarily.

For the next 250 years, the Okinawan Kingdom, as such, was in the unenviable position of having to acknowledge both Chinese and Japanese suzerainty at the same time. Finally, in May 1875, Japan forbade the islanders to send any more tribute to China, whose right to invest the Okinawan kings was now ended. In the face of mounting Okinawan protests against this arbitrary action, Japan followed its decree by dethroning the king in March 1879; he was reduced in rank, becoming a marquis of Japan. Okinawa and its neighboring islands were then incorporated within the

[4] Unless otherwise noted, the material in this section is derived from: JICPOA Bul 63–44, Nansei Shoto, dtd 15May44; CinCPac-CinCPOA Bul 161–44, Okinawa Gunto, dtd 15Nov44; EngrSec, FMFPac, EngrIntelInfo, Okinawa Gunto, n.d. The terrain description presented in this section is that of Okinawa as it existed at the time of the American invasion and before engineer and Seabee units performed their earth-transforming feats.

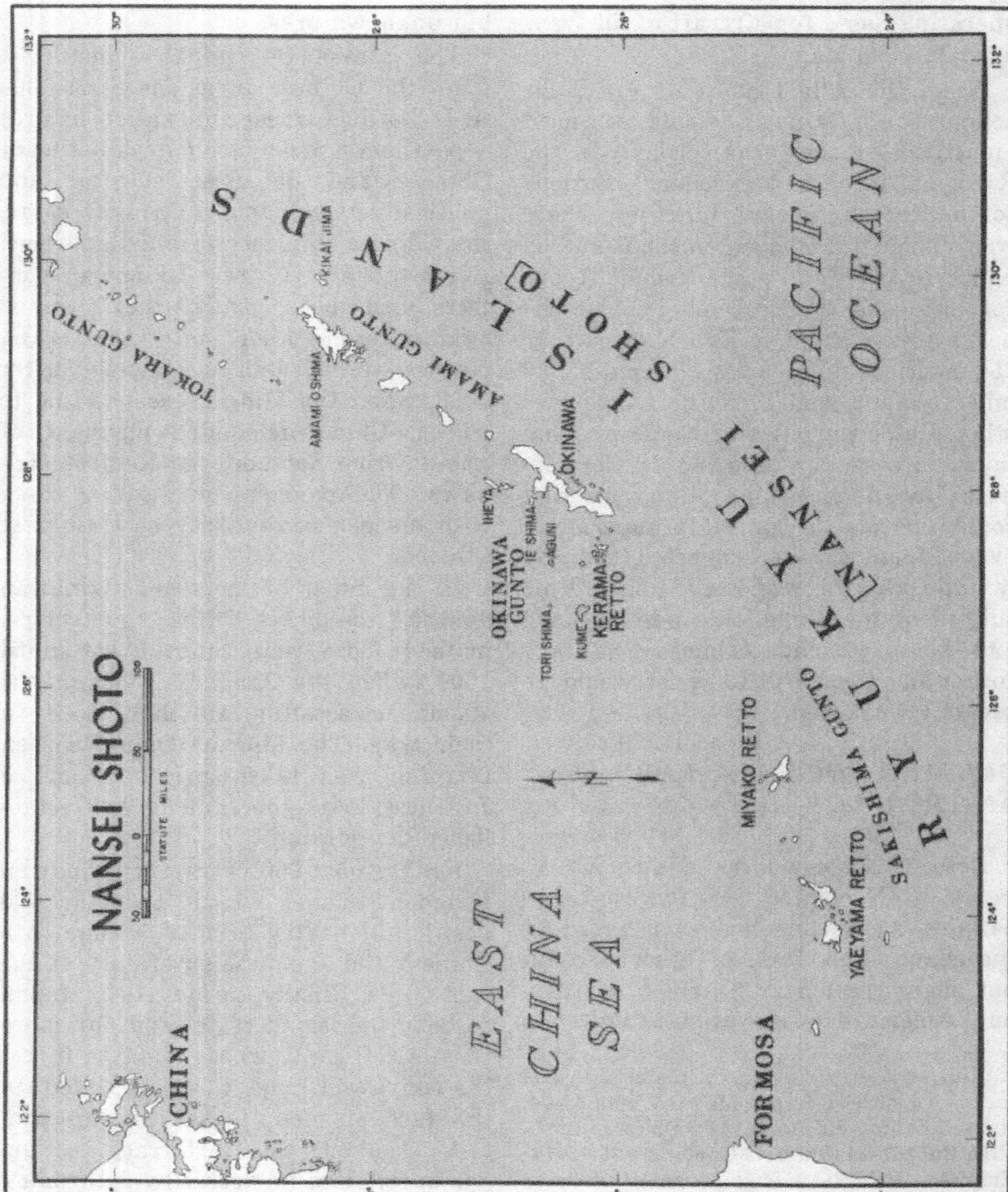
NANSEI SHOTO
STATUTE MILES
EAST CHINA SEA
CHINA
FORMOSA
PACIFIC OCEAN
RYUKYU ISLANDS
[NANSEI SHOTO]
TOKARA GUNTO
AMAMI GUNTO
AMAMI OSHIMA
KIKAI JIMA
OKINAWA
IHEYA
IE SHIMA
AGUNI
OKINAWA GUNTO
TORI SHIMA
KUME
KERAMA RETTO
MIYAKO RETTO
SAKISHIMA GUNTO
YAEYAMA RETTO
T L RUSSELL

MAP I

Japanese political structure as the Okinawan Prefecture. Over the years, China remained restive at this obvious encroachment, until the question was one of many settled in Japan's favor by its victory in the Sino-Japanese War of 1894.

The islands to which the Japanese successfully gained title, the Ryukyu Retto, were in the southernmost of two groups which make up the Nansei Shoto. The shoto is a chain of islands which stretch in a 790-mile-long arc between Kyushu and Formosa, separating the Pacific Ocean from the East China Sea. One of the groups which make up the Ryukyu Retto is the Okinawa Gunto.[5] The other four major island groups in the retto are Osumi, Tokara, Amami, and Sakishima. Okinawa Gunto is located at the half-way point in the arc and consists of Okinawa and numerous smaller islands. These include Kumi Shima, Aguni Shima, Ie Shima, and the Kerama Retto in the west; Iheya Retto and Yoron Shima in the north; and a group of small islands, named the Eastern Islands by the Americans, roughly paralleling the east central coast of Okinawa.

The island of Okinawa is narrow and irregularly shaped throughout its 60-mile length. (See Map 2.) In the north, the Motobu Peninsula juts out into the East China Sea and extends the island to its maximum breadth, 18 miles; immediately to the south is the narrowest part, the two-mile-wide Ishikawa Isthmus. The coastline of the island ranges in nature from a precipitous and rocky shore in the north, through a generally reef-bound lowland belt just below the isthmus, to an area of sea cliffs and raised beaches in the south. Landing beaches suitable for large-scale amphibious operations were neither numerous nor good. The most extensive flat areas and largest beaches on the east coast were found along the shores of Nakagusuku Wan (or bay) and, on the west coast, in the area between Zampa Misaki (or point) and Oroku Peninsula. Two major fleet anchorages existed, both on the eastern side of the island: Nakagusuku Wan (later named Buckner Bay by the Americans in honor of the Tenth Army Commander) and Chimu Wan. The leading port of the Okinawa Gunto was on the west coast at Naha, the major city of the island group. Port facilities elsewhere were limited to small vessels.

Okinawa is easily divisible into three geographical parts, each one physically different from the other. The territory north of the Ishikawa Isthmus, constituting about two-thirds of the island area, is largely mountainous, heavily wooded, and rimmed with dissected terraces—or one-time flatlands which became deeply ravined by the ravages of erosion. About 80 percent of the north is covered with a dense growth of live oak and conifers, climbing vines, and brush. The highlands, rising to rugged peaks, 1,000 to 1,500 feet in height, dominate the area. Small, swift streams drain the clay or sandy-loam topsoil of the interior which is trafficable under most conditions. Cross-country movement is limited mainly by the steepness of the hills and the lush

[5] Gunto, retto, shoto all are Japanese geographic terms for group or chain of islands; jima or shima is translated as island.

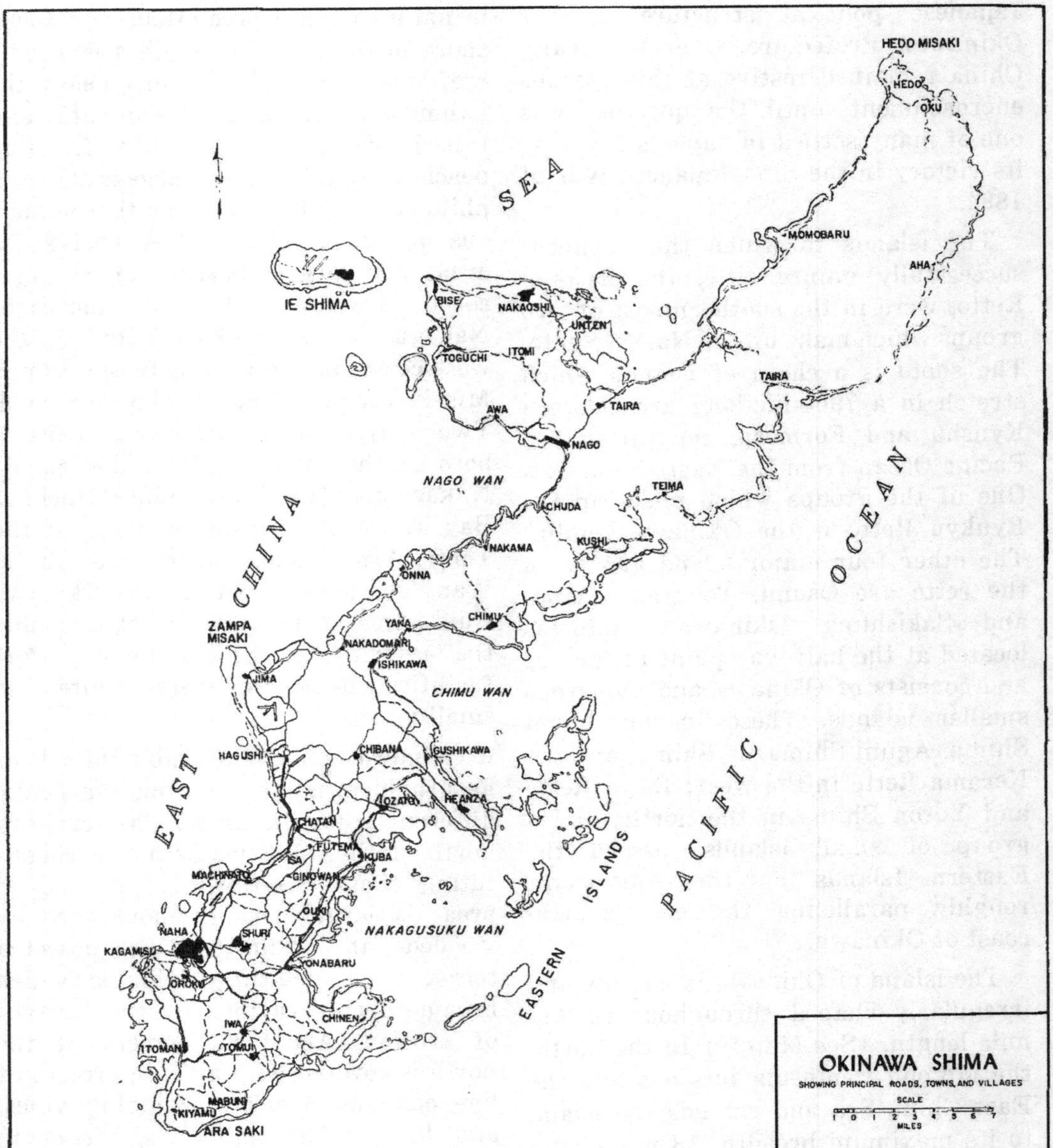

MAP 2 T.L. RUSSELL

vegetation. The few roads that existed in 1945 were mostly along the coast.

The middle division, consisting of that area lying between Ishikawa Isthmus and an east-west valley running between the cities of Naha and Yonabaru, is broadest in its northernmost part. Just south of the isthmus is an area resembling northern Okinawa, but the rest of the sector is, for the most part, rolling, lightly wooded country interrupted by steep cliffs and ravines. The few streams, flowing through hills which rarely exceeded a height of 500 feet, are generally narrow and shallow, so they could be easily bridged or forded.

The southernmost tip of the island, triangular in shape, is extremely hilly and was dominated by extensive limestone plateaus, some reaching over 500 feet in height. At each angle of the base of the triangle is a peninsula, Oroku on the west, and Chinen on the east.

The primary roads built by the Japanese were little more than coral- or limestone-surfaced trails, varying in width from 12 to 16 feet, on a sand and clay base. Use of these roads depended largely upon the weather, since rain reduced them to sticky and slow-drying morasses. In the dry season, the slightest movement on the roads threw up dense clouds of dust. The major arteries threaded along the coastlines, branching off into a few cross-island roads which then broke down into a capillary system of trails connecting the small villages, settlements, and individual farms. The central sector, the densely populated part of the island, contains an intricate network of roads. Only one, the broad stone-paved highway connecting the cities of Shuri and Naha, could support two lanes of traffic. In this area, the road net was augmented by a narrow gauge railway, with approximately 30 miles of track. This system provided the major trans-island communications net, running from Naha to Yonabaru on the east coast, via the towns of Kobakura and Kokuba, while trunk lines linked Kobakura and Kokuba with the west coast towns of Kadena and Itoman, respectively.

Okinawa's climate is tropical, with moderate winters, hot summers, and high humidity throughout the year. The annual temperature range is from a minimum of 40 degrees to a mean maximum of 95 degrees in July. The months of May through September are marked by a heavy and erratic rainfall. During the typhoon season (July-November), torrential rains and winds of over 75 miles-per-hour have been recorded.[6] During the rest of the year, except for brief downpours, good climatic conditions generally prevail.

The inhabitants of Okinawa in 1945 were heirs to a complex racial mixture. The original population is believed to have been a branch of the hairy Ainu and Kumaso stock which formerly inhabited Kyushu and other Japanese islands. A Mongoloid strain was introduced when Japanese pirates, who made Okinawa their headquarters, engaged in their time-honored habit of kidnapping women from the Chinese mainland.

---

[6] A typhoon in late September 1945 recorded winds of up to 120 miles per hour. LtGen Edward W. Snedeker ltr to Hd, HistBr, G-3 Div, HQMC, dtd 15Oct65, hereafter *Snedeker ltr 1965*.

Malayan blood was infused into this melting pot through intermarriage, immigration, and invasion. This evolution produced a people with the same basic characteristics as those of the Japanese, but with slight physical differences. The Okinawans are shorter, darker, and are inclined to have more body hair.

The 1940 census gave an estimate of slightly over 800,000 people in the Nansei Shoto as a whole, with nearly a half-million of these on Okinawa proper. Farmers constituted the largest single population class, with fishermen forming a smaller, but important, group. Approximately 15 percent of the Okinawa populace lived in Naha, and within this community were most of the higher officials, businessmen, and white collar workers—most of them Japanese who either had emigrated or been assigned from the Home Islands.

During the period of the Okinawan monarchy, there was an elaborate social hierarchy dominated by nobles and court officials. After Japanese annexation, the major social distinctions became those that existed between governing officials and natives, between urban and rural inhabitants, and between the rich and the poor—with the latter in the majority. Assimilation of the Japanese and Okinawan societies was minimal, a situation that was further irritated by the preferential treatment tendered by the Japanese to their fellow-countrymen when the more important administrative and political posts were assigned.

Another chasm separating the Japanese and Okinawan was the difference in languages. Despite a common archaic tongue which had branched into the language families of both Okinawa and Japan, there were at least five Ryukyuan dialects which rendered the two languages mutually unintelligible. The Japanese attempted to reduce the language barrier somewhat by directing that standard (Tokyo) Japanese was to be part of the Okinawan school curriculum. Several decades of formal education, however, failed to remove the influence of many generations of Chinese ethnic features which shaped the Okinawan national characteristics. The Chinese imprint on the island was such that one Japanese soldier noted that "the houses and customs here resemble those of China, and remind one of a Chinese town." [7] The natives retained their own culture, religion, and form of ancestor worship. One outward manifestation of these cultural considerations were the thousands of horseshoe-shaped burial vaults, many of impressive size and peculiar beauty, which were set into the sides of numerous cliffs and hills throughout the island.

The basic Okinawan farm settlement consisted of a group of farmsteads, each having the main and other buildings situated on a small plot of land. The farmhouses were small, thatch-roofed, and set off from the invariably winding trailside by either clay or reed walls. The agricultural communities generally clustered around their own individual marketplaces. Towns, such as Nago and Itoman, were outgrowths of the villages, differing only in the fact that these larger settlements had several modern business and government struc-

---

[7] "Diary of a Superior Private, 272d IIB," in CinCPac-CinCPOA Bul 147–45, Translations and Interrogations No. 32, dtd 16Jun45, p. 48.

tures. The island's cities, Naha and Shuri, were conspicuous by their many large stone and concrete public structures and the bustle that accompanies an urban setting. Shuri was the ancient capital of the Ryukyuan kingdom and its citadel stood on a high hill in the midst of a natural fortress area of the island.

The fundamentally agrarian Okinawan economy was dependent upon three staple crops. About four-fifths of southern Okinawa was arable, and half of the land here was used for the cultivation of sweet potatoes, the predominant foodstuff of both men and animals. Sugar cane was the principal commercial crop and its cultivation utilized the second largest number of acres. Some rice was also grown, but this crop consistently produced a yield far below local requirements. Since rice production was sufficient to satisfy only two-thirds of the population's annual consumption needs, more than 10 million bushels had to be imported annually from Formosa.[8]

Industrial development on the island was rudimentary. The Naha-Shuri area was the leading manufacturing center where such items as alcoholic beverages, lacquerware, and silk pongee were produced. Manufacturing was carried out chiefly in small factories or by workers in their homes. The only relatively important industry carried on outside of the Naha-Shuri complex was sugar refining, in which cattle supplied the power in very primitive mills. The fishing trade, of some importance, centered around Naha and Itoman. There were also small numbers of fishing craft based at all of the other usable harbors on the island; however, lack of refrigeration, distance to the fishing grounds, and seasonal typhoons all hindered the development of this industry and prevented its becoming a large source of income for the Okinawans.

From the very beginnings of the 1879 annexation, the Japanese government made intensive efforts to bring the Ryukyuan people under complete domination through the means of a closely controlled educational system, military conscription, and a carefully supervised system of local government. The prefectural governor was answerable only to the Home Minister in Tokyo. Although the elected prefectural assembly acted as the gubernatorial advisory body, the governor accepted, rejected, or ignored their suggestions as he saw fit. On a local level, assemblies elected in the cities, towns, and townships in turn elected a mayor. All local administrative units were, in effect, directly under the governor's control, and their acts or very existence were subject to his pleasure.

In every aspect—social, political, and economic—the Okinawan was kept in a position inferior to that of any other Japanese citizen residing either on Okinawa or elsewhere in the Empire. This did not prevent the government from imposing on the Okinawan a period of obligated military service.[9] The periodic

[8] MilHistSec, Japanese Research Div, HQ-USAFFE, Japanese Monograph No. 135, Okinawa Operations Record, revised edition dtd Nov49, p. 60, hereafter *Okinawa Operations Record.*

[9] Military conscription was first enforced on Okinawa in 1898, although in Japan proper, conscription had been in effect since 1873. *War History Office Comments.*

call-ups of age groups was enforced equally upon the natives of Okinawa and the Ryukyus as on the male inhabitants of Japan proper. This provided Japan with a reservoir of trained reservists from which it could draw whenever necessary.

With the exception of those drafts of reservists leaving for active duty elsewhere, Okinawa, for all practical purposes, was in the backwash of the early stages of World War II. The island remained in this state until April 1944, when Japan activated the *Thirty-second Army*, set up its headquarters on Okinawa, and assigned it responsibility for the defense of the island chain.

THE JAPANESE FORCES [10]

Following the massive and devastating United States naval air and surface bombardment of Truk, 17–18 February 1944, and the breaching of the Marianas line shortly thereafter, the Japanese *Imperial General Headquarters* awakened to the obviously weak condition of the Ryukyus' defenses. Prior to 1944, little attention had been paid to the arming of the Nansei Shoto. The island group boasted two minor naval bases only, one at Amami O-Shima and the other at Naha, and a few small Army garrisons such as the *Nakagusuku Wan Fortress Artillery Unit* on Okinawa.[11] Acting with an alacrity born of distinct necessity, *IGHQ* took steps to correct this weakness in the Empire's inner defensive positions by expediting and intensifying:

> . . . operational preparations in the area extending from Formosa to the Nansei Islands with the view of defending our territory in the Nansei area and securing our lines of communication with our southern sector of operations, and thereby build a structure capable first, of resisting the enemy's surprise attacks' and, second, of crushing their attempts to seize the area when conditions [change] in our favor.[12]

In order to improve Japanese defenses in the Ryukyus, *IGHQ* assigned this mission on 22 March 1944, to the *Thirty-second Army*, the command of which was assumed formally on 1 April by Lieutenant General Masao Watanabe. At Naha, headquarters of the new army, staff officers hoped that enough time would be available for adequate fortification of the island. All planning was tempered by memories of the immediate past which indicated that "an army trained to attack on any and every occasion, irrespective of conditions, and with no calculation as to the real chances of success, could be beaten soundly." [13] Added stimuli to Japanese

[10] Unless otherwise noted, the material in this section is derived from: Tenth Army IntelMono, dtd Aug45, pt I, secs A and B, hereafter *IntelMono;* Tenth Army G–3 POW Interrogation Summaries Nos. 1–19, Jul-Aug45, hereafter *POW InterSum;* Tenth Army G–2 Interrogation Rpt No. 27, Akira Shimada, dtd 24Jul45, hereafter *Shimada Interrogation;* Tenth Army Interrogation Rpt No. 28, Col Hiromichi Yahara (Senior Staff Officer, Thirty-second Army), dtd 6Aug45, hereafter *Yahara Interrogation;* Hayashi and Coox, *Kōgun; Okinawa Operations Record.*

[11] Actually, there were three garrisons located in Nansei Shoto: one on Amami O-Shima founded in 1922; and two founded in August 1941—Funauki Fortress at Nishi Omote-Jima and the Nakagusuku Wan Fortress noted above. *War History Office Comments.*

[12] *Okinawa Operations Record*, p. 1.

[13] *IntelMono*, pt I, sec A, p. 3.

preparations were the American invasions of Peleliu and Morotai on 15 September 1944. By this time, the Japanese high command became quite certain that either Formosa, the Ryukyus, or the Bonins, or all three, were to be invaded by the spring of 1945 at the latest. Initially, Japanese Army and Navy air forces were to blunt the assaults in a major air counteroffensive. The establishment of Allied air superiority and demonstrated weaknesses of Japanese air forces, however, caused the military leaders in Tokyo to downgrade the aviation role in the coming struggle for the defense of the Home Islands. The ground forces, then, would carry the major burden.

The *Thirty-second Army* staff planners wasted no time in organizing the ground defenses of Okinawa. They had learned by the cruel experiences of Japanese forces on islands which had been invaded by the Americans that a stand at the shoreline would only result in complete annihilation and that their beach positions would be torn to pieces in a naval bombardment. It became apparent, therefore, that the primary defensive positions had to be set up inland. Then, should the invaders escape destruction at sea under the guns and torpedoes of Japanese naval forces, or at the beachhead under the downpour of artillery shells, the death blow would be administered by the ground forces' assumption "of the offensive in due course." [14] To steel the troops' determination to fight and to keep their morale at a high peak, army headquarters devised the following battle slogans:

One Plane for One Warship
One Boat for One Ship
One Man for Ten of the Enemy
or One Tank.[15]

The command of the *Thirty-second Army* was assumed by Lieutenant General Mitsuru Ushijima in August 1944, when General Watanabe was forced to retire because of a continuing illness. Because of the importance of the impending Okinawa battle, *IGHQ* assigned General Ushijima one of the most competent officers of the Japanese Army, Major General Isamu Cho, as his chief of staff. On 21 January, army headquarters was split into two groups. Ushijima's operations staff moved to Shuri where the general was to direct his army for the major portion of the campaign. A "rear headquarters" composed of the ordnance, veterinary, judicial, intendance,[16] and the greater part of the medical staff set up near Tsukasan, south of Shuri.

Lieutenant Generals Ushijima and Cho [17] complemented each other's military qualities and personality, and formed a command team that reflected mutual trust and respect. They were ably abetted by the only holdover from

---

[14] Hayashi and Coox, *Kōgun*, p. 116.

[15] "Thirty-second Army Battle Instructions, dtd 15Feb45," in CinCPac-CinCPOA Bul 122-45, Translations and Interrogations No. 30, dtd 1Jun45.

[16] The intendance service, which had no exact U. S. military equivalent, controlled clothing, rations, forage, pay, and the upkeep of Army buildings. In effect, it combined the functions of the U. S. Army Quartermaster Corps and Finance Department. WD, *Handbook on Japanese Military Forces*, TM-30-480 (Washington, 1Oct44), p. 50.

[17] On 1Mar45, at the age of 51, Cho was promoted to lieutenant general.

the old staff, Colonel Hiromichi Yahara, who retained his billet as Senior Officer in Charge of Operations,[18] and Major Tadao Miyake as the logistics officer.[19] Ushijima, a senior officer slated for promotion to general in August 1945, was reputedly a man of great integrity and character who demonstrated a quiet competence which, in turn, inspired great confidence, loyalty, and respect from his subordinates. Cho, in comparison, was a fiery, ebullient, and hard-driving individual with a brilliant, inquiring mind. He spared neither himself nor his staff. His abounding energy was effectively counterbalanced by his senior's calm outward appearance. This combination of personalities was served by comparatively young and alert staff members who were allowed a great latitude of action and independence of thought.

The new commander of the *Thirty-second Army* inherited a combat organization which had been specially established for the expected invasion of Okinawa. Many independent artillery, mortar, antiaircraft artillery (AAA), antitank (AT), and machine gun groups supplemented the fire power of the basic infantry units assigned to the army. As a result of the *IGHQ* decision in June 1944 to reinforce the Okinawa garrison, nine infantry and three artillery battalions were to be sent to augment the force already on the island.[20] The majority of the reinforcements arrived from their previous stations in China, Manchuria, and Japan between June and August 1944.

The veteran *9th Infantry Division*, first to arrive, possessed battle honors dating from the Russo-Japanese War of 1904–5. Coming directly from Manchuria, and scheduled by the high command as the backbone of the defense force, the *9th's* stay on Okinawa was short-lived. The critical situation on Leyte required the assignment of the *9th* there, and Ushijima, ". . . in accordance with orders of Imperial General Headquarters, decided on 17 November to redeploy the 9th Division in order to send an elite unit with a proud and glorious war record to a battlefield where the Imperial Army would engage in a decisive battle." [21]

Probably, the most important of all of the factors which may have influenced the course of the coming battle for the Japanese, and favored an Allied victory, was the loss of this division and the fact that it was never replaced. It left in late December for the Philippines by way of Formosa where it sat out the rest of the war, prevented by Allied submarines and airplanes—and MacArthur's landing on Luzon in January—from either continuing on to its destination or returning to Okinawa.[22]

---

[18] Colonel Yahara's professional background included a preponderence of staff assignments. He had been a military attache in the United States and Thailand, returning to the latter country as a participant in the Burma campaign. MIS, HistDiv, GHQ, FEC, *Personal History Statements*, n.d., p. 56.

[19] *War History Office Comments.*

[20] "Empire Defense Operations Plan, dtd 15Jun44," in CinCPac–CinCPOA Bul 194–45, Special Translation No. 86, dtd 7Aug45, pp. 5–6.

[21] *Okinawa Operations Record*, p. 33.

[22] MIS, WD, *Order of Battle for the Japanese Armed Forces*, dtd 1Mar45, p. 32.

THIRTY-SECOND ARMY OFFICERS *sit for a formal portrait in February 1945. Numbers identify: (1) Rear Admiral Minoru Ota, Commander, Naval Base Force; (2) Lieutenant General Mitsuru Ushijima, Commanding General, Thirty-second Army; (3) Lieutenant General Isamu Cho, Army C/S; (4) Colonel Hiromichi Yahara, Army Senior Staff Officer. (Photograph courtesy of OCMH, DA)*

NORTH BANK *of the Bishi Gawa shows the typical integrated tomb-cave-dugout defenses which characterized Japanese organization of Okinawan terrain. (USA SC183743)*

Since the *9th Infantry Division* was no longer available to the *Thirty-second Army,* and in order to carry out his defensive plans, Ushijima asked for replacements. He was notified by *IGHQ* on 23 January 1945 that the *84th Division* in Himeji would be sent to Okinawa. This notification was cancelled that same day with the explanation that the greatest possible supply of munitions would be sent, but replacements neither could nor would be sent to the army.[23] This, in effect, put Ushijima on notice that the means to improve his situation had to be found locally.

In June 1944, the *Thirty-second Army* was to have been reinforced by Major General Shigeji Suzuki's *44th Independent Mixed Brigade (IMB),* a unit of approximately 6,000 men organized that very month on Kyushu. It was originally composed of the *1st* and *2d Infantry Units* (each essentially of regimental size) and attached artillery, engineer, and signal units. While en route to Okinawa, the *Toyama Maru,* the ship carrying the brigade, was torpedoed by an American submarine off Amami O-Shima on 29 June. More than 5,000 men were lost and only about 600 survivors of the ill-fated brigade landed on Okinawa; these were used as the nucleus of a reconstituted *2d Infantry Unit.* Other replacements were obtained from Kyushu as well as from the ranks of conscripted Okinawans, but the reorganized unit was never fully re-equipped. As a result, this lack of basic infantry equipment caused the *2d Infantry Unit* to be known among other soldiers on the island as the *Bimbo Tai* or "have-nothing-unit."[24] The *1st Infantry Unit* was never rebuilt and existed merely as a headquarters organization. Instead, the *15th Independent Mixed Regiment (IMR),* a unit newly raised in Narashino, Chiba-ken, was flown directly to Okinawa during the period 6–11 July and added to the *44th IMB* in September, bringing its strength up to about 5,000 men.

The next unit of importance to arrive was the *24th Infantry Division* which landed in August. Since its initial organization as part of the *Kwantung Army* in October 1939, the *24th* had been responsible for the security of the eastern boundaries of Manchuria. The division, commanded by Lieutenant General Tatsumi Amamiya, was well-equipped and well-trained, but not battle-proven, before it joined the *Thirty-second Army.* The *24th* was a triangular division which had been stripped of its infantry group headquarters, one battalion from each infantry regiment, an artillery battalion, and an engineer company, all of which had been added to expeditionary units sent from Manchuria to the Central Pacific in early 1944. Until a general *Thirty-second Army* reorganization in February 1945, the *24th's* infantry regiments (*22d, 32d,* and *89th Infantry*) functioned with only two battalions each. The division set up its headquarters at Kadena, and in October, it assigned 300 Okinawan conscripts, received from the *Thirty-second Army,* to each of its infantry regiments for training and retention later by the training unit. The February reorganization brought the *24th* nearly up to its

[23] *Okinawa Operations Record,* pp. 44–45.

[24] *IntelMono,* pt I, sec B, chap 3, p. 9.

original strength and made it the largest tactical unit in the *Thirty-second Army,* with more than 14,000 Japanese troops and Okinawan draftees assigned to infantry, artillery, reconnaissance, engineer, and transport regiments, and divisional troops.

The final major unit assigned to General Ushijima's command was the *62d Infantry Division,* commanded by Lieutenant General Takeo Fujioka. This was a brigaded organization which had seen action in China following its activation there in June 1943. Its table of organization, considerably different from the *24th Division's,* was similar to that of like units in the *Chinese Expeditionary Army.* Both of the *62d's* brigades had served as independent commands in China since 1938, while the division as a whole fought in the April–June 1944 campaigns in northern Honan Province. Each brigade had four independent infantry battalions (*IIB*s); the *63d Brigade* had the *11th, 12th, 13th,* and *14th IIBs,* while the *15th, 21st, 22d,* and *23d IIB*s were assigned to the *64th Brigade.* In 1944, two additional *IIBs* were sent to Okinawa as reinforcements and attached on 15 December to the division which, in turn, assigned them to the brigades. The *272d IIB* went to the *64th Brigade,* while the *273 IIB* went to the *63d.*

The *62d Division* lacked organic artillery and had few other supporting arms. It never attained a strength greater than 12,000 troops, the largest proportion of whom were infantrymen. The infantry battalions of the *62d* were the strongest units of their type on Okinawa, as each battalion mustered a total of 1,200 men organized into five rifle companies, a machine gun company, and an infantry gun company armed with two 75mm guns and two 70mm howitzers. The *272d* and *273d IIB*s were reported later as having a strength of 700 men each, but with one or two less rifle companies per battalion.

Some variance in strength was found in the infantry components of the other two major fighting organizations of the *Thirty-second Army.* The *2d Infantry Unit* and *15th IMR* of the *44th IMB* had in common three rifle battalions, an antitank company (four 37mm or 47mm AT guns), and a regimental gun company (four 75mm guns). Each of the battalions listed a total strength of 700 men who were assigned to three rifle companies, a machine gun company, and an infantry gun unit (two 70mm howitzers). The *24th Division* regimental organization was similar except for the replacement, in one battalion of each regiment, of the 70mm howitzers by a mortar platoon manning four 81mm mortars.

Since the Japanese high command envisioned the coming battle for Okinawa as developing into one of fixed position defense, the defenders were not assigned any appreciably strong armored force. The entire Japanese tank strength, given to the *Thirty-second Army* in July, consisted of the *27th Tank Regiment,* organized originally in Manchuria in April 1944, from elements of the *2d Armored Division.* It was a regiment in name only, as one of its medium tank companies was sent to the garrison at Miyako Jima. What remained was an armored task force with a strength of 750 men who filled the ranks of one light and one medium tank company, a tractor-drawn artillery battery, an infantry company, a maintenance company, and an engineer platoon. The regiment's heavy weapons included 14 medium and 13 light tanks, 4 75mm guns, 2 47mm AT guns, and 10 machine guns. The heaviest tank-mounted weapon was the 57mm gun on the medium tanks.

As the Japanese position in the Philippines became hopeless, shipments of weapons to be sent there were diverted by *IGHQ* to Okinawa. The result was that the *Thirty-second Army* possessed a heavier concentration of artillery power, grouped under a single command, than had been available to any Japanese force in previous Pacific campaigns. The total artillery strength on Okinawa, with the exception of the *24th Division's* organic *42d Field Artillery Regiment,* was grouped within Major General Kosuke Wada's *5th Artillery Command.* Besides the comparatively weak *7th Heavy Artillery Regiment* (formerly the *Nakagusuku Wan Fortress Artillery Unit*), General Wada's command included two medium regiments, a heavy battalion, and the artillery units of the *44th IMB* and *27th Tank Regiment.* Combat-tested at Bataan in the Philippines, the *1st Medium Artillery Regiment* had one of its two battalions assigned to Miyako Jima upon arrival from Manchuria in July. The other medium regiment was the *23d* which, until its departure for Okinawa in October, had been stationed in eastern Manchuria from the time of its activation in 1942. The two medium artillery regiments together mustered a total of 2,000 troops who manned 36 150mm howitzers. The artillery command also contained the *100th Independent Heavy Artillery Battalion.* This unit was formed in June of 1944 in Yokosuka and sent to Okinawa in July with 500 men and 8 150mm guns.

Besides artillery units, General Wada's troop list included a mortar regiment and two light mortar battalions. The *1st Independent Heavy Mortar Regiment's* 320mm spigot mortars were an unusual type of weapons which Marines had first encountered on Iwo Jima.[25] These awesome weapons, firing a 675-pound shell dubbed a "flying ashcan" by Americans, were the basic armament of this unit. Only half of its six batteries were on Okinawa, as the other three had been sent to Burma in mid-1942. Although the 96 81mm mortars of the *1st* and *2d Light Mortar Battalions* were nominally under the command of General Wada, actually they were assigned in close support of the various infantry units and usually operated under the direction of their respective sector defense commanders.

The infantry was strengthened with other types of artillery weapons from antiaircraft artillery, antitank, and automatic weapons units which were attached to them during most of the campaign. A dual air-ground defense role was performed by the 72 75mm guns and 54 20mm machine cannon in 4 independent antiaircraft artillery, 3 field antiaircraft artillery, and 3 machine-cannon battalions. In addition, 48 lethal, high-velocity, flat trajectory 47mm guns (located in 3 independent antitank battalions and 2 independent

---

[25] LtCol Whitman S. Bartley, *Iwo Jima: Amphibious Epic* (Washington: HistBr, G-3 Div, HQMC, 1954), pp. 13-14, hereafter Bartley, *Iwo Jima.*

companies) were added to the defense. Completing the infantry fire support of the *Thirty-second Army* were 4 independent machine gun battalions which had a total of 96 heavy machine guns.

The rest of General Ushijima's army consisted of many diverse units and supporting elements. The departure of the *9th Division* created a shortage of infantry troops which had to be made up in as expeditious a manner as possible. The reserve of potential infantry replacements on the island varied in quality from good, in the two shipping engineer regiments, to poor, at best, in the various rear area service organizations. The *19th Air Sector Command*, whose airfield maintenance and construction troops were stationed at the Yontan, Kadena, and Ie Shima airstrips, provided the largest number of replacements, 7,000 men.

Another source of troops to fill infantry ranks was found in the sea-raiding units. These organizations, first encountered by American forces in the Philippines, were designed for the destruction of amphibious invasion shipping by means of explosive-laden suicide boats. There were a total of seven sea-raiding squadrons in the Okinawa Gunto, three of which were based at Kerama Retto. Each of the squadrons had assigned to it 100 hand-picked candidates for suicide and martyrdom, whose caliber was uniformly high since each man was considered officer material. When one of these men failed to return, it was presumed that his had been a successful mission and, reportedly, he was therefore given a posthumous promotion to second lieutenant.

Japanese naval base activities on Okinawa were under the command of Rear Admiral Minoru Ota. Admiral Ota was commander of the *Naval Base Force* for the Okinawa area, commander of the *4th Surface Escort Unit*, and also was in charge of naval aviation activities in the Nansei Islands. Army-Navy relations and the chain of command on Okinawa were based locally on mutual agreements between the *Thirty-second Army* and the *Naval Base Force*.[26]

Admiral Ota directed the activities of approximately 10,000 men, of whom 3,500 were Japanese naval personnel and the other 6,000–7,000 were civilian employees belonging to sub-units of the *Naval Base Force*. Of the total number of uniformed naval troops, only about 200 were considered to have received any kind of infantry training. Upon the activation of the base force on 15 April 1944, a small number of naval officers and enlisted men, and most of the civilians, were formed into maintenance, supply, and construction units for the large airfield on Oroku Peninsula and the harbor installations at Naha. At Unten-Ko, on Motobu Peninsula in the north, were stationed a torpedo boat squadron and a midget submarine unit.

In organizing for the defense of the island, the greater portion of regular naval troops were formed into antiaircraft artillery and coastal defense bat-

[26] *Okinawa Operations Record*, pp. 57–59; for a discussion of the Japanese high command setup, see Henry I. Shaw, Jr. and Maj Douglas T. Kane, *Isolation of Rabaul—History of U. S. Marine Corps Operations in World War II*, v. II (Washington: HistBr, G–3 Div, HQMC. 1963), hereafter Shaw and Kane, *Isolation of Rabaul*, especially that section in chap I entitled "Japanese Strategy."

teries. These were broken down into four battery groups which were emplaced mainly in the Naha-Oroku-Tomigusuku area. The antiaircraft units manned 20 120mm guns, 77 machine cannon, and 60 13mm machine guns, while the 15 coast defense batteries, placed in strategic positions on the coastline under the control of Army local sector commanders, stood ready by their 14cm and 12cm naval guns. Although the total strength in numbers was impressive, the *Okinawa Naval Base Force* did not have a combat potential commensurate with its size.

Continually seeking means to bolster his defenses, General Ushijima received permission to mobilize a home guard on the island. In July 1944, the Okinawa Branch of the Imperial Reservists Association formed a home guard, whose members were called *Boeitai.* They were organized on a company-sized basis by town or village and were mainly comprised of reservists. Since the *Boeitai* represented a voluntarily organized group, it did not come under the Japanese Military Service Act, although their training and equipment came from the regular forces into whose ranks they were to be integrated when the battle was joined. The total number of *Boeitai* thus absorbed by the *Thirty-second Army* has been estimated between 17,000 and 20,000 men.

On Okinawa there were certain units which have often been confused with the *Boeitai.* These were the three *Special Guard Companies* (*223d, 224th,* and *225th*) and three *Special Guard Engineer Units* (*502d, 503d,* and *504th*) which were special components of the *Thirty-second Army.* During peacetime, each unit had a cadre of several commissioned and noncommissioned officers. When war broke out, certain designated reservists reported to the above units to which they had been previously assigned.[27]

Even the youth of the island were not exempt from the mobilization. About 1,700 male students, 14 years of age and older, from Okinawa's middle schools, were organized into volunteer youth groups called the *Tekketsu* (*Blood and Iron for the Emperor Duty Units*). These young boys were eventually assigned to front-line duties and to guerrilla-type functions for which they had been trained. Most, however, were assigned to communication units.

It has not been conclusively determined how many native Okinawans were actually added to the forces of the *Thirty-second Army,* or to what extent they influenced the final course of battle. What is known, however, is that their greatest contribution was the labor they performed which, in a period of nine months, transformed the island landscape into hornets' nests of death and destruction.

THE JAPANESE DEFENSES[28]

Continuing American successes in the conduct of amphibious operations forced the Japanese to recognize the increasing difficulties of defending against assaults from the sea. The loss

[27] *War History Office Comments.*

[28] Unless otherwise noted, the material in this section is derived from: *IntelMono; POW InterSum; Shimada Interrogation; Yahara Interrogation;* Hayashi and Coox, *Kōgun.*

of some islands in 1944 reportedly caused Japanese garrison units at other Imperial bases in the Pacific to lose confidence in themselves and their ability to withstand an American seaborne invasion. The Japanese high command hastily published the "Essentials of Island Defense," a document which credited Americans with overwhelming naval and air power, and emphasized that the garrisons should "lay out and equip positions which can withstand heavy naval and aerial bombardment, and which are suitable for protracted delaying action . . . diminish the fighting effectiveness of landing units . . . seize opportunities to try to annihilate the force in one fell swoop." [29]

This document may have influenced General Ushijima's decisions when he settled on a final defense plan, although his particular situation was governed primarily by the strength of the *Thirty-second Army* and the nature of the area it was to defend. Captured on Okinawa were a set of instructions for the defense of Iwo Jima, which were apparently a blueprint also for the defense of critical areas on the coasts of the islands of Japan. It is assumed that Ushijima may have seen these instructions, for they bore directly on his problem:

> In situations where island garrisons cannot expect reinforcements of troops from rear echelons, but must carry on the battle themselves from start to finish, they should exhaust every means for securing a favorable outcome, disrupting the enemy's plans by inflicting maximum losses on him, and, even when the situation is hopeless, holding out in strong positions for as long as possible.[30]

In order to deceive the assaulting forces as to Japanese intentions, a *Thirty-second Army* battle instruction warned the troops to "guard against opening fire prematurely." [31] A later battle instruction explained that "the most effective and certain way of [the Americans'] ascertaining the existence and organization of our firepower system is to have us open fire prematurely on a powerful force where it can maneuver." [32]

These instructions were a forewarning that, rather than forcing the issue on the beaches, "the Japanese soldier would dig and construct in a way and to an extent that an American soldier has never been known to do." [33] Japanese organization of the ground paralleled that which assault troops had discovered on Biak, Saipan, and Peleliu in 1944 and Iwo Jima in 1945.[34] General Cho, a strong advocate of underground and cave fortifications, took an active

[29] Hayashi and Coox, *Kōgun*, pp. 115–116.

[30] "Land Defense Doctrine (Provisional), dtd 1Dec44," in CinCPac-CinCPOA Bul 147–45, Translations and Interrogations No. 32, dtd 16Jun45, p. 4.

[31] "Thirty-second Army Battle Instructions No. 3, dtd 15Feb45," in CinCPac-CinCPOA Bul 122–45, Translations and Interrogations No. 30, dtd 1Jun45, p. 2.

[32] "Thirty-second Army Battle Instruction No. 8, dtd 8Mar45," in *Ibid.*, p. 7.

[33] *IntelMono*, pt I, sec A, p. 5.

[34] For a discussion of Japanese defenses on these islands, see Robert R. Smith, *The Approach to the Philippines—U. S. Army in World War II* (Washington: OCMH, DA, 1953), pp. 300–302; Maj Frank O. Hough, *The Assault on Peleliu* (Washington: HistDiv, HQMC, 1950), pp. 192–197; Bartley, *Iwo Jima*, pp. 5–18.

part in designating where defensive positions were to be placed. The most favorable terrain for the defense was occupied and honeycombed with mutually supporting gun positions and protected connecting tunnels. Natural and man-made barriers were effectively incorporated to channel attackers into prepared fire lanes and pre-registered impact areas. The reverse as well as the forward slopes of hills were fortified, while artillery, mortars, and automatic weapons were emplaced in cave mouths, with their employment completely integrated within the final protective fire plan.

Each unit commander, from brigade down to company level, was made responsible for the organization of the ground and fortification of the sector assigned to him. The need for heavy construction was lessened, in some cases, by the abundance of large caves on Okinawa which required but slight reinforcement to enable them to withstand even the heaviest bombardment. Once improvements were made, these natural fortresses served either as hospitals, barracks, command posts, or all of these combined when the size of the cave permitted. There were generally two or more entrances to the caves, which sometimes had more than one level if time and manpower was available for the extensive digging necessary. Tunnels led from the caves to automatic weapons and light artillery positions which, in conjunction with the pillboxes and rifle pits in the area, dominated each defense zone. The approaches and entryway to each cave were invariably guarded by machine guns and, in addition, by covering fire from positions outside the cave.

Integrated within the whole Japanese defensive system, these cave strongholds were, in turn, centers of small unit positions. Item Pocket, one of the most vigorously defended sectors on Okinawa, was typical of the ones American forces ran into. (See Map I, Map Section.) The area encompassed by this position, roughly 2,500 by 4,500 yards in size, was in the vicinity of Machinato Airfield. Both the 1st Marine and 27th Infantry Divisions fought bitterly to gain it. Disposed within the caves and bunkers of the pocket was a reinforced infantry battalion which manned approximately 16 grenade launchers, 83 light machine guns, 41 heavy machine guns, 7 47mm antitank guns, 2 81mm mortars, 2 70mm howitzers, and 6 75mm guns. A minefield and an antitank trench system completed the defenses. This sector was so organized that there were no weak points visible to the attacker. Any area not swept by automatic weapons fire could be reached by either artillery or mortars. These defensive positions formed a vital link in the chain of the tough outer defenses guarding Shuri.

Based on the dictum that "the island must be divided into sectors according to the defense plan so that command will be simplified," [35] each combat element of the *Thirty-second Army* was assigned a sector to develop and defend as it arrived on Okinawa. By August 1944, the *44th IMB's 2d Infantry Unit* (400 troops) under Colonel Takehiko

---

[35] "Combat Regulations for Japanese Garrison Units," n.d., in CinCPac-CinCPOA Bul 115–44, dtd 18Aug44.

Udo had occupied its assigned area, Kunigami Gun (County), and had assumed responsibility for all of the island north of the Ishikawa Isthmus, and also for Ie Shima and its airfields. Upon its arrival on Okinawa, the *24th Division* had begun to construct field fortifications around Yontan and Kadena airfields in an area bounded by Ishikawa Isthmus in the north and a line from Sunabe to Ozato in the south. Below the *24th's* zone of defense, the *62d Division* was unflagging in its efforts to alter the ridges, ravines, and hillsides north of Shuri. Responsibility for the entire southern portion of Okinawa below Shuri had been assumed by the *9th Division* commander.

The receipt of orders in November for the transfer of the *9th Division* forced a redeployment of *Thirty-second Army* troops and strained a defense that was already dangerously weak. The *24th Division* began moving south to take over some *9th Division* positions while the *44th IMB*, leaving two reinforced battalions of the *2d Infantry Unit* behind on Ie Shima and Motobu Peninsula, occupied an area which reached from Kadena airfield southward to Chatan. The *62d Division* positions were likewise affected by the withdrawal of the *9th's* 14,000 combat troops, as the northern divisional boundary of the *62d* dropped to the Chatan-Futema line. In the south, the *62d* zone of responsibility was increased tremendously to include all of Naha, Shuri, Yonabaru, and the entire Chinen Peninsula.

Although the construction of fortifications, underground positions, and cave sites had been going on since the spring of 1944, the urgency of the war situation and the expectance of an imminent invasion compelled the defenders to reevaluate their plans of deployment for blunting the assault. The exact date of the new *Thirty-second Army* plan is not known, but a reasonable assumption is that the loss of the *9th Division* in November which triggered the shuffling of units also forced a decision on a final defense plan. At the end of the month, General Ushijima and his staff pondered the following alternatives before settling on the one which they believed would guarantee the success of their mission:

> Plan I: To defend, from extensive underground positions, the Shimajiri sector, the main zone of defenses being north of Naha, Shuri, and Yonabaru. Landings north of these defenses were not to be opposed; landings south of the line would be met at the beaches. Since it was impossible to defend Kadena airfield [with available troops], 15cm guns were to be emplaced so as to bring fire on the airfield and deny the invaders its use.
>
> Plan II: To defend from prepared positions the central portion of the island, including the Kadena and Yontan airfields.
>
> Plan III: To dispose one division around the Kadena area, one division in the southern end of the island, and one brigade between the two divisions. To meet the enemy wherever he lands and attempt to annihilate him on the beaches.
>
> Plan IV: To defend the northern part of the island, with Army Headquarters at Nago, and the main line of defense based on Hill 220, northeast of Yontan airfield.[36]

Realistically appraising the many factors which might effect each one of the alternate plans, the Japanese settled on Plan I. Plan III was abandoned simply because the *Thirty-second Army*

---

[36] *IntelMono*, pt I sec A, pp. 1–2.

did not have the strength adequate to realize all that the plan encompassed. Plan IV was rejected because it conceded the loss of the militarily important south even before the battle had been joined. Plan II, the one which American staff planners feared as offering the greatest threat to a successful invasion, was regretfully relinquished by the Japanese. Ushijima, recognizing his troops' capabilities and limitations, realized that his forces, in the main, had not been trained to fight this type of delaying action which would prolong the battle, bloody the invaders, and permit the bulk of his army to withdraw to the more heavily fortified southern portion of Okinawa. Yet, in effect, this is exactly the strategy he was forced to employ after the initial American landings.

Placing Plan I into effect, the Japanese centered the main battle position in the Shuri area, where the rugged terrain surrounding the ancient capital was developed with the strongest installations oriented north toward the Hagushi beaches. (See Map 3.) The Hagushi region, coincidentally, evolved as a secondary target to the Japanese and a primary target to American staff planners. In addition, "handicapped by their lack of ability to make a logistics estimate for a landing operation," [37] the Japanese believed that the major effort would be made in the southeast with an assault across the Minatogawa beaches. Overlooking both the Minatogawa and Nakagusuku Wan beaches, Chinen Peninsula heights presented the defenders with the most favorable terrain of its type on Okinawa and, as such, it was hoped that the invaders could be met and defeated here. Since, from the standpoint of actual manpower, the Chinen sector was the weakest area in the final defense plan, a goodly portion of the artillery and infantry strength of the *Thirty-second Army*—which could have been better employed in reinforcing Shuri positions—was diverted to the peninsula, remaining there out of action during the first weeks of the campaign.[38]

Among Ushijima's most pressing needs were additional troops and time in which to train them. Extra time was needed also to provide for expanding and strengthening existing fortifications as well as the communications net. With the exception of a drastic fuel shortage, the army was in good logistical shape. Although the *Thirty-second Army* itself had no provisions in reserve, enough had been distributed to subordinate units, and stored by them in caves near troop dispositions, to last until September 1945. This system was satisfactory in that the strain on the overworked transportation facilities was removed, but when an area was overrun by Americans and the Japanese were

[37] *Ibid.*, p. 3.

[38] In the *Thirty-second Army* staff there was sharp disagreement as to the probability of this additional landing. Colonel Yahara, senior staff officer, insisted that a diversionary landing, possibly the principle one, would be made in the Minatoga region. Major Yakamaru, the intelligence officer, held that the only American landing would be in the Hagushi area. Prestige and seniority won the argument. 'Yakamaru, bitterly disappointed at the final decision, went off the next few days to inundate his sorrows in prolonged draughts of expensive *sake.*' *Shimada Interrogation.*

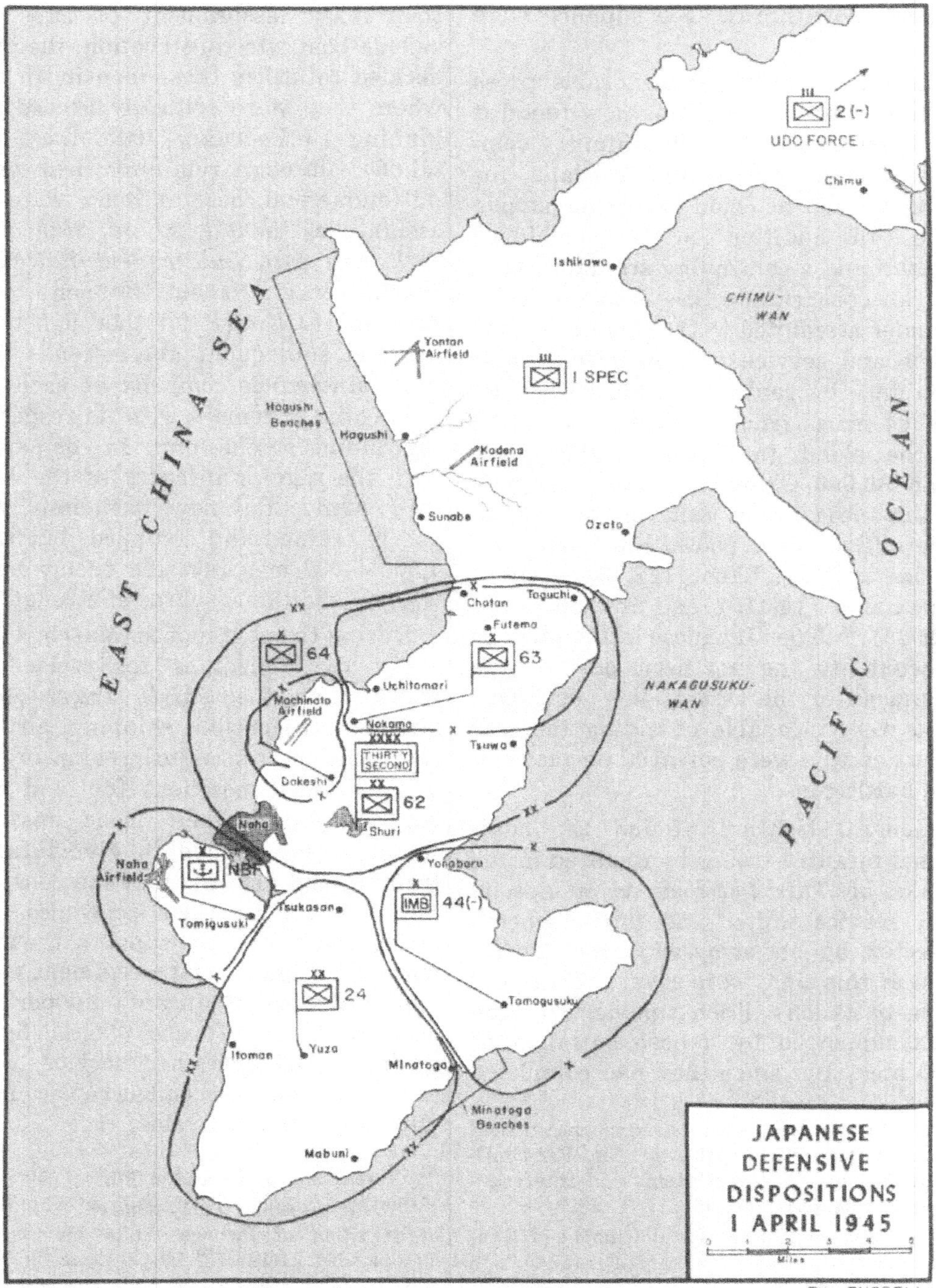

MAP 3

T. L. RUSSELL

forced to withdraw, the supplies were lost.

Unable to halt the inexorable press of time, General Ushijima now found it imperative to beef-up his infantry component from sources on the island, for he knew that he could expect no outside help. In addition to the mobilized *Boeitai* and a continuing stream of Okinawan conscriptees, the Japanese commander attempted to free his uniformed labor and service personnel for frontline duty by replacing them with able-bodied males from the large population of the island. In February 1945, more than 39,000 Okinawans were assigned to Japanese Army units on the island. The natives were placed into such categories as Main Labor (22,383), Auxiliary Labor (14,415), and Student Labor (2,944).[39] The Japanese attempted to evacuate to the northern part of the island all of the rest of the population who were incapable of aiding the war effort or who were potential obstacles in the battle zone.[40]

General Ushijima found the additional infantry troops he required in the ranks of *Thirty-second Army* special and service units. The first elements affected by an army-wide reorganization at this time were seven sea-raiding base battalions. Each suicide squadron was supported by a base battalion of 900 men, and since they had completed their basic assignment of cave and suicide boat site construction, the army decided to utilize these men in an area where they were critically needed. Beginning 13 February 1945, these battalions, although retaining their original numerical designations, were reassigned as the *1st, 2d, 3d, 26th, 27th, 28th,* and *29th Independent Battalions* (each averaging about 600 men) to the *24th* and *44th IMB* for thorough training and subsequent absorption.[41] Only the maintenance company of each battalion was to remain with its respective sea-raiding suicide unit. In comparison with the regular infantry of the Japanese Army, the new battalions were poorly trained and equipped, but these 4,500–5,000 men invested enemy forces with an additional source of strength.

During the next month, March, a final army reorganization took place, at which time the *Thirty-second Army* directed "the various shipping, air, and rear echelon forces [to] set up organizations and dispositions for land combat."[42] Besides their basic missions, these units now had to give infantry training and field fortification construction priority in their schedules. The March reorganization supplied the army with two brigades and a regiment which appeared more significant on paper than actually was the case. These lightly equipped and untrained service troops could serve only as combat replacements with slight tactical value.

[39] "Thirty-second Army Assignment of Conscript Labor, dtd Feb45," in CinCPac-CinCPOA Bul 161–45, Translations and Interrogations No. 34, dtd 27Jun45.

[40] "Standards for the Establishment of Nansei Shoto Garrison Plans, dtd 1Jan45," in *CICAS Trans* No. 83; *Okinawa Operations Record,* pp. 59–60.

[41] *IntelMono,* pt I sec A p. 10.

[42] "Thirty-second Army OpOrd A–113, Reorganization of Service Units into Combat Troops, dtd 21Mar45," in CinCPac-CinCPOA Bul 107–45, Translations and Interrogations No. 28, dtd 14May45, p. 15.

Units from the *19th Air Sector Headquarters* were funneled into the *1st Specially Established Regiment* which, under *62d Division* control, was responsible for the defense of the areas in the vicinity of Kadena and Yontan airfields. Support positions in the Naha-Yonabaru valley were assumed by the *1st Specially Established Brigade*, composed of three regiments and formed from *Thirty-second Army* transport, ordnance, construction, and supply troops formerly within the *49th Line of Communications Headquarters* command. A *2d Specially Established Brigade* of three regiments, culled from the *11th Shipping Group Headquarters* shipping, sea transport, and engineer rosters, was deployed in support of the *24th Division* mission—the defense of southernmost Okinawa. "Army rear echelon agencies not included in this order and their personnel will be under command of the front line unit in the vicinity where their duties are carried on, and will reinforce it in combat," stated the all-inclusive 21 March order which put the entire *Thirty-second Army* in a status of general mobilization for combat.[43]

By 26 March, *Okinawa Base Force* naval and civilian personnel had been formed into the same type of jerry-built, poorly equipped, and undertrained defense units as had been the service troops of the *Thirty-second Army*. On Oroku Peninsula, naval lieutenants commanded those units designated as battalions while lieutenants (junior grade) became company commanders. Admiral Ota's 13mm and 25mm antiaircraft batteries were re-equipped and transformed into an 81mm mortar battery and two independent machine gun battalions and, thus armed, were the only adequately weaponed units in the naval garrison.

In less than two months after the first reorganization order had been published, General Ushijima had nearly doubled the potential combat strength of his army by the addition of approximately 20,000 *Boeitai*, naval, and service troops. Hurriedly, the concerted efforts of this determined Japanese force converted the Shuri area into what was to be an almost impregnable bastion, for the final defensive plan was strengthened by the defenders' determination to hold Shuri to the last man.

Concurrent with the February army reorganization, the troops were deployed in their final positions. General Ushijima's main battle force was withdrawn to an outpost zone just north of Futema, while elements of the *1st Specially Established Regiment* were loosely disposed in the area immediately behind the Hagushi beaches. Although this was the least likely place where the Americans were expected to land, the Japanese troops defending this area were to fight a delaying action in any such eventuality, and then, after destroying the Yontan and Kadena airfields, were to beat a hasty retreat to the Shuri lines.

In the suspected invasion area, the Minatogawa beaches, the bulk of the Japanese infantry and artillery forces were positioned to oppose the landings. The *5th Artillery Command* observation post was established near Itokazu in control of all of its major components, which had been emplaced in defense of

[43] *Ibid.*

the Minatogawa sector. Since landings further north on Chinen Peninsula would give the invaders a relatively unopposed, direct route into the heart of the major Japanese defense system, the *44th IMB* was assigned control of the rugged heights of the peninsula. The *24th Division,* taking over the defense works begun by the *9th Division,* occupied the southern portion of Okinawa from Kiyan Point to an area just north of Tsukasan. The whole of Oroku Peninsula was assigned to Admiral Ota's forces, who were prepared to fight the "Navy Way," contesting the invasion at the beaches in a manner reminiscent of the Japanese defense of Tarawa.[44]

Since the heart and soul of the Japanese defenses were located at Shuri, the most valuable and only battle-tested organization on the island, the *62d Division,* was charged with the protection of this vital area. The Japanese had shrewdly and industriously constructed a stronghold centered in a series of concentric rings, each of which bristled with well dug-in, expertly sited weapons. Regardless of where the Americans landed, either at Hagushi or Minatogawa or both, the plans called for delaying actions and, finally, a withdrawal into the hard shell of these well-disguised positions.

[44] Although the *Base Force* was under command of the *Thirty-second Army* and Admiral Ota sincerely attempted to cooperate with the army, classic interservice rivalry apparent in many Japanese Pacific operations, on a lower echelon in the case of Okinawa, hampered the naval commander's desires. "Naval Units on Okinawa," in *POW InterSum* No. 16, dtd 28Jul45.

The isolated north was defended by the *Udo Force,* so-called after its leader and commanding officer of the *2d Infantry Unit*—Colonel Takehiko Udo. Its mission was twofold, defense of both Motobu Peninsula and Ie Shima. The reinforced battalion on Ie Shima was assigned secondary missions of destroying the island's airfield and assisting in the transfer of aviation materiel to the main island. Upon completion of these duties, the unit was then to return to Okinawa where it would be assigned to the control of the *62d Division.* Udo's battalion on Motobu Peninsula, in expectation of an invasion of Ie Shima followed by a landing on the peninsula, was disposed with its few artillery pieces so placed as to make its positions and positions on Ie Shima mutually supporting. As a result of its detachment earlier from the larger portion of the *Thirty-second Army,* Udo's command was destined to fulfill a hopeless undertaking to the very end.

Air defense was not included in the *Thirty-second Army* plan, nor was any great aviation force available to Ushijima. He had expected that approximately 300 airplanes would be sent to Okinawa, but feared that their projected time of arrival, April, would be too late to influence the local situation. The American preinvasion air and naval bombardments in March, combined with planned Japanese destruction efforts, had rendered the Ie Shima, Yontan, Kadena, and Oroku airfields unusable.

The army did expect, however, that its exertions would be complemented by the combat activity of its organic suicide sea units. The sea-raiding squadrons located at positions in Kerama Retto and along the Okinawa coast, would "blast to pieces the enemy transport groups with a whirlwind attack in the vicinity of their anchorages." [45] Unfortunately for the Japanese, their midget submarines and motor torpedo boats at Unten-Ko could not join this offensive endeavor, for, by the day of the American invasion, they had all been destroyed by American carrier strikes or scattered in the aftermath of an unsuccessful attack on the destroyer *Tolman* of Task Force 52.[46]

The significance of *Thirty-second Army* deployments and redeployments, the frenzied last-minute preparations, and the general air of expectancy were not lost upon even the lowest ranks. One private wrote as early as February, "it appears that the army has finally decided to wage a decisive battle on Okinawa." [47] Another soldier noted that "it's like a frog meeting a snake, just waiting for the snake to eat him." [48]

Between 20 and 23 March 1945, the Japanese command on Okinawa made an even more realistic estimate than had the troops of what the future held for the garrison. The Japanese reacted to news of a conference held in Washington between Admirals King and Nimitz in early March by placing a general alert into effect "for the end of March and early April," since statistics demonstrated "that new operations occur from 20 days to one month after [American] conferences on strategy are held." [49] This estimate of when the Americans were expected was reduced three days after its publication following receipt of reports of increased shipping in the Marianas, and when repeated submarine sightings and contacts were made. All of this enabled the Japanese intelligence officers to predict without hesitation that the target was to be "Formosa or the Nansei Shoto, especially Okinawa." [50]

[45] "Thirty-second Army OpOrd A 115, dtd 23Mar45," in *CICAS Trans* No. 321.

[46] "Naval Units on Okinawa," in POW InterSum No. 16, dtd 28Jul45; CTF 52 AR, Okinawa Gunto, 21Mar-20Apr45, dtd 1May45, chap III, p. 9, hereafter *CTF 52 AR*.

[47] "Diary of a superior private, 272d IIB," in CinCPac-CinCPOA Bul 147-45, Translations and Interrogations No. 32, dtd 16Jun45, p. 48.

[48] "Diary of a radioman, 5th Harbor Base Unit," in CinCPac-CinCPOA Bul 161-45, Translations and Interrogations No. 34, dtd 27Jun45, p. 50.

[49] "Thirty-second Army Estimate of the Situation, dtd 20Mar45," in CinCPac-CinCPOA Bul 140-45, Translations and Interrogations No. 31, dtd 7Jun45, p. 5.

[50] "62d Division IntelRpt on Findings Since the 20Mar Rpt, dtd 23Mar45," in *Ibid.*, p. 6.

Project ICEBERG

THE TASK DEFINED [1]

Three weeks after receipt of the Joint Chiefs of Staff directive ordering the Okinawa invasion, Admiral Nimitz' headquarters published and distributed the ICEBERG Joint Staff Study. This study served as a planning guideline for the units assigned to the campaign and defined for them the objectives, the allotment of forces, and roughly outlined the scheme of maneuver ashore.

Although Operation CAUSEWAY, the invasion of Formosa, had been cancelled in favor of ICEBERG, the principal commanders for CAUSEWAY were retained for the Okinawa landing and redirected their staffs' efforts towards planning for the assault on the newly assigned target. Admiral Raymond A. Spruance, whose Task Force 50 (TF 50) contained the Fifth Fleet and the Central Pacific Task Forces, was made responsible for the Ryukyus operation. His staff, previously charged with preparing plans for the Iwo Jima invasion scheduled for 20 January 1945, was now given the concurrent assignment of planning for Okinawa.

Certain assumptions governed task planning and the assignment of assault and garrison forces for ICEBERG. Adherence to the scheduled 1 March 1945 invasion date (L-Day) for Okinawa was based on the presupposed seizure of Iwo Jima at a date early enough to permit release of naval gunfire and air support units for the second operation. It was further assumed that ICEBERG commanders would be able to secure the prompt release from General MacArthur of assault shipping, support shipping, supporting naval forces, and Army troops assigned to the Philippines operation which had been earmarked for use later at Okinawa. Finally, before Okinawa was invaded, Allied air and surface superiority had to be gained in the target area.

This last point was one of the most important in the overall concept of the operation, for it was believed that air attacks on Japan, together with the conquest of Iwo Jima, would force a concentration of Japanese air strength on the bases which ringed the Home Islands. It would be necessary, therefore, to destroy enemy air installations at Japanese staging areas in Kyushu and Formosa, and neutralize those at Okinawa, since it was a basic assumption that enemy aircraft would vigorously oppose any invasion attempt. For this reason, the scheme of maneuver ashore included plans for the

[1] Unless otherwise noted, the material in this section is derived from: *ICEBERG Study;* CinCPac-CinCPOA OPlan 14–44, dtd 31Dec44, hereafter *CinCPOA OPlan 14–44; USAFMidPac G–5 Hist;* ComFifthFlt OPlan 1–45, dtd 3Jan45, hereafter *ComFifthFlt OPlan 1–45;* CG, Tenth Army TntvOPlan 1–45, ICEBERG, dtd 6Jan45, hereafter *Tenth Army TntvOPlan 1–45;* Samuel Eliot Morison, *Victory in the Pacific, 1945—History of United States Naval Operations in World War II,* v. XIV, (Boston: Little, Brown and Company, 1960), hereafter Morison, *Victory in the Pacific.*

early securing of airfields on Okinawa and their equally early use by Allied land-based aircraft. Japanese sea communications in the Ryukyus area were to be severed before the operation by surface and air attacks on enemy shipping and by a maximum effort mounted by American submarines.

According to the ICEBERG staff study, operations ashore were to be conducted in three phases. To be accomplished in the first phase were the capture of the southern portion of Okinawa and small adjacent islands and the initial development of base facilities. In Phase II, Ie Shima and the remainder of Okinawa were to be seized and the base build-up continued with the construction of installations in favorable locations designated in the development plan. Phase III required the exploitation of Allied positions in the Nansei Shoto and, when Admiral Nimitz directed, the seizure and development of additional positions with forces then locally available. (See Map 4.)

It was envisoned that an army of two corps, each composed of three reinforced infantry divisions, would be required in the initial assault. In addition, two divisions were to be assigned as area reserve. Okinawa's proximity to the heart of the Empire as well as to other major Japanese bases, and the expectation of fanatic resistance by enemy troops on a battleground of such large dimensions, presaged a prolonged period of fierce combat. For these reasons, a new command relationship was established for the Okinawa operation differing, in some respects, from that which had been effective in previous Pacific campaigns.

As strategic commander of the invasion forces, Admiral Nimitz directed that the chain of command would descend to Admiral Spruance, thence to Vice Admiral Richmond K. Turner who would command Task Force 51 (Joint Expeditionary Force), and then to Lieutenant General Simon B. Buckner, Jr., USA, who would command the Army, Navy, and Marine units comprising the Expeditionary Troops. When Spruance had determined that the amphibious phase of the invasion had ended, he would pass the command of all forces ashore to Buckner. As Commanding General of the Tenth Army, Buckner would assume responsibility for the defense and development of positions captured on the island. When the situation permitted, he would also relieve Admiral Spruance of the responsibility for the defense and development of the Ryukyus as a whole and, at that time, he would be directly responsible to CinCPOA for the captured island positions and for the waters within a 25-mile radius. Concurrently, responsibility for the establishment of an Island Command and a military government on Okinawa would be General Buckner's also.

ALLIED COMMANDERS AND FORCES [2]

Many units of Admiral Nimitz' command not directly assigned Task Force 50 were to support the Okinawa landing

---

[2] Unless otherwise noted the material in this section is derived from: *ICEBERG Study; CinCPOA OPlan 14-44; USAFMidPac G-5 Hist; ComFifthFlt OPlan 1-45; Tenth Army TntvOPlan 1-45;* Morison, *Victory in the Pacific.*

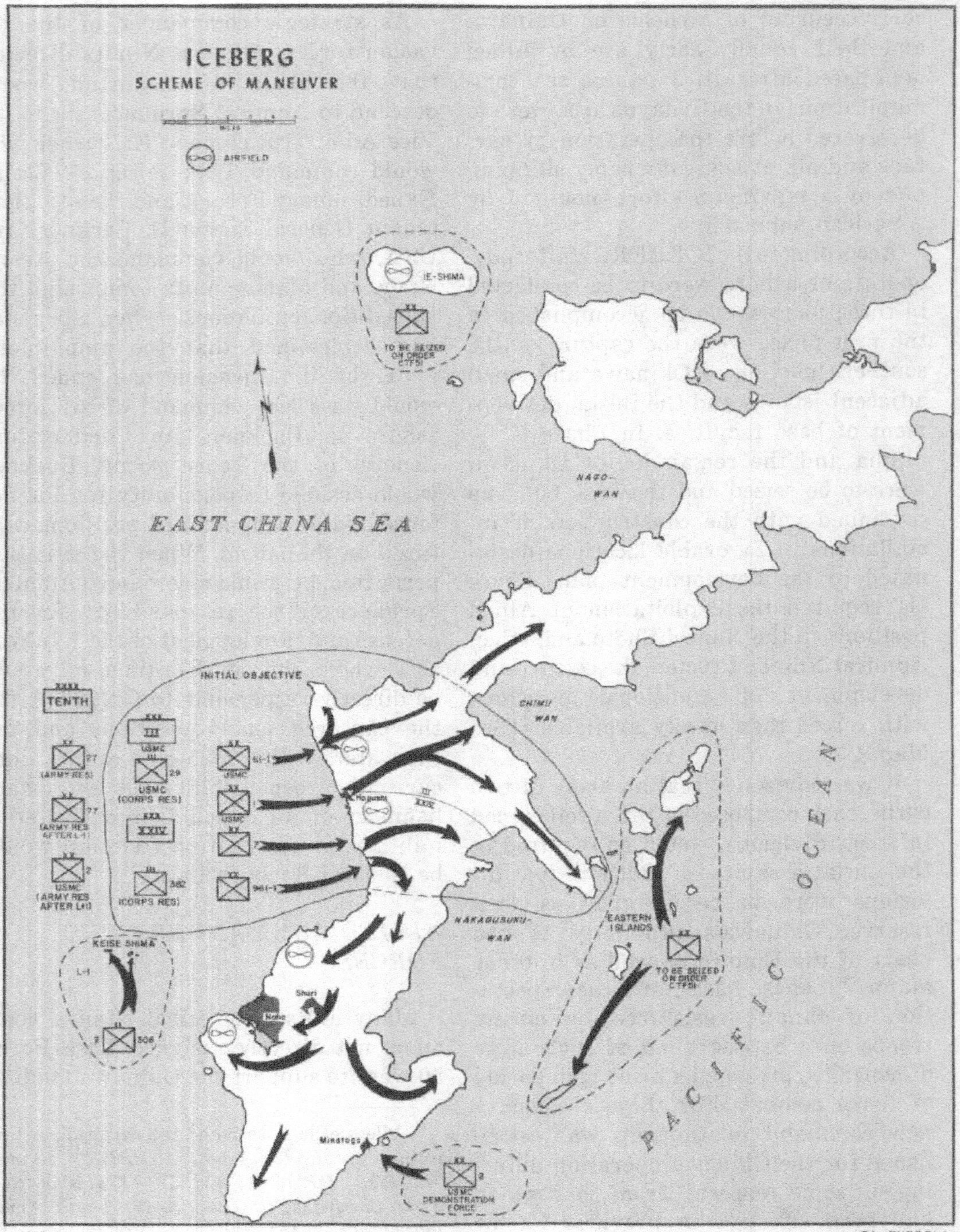

MAP 4 T.L. RUSSELL

from bases widespread in the Pacific Ocean Areas. Additionally, from their airdromes in China and the Southwest Pacific, Army Air Forces elements were to assist the ICEBERG effort, both prior to and during the course of the campaign. In all, about 548,000 men of the Marine Corps, Army, and Navy, together with 318 combatant and 1,139 auxiliary vessels—exclusive of numerous small personnel craft of all types[3]—and a profusion of strategic and tactical aircraft were to strike some of the last blows dooming the Japanese attempts to gain supremacy in Asia and the Pacific.

In the Fifth Fleet were the Covering Forces and Special Groups which included the Fast Carrier Force (TF 58, Vice Admiral Marc A. Mitscher) and the British Carrier Force (TF 57, Vice Admiral Sir H. Bernard Rawlings, RN). These two forces were to conduct air strikes and neutralize Japanese air power prior to the landing, and prevent enemy air and surface interference with the Allied landing and subsequent occupation of Okinawa.

The units more directly concerned with the landing were components of Turner's Task Force 51. Its complex composition reflected its many assignments incident to the capture, occupation, and defense of Okinawa. Any enemy attempt to disrupt the movement to the target or landing on the beach would be handled by the force's support elements. These naval units would also undertake air support and minesweeping operations once the beachhead had been gained. Assignments for these tasks were allocated, in turn, to the Amphibious Support Force (TF 52, Rear Admiral William H. P. Blandy) which provided direct air and naval support, and to the Gunfire and Covering Force (TF 54, Rear Admiral Morton L. Deyo).[4] The Northern Attack Force (TF 53, Rear Admiral Lawrence F. Reifsnider) and the Southern Attack Force (TF 55, Rear Admiral John L. Hall, Jr.,) contained the transports which were to lift the assault troops to the objective and the tractor units which were to land them on L-Day.

The assault of Okinawa and its surrounding islands was to be accomplished by the landing forces of Buckner's Expeditionary Troops (TF 56). The assault force of the Northern Attack Force was Major General Roy S. Geiger's III Amphibious Corps (IIIAC), composed of the 1st Marine Division (Major General Pedro A. del Valle) and the 6th Marine Division (Major General Lemuel C. Shepherd, Jr.). The Army XXIV Corps (Major General John R. Hodge) would be lifted by the Southern Attack Force and would consist of the 7th Infantry Division (Major General Archibald V. Arnold) and the 96th Infantry Division (Major General James L. Bradley).

One other major Marine echelon in the Tenth Army was Major General Francis P. Mulcahy's joint air task

[3] *War Reports*, p. 664.

[4] Vice Admiral Jesse B. Oldendorf, originally the commander of TF 54, was injured at Ulithi shortly before the operation. "Fortunately, Rear Admiral Morton L. Deyo, veteran gunfire support commander in Operation OVERLORD [Normandy invasion] and DRAGOON [invasion of southern France], was available to relieve Admiral Oldendorf." Morison, *Victory in the Pacific*, p. 109.

command, Tactical Air Force (TAF), which was to provide land-based air support for the operation once its squadrons were ashore. The elements initially assigned to TAF were to come primarily from the 2d Marine Aircraft Wing (2d MAW). Although TAF was established under the Tenth Army on 21 November 1944, its staff was not really organized until late in December. By that time, much of the earlier, basic, and important preinvasion planning had been completed without TAF participation. As a matter of fact, the last of the personnel assigned to TAF staff did not even report until after the assault echelon had already left for the target.[5] Although he had not taken part in ICEBERG planning, General Mulcahy was kept fully abreast of Tenth Army activities and decisions by his chief of staff, Colonel Perry O. Parmelee, who daily visited Buckner's headquarters and attended briefings and conferences there.[6]

A most important element of TAF was its fighter arm, the Air Defense Command (ADC), headed by Brigadier General William J. Wallace who had formerly been AirFMFPac Chief of Staff. Wallace's squadrons were to begin operations from previously designated airfields on Okinawa as soon as they had been captured by the ground troops. Initially, General Wallace's command consisted of a headquarters squadron and a service squadron, and three MAGs with a total complement of nine fighter, two night fighter, and four air warning squadrons. The radar installations of the units last named would give early warning of enemy air attacks. An Army Air Forces fighter wing was also part of ADC, but only one group was to join TAF before the campaign was brought to a close.

General Mulcahy's Bomber Command was made up wholly of AAF flight and support elements, none of which arrived on Okinawa before the beginning of June. Photographic coverage of enemy installations, interpretation of the pictures thus obtained, and an aerial photographic survey of the island for mapping purposes were to be the missions of an AAF photo-reconnaissance squadron which was also part of the TAF organization.

Rounding out the Tenth Army air force were two Marine torpedo-bomber squadrons which were to conduct anti-submarine warfare operations together with the carrier-based naval aircraft at the target. The Marine squadrons were also prepared to conduct bombing attacks on ground targets and any other missions when the need for them arose.

Marine aviation, other than that which was organic to TAF, was to play an important part in the invasion. Artillery spotting was the assigned mission of Marine observation squad-

---

[5] *Tenth Army AR*, chap 11, sec IV, pp. 2–3; TAF WarDs, 29Oct44-31Jan45; Tactical Air Force, Tenth Army, AR, Ph I, Nansei Shoto, 8Dec44-30Jun45, dtd 12Jul45, pt I, chap 3, p. 2, hereafter *TAF AR*.

[6] MajGen Ford O. Rogers interview with HistBr, HQMC, dtd 14Apr66, (Oral History Collection, HistBr, HQMC), hereafter *Rogers interview*. General, then Colonel, Rogers, who was the Air Defense Command Chief of Staff, did not arrive at Pearl Harbor to take over his new job until early January 1945. In commenting on the fact that TAF and the Air Defense Command had not participated in planning for the invasion, General Rogers said that it had no serious or visible effect on the conduct of Marine air operations at Okinawa.

rons attached to the Marine divisions and corps. Scheduled to control all aircraft in support of the ground forces were Colonel Vernon E. Megee's Landing Force Air Support Control Units (LFASCUs). When directed by Admiral Turner, LFASCUs, set up ashore at the headquarters of Tenth Army and its two corps, would take over control from their shipboard naval counterparts.

In addition to the tactical units assigned to the Tenth Army for the assault and consolidation phases of the operation, General Buckner was to have direct command of the defense and service troops assigned for the garrison phase. Major General Fred C. Wallace, USA, was designated Island Commander, Okinawa,[7] while the Naval Forces, Ryukyus, were to be commanded by Rear Admiral Calvin H. Cobb, who would assume his command upon completion of the amphibious phase of the operation. Although strategic air force and naval search squadrons were to be based on Okinawa, they would remain under the operational control of the Commanding General, Army Air Forces, Pacific Ocean Area, and Commander, Fifth Fleet, respectively.

Infantry units were assigned also to the Western Islands Attack Group (TG 51.1, Rear Admiral Ingolf N. Kiland) which had the 77th Infantry Division (Major General Andrew D. Bruce) as its landing force; the Demonstration Group (TG 51.2, Rear Admiral Jerauld Wright) whose landing force was the 2d Marine Division (Major General Thomas E. Watson); and the Floating Reserve Group (TG 51.3, Commodore John B. McGovern) which carried 27th Infantry Division (Major General George W. Griner, Jr.).

JOINT PREPARATIONS AND PLANNING[8]

Intensive joint planning attested to the immensity of the future operation. Smooth Army, Navy, and Marine Corps coordination of operational, logistical, and administrative matters was imperative. Since the Tenth Army, under CinCPOA, would consist of an Army corps and a Marine amphibious corps, and a large naval contingent, General Buckner believed that it was important for him to have a joint staff. He therefore requested Admiral Nimitz to authorize a Marine and naval augmentation of his staff. When this request was granted, approximately 30 Marine and 30 Navy officers, and enlisted assistants from each of these services, were assigned and integrated within the Tenth Army staff. "There was no Marine or naval section of the staff." [9] One of the Marine officers was Brigadier General Oliver P. Smith, who became the Marine Deputy Chief of Staff; he had been the Assistant Division Commander of the

[7] This command also was to include the Ie Shima garrison, and, on Okinawa, the Naval Operating Base and the Naval Air Bases.

[8] Unless otherwise noted the material in this section is derived from: *Tenth Army AR;* 1st MarDiv SAR, Nansei Shoto, 1Apr-31Jun45, dtd 1Jul, hereafter *1st MarDiv SAR;* 6th MarDiv SAR, Okinawa Operation, Phases I and II, dtd 30Apr45, hereafter *6th MarDiv SAR, Ph I & II.*

[9] Gen Oliver P. Smith, Personal Narrative, n. d., p. 4, hereafter Smith, *Personal Narrative.*

1st Marine Division in the Peleliu campaign. His counterpart on the Tenth Army staff was Brigadier General Lawrence E. Schick, who filled the billet of Army Deputy Chief of Staff.

When General Smith arrived at Tenth Army headquarters, he found that CinCPOA had already approved the Marine augmentation for the Army staff. The Marine general believed that this augmentation was overly large, for:

> This padding would result in Marine officers doing clerical duty at Army Headquarters as there were manifestly not enough bona fide billets to take care of all the Army officers on the staff as well as the Marine and naval officers.[10]

After considerable discussion with the Army officer responsible for the assignment of staff billets, General Smith managed to have the number of Marine officers on the Tenth Army staff reduced by nearly 30 percent.[11]

A tactical concept based upon the directive stated in the ICEBERG joint staff study, and later incorporated in the TF 50 operation plan, required "early use of sufficient airdrome capacity in Okinawa, together with unloading facilities adequate to support its development and to maintain positive control of the air in the area." [12] In a study of all landing beach areas in southern Okinawa, those beaches on the west coast which lay north and south of Hagushi were deemed to be best suited to support the ICEBERG landing. Admiral Turner's operation plan assumed that there would be bitter Japanese air reaction to the Okinawa invasion; that enemy submarines would be very active in the target area; that the Japanese surface fleet might possibly sortie out from its bases in Japan; and, that attempts might be made to reinforce the garrison on Okinawa. The first three assumptions proved correct; the fourth was not tested because, in accordance with the JCS directive ordering the invasion of Okinawa, Allied air and surface superiority had been gained prior to L-Day.

Based on Admiral Turner's plan, the Tenth Army staff drew up Plan Fox, which committed the assault forces to a landing on the west coast. Plan Fox also included the pre-L-Day capture of Keise Shima, since a study of this small island indicated the feasibility of its use as a fixed emplacement for artillery which would first augment the naval and air bombardment of the main objective before the landing, and afterwards provide support during the land campaign. This plan, approved by Buckner, was presented to Turner at the initial joint conference held at Pearl Harbor on 1 November 1944.

Following this presentation, Turner stated his views of the operation and outlined what would be the requirements of the Navy during the course of ICEBERG. He believed that, prior to the landings on Okinawa, the adjacent islands had to be neutralized. Once this had been done, the major landings on Okinawa would be more secure and the fleet could be replenished in a safe an-

---

[10] *Ibid.*, p. 25.

[11] A list of the names of Marine officers on the Tenth Army staff when ICEBERG forces landed on Okinawa is located in *Ibid.*, pp. 25–26.

[12] *ComFifthFlt OPlan 1–45*, Anx B, p. 2.

chorage without danger from enemy surface vessels or submarines.

Two provisions of Plan Fox particularly concerned the Fifth Fleet commander. Because of the suspected presence of Japanese mines and submarines immediately west of Okinawa, should the Hagushi beaches be used for the invasion, the landings here would perforce require the fleet to steam into a hazardous area. The second apprehension arose because 1 March had been scheduled as L-Day. He feared that unfavorable weather conditions, which generally prevailed in March, might possibly affect the conduct of the landings and unduly prolong the unloading of supplies on exposed beaches. Available meteorological data justified this concern, for from October to March the Ryukyus experienced strong northerly winds with a mean velocity of 17-19 miles-per-hour as well as frequent gales. A generally moderate wind, averaging 11 miles-per-hour, marked the beginning of the summer monsoon period and characterized the weather of Okinawa in April, which was a more suitable time for the invasion.[13] In any case, Turner requested that the possibility of landings along the east coast be restudied. At the same time, he suggested that the value of a feint landing be determined and, if valid, should be incorporated in the plan finally adopted for ICEBERG.

After a lengthy discussion of the problems inherent in the proposed plan, the conferees concluded that a landing on the western beaches on 1 March was fraught with considerable risk. The alternatives were either a 30-day delay of the operation or a landing on the southeast coast on the date originally scheduled for the assault. All other possible courses of action were re-examined, with the result that the Hagushi beaches were recommended again as the site for the landings. Final approval was withheld by Turner because he retained doubts as to the practicality of landing and supporting the proposed assault force of four divisions over the Hagushi beachhead. In spite of the objections of Admiral Turner, the Plan Fox estimate was distributed on 5 November.[14] When completed on 9 November, another detailed study upheld the original contention that Hagushi held the only beaches in southern Okinawa adequate to receive four divisions abreast and, subsequently, to handle sufficient logistical support for the operation.

In the face of these convincing arguments, Admiral Turner accepted the plan with the proviso that both Kerama Retto and Keise Shima were to be captured prior to the main landing. With minor exceptions, General Buckner concurred with these modifications, and the revised plan was forwarded to Turner on 11 November. The original target date of 1 March was changed twice within the next month, first to 15 March and finally to 1 April. The first change was made on 19 November in anticipation of bad weather at the target at the beginning of March. On 7 December, Admiral Nimitz advanced L-Day two more weeks when doubts arose as to whether the shipping assigned to General MacArthur's Lingayen Gulf opera-

[13] *HQUSAFPOA Study*, sec XX, p. 1.

[14] Smith, *Personal Narrative*, p. 13.

tions could be returned in time to permit its reemployment at Okinawa.[15]

Since the naval planning staff recommended a sustained seven- or eight-day bombardment of the assault beaches, the resulting expenditure of Navy supplies and ammunition would force the bombardment group to either withdraw from the area for resupply and refueling or to conduct these operations under dangerous conditions in the open sea offshore of the objective. Basically, it was this consideration that prompted Turner's insistence on the pre-L-Day capture of the entire Kerama group. At first, these islands appeared to be only worthy as targets for amphibious raids in which the raiding parties would retire after destroying enemy coastal artillery. Later plans for their capture grew out of Admiral Turner's proposal that, once taken, the Keramas provide a protected anchorage for the establishment of a small-boat pool and a seaplane base.

Because the Kerama assault was now to be a full-scale invasion instead of a raid, the assignment of a larger force was indicated and Major General Thomas E. Watson's 2d Marine Division was chosen initially. This unit, designated IIIAC Reserve, had been slated for early commitment in support of operations on Okinawa, and so the task of capturing the Keramas was given instead to the 77th Infantry Division while the Marine division was assigned tentatively to a feint landing off southeastern Okinawa.[16]

As the scope and importance of preliminary operations grew, the reserves which had been made available to General Buckner originally decreased in number, and it was found necessary to secure from CinCPOA release of the area reserve division (27th Infantry Division). This unit was then designated as the Tenth Army floating reserve and was replaced by the 81st Infantry Division which remained in New Caledonia under Admiral Nimitz' control.

The alternate plan for the operation, Plan Baker, was approved on 3 January 1945. It envisioned first the capture of

[15] *USAFMidPac G-5 Hist*, pp. 183, 201. "The deferment of the target date to April 1 was most fortunate from the logistic angle. Under CinCPOA procedures, all maintenance supplies for Okinawa were to be shipped from the West Coast to the control point at Ulithi (3d and subsequent echelons were staged through Eniwetok) for call forward as required. Requisitions for these supplies had to be in the hands of West Coast supply agencies 60 days prior to sailing date of the shipment. Due to the sailing time required, requisitions for the first maintenance shipment to support a 1 March target date had to be on the West Coast by 20 November. With no firm tactical plan until after the conference with Admiral Turner on 9 November, and lacking a firm troop basis, the determination of supply requirements had to be based on very rough estimates. The 30-day delay in target date enabled supply agencies to make a more careful estimate of the supply requirements of the assault force. This delay also enabled critical supplies and augmentation personnel, required for the assault, to be shipped to mounting points of the divisions (some had to be shipped by air) prior to mounting date." BGen David H. Blakelock, USA ltr to CMC, dtd 3Oct54, hereafter *Blakelock ltr.*

[16] *Tenth Army AR*, chap 3, pp. 11–12. The original concept of the operation anticipated that the 2d Marine Division would come out of army reserve, pass through the 1st Division, and take the Katchin Peninsula to the southeast of the latter's zone. *1st MarDiv SAR*, chap III, OperAnx, p. 1.

Kerama Retto, followed by a sweep of the Eastern Islands by General Watson's Marines. Both of these actions were to be conducted prior to the assault of Okinawa itself. A mixed Marine and Army corps artillery group was to support both the XXIV and III Amphibious Corps assault of the east coast.

On L-Day, General Geiger's Marines would land between Chinen Point and Minatoga, secure the high ground behind the beaches, and, following the Army landing two days later, tie-in with XXIV Corps at Yonabaru. After effecting this juncture, both corps were to make a rapid advance across the island during which time the Marines were to take the airfield on Oroku Peninsula and the Army was to capture the unfinished field at Yonabaru. Included in the alternate plan were provisions for the capture of Ie Shima, feints against Chimu Wan on L plus 3 or 4, and, overall, the maintenance of flexibility of action in the commitment of Army reserves to either of the corps zones or for the protection of XXIV Corps' northern flank.

Although the principal advantages of Plan Baker were that the approach to the east coast of Okinawa was more direct and the weather here was vastly superior to that of the west coast, they were outweighed by the disadvantages. These included: (1) the difficulty of providing optimum naval gunfire support because of the interposition of the Eastern Islands and off-shore islets, (2) the paucity of good beaches, (3) the length of time it would take to uncover airfields, located, for the most part, on the west coast, and, (4) because of Plan Baker landing zone assignments, the possibility that Japanese forces might be able to concentrate considerable strength against IIIAC troops before they could even contact the XXIV Corps. General Smith was convinced at this time that "in the advent of bad weather on the west coast, landings would have been delayed rather than resort to the east coast landing as provided in the alternate plan." [17]

General Geiger became involved in the planning for ICEBERG in November 1944, when he was directed to report to General Buckner for planning purposes. Upon receipt of this order, the IIIAC commander immediately reported by dispatch. Shortly thereafter, IIIAC headquarters received a copy of the tentative Plan Fox together with all available intelligence on the prospective target, and a request that Geiger prepare a tentative corps operation plan.

When the IIIAC plan was completed, and at the request of Buckner, Geiger, accompanied by his chief of staff, Colonel Merwin H. Silverthorn,[18] his G–2, Lieutenant Colonel Sidney S. Wade, his G–3, Colonel Walter A. Wachtler, his G–4, Colonel Francis B. Loomis, Jr., and other members of his staff, departed Guadalcanal for Pearl Harbor, arriving at Schofield Barracks on 9 December. After personally contacting their opposite numbers on the Tenth Army staff, the IIIAC staff officers prepared to present their plan to General Buckner.

Geiger planned to employ the 1st and 6th Marine Divisions in the assault, with General del Valle's division on the right or south flank. The choice of these

[17] Smith, *Personal Narrative*, p. 30.

[18] Silverthorn was promoted to the rank of brigadier general on 19 December while still in Hawaii.

divisions was logical since they were both located in the Solomons and there would be no problem in establishing liaison. The 2d Division, based on Saipan, would be the floating reserve of the army, according to the IIIAC plan. The question then arose regarding what steps would be taken if the Japanese were encountered in strength as IIIAC advanced eastward across Okinawa, for there was no doubt that an additional division would have to be inserted in the line before the east coast was reached. General Smith took this question up with the Tenth Army commander, who agreed that IIIAC would have first call on the 2d Marine Division.[19]

General Watson's division was scheduled to make the feint landings on the southeast coast of Okinawa on L-Day and L plus 1, and it was not contemplated that Geiger would need it before the third day of the operation. The IIIAC staff presented their plan orally to General Buckner on 19 December, when it was approved. According to General Smith, who was present on this occasion, Geiger's staff members "did a very creditable job. . . ." [20]

SCHEME OF MANEUVER [21]

Basically, the scheme of maneuver ashore was designed to attain early use of the airfields so that land-based air supremacy over the target could be gained and held. An additional dividend derived from the capture of the airfields would be their use as staging bases for continuing mass air raids on both Japan and those areas within flying range of Okinawa under enemy control. As in the case of earlier amphibious landings in the Pacific, certain preliminary softening-up steps had to be taken before the main assault was launched.

Kerama Retto was to be seized by the 77th Infantry Division (Reinforced) on 26 March 1945, or six days before L-Day. Following the first day of operations in the Kerama Retto and beginning the night of the 26th, Marines of the FMF Amphibious Reconnaissance Battalion were to reconnoiter the reef islets of the island group. First they were to investigate Keise Shima for the presence of enemy troops, and in the following days and nights prior to L-Day, they were to land on Aware Shima, Mae Shima, and Kuro Shima. To support the landing on Okinawa, a field artillery group of XXIV Corps Artillery was to land and be emplaced on Keise Shima prior to L-Day. While these operations were underway, Okinawa would receive increased air and naval gunfire bombardment which would mount in intensity until the first assault waves neared the beaches. At this time, the fire would lift from the beach area and continue inland.

The Army and Marine divisions were to land on the Hagushi beaches, General Geiger's corps on the left. The mouth of the Bishi Gawa marked the beginning of the corps boundary, which

---

[19] Smith, *Personal Narrative*, p. 28.

[20] *Ibid.*

[21] Unless otherwise noted the material in this section is derived from: *Tenth Army TntvO-Plan 1-45;* IIIAC OPlan 1-45, dtd 1Feb45, hereafter *IIIAC OPlan 1-45;* 1st MarDiv OPlan 1-45, ICEBERG, dtd 10Feb45, hereafter *1st MarDiv OPlan 1-45;* 6th MarDiv OPlan 1-45, dtd 10Feb45, hereafter *6th MarDiv OPlan 1-45.*

310-224 O - 69 - 6

roughly followed the course of the river to a point just north of Kadena; here, the line headed almost due east to bisect the island. (See Map 4.)

Once landed north of Hagushi town, the Marine assault divisions were to move rapidly inland, coordinating their advance with that of XXIV Corps. On the Marine left flank was the 6th Division; the 22d Marines on the left and the 4th Marines, less its 2d Battalion in division reserve, on the right. The 29th Marines, the third infantry regiment of the 6th Division, was corps reserve and was to be ready to land on any of the beaches. It was also to be prepared to revert one battalion landing team to the 6th Division on order. General Shepherd's initial mission was the capture of Yontan airfield while protecting the northern flank of the Tenth Army.

General del Valle's division, landing to the right of the 6th, was to assist in the capture of Yontan by quickly seizing the high ground northeast of China. The attack was then to continue, with major emphasis placed on maintaining contact with General Hodge's corps and assisting his advance. The 1st Marine Division scheme of maneuver placed the 5th and 7th Marines in the assault, 7th on the left, and the 1st Marines in division reserve.

Adjoining the 1st Marine Division was to be the 7th Infantry Division, with one regiment in division reserve but under the operational control of XXIV Corps. The other Army assault division was to be 96th, which was to land with two regiments abreast and a third in corps reserve.

Artillery support for the Marines was to come from IIIAC Corps Artillery and those artillery units organic to the divisions. General Geiger's guns were to land on his order to support the attack and, once ashore, corps artillery would coordinate all supporting arms in the Marine sector. XXIV Corps Artillery, less the group on Keise Shima, would land on General Hodge's order and support the attack with long-range interdiction, counterbattery, and harassing fires.[22]

Following the initial landing, operations were designed to isolate the Phase I objective, which consisted of that part of the island lying south of a general line drawn across the Ishikawa Isthmus, through Chimu, and including the Eastern Islands. In order to prevent enemy reinforcement from the north and to fulfill its assignment in Phase I, IIIAC was to gain control of the isthmus as swiftly as possible. To seal off the Japanese in the south, General Hodge's troops were to drive across the island, his right flank units holding a line that ran through Futema to Kuba Saki. Once the central portion of the island had been captured and secured, the direction of attack would be faced to the south and continued until all of the objectives of the first phase had been achieved.

Phase II, the seizure of northern Okinawa and the capture of Ie Shima, was to be executed with Tenth Army troops locally available when Buckner was satisfied that Phase I had been accomplished. The first major military objec-

---

[22] XXIV Corps FldOrd 45, dtd 8Feb45, pp. 4–8.

tive in the north was Motobu Peninsula, which was to be taken by means of simultaneously launched attacks from sea and land. Once the peninsula had been gained, a shore-to-shore assault would be made against Ie Shima. The end of Phase II would be signalled when the rest of northern Okinawa had been captured.

While higher echelon air planning for ICEBERG detailed both strategic and tactical missions, the Tenth Army was more immediately concerned with the latter. Carrier-based tactical aviation, aboard the TF 52 escort carrier group (TG 52.1, Rear Admiral Calvin T. Durgin), was to provide the invasion force with air support until General Mulcahy's squadrons were established ashore and could take over. At this time, TAF would also be responsible for overall air defense.

When this responsibility was assumed, TAF operations would be based on the following order of priority: (1) attainment of air superiority by annihilation of enemy aircraft in the air and on the ground, and destruction of enemy air installations; (2) interdiction and destruction of enemy troop and supply movements immediately within or heading towards the target area; and (3) execution of combined air-ground attacks on specific frontline objectives. The importance of the first priority lay in Tenth Army recognition of the yet-existing Japanese air strength and the threat it posed to the invasion force.

As soon as Air Defense Command fighter squadrons were established ashore on captured airfields, they were to begin fulfilling their assigned missions. From these fields, ADC was to provide air defense to ground units on the island and naval forces in its environs. Combat air patrols, close air support, and other related flight missions were considered the means by which the defense was to be maintained. Although it was a function of ADC, close air support is not normally a part of air defense; it is more closely associated with a ground offensive concept. Despite this fact, however, Okinawa's terrain and the nature of the Japanese defenses were to provide Marine aviators of the Air Defense Command with ample opportunities to display close air support techniques born of experience accumulated in earlier Pacific campaigns.

LOGISTIC SUPPORT PLANNING [23]

Fortunately for those preparing ICEBERG, much in the logistical plans for the cancelled Formosa operation could

[23] Unless otherwise noted the material in this section is derived from: *ICEBERG Study;* CTF 51 OPlan Al-45, dtd 16Feb45, hereafter *CTF 51 OPlan A1-45;* CTF 51 General Action Report, Capture of Okinawa Gunto, Phases I and II, 17Feb-17May45, dtd 25Jul45, hereafter *CTF 51 AR; CinCPOA OPlan 14-44; Tenth Army TntvOPlan 1-45; Tenth Army AR; USAFMidPac G-5 Hist;* IIIAC, AR, Ryukyus Operation, Phases I and II (Okinawa), dtd 1Jul45, hereafter *IIIAC AR;* Tactical Air Force, Tenth Army, OPlan 1-45, dtd 10Feb45, hereafter *TAF OPLAN 1-45; TAF AR;* IsCom OPlan No. 1, LEGUMINOUS [code name assigned to island of Okinawa], dtd 1Feb45, hereafter *IsComOPlan No. 1;* IsCom, Okinawa, AR, 13Dec44-30Jun45, dtd 30Jun45, hereafter *IsComAR; War Reports; 1st MarDiv SAR; 6th MarDiv SAR, Ph I & II.*

be salvaged and adapted for the invasion of Okinawa with but few changes. Without competent logistics planning of the highest order, and utilization of a resupply and shipping support schedule designed to function with clockwork precision, the target date for the Okinawa operation could not have been met. This would have caused all related planned strategy to have been either nullified or advanced to a later date.

The logistics plan for Okinawa "was the most elaborate one of its kind developed during World War II, involving prearranged movement of both assault and cargo shipping over vast ocean distances."[24] The plan required establishment of a 6,000-mile-long supply line, stretching across the Pacific, with 11 different ports-of-call,[25] to support the mounting of 182,821 troops encumbered with some 746,850 measurement tons[26] of cargo loaded into 434 assault transports and landing ships.

A great limitation imposed upon preinvasion logistical planning was the shortage of shipping and the delay in the return from the Philippines of the vessels which were to be used for Okinawa. Seeking a solution to lift and timetable problems was not the only concern of the Tenth Army logistics staff, "for the mere loading of more ships led only to congestion at the receiving end unless the development of unloading facilities kept pace."[27]

It had been decided that the Hagushi beaches were sufficiently large to handle the supply tonnage required by the assault echelon of two corps and their support troops; however, it was impossible to prophesy exactly how soon after the landings the beachhead would be secured and the advance continued inland, or how soon thereafter base development could begin and the supplies for this aspect of Phase I would be required and available. Nor was it possible to forecast the possibility that Phase II would be completed before the accomplishment of Phase I. Nonetheless, estimates of troop progress had to be made in order to prepare a logistics plan at all.

The main features of the ICEBERG logistics plan required an initial supply level to be taken to Okinawa by the assault troops who were mounted at such distantly scattered points as Leyte, Guadalcanal, Espiritu Santo, Banika, Pavuvu, Saipan, Eniwetok, Oahu, and the west coast of the United States. Upon completion of the assault phase of the landing, a staggered series of supply shipments would replenish the Tenth Army in accordance with a schedule established earlier. This timetable had been based on the estimated time required to conduct combat operations ashore and, in turn, on how quickly the beach and port capacity could be expanded.

---

[24] Robert W. Coakley and Richard M. Leighton, "Global Logistics and Strategy, 1943–1945 —U. S. Army in World War II" (unpublished MS, OCMH), pt VI, "Shift to a One-Front War," chap XXIV, p. 26.

[25] Guadalcanal, Espiritu Santo, New Georgia, Makin, Kwajalein, Saipan, Ulithi, Manus, Milne Bay, Finschhafen, and Hollandia.

[26] Measurement ton is defined as a unit of carrying capacity of a ship, usually equal to 40 cubic feet; it is sometimes designated a freight ton.

[27] Coakley and Leighton, *op. cit.*, pt V, "The War Against Japan, 1943–1944," chap XVIII, p. 3.

Beginning on 20 February 1945, ICEBERG replenishments were to leave the west coast every 10 days for regulating points at Ulithi, Eniwetok, and Saipan, the first shipments to arrive at each place on L minus 5 (27 March). The supplies would remain at these points until they were called-up by General Buckner. It was planned to continue these automatic resupply shipments for a period of 210 days beyond L-Day. The Tenth Army was also to have emergency reserves located at Saipan, Tinian, and Guam.

The prediction of supply requirements depended upon completed tactical plans, a firm troop basis, and other necessary items of information which either were nonexistent or had not yet been made available to the logistics planners. Adding to the logistics dilemma was the factor of time, for it look 120 days for supplies to be requisitioned, procured, and shipped from the Pacific Coast of the United States to the objective.

To facilitate the preparation and shipment of resupply items in accordance with the scheduling of the various invasion echelons, Army commanders established a standard unit of supply, or "block requisitions," tailored specifically to the organization of each of the support and assault elements. The composition of the individual block requisition was determined by estimating the logistic support required by a particular unit for a given number of days regardless of the combat situation.

In contrast to this approach, Marine supply agencies, drawing on their experience, felt that the combat situation as envisioned in the planning stages should govern the nature of the supplies requisitioned, and the number, types, and frequency of shipments. Tenth Army considered the Marine system to be more flexible than the Army's because the requisitioning agencies were better able to make the several automatic resupply shipments conform to their view of how the campaign would progress.

Each service was responsible for initial support of its own elements in the Okinawa task force, with the exception of troops mounting in the South and Southwest Pacific. Area commanders there would be charged with logistical support of units assigned to ICEBERG. After the landing had been accomplished, and when directed by Admiral Turner, Island Command would take over as the Tenth Army central support agency charged with funneling supplies to all of the assault forces.

Early in January [28] it became obvious that ICEBERG had been allocated insufficient shipping to accomplish the tactical mission, to support base development, and to lift to the target those air units which were to be committed early in the campaign. An inadequate transport quota for engineer units, whose services would be needed in the

---

[28] Actually, questions regarding the allocation of shipping had appeared earlier, for on 24 November 1944, Colonel Francis B. Loomis, Jr., the G–4 of IIIAC, reported in at the Tenth Army headquarters for a short tour of temporary duty. General Geiger apparently was concerned about the shipping problem and believed it necessary to have a G–4 representative with the army. Smith, *Personal Narrative*, p. 9.

early development of airfields, roads, and waterfront facilities, was improved slightly by scheduling the immediate return of assault LSTs to Saipan after the initial landings to shuttle eight naval construction battalions (Seabees) to the target. In the same manner, other LSTs would be sent to Leyte to pick up any XXIV Corps equipment not carried in assault shipping.[29]

Of the overall inadequate shipping situation and its effect on the combat divisions, the former G–4 of the Tenth Army recalled that, if needed, ICEBERG was to get all shipping available in the Pacific, because:

> the amount of assault shipping assigned for the operation was far below that required to properly lift the assault elements of the Tenth Army. This resulted in [the Tenth Army being given] authority to modify Combat Loading Doctrine so that the most essential equipment and supplies could accompany the assault echelon. Additional items that should have been in the assault echelon were loaded in a subsequent shipping echelon.[30]

The overall assault lift was augmented by other means also. Vessels to be used for the Luzon and Iwo Jima landings were made available later for Okinawa through adherence to a stringently monitored and thoroughly regulated shipping schedule. Additional space for Tenth Army troops was gained by reducing the tonnage requirements of IIIAC, substantially at the expense of the 2d Marine Division. It was reasoned that since the division was not going to be committed immediately, it could acquire whatever additional shipping it needed within a short time following the initial assault. Further lift capacity was gained by loading landing ships to their rated limits, by the addition to the invasion flotilla of newly constructed attack transports (APAs) with greater cargo-carrying characteristics, and by an increased allocation of landing ships, tank, (LSTs) and landing ships, medium (LSMs).

The shipping allocation for the garrison forces was governed by the estimated capacity of Okinawan beach and port unloading facilities. Past experience, however, resolved the size of the lift necessary to transport an assault echelon of three reinforced Marine divisions, three reinforced Army divisions, a Marine amphibious corps headquarters and corps troops, and an Army corps headquarters and corps troops. Thus, the required assault tonnage was a firm figure from the beginning and was deducted from that allotted to the ICEBERG forces overall. The remainder was assigned as the lift for Tenth Army support troops, which included air, naval, and airfield construction units.

After the Marianas and Palau operations, it was found that one transport group (12 APAs and 3 cargo ships, attack (AKAs)), made up of three transport divisions, had sufficient lift capacity for a combat-loaded reinforced infantry division. For the ICEBERG lift, however, a new shipping echelon, the transport squadron (transron) was formed to carry a proportionate share of assault forces, corps troops, and ele-

---

[29] *Blakelock ltr.*

[30] BGen David H. Blakelock, USA, ltr to Asst G–3, HQMC, dtd 6Nov65, hereafter *Blakelock ltr 1965.*

ments from corps and army headquarters. The transron was nothing but the old transport group augmented by three APAs and three AKAs.

Each transron was to be accompanied by one APH, which was a troop transport specially rigged as a hospital and equipped to treat casualties and then evacuate them from the battle zone.[31] There were to be six hospital ships (AHs) assigned to ICEBERG; one was to be on station L minus 5 with the Kerama Retto invasion group, three were assigned to the main attack forces and were to arrive off Hagushi on L plus 1, while the other two were scheduled to reach Okinawa three days later.

Improved casualty evacuation was planned for this invasion by assigning four hospital landing ships (LST(H)s) to each of the two naval attack forces in the major assault. Assigned to each vessel was a naval medical officer who functioned as an evacuation control officer and, as such, was responsible for screening the wounded as they arrived, giving treatment and classifying them with reference to their estimated recovery time, and transferring the casualties in accordance with the provisions of a system related to their recovery classification. Accordingly, hospital ships would evacuate those men wounded seriously enough to require hospitalization for two months or more. Casualties requiring treatment for a minimum of two and a maximum of eight weeks would be evacuated in APHs during the initial assault phase and, after that, would receive further treatment in hospitals established on Okinawa. Those men who could be returned to duty within two weeks after being wounded would be treated and held in the hospital transports or landing ships until they had fully recovered or until the land-based hospitals had been established.

The LST(H)s were to remain on station until released by Admiral Turner, at which time the medical officers aboard would land and assign casualties directly to the ships from aid stations set up on the beaches. When General Buckner assumed command ashore, he would become responsible for the establishment and administration of medical services on the island, and for air evacuation of casualties, when airfields became operational.

The equipment and supplies to be taken to Okinawa by the corps and the divisions had been specifically designated by Tenth Army order. After cargo space in assigned shipping had been allocated to this material, any other available space would be filled by additional items which the corps and division commanders had decided the troops could carry. Logistical planning on the division level was influenced by the supposition that the beaches would be heavily defended and that the inland advance stubbornly resisted. As a result, only "hot cargo," predetermined blocks of high-priority supplies, was to be landed on L-Day. Included in a block of cargo

[31] The APH should not be confused with the better-known hospital ship (AH), which is unarmed and protected only by international recognition of the provisions of the Geneva Convention.

were one CinCPOA unit of fire[32] for all weapons and rations and water for one day. Moreover, all organic division motor transport would be taken to the target in available shipping space because the prospect of prolonged operations over a relatively large land mass envisioned wide-spread use of vehicles.[33]

To assist in Marine logistical planning and preparations, Fleet Marine Force, Pacific, established the 2d Field Service Command on Guadalcanal. Here relatively close liaison could be maintained with Marine ICEBERG elements mounting from the Solomons. This service command was empowered to coordinate the efforts of the supply agencies of both the 1st and 6th Marine Divisions and to deal with Army and Navy sources of supply directly. In the same manner, the Marianas-based 1st Field Service Command assisted the 2d Marine Division. Re-equipment of General del Valle's division on Pavuvu was relatively simple since its primary supply source, the 4th Base Depot, under the 2d Field Service Command, was on the other major island in the Russells, Banika. General Shepherd's division experienced some difficulties, however, because its supply source was a transfer rather than a stocking agency and had to obtain its requisitioned items from the 4th Base Depot. As a result of the cumbersome and time-consuming administrative procedures involved in processing requisitions through the several service echelons in the area, the 6th Division experienced many delays in the delivery of much of its needed equipment and supplies.[34] Both assault divisions, however, embarked for the target with but few shortages, none of which affected combat readiness and efficiency.

By the time that the TAF logistics section had been activated, AirFMFPac had already issued warning orders and was in the process of preparing subordinate units for the impending campaign. The basis for logistic support of Marine aviation units was different, in certain ways, from that of Marine ground elements. While items peculiar to the Marine Corps were drawn by both ground and air units from the same sources, all technical aviation materiel was received through Navy supply channels or, in some cases, from the Army. Since this was the case, the TAF logistics staff established liaison with representatives of Commander, Aircraft, Pacific Fleet (ComAirPac), the agency responsible for fulfilling the fuel and installation requirements at the Okinawa air fields the TAF units were to occupy. The supply section of Commander, Naval Air Bases, Okinawa (ComNABS) was made

[32] CinCPOA Unit of Fire Table, dtd 6Dec44, included in *Tenth Army TntvOPlan 1-45*, Anx 13, App B, was based on the successful criteria established by use during the Central Pacific landings. Allocation of ammunition for various ordnance was, for example: 100 rounds for each M-1; 1,500 rounds for .30 caliber and 600 rounds for .50 caliber machine guns; 275 rounds for 60mm and 81mm mortars; 250 rounds for 105mm howitzers; 150 rounds for 155mm howitzers.

[33] *1st MarDiv SAR*, pp. 1-2.

[34] 6th MarDiv SAR, Okinawa Op, Ph III, dtd 30Jun45, pp. 5-6, hereafter *6th MarDiv SAR, Ph III*, and including the SARs of the following units: 4th, 15th, 22d, and 29th Marines, hereafter (unit) *SAR, Ph III*.

the ComAirPac type command logistics representative for these matters.

All other supply requirements were to be handled by the supply section of the Navy's Pacific service command. Liaison was also established with Army Air Forces logistics representatives to determine the nature and extent of support required by Army elements in General Mulcahy's command. Arrangements were then made to obtain special combat clothing and equipment for the AAF personnel to be assigned to TAF. Based upon the latter's recommendations, automatic resupply shipments for the Army squadrons were adjusted to coincide with the schedule established for the Marines.

The organization and general administration of the supply system on Okinawa was to be an Island Command function, in which it would receive and distribute Tenth Army supplies. The Marine groups in TAF, however, would support their own squadrons and would draw Marine Corps supplies from the 2d Wing or other designated Marine sources. Air base commanders would provide aviation fuel and lubricants to squadrons operating from their strips; all technical aviation supplies were to be requisitioned through ComNABS, Okinawa.

Service units organic to the AAF fighter and bombardment groups would support the flying squadrons of each. All supplies other than the technical items peculiar to AAF planes would be requisitioned from sources designated by the Island Commander. Until an Air Service Command Depot was established on Okinawa, the one at Guam would supply the remainder.

BASE DEVELOPMENT AND MILITARY GOVERNMENT PLANNING[35]

A second logistic mission given to ICEBERG, separate yet related to the assault effort, was the immediate development of Okinawa as an advanced air and fleet base. In order to support all of the aircraft assigned to the invasion, eight airfields and one seaplane base were to be built almost immediately and during the later phases of the operation this number would be increased. Also, two ports were to be developed—one, Nakagusuku Wan, by the Navy and the other, Naha harbor, by the Army. Since Okinawa was to serve merely as a staging base for final operations against Japan, it was not contemplated that the installations on the island were to be of permanent construction. Ie Shima was included in the base development program as the island was to hold four airfields and to garrison ground and antiaircraft artillery defense troops.

Base development would proceed right on the heels of the assault troops as two of Okinawa's airfields were to be seized, improved, and made operational by L plus 5, while two more fields were to be available by L plus 20. The preparation of Okinawa as a mounting and staging point was to be undertaken concurrently. First priority was given the early development and activation of airfields; next in order of importance was the construction of bulk fuel storage facilities; and the third most important mat-

[35] Unless otherwise noted the material in this section is derived from: *ICEBERG Study; USAFMidPac G-5 Hist; War Reports; Tenth Army AR; IsCom OPlan No. 1; IsCom AR.*

ter was the development of waterfront installations. Reflecting the urgency of these tasks, every effort was made to schedule the shipments of supplies required to support base improvement so that they would arrive at the island when they were needed. Accordingly, garrison troops and the materials which they were to employ were to arrive in 17 successive echelons. The timing of their arrival was governed not only by the preplanned work schedule but also by the projected unloading capacity of the captured beaches.

To establish this schedule, a series of echelonment conferences were held between the staffs of the Tenth Army and the different type commanders who were furnishing troops for the operation. In any large amphibious operation, it is neither possible nor feasible, because of shipping limitations, to transport to the target in the assault convoy both those troops required to undertake the campaign to its end and the troops, equipment, and supplies required to develop the captured base. Even if all required shipping had been made available for an operation of the size of Okinawa, it would have been patently undesirable to schedule the simultaneous arrival at the target of both assault and garrison troops. Until the assault forces had landed, unloaded their shipping, and gained enough room on the beaches for the landing of the garrison elements and equipment, the shipping in which garrison troops were embarked would have had to lie off Okinawa, where it would have been vulnerable to enemy submarines and aircraft. For these reasons, it was imperative that echelonment plans covering the movement of thousands of assault, service, and construction troops had to be precise.

In addition to its other functions, Island Command was also to establish a military government on Okinawa. Since this was to be the first Pacific operation in which large numbers of enemy civilians would be encountered by combat troops, it was expected that the island would serve as a valuable testing ground of civil affairs and military government procedures which would be applied later when Japan itself was occupied.

In 1943, the JCS gave the Navy basic responsibility for establishing military government on certain outlying islands of the Japanese Empire, once they had been captured. Included in this group were the Ryukyus. Because the Tenth Army would be in overall control of the Okinawa land campaign, Admiral Nimitz believed that General Buckner should be responsible for military government on the island. Accordingly, once the War Department concurred in this transfer of authority, CinCPOA was able to get the 1943 JCS order reversed.

Because of its European commitments, the Army was unable to furnish all of the civil affairs personnel needed to round out the entire Tenth Army military government component. Therefore, the Navy supplied Brigadier General William E. Crist's command with naval officer and enlisted personnel so that Military Government would have well-balanced teams.

Direct naval participation in military government planning for Okinawa began in July 1944, when work was begun in New York City by the research staff of the Chief of Naval Operations' military government section. The pooled efforts of the staff resulted in the *Civil Affairs Handbook for the Ryukyu Islands*, a publication which proved to be of inestimable value to Tenth Army civil affairs administrators during both the ICEBERG planning phase and the rehabilitation period after Okinawa had been secured.[36]

The ICEBERG joint staff study originally anticipated that, within the Okinawan population to come under Tenth Army control, a small element would be "antipathetic" and would have to be "placed under detainment pending screening and probable internment." No figures were available to determine how many mainland Japanese civilians on Okinawa might possibly be captured, but preparations had to be made for the construction of an internment camp whose facilities were flexible enough to provide for upwards of 10,000 island natives and Japanese civilian internees. It was expected that by L plus 40 this number would skyrocket to an approximate total of 306,000 captured civilians, whose food, clothing, and housing would have to come from captured stocks of salvagable material, since there was no room aboard assault ships for supplies of this nature. By the time ICEBERG had reached the garrison phase, 12 military government camps were to be in operation, each unit staffed and equipped to handle 2,500–10,000 civilians.

[36] Deputy Commander for MilGovt ltr to ComNOB, Okinawa, and Chief MilGovt Officer, Ryukyus, dtd 1Jul45, Subj: Rpt of MilGovt Activities for Period from 1Apr45 to 1Jul46, hereafter *MilGovt AR*.

Assigned to General Crist's jointly staffed military government section were such varied Army and Navy units as a military police battalion, a truck company, 20 Navy dispensaries, and 6 Navy hospital units. In addition to these and some purely administrative elements, 350 officer and 890 enlisted civil affairs personnel were organized into four types of teams, each of which had been tailored for specific functions. One of the teams was assigned to each of the assault divisions and, after landing, was to conduct preliminary reconnaissance missions relating to military government as the attack advanced. Teams in another group, attached to the two corps and all divisions also, were to take charge of civil affairs behind the front lines as civilians were encountered by the combat forces. A third type of team was made up of refugee camp administrators, while in the fourth category there were six teams, each of which was to take charge of one of the six military government districts into which Okinawa was to be divided.

The Chief Military Government Officer was to be directly subordinate to the Island Commander and would function as his deputy. The importance of this close relationship and the emphasis placed on intensive civil affairs planning was justified later during the campaign, when, by 30 April, there were approximately 125,000 civilians under military government jurisdiction on Okinawa. This figure climbed steadily following this date, reached 147,829 by 31 May, 172,670 by 15 June, and totaled 261,115 on 30 June.[37]

[37] *Tenth Army AR*, chap 11, sec XXVII, p. 4.

INTELLIGENCE PLANNING [38]

In October 1944, the statement that "information as to enemy defensive installations on Okinawa Jima is meager," [39] was indisputable. Despite the early lack of information concerning the island, the various intelligence gathering and processing agencies in the Pacific, as well as those in the United States, began to sift through available material and soon were able to clarify the enemy situation for ICEBERG forces. In keeping with the established principle of coordinated planning, the corporate activities of all intelligence agencies in the various Pacific commands quickly resulted in the production of urgently needed basic intelligence.

Currently valid military information of the Japanese situation was difficult to obtain because of the location of Okinawa within the Empire's well-protected, strategic, inner defense line. For the most part, captured documents, interrogations of prisoners as well as of former island inhabitants, and old Japanese publications provided the basis for the intelligence estimates initially issued.[40] In addition, the Navy was able to make use of both captured and previously available hydrographic charts for navigational studies of the waters surrounding Okinawa.

For a terrain study, a determination of the location and nature of enemy defenses, and an estimate of enemy strength, most of the data at hand was inadequate and an aerial photographic mission over the target had to be laid on. In conjunction with other information of the enemy received right up to L-Day, the thorough interpretation and evaluation of these photographs enabled Tenth Army to issue detailed intelligence studies which contained an accurate estimate of the Japanese situation.

Aerial photos were required also for use in the production of a map of the target. It was difficult to obtain adequate photographic coverage at first because of the distance of Okinawa from the closest Allied air base, some 1,200 nautical miles. This factor limited the conduct of such missions to either carrier aircraft, whose ships could carry them close to the target, or B–29s. Other obstacles to the amassing of a complete intelligence picture of Okinawa were the notoriously poor weather over the target, the vastness of the land mass to be photographed, and the schedule of carrier strikes against the target—few of which were timed to coincide with immediate Tenth Army intelligence requirements.

On 29 September 1944, the first ICEBERG photographic mission was flown by B–29s. While they covered all of Okinawa, and the outlying islands to a degree, the results of this flight were

---

[38] Unless otherwise noted, the material in this section is derived from: *ICEBERG Study; CTF 51 OPlan A1–45; CTF 51 AR; Tenth Army TntvOPlan 1–45; Tenth Army AR; IIIAC OPlan 1–45; IIIAC AR; 1st MarDiv OPlan 1–45; 1st MarDiv SAR; 6th MarDiv SAR, Ph I & II.*

[39] *ICEBERG Study*, App A (Enemy Strength and Capabilities), p. 19.

[40] Admiral Turner's intelligence staff discovered the existence in the United States of an American civilian who, having spent many years in Japan and Okinawa, was able to divulge much valuable information. "He worked with this section and later worked with both the Army and the Navy at the objective." *CTF 51 AR*, pt V, sec A, p. 1.

limited by clouds which obscured about half of the area photographed, mainly the northern portion of the major island. Because of this inadequate photo coverage, the first map produced and distributed had many blank portions in which there was little or no topographic detailing. Modifications of this first map were made later in the campaign, when captured Japanese maps provided more thorough contouring information.

During the first fast-carrier strikes on Okinawa Gunto of 10 October 1944, large scale vertical and oblique aerial photographs were acquired, giving 90 percent coverage of the area. From 29 September 1944 to 28 March 1945, a total of 224 photo-reconnaissance sorties were flown over the target. Information gained from these photographs was collated and analyzed, and the resultant intelligence summaries were distributed to Tenth Army units.

In the week preceding L-Day, escort carrier-based photographic aircraft flew daily missions over the island. Careful interpretation of the photos thus obtained permitted bomb damage assessments and, at the same time, comparison of these photos with ones taken earlier enabled the interpreters to locate many enemy installations previously concealed by effective camouflage. From a close study of successive sorties, it was possible to determine each displacement of the enemy's defensive positions, to hazard guesses of his relative strength, and to compile a preliminary target information list for distribution to artillery units.[41]

[41] IIIAC Artillery Action Report, Phase I, Nansei Shoto, 1Apr-30Jun45, dtd 25Jul45, p. 7, hereafter *IIIAC Arty AR.*

After L-Day and while the fighting was still in progress, the island was completely rephotographed, the results of which enabled a more accurate map to be printed and distributed.[42] A scale of 1:25,000 was used for the basic map originally issued from which maps of the initial zones of action, scaled at 1:10,000, were produced for the use of the lower echelon assault units. At the same time, smaller scale maps were reproduced for use as road maps in traffic control planning.

The Tenth Army made rubber relief maps on a scale of 1:10,000, which were issued to General Geiger's troops in sufficient quantity to permit distribution down to and including assault battalions. The mapping sections of IIIAC, and the 1st and 6th Marine Divisions worked together to produce plastic terrain models of the corps zone of action. Made to a scale of 1:5,000 and constructed with a 2:1 vertical exaggeration, these models facilitated the briefing of commanders and their troops for the prospective operation. Wholesale distribution of these relief models was made soon after the troops embarked for the target, at which time some 600 copies of a 1:5,000 map of the landing beaches, specially prepared by the 1st

[42] ". . . the absence of an adequate one over twenty five thousand map during the planning phase, and even during the early phases of the operation, served greatly in influencing everything the landing force did. It was often a critical impediment. As a matter of fact, the area in which the 6th Division operated as early as L-plus 4 had large blank segments on the map." CMC [Gen. Lemuel C. Shepherd, Jr.] Memo to G–3, dtd 18Jan55, hereafter *Shepherd memo I.*

Marine Division, were issued to its assault units.

To supplement aerial photographs, the USS *Swordfish*, a specially equipped submarine, was dispatched to Okinawa from Pearl Harbor on 22 December 1944 with the mission of photographing Okinawa's beaches and Japanese defensive installations on the island coasts. After making her last known radio transmission on 3 January 1945, the submarine was never heard from again and was reported missing. As a result, no beach photographs were taken before L-Day for, as succinctly stated in the TF 51 AR, "no information from submarine reconnaissance was available." [43]

In October 1944, enemy strength on Okinawa was set at 48,600. It was estimated that two well-trained and experienced infantry divisions, and a tank regiment, comprised the major defense force on the island. At this time, it was recognized that an additional threat to the landings was posed by the size of the civilian population located in southern Okinawa. This manpower potential of more than 300,000 individuals would swell the enemy strength figure if they were used to form a home guard or militia, or to conduct guerrilla activities. In January 1945, the Tenth Army estimate assumed that the Japanese reinforcement capability could increase the regular force figure to 66,000 by L-Day, at which time enemy defense forces on Okinawa would be two and a half infantry divisions. If the enemy exerted his maximum reinforcement capability, he could then oppose the landing with four infantry divisions constituting the principal combat elements of the defense. Total Japanese strength would then be 87,000 men.

All possible Japanese courses of action were considered, and troop dispositions for each course were analyzed in light of what was known of current Japanese tactical doctrine and its evolution to date. All indications pointed to the fact that the enemy would most likely organize the southern third of Okinawa for a defense in depth while the bulk of his troops were withheld as a mobile reserve. This course of action would present a potentially more dangerous situation to the landing force than would the more commonly experienced alternative of a determined defense of the beaches.

An interpretation of aerial photographs in February revealed that the enemy force on Okinawa comprised two infantry divisions and an independent mixed brigade, service and support troop reinforcements for the infantry, all totaling an estimated 56,000–58,000 men. It also appeared that, while the far northern sector was defended by a single battalion only, the main force was disposed in the south in the projected XXIV Corps area. In the III Amphibious Corps zone of action, it was estimated that two infantry regiments defended. Conceivably, these six or seven thousand men could be reinforced by local auxiliaries.

While the small garrison in the north was given the capability of mounting counterattacks against the invader left flank, it was expected that the most

[43] Theodore Roscoe, *United States Submarine Operations in World War II* (Annapolis: United States Naval Institute, 1949), pp. 446–447; *CTF 51 AR*, pt V, sec A, p. 1.

violent enemy reaction would come from the heavily defended south, on the XXIV Corps' right flank, where the Japanese mobile reserve would be maintained in considerable strength. It was anticipated that, as soon as the Japanese had appraised the landing force's dispositions, a counteroffensive in force would be mounted by the enemy reserve.

The estimate of Japanese strength was again revised in mid-February, this time downwards to 37,500–39,500, when information was received that a full division had been withdrawn from Okinawa. In view of this reduction, and supplemented by indications that the enemy was concentrating in the Nakagusuku Bay area, it was presumed that the two Marine divisions would be opposed in their zones by no more than one infantry regiment deployed in position, and that the total number of Japanese troops in the overall sector would be more than 10,000.

This numbers guessing game continued when, a month later, the estimate of Japanese defense forces was revised upwards to 64,000. It appeared that the enemy had been able to reinforce the garrison with an understrength infantry division as well as with some miscellaneous units of unknown origin, in all about 20,000 men. It was believed that an additional force of 4,000–6,000 men had arrived in March, having been lifted by shipping which successfully evaded the Allied blockade. The Tenth Army assumed that, if the March enemy reinforcements were the advance elements of another division, it was reasonable to assume further that by 1 April the landing force would be opposed by at least 75,000 men. In the week preceding L-Day, while the assault elements sortied for the target, still another estimate of enemy strength in the IIIAC zone was issued. In this supplementary revision, it was stated that the principal Japanese opposition now would come from two reinforced infantry regiments with a strength of 16,000 men.

Air and naval capabilities assigned to the Japanese remained relatively unchanged all during the planning phases of ICEBERG. At all times it was expected that the enemy would be capable of mounting heavy and repeated air attacks against invasion shipping. It was expected that this vigorous air effort would include continued employment and intensification of the suicide bombing tactics which first had appeared during the invasion of Leyte in October 1944. The Japanese were credited with an air strength of approximately 3,000 planes which were based within range and capable of blunting the Okinawa landing. Along with this air capability, the enemy was believed able to mount an airborne counterattack, for "as air action is practically the only assistance he can give the Okinawa garrison from outside [the island], he may expend considerable aircraft and endeavor to land several thousand troops within our beachhead." [44]

It was known that the Japanese had suicide motor torpedo boat units at Okinawa and it was assumed that midget submarines were based there also. Added to the possible tactical employment of these suicide organizations was the potential use of suicide swimmers

---

[44] *Tenth Army TntvOPlan 1–45*, anx 3, sec V, p. 14.

whose mission was also disruption of the invasion fleet at anchor off the objective. Although the Japanese Navy was a mere shadow of its former self, it still retained operational forces strong enough to pose a threat to the landing's success. For that reason, it was deemed necessary to maintain a strong surface cover at the objective.

While the southern part of Okinawa was ideally suited for the tactical use of tanks, the enemy was not given an armored capability. This was because the relation of estimated tank strength to the total estimated garrison strength was too low, and it was not felt that this support arm would offer any great opposition.

Three months after Admiral Nimitz had received the JCS directive for Okinawa's invasion, General Buckner issued the initial operation order setting the ICEBERG juggernaut's wheels into motion. During the course of this planning period, each Tenth Army general and special staff section prepared that portion of the operation order for which it was responsible while maintaining liaison with the subordinate units which were preparing to put words into action. Although most of the ICEBERG assault, support, and garrison forces did not issue their own operation orders until January 1945, warning orders had already alerted them to the impending invasion.

Assault Preparations

TRAINING AND REHEARSALS[1]

The Pacific-wide dispersion of troops and shipping assigned to ICEBERG prevented the Tenth Army from conducting either training or rehearsals as a cohesive unit. Because of the vast distances separating General Buckner and his corps and division commanders, the latter were invested with the responsibility for training their respective organizations along the lines of Tenth Army directives. With these orders as a guide, all Marine units committed to the operation were trained under the supervision of FMFPac.

Assault preparations of ICEBERG Army divisions were hindered by the limited time available for their rehabilitation, reorganization, and training. This was especially true in the case of XXIV Corps units already in combat in the Philippines. Many of the garrison and service units which were to be attached to the various assault forces were also handicapped by the time factor because they, too, were either fighting or heavily committed in support of operations in the Philippines. In order that Tenth Army staff planners could better evaluate the combat readiness of all organizations within the command, each of General Buckner's commanders submitted a monthly training status report to ICEBERG headquarters on Oahu.[2] Since the reports lacked what an inspection at first-hand could provide, Buckner and some of his principal staff officers made a series of flying trips to each of the corps and divisions. These personal visits at the end of January 1945 "did much to weld the far-flung

[1] Unless otherwise noted, the material in this section is derived from: *CTF 51 AR; Tenth Army AR; TAF AR;* XXIV Corps AR, Ryukyus, 1Apr-30Jun45, n.d., hereafter *XXIV Corps AR; IIIAC AR; IIIAC Arty AR; IsCom AR; MilGovt AR; 1st MarDiv SAR;* 2d MarDiv AR, Phase I, Nansei Shoto, dtd 15Apr45, hereafter *2d MarDiv AR; 6th MarDiv SAR, Ph I & II;* 7th InfDiv OpRpt, Ryukyus Campaign, dtd 30Jul45, hereafter *7th InfDiv OpRpt;* 27th InfDiv OpRpt, Phase I, Nansei Shoto, 1 Jan-30Jun45, dtd 19Jul45, hereafter *27th InfDiv OpRpt;* 77th InfDiv OpRpt, Phase I (in 3 parts—Kerama Retto, Keise Shima; Ie Shima; Okinawa), n.d., hereafter *77th InfDiv OpRpt* and appropriate part; 96th InfDiv AR, Ryukyus Campaign, dtd 28Jul45, hereafter *96th InfDiv AR.*

[2] The voluminous training status reports were in reality check-off lists for newly formed divisions. When the Marine Deputy Chief of Staff of the Tenth Army learned that the IIIAC assault divisions, whose troops had recently been in combat and were in an advanced state of training, had to submit these reports, he pointed out that preparation "of these reports merely harassed the divisions and served no useful purpose." Once General Buckner "saw the training being engaged in by the 1st and 6th Marine Divisions . . . the word was passed to [Brigadier General] Silverthorn to forget about the submission of the Status Reports." Smith, *Personal Narrative,* p. 43.

units of the Tenth Army into a unified whole."[3]

One determinant forcing the postponement of the Formosa-South China invasion in favor of the Okinawa assault had been the shortage of service and support troops, a shortage that still existed when the Tenth Army began its final training and rehearsal phases.[4] Many of these specialist units were slated to reinforce corps and divisions for the assault and then to augment Island Command during the initial base development. Because they were too deeply involved otherwise, often with primary missions related to the buildup for the operation, the support troops could not train with the assault units they were to reinforce. The time borrowed for training would seriously disrupt the mounting and staging efforts for ICEBERG. Nevertheless, support unit commanders carried out adequate individual weapons' qualification and physical conditioning programs which met Tenth Army training requirements. Although they were released to General Buckner's control only a few days before mounting for the target, a number of garrison organizations were able to conduct limited training with the combat outfit to which they were attached.

The major assault components of the Tenth Army were battle-experienced for the most part, but they needed to undertake an intensive training schedule in order to bring veterans and newly absorbed replacements alike to peak combat efficiency. To accomplish this task, Army and Marine Corps units in the South Pacific, and the 2d Marine Division on Saipan, conducted extensive programs which fulfilled the training requirements stipulated by General Buckner's directives. General Hodge's XXIV Corps, however, was engaged in operations on Leyte, and his divisions were not released to Tenth Army by General MacArthur until 10 February 1945,[5] just two months before L-Day.

After the extended period of bitter fighting in the Philippines, however, the combat units slated for ICEBERG were understrength. General Hodge's problems were further aggravated when his infantry divisions were required to fur-

[3] *Blakelock ltr.* Of Buckner's visit to his Marine units in the Solomons, General Geiger wrote: "General Buckner and five of his staff spent about three days with us the latter part of January. You know how difficult it is to make a very favorable impression in the mud down here, especially in cleanliness of equipment; but I believe he and his staff were satisfied with what they saw. They spent a day with each division observing training. We took them over to the 1st Division in a PBY-5A." MajGen Roy S. Geiger ltr to LtGen Alexander A. Vandegrift, dtd 2Feb45 (Vandegrift Personal Correspondence File, HistBr, HQMC).

[4] General Oliver P. Smith recalled that in order to obtain at least a minimum of additional service troops, the Tenth Army "had to comb the entire Pacific Ocean Area and resort to considerable improvisation. For example, a veteran tank battalion was broken up to make Quartermaster Truck Companies. I happened to talk to the battalion commander of this tank battalion, who was heartbroken over the matter. . . ." Smith, *Personal Narrative*, p. 12.

[5] In 1944 when the JCS approved the recommended invasion of Leyte, the XXIV Corps was fully combat loaded, already at sea, and en route for the invasion of Yap. As a result of the decision to land at Leyte, the Yap landing was cancelled and the XXIV Corps was reassigned to General MacArthur, who ordered it to invade Leyte on 20 October 1944.

nish the Leyte Base Command[6] with large working parties as soon as the troops returned from mopping-up operations at the front. The servicing, crating, and loading of organic division equipment siphoned off the services of other infantrymen as well as making it impossible to impose a major training program on any of the divisions. Finally, as one command reported, the "deterioration of the physical and mental condition of combat personnel after 110 days of continuous contact with the enemy made it plain that rigorous field training in the wet and muddy terrain would prove more detrimental than beneficial."[7]

Besides undertaking the many other incidental duties preparatory to mounting for Okinawa from Leyte, some Army units had to construct their own camps and make their own billeting arrangements as soon as they arrived in the rear area from the front lines. What little time was available to the Southern Landing Force before L-Day was divided between training in small-unit tactics and practice for breaching and scaling operations, in anticipation of the conditions to be found at Okinawa beaches. Because of the large influx of raw replacements into the divisions, great emphasis was placed on developing the teamwork of riflemen and their supporting weapons.

Of the three divisions in XXIV Corps, the 96th was the most fortunate in that some of its new troops arrived during mopping-up stages on Leyte. At that time, the replacements were given an opportunity to take an "active part in combat and reconnaissance patrols, gaining valuable battle indoctrination through physical contact and skirmishes with small isolated groups of Japanese."[8]

According to the Tenth Army Marine Deputy Chief of Staff, General Smith:

> The conditions of the Army divisions on Leyte gave General Buckner considerable concern. This was not the fault of the divisions; they were excellent divisions. However, they had been in action on Leyte for three months and two of the divisions were still engaged in active operations. The divisions were understrength and adequate replacements were not in sight. There were [numerous men suffering from] dysentery and skin infections. Living conditions were very bad. A considerable number of combat troops had been diverted to Luzon and converted into service troops. There was some doubt as to whether reequipment could be effected in time.[9]

The fighting record of the XXIV Corps on Okinawa indicates how well it overcame great obstacles in preparing for its ordeal. Once they had reconstituted their combat organizations, trained their fresh replacements, and attended to the many details incident to mounting for the target, the veteran units of this corps were able to give good accounts of themselves against the enemy.

In the South Pacific and the Marianas, Tenth Army units were not as heavily committed as the units of the Southern Landing Force, and completed a more comprehensive training pro-

[6] This logistical organization suffered from a chronic shortage of service troops which threatened the dual mission of the command of supporting both the Luzon and Okinawa invasions.

[7] *7th InfDiv OpRpt*, p. 28.

[8] *96th InfDiv AR*, chap V, p. 1.

[9] Smith, *Personal Narrative*, p. 47.

gram. The 27th Infantry Division, ICEBERG floating reserve, arrived at Espiritu Santo in the New Hebrides from Saipan during September and October 1944. The division was undivided in its opinion that this base was a "hellhole," unfitted for the division's rehabilitation and training because of the island's torrid climate, its topography, and lush, tropical vegetation.[10]

Upon receipt of advance information that it was to take part in the Okinawa assault, the 27th instituted an accelerated combat training program which was calculated to qualify it, by 30 January 1945, for a period of prolonged operations against the enemy. Launched on 23 October, the level of the program advanced progressively from individual schooling to combined company and battalion exercises and, finally, to a two-week stretch of regimental combat team (RCT) [11] maneuvers. During this staging period, in which 2,700 replacements arrived and were assigned, the division stressed training for offensive and defensive night operations.

Most Marines in IIIAC assault divisions had recently been in combat, yet their training programs were stringent and comprehensive. Like all other veteran ICEBERG forces, the Marine divisions were confronted with the need to obtain, integrate, and train replacements. Marine training overall emphasized the development of a tank-infantry-artillery team and focused attention on tactical innovations such as the use of the armored amphibian's 75mm howitzer for supplementary artillery support. While other Tenth Army units were required to undertake amphibious training, General Geiger's troops did not have to, since General Buckner considered his Marine divisions eminently qualified in this aspect of warfare.

Following the Peleliu campaign, General del Valle's 1st Marine Division had returned to Pavuvu for rest and rehabilitation. The division was first based on the island in April 1944, arriving there after completion of the New Britain operation. At that time, and with some difficulty, the Guadalcanal and Cape Gloucester veterans converted the overrun coconut groves into some semblance of a habitable cantonment. Because of its small size, Pavuvu was not particularly suited for training as large a unit as a division; its terrain limited the widespread construction of machine gun and mortar firing ranges.[12] All artillery firing had to be conducted on Guadalcanal.

[10] Capt Edmund G. Love, *The 27th Infantry Division in World War II* (Washington: Infantry Journal Press, 1949), p. 521, hereafter Love, *27th InfDiv Hist.*

[11] An RCT in an amphibious operation was an infantry regiment reinforced by supporting arms, *i.e.*, artillery, tanks, engineers, etc., which made up a balanced team for specific missions and whose services were required for initial operations ashore.

[12] During the division training phase, physical conditioning hikes were made on the shore road which encircled only that part of the island occupied by the division. Both sides of the road were used as units followed one another, the group on the inside track marching in a clockwise direction, while the outer group hiked counterclockwise—both groups passing each other several times as they crowded the limited road net. At mealtime, the unmistakable smell of New Zealand-grown sheep being cooked filled the air, and the Marines, as if one, would curse, "Mutton again, dammit."

After its return from the Palaus, the ranks of the 1st held some 246 officers and 5,600 enlisted Marines who had already served overseas nearly 30 months. Within that time, the division had made three assault landings and it was now to make a fourth. If the division was to go ashore at full strength, it appeared, at first, that it would be necessary for the veterans of Guadalcanal, Cape Gloucester, and Peleliu to fight at Okinawa too. A potentially serious morale problem was alleviated when the division received four replacement drafts by 1 January. These drafts, plus a steady flow of individual replacements, brought officer strength to slightly above the authorized figure and exceeded the authorized enlisted strength figure by more than 10 percent. As a consequence, all eligible enlisted Marines were able to return to the States.

At the same time, an extensive leave program was established for officers who, though eligible, could not be spared for rotation. Fifty-three of them were permitted to take 30 days leave in the United States, after which they were to return to Pavuvu. "In addition, six went to Australia and one to New Zealand. Some fifty key enlisted men [eligible for rotation] also elected to take leave in Australia in order that they could continue to serve in the First Marine Division." [13] By the time the division embarked for Okinawa, approximately one-third of its Marines had been in two invasions, one-third had faced the enemy once, and the remainder were men who had seen no combat whatsoever. The majority of the latter were replacements who had arrived at Pavuvu while the 1st was at Peleliu.

As soon as the training cycle of General del Valle's infantry units reached the regimental level and outgrew Pavuvu's facilities, each RCT was rotated to Guadalcanal, about 65 miles to the southeast, for two weeks of more intensive combined-arms training. Special emphasis was given to preparing the division for warfare of a type and on a scale differing in almost every respect from that which it experienced in the tropical jungles of Guadalcanal and New Britain, and on the coral ridges of equatorial Peleliu. As an integral part of a much larger force, this division was to invade, for the first time, a land mass "which contained extensive road nets, large inhabited areas, cities and villages, large numbers of enemy civilians, and types of terrain" [14] not found in the South Pacific. Besides being schooled to fight under the conditions anticipated at Okinawa, the troops were trained to defend against paratroop attack and indoctrinated in the techniques of dealing with hostile civilians.

In commenting on the personnel situation of his regiment during its training period, the former commanding officer of the 11th Marines stated:

> The heavy casualties suffered at Peleliu, plus the rotation without immediate replacement of all officers and men with 30 months' service in the Pacific after that battle, posed a severe problem. Only one battalion commander remained of the four who went to Peleliu. There were only eight field officers in the regiment including myself and the [naval gunfire] officer. Fourteen captains with 24 months' Pacific

---

[13] *1st MarDiv SAR*, chap III (Personnel Anx), pp. 2–3.

[14] *1st MarDiv SAR*, chap IV, p. 1.

service were allowed a month's leave plus travel time in the United States, and they left Pavuvu at the end of November and were not available for the training maneuver at first. I recall that the 4th Battalion (LtCol L. F. Chapman, Jr.) had only 18 officers present including himself. He had no captains whatever. The other battalions and [regimental headquarters] were in very similar shape. The 3d Battalion had to be completely reorganized due to heavy casualties on Peleliu and was the only one with two field [grade] officers. But it had only about 20 officers of all ranks present.[15]

General Shepherd's 6th Marine Division was activated on Guadalcanal in September 1944, and was formed essentially around the 1st Provisional Marine Brigade. This unit had taken part in the Guam invasion and had been withdrawn from that island late in August. The infantry components of this new division were, with a few exceptions, veterans of the Pacific fighting. The 4th Marines was made up of the disbanded Marine raider battalions, whose troops had fought on Guadalcanal, New Georgia, and Bougainville; the infantry regiment as a whole had landed on Emirau and Guam. The 22d Marines had participated in the Eniwetok and Guam campaigns, and the 1st Battalion, 29th Marines had augmented the 2d Marine Division for the Saipan assault. After its relief on Saipan, 1/29 was sent to Guadalcanal to await the arrival from the United States of its two sister battalions, and eventual assignment to the 6th Division.

At the time of the 6th Marine Division activation, the division was some 1,800 men understrength and, as was the case with other IIIAC units, in very few instances did the classification of the replacements received by General Shepherd correspond to his actual needs. Paralleling other instances, where the composition of stateside-formed replacement drafts did not satisfy critical shortages in specific specialist fields, the 15th Marines was assigned and forced to retrain antiaircraft artillerymen from disbanded defense battalions whose previous experience and training was not considered the same as that needed by field artillerymen.

Most of the men in the 6th Division had fought in at least one campaign, while others were Pacific combat veterans who were now beginning a second tour of overseas duty. The division was based on Guadalcanal, where kunai grass and steaming tropical jungle provided an excellent environment in which General Shepherd's men could fulfill a rugged training schedule. The program began on 1 October and proceeded from small-unit exercises through large-scale combined-arms problems employing battalion landing teams (BLTs) [16] and RCTs; all training culminated in an eight-day division exercise in January 1945. Anticipating how the division was to be employed on Okinawa, General Shepherd emphasized the execution of large-unit maneuvers, swift movement, and rapid troop deployment.

The IIIAC Artillery faced the same replacement retraining problems that plagued the 15th Marines. When the 6th 155mm Howitzer Battalion and the

[15] MajGen Wilburt S. Brown ltr to CMC, dtd 10Oct54, hereafter *Brown ltr.*

[16] The BLT was reinforced for the assault in a manner similar to that of an RCT, but on a lesser scale.

Headquarters Battery, 2d Provisional Field Artillery Group, were formed in October and November 1944, their cadres were withdrawn from existing units of corps artillery. The latter was further drained when 500 combat veterans, mostly valuable noncommissioned officers, were rotated home in November. There were few experienced artillerymen in the group replacing them.

At the same time that rehearsals were being conducted for the coming operations, Brigadier General David I. Nimmer's Corps Artillery battalions were forced to conduct training sessions (retraining classes in the case of radar technicians and antiaircraft artillerymen coming from disbanded defense battalions) in order to ensure that all firing battery personnel would be completely familiar with the weapons to which they were newly assigned. Another matter adversely affecting the artillery training program was the delay, until 15 November and 10 December respectively, in the return of the 3d 155mm Howitzer and the 8th 155mm Gun Battalions from the Palaus operation. General Nimmer's organizational and personnel problems were complicated further by the fact that approximately 10 percent of his unit strength joined after active training ended in February, while 78 communicators and 92 field artillerymen did not join until after Corps Artillery had embarked for Okinawa.

VMO–7, the Marine observation squadron assigned to Corps Artillery, did not arrive before General Nimmer's units mounted out, but joined them later at the target. Three days before embarkation, the commanding officer of the 2d Provisional Field Artillery Group joined. Despite these hitches to IIIAC Corps Artillery pre-combat preparations, General Nimmer considered all of his embarked artillery units ready, although "both individual and unit proficiency were not up to the standards that could have been obtained under more favorable circumstances." [17]

As soon as General Geiger's staff began planning for the Marine Corps role in ICEBERG, the commanders of the IIIAC Corps Artillery and the 11th and 15th Marines established liaison with one another in order to coordinate their unit training programs. These senior Marine artillery officers "resolved that in this operation we would take advantage of all previous experience, good and bad, and give a superior performance. Accordingly, great care was given to . . . the ability to rapidly mass fires of all available guns at any critical point." [18]

Artillery training was directed toward attaining this capability. General Nimmer's staff devised and wrote the standard operating procedures to be used by all Marine artillery units assigned to ICEBERG. These procedures established the techniques to be used for requesting and the subsequent delivery of reinforcing fires. During the training period, firing batteries constantly put the new doctrine into practice.

With the exception of the 12th Marines, the 2d Marine Division artillery regiment, all other Marine artillery units in the Tenth Army conducted a

[17] *IIIAC Arty AR*, p. 10.

[18] Col Frederick P. Henderson ltr to CMC, dtd 11Mar55, hereafter *Henderson ltr.*

combined problem on Guadalcanal, 11–13 January. A majority of the firing missions were spotted by aerial observers. Conditions anticipated on Okinawa were simulated as closely as possible, although the large military population and the consequent profusion of various installations on Guadalcanal necessarily limited the size of the artillery ranges available for the big guns. By the end of the combined problem, when a firing mission was called in, the Marines "were able to have all artillery present, laid and ready to fire in an average of five minutes from the time it was reported." [19]

General Watson's 2d Marine Division, reserve for IIIAC and its third major element, was in garrison on Saipan where a division-wide training program was effectively integrated with mopping-up operations against enemy forces remaining at large on the island. More than 8,000 Marine replacements received valuable on-the-job experience routing Japanese holdouts during the first months of the division training program which began 15 September. Saipan's rapid build-up as a supply center and an air base restricted the training efforts of the division, however, and maneuver room and impact areas were soon at a premium.

In the course of his inspection trip to Tenth Army units, General Buckner visited the 2d Marine Division. On the morning of 3 February, he trooped the line of the 8th Marines and then inspected the regimental quarters and galleys. It seemed to General Smith that the men of the 2d Division looked very fit, and that they had made a tremendous impression on the Tenth Army commander. Buckner was particularly impressed with the battalion commanders, and told his deputy chief of staff that "he had never before had the privilege of meeting such an alert group. . . ." [20]

A lack of suitable beaches on Saipan confined final division rehearsals to simulated landings only. Because of the indefinite nature of its employment once it had made the feint landings on L-Day and L plus 1, the 2d Division had to select an arbitrary landing scheme of two RCTs abreast for the rehearsal pattern. Bad weather prevented LVT launchings on two days, neither air nor naval gunfire support was available, and, finally, on 19 March—the last day of the exercises—only the naval portion of TG 51.2 (Demonstration Group) was able to participate in the demonstration rehearsal.

On Espiritu Santo, the Tenth Army's other relatively isolated unit—the 27th Infantry Division—conducted rehearsals from 20 to 25 March while its transport squadron was being loaded. This division was in the same position as General Watson's in that it faced a profusion of potential missions. The rehearsals of both reserve divisions were

---

[19] *Henderson ltr.* After the 11th Marines arrived on Guadalcanal, 15 December 1944, it joined the 15th Marines to train with IIIAC Artillery for seven straight weeks with only one break, Christmas Day. In this joint training effort, great stress was placed on such artillery tactics as proper conduct of fire, with the battalions registering, firing missions, and displacing several times a day "to overcome the improvised jungle methods heretofore used by the division in previous campaigns." *Brown ltr.*

[20] Smith, *Personal Narrative*, p. 52.

based, therefore, on a number of hypothetical landing assignments.

Satisfactory practice landings were made by all of the other Tenth Army asault divisions. IIIAC rehearsals took place off the Cape Esperance-Doma Cove beaches on Guadalcanal from 2 through 7 March. Although reefs do not exist here, a transfer line was simulated 200 yards from the shore in an attempt to duplicate actual landing conditions in the corps zone on Okinawa. During the six days of rehearsals, Rear Admiral Reifsnider's staff officers made certain that assault wave control was emphasized and that the training of communications elements was intensified at all command levels.

Because naval gunfire and air-support units assigned to ICEBERG were committed elsewhere at this time, the token prelanding bombardment furnished by vessels in the area, and the air support supplied by F6Fs (Hellcats) and TBMs (Avengers), flying in from Henderson Field and nearby carriers, only approximated the tremendous volume of fire to be laid on the Hagushi beaches. Practice landings were made by IIIAC units on 3 March, followed the next day by a critique aboard the TF 53 flagship, USS *Panamint*. Other preliminary landings on the 5th preceded the landing of the entire IIIAC assault echelon on 6 March. Corps and division command posts were set up ashore, a primary communications net was established, and some equipment was unloaded. On 7 March, the reserve regiments—the 1st Marines for the 1st Division and the 29th Marines for the IIIAC—climbed down the nets into invasion craft, which were formed into boat waves, and then landed on the beaches.

General Geiger's corps artillery units did not participate in these final rehearsal exercises except to land battalion, corps, and group headquarters reconnaissance parties. The shortage of time prevented the landing of any of the artillery pieces which were to go ashore at the target.

Nearly 3,000 miles away from Guadalcanal, in the Philippines, assault elements of XXIV Corps conducted rehearsals in Leyte Gulf from the 15th to the 19th of March under the watchful eyes of Admiral Hall and his attack force staff officers. Because the missions assigned XXIV Corps divisions varied so widely, the nature and conduct of their rehearsals tended to reflect this variance.

The 77th Infantry Division was to make the initial ICEBERG assault, the landing on Kerama Retto. In order to familiarize the troops with conditions at their impending target, practice landings were made in southeastern Leyte's Hinunangan Bay on islands that closely resembled some of those in the Keramas. For two days, 14 and 15 March, adverse weather conditions and heavy swells prevented any landings at all, but adherence to any firm rehearsal schedule was not considered necessary since the mission of the 77th involved several landings independent of each other.

Poor weather on the 15th forced the cancellation of a planned rehearsal for the Ie Shima invasion, while only the division reserve (307th Infantry) made any practice landings on the 16th. Although General Bruce was satisfied with the rehearsals since "all elements scheduled for a specific mission satisfactorily executed a close approximation of their mission,"[21] Admiral Kiland was not so confident. The Western Islands Attack Group Commander felt that "considering the complexity of the operation and the relative inexperience of naval personnel involved, the curtailment of these exercises by weather conditions made the training provided entirely inadequate."[22]

On 16 March, the 7th and 96th Divisions landed under perfect weather conditions and on the 18th held unit critiques, in which certain basic discrepancies and difficulties discovered in the first exercise were ironed out. The following day, the two divisions landed again. A high-level critique was held on the 21st for the major Army and Navy commanders on Admiral Hall's flagship, USS *Teton*. Also present were Admiral Turner and General Buckner. At this time, all of the XXIV Corps rehearsals were evaluated, and efforts were made to ensure that the actual landing would be better coordinated.

As the normal duties of most of the flying squadrons assigned to TAF constituted their combat training, and since they would not begin operations at Okinawa until after the landing, when the airfields were ready, they were not required to conduct rehearsals for ICEBERG. TAF ground personnel scheduled to travel to the target with the assault echelon, participated in the landing rehearsals that were held at Guadalcanal and Leyte. Their troop training, for the most part, was conducted aboard ship en route to the staging areas, and consisted of familiarization lectures about the enemy, his tactics, and his equipment.

Like the other Okinawa-bound Tenth Army units mounting from Pearl Harbor, Island Command troops conducted individual and unit training programs which consisted of specialist as well as combat subjects. The Island Command assault echelon was composed chiefly of headquarters personnel who were to initiate the base development plan as soon as practicable after the landing. Within this echelon also were shore party, ordnance, ammunition, supply, signal, quartermaster, truck, and water transportation units, whose support services would be required immediately after the initial assault.

At Fort Ord, California, officers to staff military government teams began assembling in late December 1944. A number of these officers had already received approximately three months of military government training at either Princeton or Columbia Universities. In California and at the staging areas where they joined the assault forces, these Army and Navy officers received instructions pertinent to the ICEBERG military government plan. Many in the Navy enlisted component in the military government section had never received any specialized civil affairs training before they arrived at Fort Ord, where they were assembled just in time to embark with the teams to which they were assigned.[23]

---

[21] *77th InfDiv OpRpt*, Kerama Retto, Keise Shima, p. 20.

[22] CTG 51.5 AR, Capture of Okinawa Gunto, Phases 1 and 2, 9Mar-2Apr45, dtd 26May45, chap II, p. 2, hereafter *CTG 51.1 AR*.

[23] *MilGovt AR*, p. 3.

By 1945, the roll-up of enemy positions in the Pacific had progressed to the point where some Tenth Army units were able to mount and stage on the threshold of Japan. XXIV Corps prepared for Okinawa in the Leyte Gulf area, only 1,000 miles from the Ryukyus, while in the Marianas, just slightly farther away from the target, other ICEBERG forces made ready for the attack. Northern Attack Force units, however, had a considerably longer journey to the Ryukyus as they prepared in the Solomons.

MOUNTING AND STAGING THE ASSAULT [24]

Each attack force of the Joint Expeditionary Force was organized differently for loading, movement, and unloading at the target. The nine transport divisions in the three transrons of Admiral Hall's Southern Attack Force were reorganized and expanded to number 11 transport divisions (transdivs). Assigned to these two additional transdivs were those ships slated to lift XXIV Corps troops at Leyte and those which were to load Tenth Army and Island Command forces waiting on Oahu. The Northern Attack Force, which was to carry IIIAC troops, was not so augmented. General Geiger was so impressed with how well the reorganization of Admiral Hall's transport force had eased movement control and increased the efficiency of loading and unloading operations, that he requested the formation of a similar corps shipping group for future IIIAC operations.[25]

The commanding generals of the 1st and 6th Marine Divisions were made responsible for the loading and embarkation of their respective organic and attached units, while IIIAC itself supervised the loading of corps troops. In addition, General Geiger was responsible for embarking Marine Air Group 33 (MAG–33) of TAF, which was based on Espiritu Santo, 555 miles southeast of Guadalcanal.

Although some Northern Attack Force vessels were partially combat-loaded before the rehearsal period, all required additional time off the Guadalcanal, Banika, and Pavuvu beaches to take on vital cargo and to top-off water and fuel tanks. The Northern Tractor Flotilla was the first increment of TF 53 to leave the Solomons for the staging area at Ulithi. Departing on 12 March, the holds and above decks of the landing ships in the convoy were solidly packed with amphibious vehicles, tanks, artillery, and various other combat gear. For this invasion, IIIAC wanted to avoid subjecting assault troops to the crowded conditions and debilitating effects of prolonged confinement aboard LSTs and LSMs.

While APAs were hardly luxurious, their accommodations were far better than those of the landing ships. To ease the first leg of the journey to Okinawa,

[24] Unless otherwise noted, the material in this section is derived from: *CTF 51 AR; Tenth Army AR; TAF AR; IIIAC AR; XXIV Corps AR; 1st MarDiv SAR; 2d MarDiv AR; 6th MarDiv SAR; 7th InfDiv OpRpt; 27th InfDiv OpRpt; 77th InfDiv OpRpt*, Kerama Retto, Keise Shima; *96th InfDiv AR.*

[25] CNO, Amphibious Operations—Capture of Okinawa, 27Mar-21Jun45 (OpNav 34–P–0700), dtd 22Jan46, chap 7, p. 20, hereafter *CNO Record.*

many assault wave Marines were embarked on the faster attack transports which, together with the rest of the Northern Attack Force transport groups, left from the mounting area on 15 March to join the ICEBERG force gathering at Ulithi.

The immense lagoon at Ulithi Atoll was the westernmost American fleet anchorage, staging base, and repair depot in the Pacific. Midway between the Marianas and the Palaus, Ulithi was captured without opposition in September 1944, and was developed immediately to support naval operations in the western Pacific as well as to serve as an advance base for the Philippines invasion. Once occupied and built up, the islets of the atoll served also as limited recreation areas where personnel of all services could regain their landlegs and participate in a somewhat restricted physical conditioning program.

On 21 March, both the transport group and the tractor flotilla of TF 53 arrived at Ulithi, anchored, and on the following day, APA-borne assault troops were transferred to the landing ships which were to carry them the remaining 1,400 miles to Okinawa. Once the transfer was completed, small boats began ferrying recreation parties ashore. Here the rigors of shipboard confinement were forgotten by a combination of organized athletics and an issue of not-too-cool cokes and beer.

For many of the troops, this stopover on the long voyage towards the unknown was made exciting by the fascinating sight of the constantly shifting fleet groupment whose makeup changed from day-to-day and hour-to-hour as carriers, battleships, cruisers, and smaller combat vessels departed for strikes against the enemy or returned from completed missions. In the midst of this activity, the scattered elements of the Expeditionary Troops filtered in to join those forces which had arrived earlier.

Despite the relaxing effect of sun, sand, and surf at Ulithi, the nightly alerts to the presence of Japanese snooper planes was a continual reminder that a war still existed. This grim fact was brought home to many men in the invasion force on the gloomy, fog-bound Saturday afternoon of 24 March when the battered carrier *Franklin* limped into the anchorage shepherded by the USS *Santa Fe*.

On the next day, a brilliantly sunlit Sunday, the bruised and battered *Franklin* could be seen more clearly as she lay at anchor. Her top rigging, aerials, and radar towers were gone or twisted completely out of shape. Her flight deck was buckled and undulating. These were the external damages wrought by the internal explosions of bombs that had penetrated to lower decks when Japanese suicide planes furiously attacked the carrier on 19 March, during TF 38 strikes against enemy shipping at Kure and Kobe. As the most heavily damaged carrier to be saved in the war, the *Franklin* was able to make the 12,000-mile trip to New York for repairs under her own power, stopping only at Pearl Harbor on the way.

The Northern Tractor Flotilla sortied from the Ulithi anchorage for Okinawa on 25 March and, two days later, the remainder of the assault echelon set forth in its wake. Saipan was the scene, on the same dates, of the Demonstration Group departure.

Loading operations of the 2d Marine Division were eased by the fact that its lift, Transron 15, had laid over briefly at Saipan in February while en route to Iwo Jima. At that time, division transport quartermasters (TQMs) obtained ships' characteristics data which proved more accurate than the information provided earlier by FMFPac. As a result, the TQMs were better able to plan for a more efficient use of cargo and personnel space.

In addition to the responsibility for loading his reinforced division, General Watson was given the duty of coordinating the loading of all ICEBERG Marine assault and first echelon forces elsewhere in the Marianas and at Roi in the Marshalls.[26]

In preparing for Okinawa, the only real problem confronting General Mulcahy's Marine air units was the coordinated loading of ground and flight elements. According to the logistical planning, planes and pilots were to be lifted to the target on board escort carriers, while ground crews and nonflying units were to make the trip in assault and first echelon shipping. As the organizations comprising the Tactical Air Force were widely dispersed, their loading and embarkation was supervised, of necessity, by local commanders of the areas where the air groups and squadrons were based.

Mounting from Oahu in the TAF assault echelon were the headquarters squadrons of the 2d Marine Aircraft Wing and MAG-43, and Air Warning Squadrons (AWSs) 7 and 8. Headquarters Squadrons 2 and 43 became the headquarters commands of TAF and General Wallace's Air Defense Command, respectively. The TAF transport quartermaster coordinated the mounting out of the Oahu-based units with his opposite numbers on the staffs of the Tenth Army and the 2d MAW. The Marines from AWS-8 and the forward echelons of Mulcahy's and Wallace's headquarters commands left Pearl Harbor on 22 February, while AWS-7 departed Pearl the same month in two increments, one on the 10th and the second on the 21st.

Colonel John C. Munn's MAG-31 embarked from Roi and Namur in the Marshall Islands. The group service squadron and ground personnel of Marine Fighter Squadrons 224, 311, and 441 boarded transport and cargo vessels which, in turn, joined the ICEBERG convoy forming at Saipan. Flight personnel and their planes went aboard the escort carriers *Breton* on the night of 22-23 March, *Sitkoh Bay* on 24 March and were staged through Ulithi where they were joined by Marine Night Fighter Squadron 542.

MAG-33 (Colonel Ward E. Dickey) mounted from Espiritu Santo in the New Hebrides. While IIIAC was responsible for the embarkation of the MAG, the group itself supervised the loading of its ground and service elements which joined the Northern Attack Force off Guadalcanal. The pilots of VMF-312, -322, and -323 flew their F4Us (Corsairs) to Manus via

[26] Units involved were: MAG-31 at Roi; 1st SepEngrBn and 16th AAA Bn at Tinian; Corps EvacHosp No. 2 and 2d AAA Bn at Guam; and 7th FldDep, 1st Prov MP Bn, and LFASCU-1 at Saipan.

Guadalcanal and Green Island. There they boarded the escort carriers *White Plains* and *Hollandia*. Already on board the latter was VMF(N)-543 which had boarded the vessel at Pearl Harbor on 11 March. Its ground personnel had departed from the same port three weeks earlier.

Outside of the TAF chain of command, but closely related to its combat functions, were Landing Force Air Support Control Units 1, 2, and 3. Two weeks after returning to its Saipan base from the Iwo Jima operation, LFASCU-1 loaded aboard ship for an immediate return engagement with the enemy at Okinawa. The other two LFASCUs were based at Ewa, T. H., where they trained for ICEBERG, and mounted for the invasion in February 1945, later staging for the target through Leyte.

As it had no need for an intermediate staging area, the XXIV Corps departed for Okinawa directly from Leyte. General Bruce's 77th Infantry Division, which was to open the Ryukyus operation with the assault on Kerama Retto, finished loading its landing ships on 18 March and its transports on the 20th, each echelon leaving for the target on the day following. The 7th and 96th Divisions conducted their own loading under the supervision of XXIV Corps TQMs, who spotted Southern Landing Force shipping at the most satisfactory point on the landing beaches. The Southern Tractor Flotilla departed Leyte during the morning hours of 24 March; the transport groups followed three days later.

By the evening of 27 March, all ICEBERG assault elements were at sea, converging on Okinawa. Soldiers and Marines aboard the transports and landing vessels had already made themselves as comfortable as possible under the crowded conditions and had settled down to shipboard routine. Officers and key NCOs reviewed their unit operation plans, examined maps and terrain models of the landing area, and held daily briefing sessions with their men. At the same time they squared away their combat gear for the invasion, most of the men of Hebrew and Christian faiths also prepared themselves for religious observances of Passover or Good Friday and Easter, all three holidays falling within a few days of each other in 1945.

NEUTRALIZING THE ENEMY [27]

After the first carrier strike of 10 October 1944, Naha's fire- and explosion-gutted ruins furnished the Japanese defenders with visual evidence of the effectiveness of American naval air power and served as an ominous portent of the future. One observer, a Japanese soldier, complained in his diary that, "the enemy is brazenly planning to completely destroy every last ship, cut our supply lines, and attack us." [28]

Okinawa was not visited again by Vice Admiral John S. McCain's Fast Carrier Force (TF 38) until 3 and 4

[27] Unless otherwise noted, the material in this section is derived from: CinCPac WarDs Jan-Mar45; *CTF 52 AR; Tenth Army AR; War Reports; Okinawa Operations Record;* Morison, *Victory in the Pacific.*

[28] "Diary of a radioman, 5th Harbor Base Unit," in CinCPac-CinCPOA Bul 147-45, Translations and Interrogations No. 32, dtd 16Jun45, p. 50.

January 1945, when, in conjunction with a heavy attack on Formosa, the Ryukyu and Sakashima Islands were also struck. Commenting on this raid, a Japanese replacement confided in his diary that "seeing enemy planes for the first time since coming to Okinawa somehow or other gave me the feeling of being in a combat zone." [29] The return of the Navy planes on 22 January reinforced his first impression and further shook his seeming complacency, as that day's diary entry implied resentment. "While some fly around overhead and strafe, the big bastards fly over the airfield and drop bombs. The ferocity of the bombing is terrific. It really makes me furious. It is past 1500 and the raid is still on. At 1800 the last two planes brought the raid to a close. What the hell kind of bastards are they? Bomb from 0600 to 1800!" [30]

During January, TF 38 struck Formosa and the Ryukyus twice, and made some uninvited calls on South China coastal ports, all while covering the Luzon landings. After its last attack, the force retired to Ulithi where reinforcing carriers were waiting to join. On 27 January, the same day that Admiral Nimitz arrived at his new advance headquarters on Guam,[31] the command of the Pacific fleet's striking force was changed and Admirals Spruance and Mitscher relieved Halsey and McCain. When Mitscher's carriers departed Ulithi on 10 February, it was in the guise of Task Force 58, which was destined to continue the work that TF 38 had begun.

As a diversion for the 19 February Marine landing on Iwo Jima, and to reduce the Japanese capability for launching air attacks against the expeditionary force, Mitscher's Fast Carrier Force struck at the Tokyo area on 16–17 February and again on the 25th. In between these attacks, Mitscher's planes and ships supported the Iwo assault from D-Day until the 23d, at which time they sortied for the 25 February Tokyo strike. As TF 58 retired to Ulithi on 1 March, planes of Task Units 58.1, 58.2, and 58.3 photographed Okinawa, Kerama Retto, Minami Daito, and Amami O Shima, and bombed and strafed targets of opportunity. These three units returned to Ulithi on the 5th.

At the same time that the fast carriers were making their forays, American submarines and naval patrol bombers ranged the western Pacific taking a steadily increasing toll of Japanese shipping. The bottom of the China Sea was littered with the broken hulls and loads of enemy transports and cargo ships which never reached their destinations. Almost complete isolation of the Okinawa garrison was accomplished by mid-February 1945 through the combined efforts of Navy air and submarine forces. It soon became apparent to General Ushijima that his Ryukyus command stood alone since "communications between the mainland

---

[29] "Diary of an unidentified superior private, 273d IIB," in *Ibid.*, p. 70.

[30] *Ibid.*

[31] Because Pearl Harbor was too far from the war area, the recently promoted Fleet Admiral Nimitz transferred his flag to Guam on 2 January in order to direct the Pacific war effort from a vantage point closer to the scene of the fighting.

of Japan and Formosa had been practically severed."[32]

The neutralization and isolation of Okinawa was furthered by the continuous series of strategic air strikes on the Japanese industrial network by Army Air Forces bombers, which mounted attacks from bases in China, India, the Philippines, the Marianas, and the Palaus. Massive raids on the factories of the main islands as well as on outlying sources of raw materials hindered Japan's ability and will to continue the war. Giant super-fortresses also rose from airfields in the southern Marianas in steadily increasing numbers to hit Tokyo, Osaka, Nagoya, and Kobe, widening the fire-swept circle around the expanse of previously devastated areas. During the interludes between carrier-plane attacks on the Ryukyus, B–29 appearances over Okinawa became so commonplace that the Japanese defenders referred to their visits as "regular runs."[33]

With the approach of L-Day, the tempo of covering operations was accelerated throughout the Pacific. For its final strike on Japan prior to the Okinawa landing. TF 58 steamed out of the Ulithi anchorage on 14 March. Four days later, carrier-launched planes interdicted Kyushu's heavily-laden airfields, and attacked installations on Shikoku and Honshu islands on the 19th. The task force did not escape unscathed this time, however, for the enemy was ready and retaliated with heavy counterstrikes during which the Japanese pilots displayed reckless abandon and a wanton disregard for their lives. Five carriers and other ships in the task force were hit hard. A temporary task group composed of the damaged carriers *Wasp*, *Franklin*, and *Enterprise*, the cruiser *Santa Fe*, and Destroyer Squadron 52 returned to Ulithi for necessary repairs. The ships remaining in TF 58, the carriers, the battleship force, and the protective screen, were reorganized into three task groups of relatively equal strength on 22 March. With this force, Admiral Mitscher then began the final run on Okinawa for the beginning of the preinvasion bombardments.

PREINVASION PREPARATIONS AND THE KERAMA RETTO LANDING[34]

The first elements of the ICEBERG force to appear at the target were the doughty sweepers of Mine Group One, which began operations off Kerama Retto and the southeastern coast of Okinawa on 24 March, just two days before the 77th Infantry Division was to land in the Keramas. After the minecraft cleared a channel outside the 100-fathom curve off the Minatoga beaches, part of Admiral Mitscher's battleship force, temporarily organized as TF 59, steamed through the swept area and bombarded Okinawa while TF 58 planes covered and neutralized enemy shore

[32] *Okinawa Operations Record*, p. 62.

[33] "Diary of an unidentified superior private, 273d IIB," *op. cit.*, p. 49.

[34] Unless otherwise noted, the material in this section is derived from: *CinCPOA OPlan 14–44; CTF 51 OPlan A1–45; CTF 51 AR; CTF 52 AR; CTG 51.1 AR; Tenth Army AR; 77th InfDiv OpRpt;* PhibReconBn, FMFPac, AR, Phase I and II, Nansei Shoto Operation, n.d., hereafter *PhibReconBn AR; War Reports;* Morison, *Victory in the Pacific*.

installations. By late afternoon, as TF 59 withdrew to rejoin the carrier force, the mine vessels finished that day's planned program of preliminary sweeps.

During these operations, the Amphibious Support Force, with elements of the Gunfire and Covering Force (Admiral Deyo serving as Officer in Tactical Command), had completed the run from Ulithi and deployed into approach formation. Two fire support units left TF 54 to begin their respective assignments—one unit to cover the sweep conducted between Tonachi Shima and Kerama Retto, and the other to cover the mine sweepers off Okinawa and to begin bombarding the demonstration beaches.

An important TF 52 element was the Underwater Demolition Flotilla, consisting of 10 underwater demolition teams (UDTs) organized into two groups, Able and Baker. On the afternoon of 24 March, the high speed destroyer transports (APDs) of Group Able and destroyers of TF 54 formed for the next day's UDT and NGF operations at Kerama. The remainder of Admiral Deyo's force was concentrated and ready to repulse all Japanese surface or air attacks.[35]

A carefully planned feature of the ICEBERG operation was this concentration of naval strength. With ample sea room and sufficient fighting power to eliminate any or all of the remnants of the Japanese Navy, TF 58 lay to the east of the Ryukyus in the Pacific Ocean. In the East China Sea, to the west of Okinawa, the majority of the combat ships of the Amphibious Support Force was concentrated, ready to stop any attempt to reinforce or evacuate the garrison. At night, the ships assigned to the bombardment of Okinawa's southeastern coast retired together so that their mission could be resumed without delay the following morning. In the event of any surface action, each of these task groups was able to operate and support itself independently.

As vaster areas surrounding Okinawa were swept clear of mines, destroyers and gunboats began patrol operations and made the beleaguered enemy's isolation a certainty. Shipborne radar picket stations, disposed from 15 to 100 miles offshore, encircled the island to protect the invasion force from the constant threat of surprise enemy air attacks. Aboard the destroyers and destroyer minesweepers serving as picket vessels were fighter-director teams which controlled the combat air patrols (CAPs) of carrier planes which orbited overhead during the hours of daylight. When Japanese flights were detected on picket radar screens, the CAP was vectored out to intercept and destroy the enemy. The bulk of the heavy losses incurred by the Navy during the battle for Okinawa was borne by the vessels comprising the radar picket fleet. The value of their services in protecting the vulnerable transport

[35] The command relationship between Admirals Blandy (CTF 52) and Deyo (CTF 54) requires clarification. Blandy, as a Senior Officer Present Afloat (SOPA), was responsible for the execution of all operations at the target while Deyo, although Blandy's senior, was responsible for the TF 54 movement and approach to the target (and for any TF 52 ships moving with TF 54), for the conduct of surface actions should there be any, and for the night deployments. *CNO Record*, chap 1, pp. 33–34.

310-224 O - 69 - 8

and service areas is measured by the large number of Japanese planes shot down before they had reached their objectives.

Although the destructive TF 58 raids on Kyushu had temporarily disrupted enemy plans for air attacks from the home islands, the Japanese managed to mount an increasing number of raids from fields in the Formosa area. Once it became apparent that Okinawa was to be invaded and that Okinawa waters held lucrative targets, forward elements of the *8th Air Division* rose from their fields in the Sakashimas to make their first *Kamikaze* attack on ships standing off Kerama Retto at dawn on 26 March.[36]

Beginning with this first, hour-long enemy air raid, the loss of lives and damage to ships mounted as Japanese bombers and suiciders made sneak attacks on the amphibious force in the dawn and dusk twilight hours.[37] As part of its planned schedule of preliminary operations supporting ICEBERG, Vice Admiral Sir H. Bernard Rawling's British Carrier Force (TF 57) struck Sakashima Gunto on the 26th and 27th. Since the carriers had blocked the use of Sakashima and Kyushu, the Japanese had to use Okinawa-based planes to attack the American invasion forces. The employment in three suicidal forays of all available aircraft, including trainers, liaison craft, and planes of a *Special Attack Unit* which managed to fly in from Kyushu, led to the complete elimination of the air strength of the Okinawa garrison by 29 March.[38]

Claims of enemy airmen who survived to return to home bases were grossly exaggerated, but their destructiveness was extensive. A summary of damages to American forces for the period 26 to 31 March reveals that six ships, including Admiral Spruance's flagship *Indianapolis*, were crashed by suicide-bent enemy pilots. Near misses accounted for damage to 10 other vessels, while floating mines sank 2 ships and an encounter with a Japanese torpedo boat gave another American ship minor damage.

Despite costly harassment from Japanese air attacks, Admiral Blandy's force proceeded with its primary task of preparing the target for the assault. Four Group Able UDTs cleared beach approaches in Kerama Retto on 26 March and began blowing Keise Shima reefs the next day. Because Okinawa's offshore waters had not been completely cleared of mines, the reconnaissance and demolition work scheduled for the 28th was delayed a day. Elements of Group Able scouted the demonstration beaches on the 29th, while Group Baker teams reconnoitered the Hagushi beaches.

During this reconnaissance of the west coast landing area, Group Baker swimmers discovered approximately 2,900 wooden posts embedded in the reef near its seaward edge and stretching for some distance on either side of the Bishi Gawa. These posts, which were on the average six inches in

[36] *Okinawa Operations Record*, "Record of the 8th AirDiv," Chart 1.

[37] More correctly known as morning and evening nautical twilights, these are the brief periods preceding sunrise and following sunset when general outlines may be visible, although the horizon cannot be distinguished, and approaching aircraft are virtually invisible.

[38] *Okinawa Operations Record*, pp. 65–66.

LANDING CRAFT *form up for the run to Kerama Retto beaches on 26 March 1945. Viewed from USS* Minneapolis. *(USN 80–G–316830)*

155MM GUNS *of the 420th Field Artillery Group are set up on Keise Shima to shell enemy main defenses prior to the Tenth Army assault landing. (USA SC205503)*

diameter and from four to eight feet in height, were generally aligned five feet apart in rows of three or four. Although some of these obstacles were loose, a few were set in concrete and the rest wedged into the coral. On 30 March they were blown up with hand-placed charges. All but 200 posts were destroyed by L-Day and it was believed that the landing would not be hindered by those that remained.[39]

Accompanying the UDTs during the beach reconnaissance and initial demolitions operations were assault troop observers, who acted as liaison and reconnaissance personnel.[40] Their primary function was to brief the UDTs on the schemes of maneuver and location of the landing areas of their respective assault units, to make certain that specific beaches were cleared, and to obtain current intelligence concerning the beaches and surrounding terrain. As soon as these preinvasion operations had been completed (29 and 30 March), the observers were returned by ship to join their parent organizations in the approaching attack groups. In general, the intelligence reports submitted by the observers favored a successful landing across the entire Tenth Army front.

Because the waters surrounding Okinawa had been heavily mined,[41] the scheduled NGF bombardment did not begin until 25 March (L minus 7) when TF 54 fire support vessels were able to close to ranges of maximum effectiveness. Carrier air was able to pound Okinawa repeatedly, however, and was met by only ineffectual and desultory fire from enemy antiaircraft defenses. In the course of the 3,095 sorties that the TF 52 Combat Air Support Control Unit (CASCU) directed against Okinawa prior to L-Day, special attention was given to the destruction of submarine pens, airfields, suicide boat installations, bridges over the roads leading into the landing area, and gun positions. After the pilots were debriefed, each day's strike results were evaluated by the CASCU on board Admiral Blandy's flagship, USS *Estes*, and considered together with damage estimates of ships' guns. The schedule of air missions and the NGF plans were revised and coordinated, and plans for the next day's sorties and shoots were then issued.

Initial target lists compiled by the Tenth Army artillery section and TF 54 intelligence section were constantly revised as analyses, based on aerial observation and photo reconnaissance, were received. As new evaluations were made of the destruction of enemy positions and installations, and new targets tabulated, cards listing the corrected data were delivered to the target information centers (TICs) of IIIAC and XXIV Corps. From the time that the bombardment of Okinawa began until L-Day, General Nimmer's TIC received copies of all dispatches sent from the objective by CTF 54. From these reports, all information relative to the discovery, attack, damage, and destruc-

[39] *CTF 52 AR*, pt V, sec G, p. 3.

[40] The allocation of Tenth Army observers to the UDTs was based on the assignment of one officer for each battalion in the assault, one for each RCT, division, and corps, and one for the army. *Tenth Army TntvOPlan* 1–45, Anx 4, App A, p. 1.

[41] TF 52 estimated that its minecraft swept and reswept over 3,000 square miles in the six days before L-Day. In this period, 257 mines were destroyed.

tion of targets in the IIIAC landing zone was excerpted and used to bring the target map and target file up-to-date.[42]

Although Admiral Blandy's bombardment force expended 27,226 rounds of 5-inch or larger-caliber ammunition on Okinawa, extensive damage was done only to surface installations, especially those in the vicinity of the airfields. As the ground forces were to discover later, the Japanese sustained little destruction of well dug-in defenses, and few losses among the men who manned them. On the day before the landing, as a result of his evaluation of the effect of air and NGF bombardment, CTF 52 could report that "the preparation was sufficient for a successful landing."[43] Admiral Blandy also stated that "we did not conclude from [the enemy's silence] that all defense installations had been destroyed. . . ."[44]

A prerequisite which Admiral Turner felt would guarantee the success of ICEBERG was the seizure of Kerama Retto and Keise Shima prior to L-Day. Because of the advantages to be gained by all ICEBERG assault and support elements, the taking of these islands was made an essential feature of the Tenth Army operation plan. Naval units, particularly, would benefit since the Keramas provided a sheltered fleet anchorage in the objective area where emergency repairs, refueling, and rearming operations could be accomplished. Once the envisioned seaplane base was established, Navy patrol bombers could range from Korea to Indochina in search and rescue missions and antisubmarine warfare operations. With the emplacement of XXIV Corps Artillery guns on Keise Shima and their registry on Okinawa, preliminaries for the main assault would be complete.

Even if the Keramas had had no value as an advanced logistics base, they would have been taken. The suspected presence of suicide sea raiding squadrons in the island chain was confirmed when the 77th Infantry Division landed, and captured and destroyed 350 of the squadrons' suicide boats. Their threat to the Okinawa landing was undeniable, for these small craft were to speed from their hideouts in the Keramas' small islands to the American anchorages. Here, "The objective of the attack will be transports, loaded with essential supplies and material and personnel . . .," ordered General Ushijima. "The attack will be carried out by concentrating maximum strength immediately upon the enemy's landing."[45] The surprise thrust into the Keramas frustrated the Japanese plan and undoubtedly eased initial ICEBERG operations at the Hagushi beaches. At the time that the 77th was poised to strike the Keramas, the islands were defended by approximately 975 Japanese troops, of whom only the some 300 boat operators

[42] "At H-Hour on D-Day it is estimated that the TIC files showed about 500 active targets suitable for attack by artillery, naval gunfire, and support aircraft, all located in the III Phib Corps zone of action for Phase One of the operations." LtCol John G. Bouker ltr to CMC, dtd 9Jan48, Subj: Operations of III Amphibious Corps Target Information Center on Okinawa, hereafter *Bouker ltr*.

[43] *CTF 52 AR*, pt V, sec C, p. 5.

[44] *Ibid.*

[45] "Thirty-second Army OperO No. 115, dtd 23Mar45," in Tenth Army Trans No. 231, dtd 6Jul45.

of the sea raiding squadrons had any combat value. The rest of the defense was comprised of about 600 Korean laborers and nearly 100 base troops.

On 26 March, the day following the sweep of Kerama waters by the minecraft and reconnaissance of its beaches by UDT personnel, Admiral Kiland's Western Islands Attack Group moved into position for the assault. A battleship, two large cruisers, and four destroyers had been assigned to provide NGF support for the landing, but only the 5-inch guns of the destroyers were used extensively. The capital ships were not called on to fire but remained on standby. As LSTs disgorged their cargo of armored amphibians and troop-laden assault tractors for the run to the beaches, carrier planes orbited the transport area to ward off Japanese suicide planes which were beginning to filter through the outer fighter screen. Aircraft bombed the beaches as the assault waves were guided toward the target by LCIs assigned to give close-in support.[46] (See Map 5.)

At 0801, the first of the four assault battalions of the 77th hit its target and in a little over an hour's time the other three had attacked their own objectives. Before noon, General Bruce saw that the rapid progress of his landing teams ashore would permit yet another landing that day, so he directed the 2d Battalion, 307th Infantry, a reserve unit, to take Yakabi Shima. Since this island's defenders offered little resistance, it was secured by 1341. By the end of the day the 77th had done quite well, having seized three Kerama islands outright and established a firm foothold on two others.

Within a six-day period, 26–31 March, at a cost of 31 killed and 81 wounded, the 77th Infantry Division completely fulfilled its mission as the vanguard of ICEBERG Expeditionary Troops. In the process of removing the threat posed by the Japanese to operations in the Kerama anchorage, General Bruce's troops killed 530 of the enemy, captured 121 more, and rounded up some 1,195 civilians. All of the enemy were not disposed of, however, for scattered Japanese soldiers remained hidden in the hills of the various Kerama islands and even occasionally communicated with units on Okinawa.[47]

Marine participation in pre-L-Day activities was confined to the operations of Major James L. Jones' FMF Amphibious Reconnaissance Battalion. For the Keramas invasion, it was split into two tactical groups, one under the battalion commander and the other under his executive officer. Only two companies of the battalion were available for anticipated missions, since Company B had been assigned to the V Amphibious Corps for the Iwo Jima landing and did not return to its parent organization until after L-Day.

On the night of 26–27 March, while the 77th consolidated its conquests and prepared for the next day's battles, Major Jones' men landed on and reconnoitered the reef islets comprising

[46] The Support Craft Flotilla was composed of two mortar, three rocket, and four gunboat divisions. The 42 LCIs and 54 LSMs in this unit were organized into teams of various types and assigned to a specific beach and landing. PhibGru 7 AR, dtd 26May45.

[47] *Okinawa Operations Record*, p. 66.

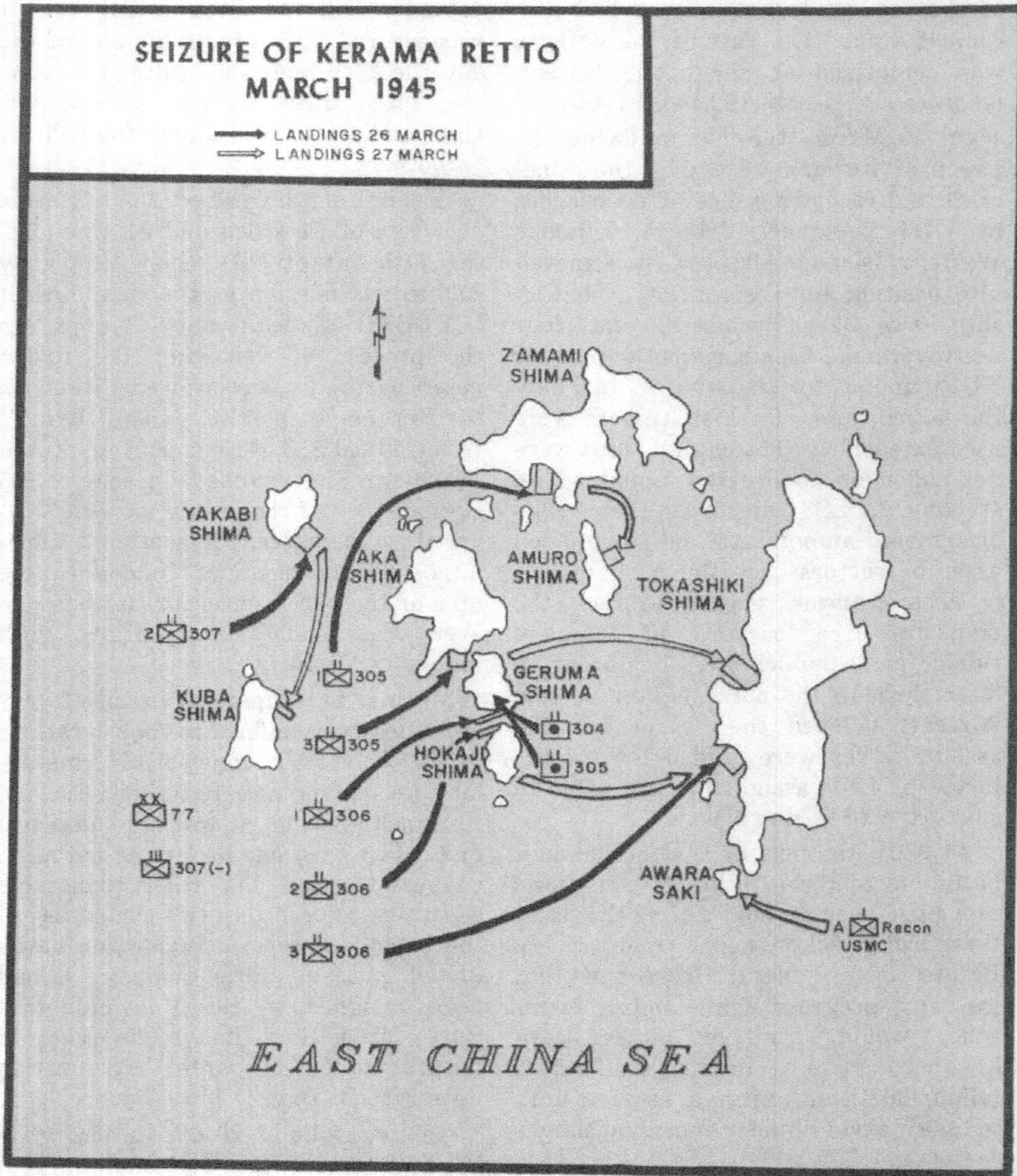

MAP 5

T.L. RUSSELL

Keise Shima.[48] When reconnaissance of the small group revealed no trace of the enemy, the Marines reboarded their APDs with information of reef, beach, and terrain conditions. Their findings were then forwarded to the 77th Division intelligence officer for evaluation and distribution to the units which were to land on Keise Shima.

During the night of 27–28 March, scouts from Company A landed on Aware Saki, a small island off the southern tip of Tokashiki Shima. Again there was no evidence of the enemy. The only encounter with the Japanese occurred on 29 March, during early morning landings on Mae and Kuro Shima, two small islands which lie between the Kerama Retto and Keise Shima. At 0630, a suicide boat, apparently manned by only one soldier, was observed heading at high speed for Mae Shima's beach from Tokashiki Shima. The one-man regatta was brought to a speedy and spectacular end as the explosive-laden craft disintegrated under a hail of machine gun fire. A reconnaissance of Kure Shima shortly thereafter indicated no troops or civilians, nor any installations.

In order to remove some of the obstacles to the landing of XXIV Corps Artillery units on Keise Shima on 31 March, UDTs blasted a path through the off-shore coral reefs early that morning, and this completed the work they had begun on the 27th. Then, after 2/306 landed unopposed and determined that no enemy had slipped back to Keise after the Marine reconnaissance, men and equipment of the 420th Field Artillery Group went ashore and immediately set up to fire. By 1935, the group's 155mm guns ("Long Toms") began registering on preselected targets in southern Okinawa. There, the Japanese later reported, they "incessantly obstructed our movements by laying an abundant quantity of fire inside our positions, the fire being directed mainly to cut off our communications." [49] Also landing on Keise were part of a team from an Army air warning squadron and an AAA (automatic weapons) platoon which, when ashore and set up, became part of the area antiaircraft defense system.

Already beset by American carrier-based strikes and by ships' gunfire which blasted Okinawa in a precise and businesslike manner, the Japanese felt that the artillery fire from Keise was overdoing it a bit. A special attack unit was formed to raid the island artillery emplacements, and the *5th Artillery Command's* 15cm guns were ordered to conduct counterbattery fire in an attempt to destroy the American Long Toms. Neither measure attained success, and the *Thirty-second Army* was never able to enforce its order to "stop

[48] Keise Shima or Keisan Sho (sho, in Japanese, is shoal or reef) is five miles WNW of Naha and consists of three sand-and pebble-islands; Kuefu Shima, Naganna Shima, and Kamiyama Shima, the last-named being two low sandy islands separated by a 100-yard-wide strait. For simplicity, and in keeping with the usage established by action reports and previously written histories of the Okinawa campaign, this small island grouping will be referred to as Keise Shima in this account also.

[49] *Okinawa Operations Record*, pp. 66–67.

the use of enemy artillery on Keise Shima."[50]

Before L-Day, the floating naval base in the Kerama Retto was functioning at a high pitch. From watery take-off lanes, seaplanes rose to harass enemy submarines and shipping in the China Sea. *Kamikaze*-damaged vessels were salvaged and repaired on an around-the-clock schedule, while the rearming, refueling, and revictualling of healthy ships kept pace. Without this frontline logistical facility, "many more ships and personnel of the service force than were available in the Okinawa area would have been required at sea to make replenishment an accomplished fact for all fleet forces."[51]

In contrast to the conspicuous prelanding operations of ICEBERG forces in the target area, the *Thirty-second Army* was able to surround its tactical dispositions with a greater degree of secrecy. Not until after the landing on Okinawa and relentless probing by the assault forces did the Tenth Army learn what the strength of the enemy was and where his positions were. Before L-Day, American knowledge of enemy dispositions was sketchy, and as late as L minus 1 (31 March), the G–3 of the 6th Marine Division was told that "the Hagushi beaches were held in great strength."[52]

The factor which tipped the scales in favor of an unopposed Allied landing on the Hagushi beaches was General Ushijima's decision to defend the southeast coast of Okinawa in strength. When the 2d Marine Division made its feint landings on D-Day and D plus 1, the Japanese commander's staff believed its earlier estimate that "powerful elements might attempt a landing [on the Minatoga beaches]" was fully justified.[53] Consequently, a substantial portion of the artillery and infantry strength of the *Thirty-second Army* was immobilized in face of a threat in the southeast that never materialized.

Although Ushijima's command had prepared for an American landing elsewhere, from the Japanese point of view the Hagushi beaches remained the most obvious target. Even while propaganda reports—mostly untrue—of successful *Kamikaze* attacks against the invasion fleet bolstered Japanese morale, the commander of a scratch force formed from airfield personnel on the island warned his men not "to draw the hasty conclusion that we had been able to destroy the enemy's plan of landing on Okinawa Jima."[54] The commander of

[50] "32d Army OperO A #127, dtd 6Apr45," in *CICAS Trans No.* 266. Although this order directed the CO of the shipping engineers to organize a raiding unit and attack Keise on the night of 6 April 1945, no contemporary records show that this operation was carried through.

[51] RAdm Worrall R. Carter, *Beans, Bullets, and Black Oil* (Washington: GPO, 1953), p. 353, hereafter Carter, *Beans, Bullets, and Black Oil.*

[52] Capt Edward F. Townley, Jr., ltr to Asst G–3, HQMC, dtd 4Dec65.

[53] *Okinawa Operations Record*, p. 67.

[54] "1st Specially Established Regt OperO No. 1, dtd 30Mar45," in CinCPac-CinCPOA Bul 107–45, Translations and Interrogations No. 28 dtd 14May45.

the *1st Specially Established Regiment*, Lieutenant Colonel Tokio Aoyangai, proved himself clairvoyant, for in less than 24 hours after his message had been distributed, the Northern and Southern Attack Forces were moving into their transport areas ready to launch the assault.

The First Days Ashore

SEIZURE OF THE BEACHHEAD [1]

Optimum weather conditions for an amphibious landing prevailed at the target on L-Day when the Central Pacific Task Forces launched the attack against Okinawa on 1 April 1945, Easter Sunday. The coming of dawn revealed cloudy to clear skies and a calm sea with but a negligible surf at the shore. Moderate easterly to northeasterly winds were blowing offshore, just enough to carry the smoke away from the beaches. To the many veteran jungle fighters among the invading troops, the 75-degree temperature seemed comfortably cool.

At the target, the major naval lift and support elements moved into their assigned areas off the Hagushi beaches. Once in position, the ships prepared to debark troops. Off the Minatoga beaches on the other side of the island, the same preparations were conducted concurrently by the shipping that carried the 2d Marine Division.

Admiral Turner unleashed his forces at 0406 with the traditional order, "Land the Landing Force," [2] and Okinawa's ordeal began with a percussive overture of naval gunfire. (See Map 6.) The enemy reacted to the landing shortly after dawn as he mounted scattered air attacks on the convoys. In the continuing belief that the main effort was directed at the Minatoga area, the few Japanese aircraft not destroyed by American carrier air or ships' antiaircraft guns disregarded the more lucrative targets off Hagushi and concentrated on Demonstration Group shipping. *Kamikazes* struck the transport *Hinsdale* and LST *884* as troops, mostly from the 3d Battalion, 2d Marines [3] and its reinforcing elements, were disembarking for the feint run into the beaches. Reported killed were 8 Marines; 37 were wounded, and 8 were

[1] Unless otherwise noted, the material in this section is derived from: *CTF 51 AR; Tenth Army AR; XXIV Corps AR; IIIAC AR; 1st MarDiv SAR;* 1st MarDiv G-3 Jnl, 28Feb-13Jul45, hereafter *1st MarDiv G-3 Jnl,* with appropriate date; 1st Mar SAR, Nansei Shoto, dtd 25Jul45, hereafter *1st Mar SAR;* 5th Mar S-3 Jnl, 15Jan-26Jul45, hereafter *5th Mar S-3 Jnl,* with appropriate date; 7th Mar SAR, Phase I and II, dtd 1May45, and Phase III, dtd 11Jul45, hereafter *7th Mar SAR; 6th MarDiv SAR, Ph I&II;* 6th MarDiv Unit Jnl, Phase I and II, 1-22Apr45, hereafter *6th MarDiv Jnl, Ph I&II,* with appropriate date. The action reports of the major component units of the 6th Marine Division are included as annexes to the division SAR and will be cited separately as *4th Mar SAR, Ph I&II; 1/4 SAR, Ph I&II; 6th TkBn SAR, Ph I&II;* etc.

[2] *CTF 53 AR,* pt III, p. 12.

[3] Throughout this campaign narrative, the designation "3d Battalion, 1st Marines" will be used interchangeably with "3/1" or "BLT 3/1," the latter number depending upon which regimental unit is involved. Also, 4th Marines or 5th Marines is synonymous with RCT 4 or RCT 5. Reinforcing troops which make an infantry regiment an RCT are considered to be included in the RCT designation. A BLT's reinforcement is likewise to be included in the BLT designation.

MAIN BATTERIES *of USS* Tennessee *hurl tons of high explosives at Okinawa as assault amtracs head for the beachhead. (USN 80-G-319325)*

AERIAL VIEW *of the Hagushi anchorage and Yontan airfield on L plus 2, looking southeast from Zampa Misaki. (USN 80-G-339242)*

listed as missing in action.[4] It is somewhat ironic that units not even scheduled to land on Okinawa on L-Day sustained the first troop casualties.

Air support arrived over the target in force at 0650 [5] and the assault forces began to debark ten minutes later. The transport areas became the scenes of purposeful activity as troops climbed down landing nets into waiting landing craft, while armored amphibians, and amphibian tractors preloaded with troops and equipment, spewed forth from the open jaws of LSTs. At the same time, tank-carrying LCMs (landing craft, mechanized) floated from the flooded well decks of LSDs (landing ships, dock). Other tanks, rigged with T–6 flotation equipment, debarked from LSTs to form up into waves and make their own way onto the beaches.[6]

In reply to the murderous pounding of the Hagushi beaches by 10 battleships, 9 cruisers, 23 destroyers, and 177 gunboats, the Japanese returned only desultory and light artillery and mortar fire. Even though the assault waves formed up in assembly areas within range of this fire, neither troops nor invasion craft were hit. During the early morning hours, the two battalions of the 420th Field Artillery Group on Keise Shima received heavy enemy counterbattery fire, which stopped American unloading operations on the reef for four hours but caused no damage.

Lying off each Okinawa invasion beach were control vessels marking the lines of departure (LD). Landing vehicles quickly formed into waves behind the LD and at 0800, when the signal pennants fluttering from the masts of the control vessels were hauled down, the first wave, composed of LVT(A)s (landing vehicle, tracked (armored)), moved forward to the beaches in an orderly manner behind a line of support craft. Following on schedule, hundreds of troop-carrying LVTs, disposed in five to seven waves, crossed the lines of departure at regular intervals and moved determinedly towards the shore.

Despite the ferocity of the prelanding bombardment, enemy artillery and mortars continued scattered but ineffectual fire on the invasion waves as they made the 4,000-yard run to the beach. On approaching the shoreline, the LVT(A)s fired upon suspected targets, while naval gunfire lifted from the beach area to hit inland targets. Carrier fighters that had been orbiting lazily over the two flanks of the beachhead began diving over the landing area and neutralized it with repeated strafing, bombing, and napalming runs.

As the assault waves hit the beaches, smoke was laid down on the hills east of Yontan to prevent enemy observation of the landing zone. On the other side of the island, Demonstration Group

---

[4] *2d MarDiv AR*, Anx A–N. In the same attack, ships' crews sustained casualties of 1 killed, 34 wounded and 10 missing.

[5] More than 500 planes of the Fifth Fleet were engaged in troop support missions between 0700 and 1000 on L-Day.

[6] Medium tanks fitted with T–6 flotation devices were able to proceed to the beach under their own power, utilizing their tracks for propulsion. The equipment consisted of flotation tanks welded to the outside of the armored vehicle, an improvised steering device, electrical bilge pumps, and electrically detonated charges to jettison the flotation tanks when they were no longer required and once the beach was reached. *6th TkBn SAR, Ph I&II*, pp. 39–41.

landing craft raced toward the Minatoga beaches only to reverse their course and retire to the transport area behind a smoke screen as the fourth wave crossed the line of departure.

Neither the reef fringing the beaches nor enemy mortar fire on the beaches themselves interfered with the successful XXIV Corps landing south of the Bishi Gawa. The eight Army assault battalions were landed by six successive waves of LVTs and moved forward without opposition.

The sea wall, which had caused some concern to XXIV Corps planners, had been breached by naval gunfire. In anticipation of the early build-up ashore, engineers, landing in the first waves, blasted additional beach exits in those portions of the wall which remained standing. Upon landing, the LVT(A)s poured through these breaches, hard on the heels of the infantry, and moved to protect the flanks, while amphibious trucks (DUKWs), preloaded with 4.2-inch mortars, and tanks rolled inland.

Off the Marine landing zone, north of the Bishi Gawa, the reef was raggedly fissured and became smoother only as it neared the beach. A rising tide floated the landing vehicles over a large portion of the reef and the boulders which fringed it. On the northern flank of the 1st Marine Division, however, the large circular section of the reef off the Blue Beaches presented difficulties to the tractors attempting to cross at that point, and delayed their arrival at the beach.

During the approach of IIIAC assault waves to assigned beaches, several instances occurred when inexperienced wave guide officers failed to follow correct compass courses or when they did not guide by clearly recognizable terrain features on the shore. Some troops were thus landed out of position. For example, Lieutenant Colonel Bruno A. Hochmuth's 3d Battalion, 4th Marines —assigned to Red Beach 1—was landed on the right half of Green 2 in the zone of the 22d Marines (Colonel Merlin F. Schneider), and on the rocky coast line between Green 2 and Red 1. (See Map 6.)

Elements of Colonel Edward W. Snedeker's 7th Marines were landed in relatively insignificant numbers on the beach of the 4th Marines (Colonel Alan Shapley). On the extreme right flank of the 1st Marine Division, however, the fourth wave of RCT–5 was diverted across the corps boundary and landed on the right flank of the 7th Infantry Division. The Marine wave consisted principally of Lieutenant Colonel Charles W. Shelburne's 1/5 reserve—Company B—and part of the battalion command post group. By 0930, a sufficient number of LVTs had been sent to pick up all but one lieutenant and two squads, who did not rejoin their parent unit until L plus 3.[7] The fifth and sixth waves of the 5th Marines landed on

[7] 1/5 SAR, Ph I&II, dtd 29Apr45, p. 5, and Ph III, dtd 9Jul45, hereafter *1/5 SAR*. Relating to this mixup in landing beaches, the 1st Division Shore Party Officer wrote, "there is a point here which should not be missed. 1st MarDiv beaches were designated Blue and Yellow and 7th Div's were Purple and Orange from left to right as you came in from the sea. Some of the ships mounting out the 7th Div did not have Purple and Orange bunting and substituted Blue and Yellow." Col Robert G. Ballance ltr to CMC, dtd 22Mar55, hereafter *Ballance ltr*.

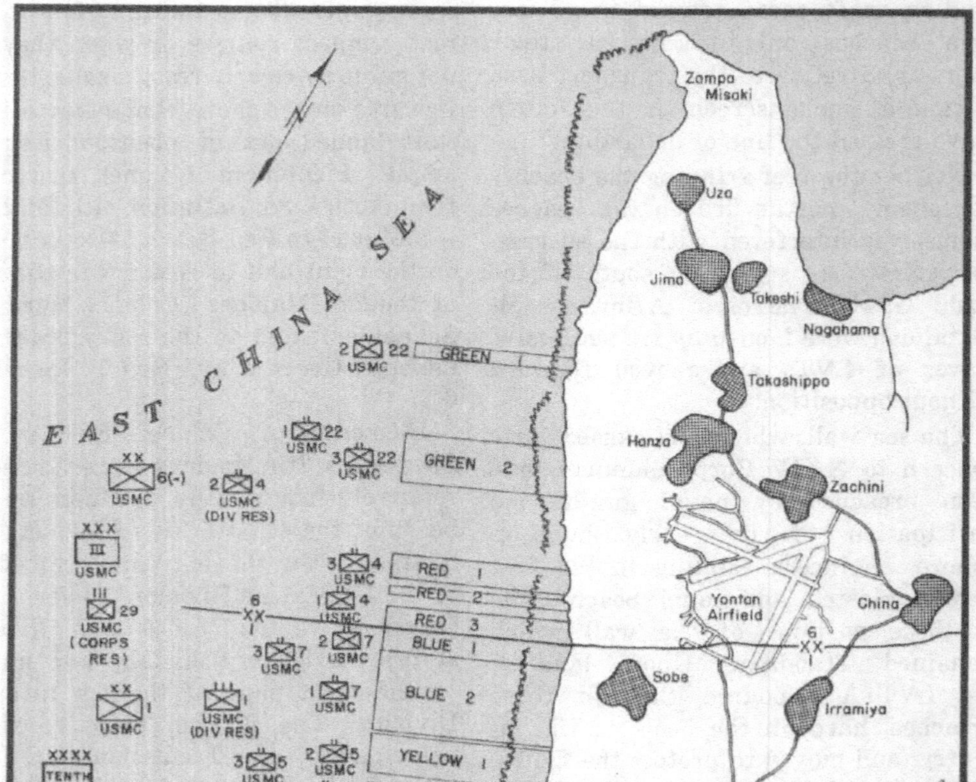

MAP 6 T. L. RUSSELL

their assigned beaches after the guide officers of these waves corrected the faulty course heading followed by the fourth wave.

Despite these unexpected deviations from the landing plan, all the LVT(A)s spearheading the IIIAC attack reached the beach by 0830 and all eight assault battalions were ashore by 0900. The beaches had not been mined and opposition to the landing consisted only of sporadic mortar and small arms fire. This resulted in but few casualties and caused no damage to the LVTs. "With utter consternation and bewilderment and with a great deal of relief the assault wave landed against practically no opposition." [8]

As the assault troops surged up the terraced slopes behind the beaches and sped inland, the center of invasion activity shifted from the line of departure to the transfer line at the edge of the reef. There, small boat, LVT, and DUKW control was established to unload support troops and artillery units on call. Supporting units continued to pour ashore during the morning as the attack progressed against only slight resistance. At the transfer line, reserve infantry elements shifted from ships' boats into the LVTs which had landed assault troops earlier. Flotation-equipped tanks made the beaches under their own power, others were landed at high tide from LCMs, and the remainder were discharged directly onto the reef from LSMs and LSTs. DUKWs brought the 75mm and 105mm howitzers of the light field artillery battalions directly ashore.

All tanks in the 1st Division assault wave, landing from LCMs and LCTs, were on the beach by noon. One exception was a tank that foundered in a reef pothole. The captain of LST *628*, carrying the six T–6 flotation-equipped tanks of the 1st Tank Battalion (Lieutenant Colonel Arthur J. Stuart), disregarded the operation plan and refused to allow them to launch until H plus 60 minutes.[9] At that time, he set them in the water 10 miles off the landing beaches. These tanks finally reached shore after being afloat for more than five hours, but two of them were hung up on the reef because of the ebbing tide. Because the LSMs carrying Lieutenant Colonel Stuart's reserve tanks had great difficulty in grounding on the reef on L-Day, the first tracked vehicle off the ramp of one was lost in an unseen pothole. Of the four LSMs employed, two finally landed their cargo late on L-Day, another at noon on L plus 1, and the last on 3 April.

Tanks were landed early in the 6th Division zone, where each of the three companies of the 6th Tank Battalion (Lieutenant Colonel Robert L. Denig, Jr.) employed a different landing procedure. Tanks equipped with flotation gear swam to the reef, easily negotiated the rugged coral, continued on to the beaches, where they jettisoned their pontoons and became operational by H plus 29 minutes. The company in LCMs

---

[8] *4th Mar SAR, Ph I&II*, pp. 6–7.

[9] In his action report, the LST captain stated that he did not hold a complete copy of the basic Fifth Fleet plan. He also commented that the plans held aboard his ship were distributed so late that there was insufficient time for "'adequate planning, preparation and training.'" *Dir, Naval Hist ltr I.*

came in at high tide (0930) and landed without incident. The third company successfully landed directly from the LSMs grounded on the reef but forded the deep water between the grounding point and the shore with difficulty.[10]

Soon after landing, the accelerated pace of the 6th Division assault to the north overextended Colonel Schneider's 22d Marines. Troops were taken from his left battalion, 2/22 (Lieutenant Colonel Horatio W. Woodhouse, Jr.), to guard the exposed flank. This, in turn, weakened the 22d attack echelon and gave it a larger front than it could adequately cover. Consequently, a considerable gap developed between the 2d Battalion and the 3d Battalion (Lieutenant Colonel Malcolm "O" Donohoo), advancing on the right. In less than 20 minutes after the landing, Colonel Schneider ordered his reserve, 1/22 (Major Thomas J. Myers), ashore. Upon landing, the 1st Battalion (less Company C, remaining afloat as regimental reserve) was committed in the center of the regimental zone.

Still meeting no opposition while it continued the rapid move inland, by 1000 the 22d Marines found its left flank unit stretched dangerously thin. As he pressed the division attack to exploit initial success, General Shepherd anticipated Schneider's request for reinforcements to cover the exposed flank, and asked corps to release one BLT of the 29th Marines.

During the unopposed 22d Marines advance on Hanza, the 4th Marines moved on Yontan airfield against light to moderate resistance. Isolated enemy pockets, built around light machine guns, slowed the regiment only slightly as it penetrated several hundred yards inland and made contact on its right with the 7th Marines at the division boundary. By midmorning, the 4th reached the airfield and found it unguarded.

Only intermittent sniper fire coming from beyond the field opposed the 4th Marines as it swept across the air facility and secured its objective by 1300. The airfield was found to be essentially intact, but all buildings had been stripped and the antiaircraft emplacements contained only dummy guns.

As this rush carried Colonel Shapley's regiment beyond adjacent units, a wide gap developed between its left flank and the right of the 22d Marines, then in the vicinity of Hanza. Shapley's regiment jumped off again at 1330 against only light resistance on its left. Tanks were called in to reduce several cave positions in this area. After these positions were neutralized, the attack continued slowly through rugged, wooded terrain. In order to maintain contact with the 7th Marines and to rectify the overextended condition of the 4th, 2/4 was released from division reserve at 1500 and immediately committed on the regiment's left to establish contact with the 22d Marines.

Because the division left flank was still dangerously exposed, General Shepherd regained 1/29 (Lieutenant Colonel Jean W. Moreau) from corps. Released by IIIAC at 1300, the battalion landed at 1500 and, with its left flank anchored on Green Beach 1, completed tying in with 22d Marines at 1700.

The 1st Marine Division, to the south of the 6th, met with the same sur-

[10] *6th TkBn SAR, Ph I&II*, pp. 15–16, 39–41.

310-224 O - 69 - 9

prising lack of resistance. By 0945 on the division left, Colonel Snedeker's 7th Marines had advanced through the village of Sobe, a first priority objective, and the 5th Marines (Colonel John H. Griebel) was 1,000 yards inland standing up. With the beaches clear, and in order to avoid losing any troops as a result of anticipated enemy air attacks against the congested transport area, the division reserve was then ordered ashore. Colonel Kenneth B. Chappell's 1st Marines embarked two BLTs, and the third was to land as soon as transportation became available.

Reserve battalions of both assault regiments were picked up by LVTs at the transfer line and shuttled to the beach before noon. The 3d Battalion, 7th Marines (Lieutenant Colonel Edward H. Hurst), landed in the center of the regimental zone of action and then moved to the rear of Lieutenant Colonel Spencer S. Berger's 2/7, the left flank unit. At 1400, 3/5 was positioned 400 yards behind 1/5 on the division right boundary. When the 5th Marines reserve was moved up behind the assault battalions, the commanding officer of 3/5, Major John A. Gustafson, went forward to reconnoiter. An hour later, at 1500, his group was fired on by a small bypassed enemy pocket and Gustafson was wounded and evacuated. His executive officer, Major Martin C. Roth, took over temporarily until 4 April, when Lieutenant Colonel John C. Miller, Jr., assumed command.[11]

Thus disposed in depth with its reserve elements echeloned to guard the flanks, the 1st Division continued its steady advance over the rolling checkerboard terrain. In addition to having developed the numerous caves that honeycombed the many hillsides in the zone, the Japanese had begun to organize other positions throughout the area and the Marines encountered innumerable field fortifications in varying stages of development. These defenses, however, were only held by small, scattered groups of service troops and home guards. According to a postwar Japanese source, these troops comprised ". . . a hastily organized motley unit . . . facing extreme hardship in trying to achieve an orderly formation." [12]

The principal bridge over the Bishi Gawa below Hiza was undamaged and standing,[13] and local defense forces had made little or no effort to destroy the narrow bridges that spanned lesser streams. What proved a greater hindrance to the advance than the desultory enemy attempts at halting it was what one observer described as "an excellent network of very poor roads." [14]

By 1530, the majority of IIIAC supporting troops and artillery was ashore. One howitzer of Colonel Robert B. Luckey's 15th Marines and three of the 11th Marines (Colonel Wilburt S. Brown) were lost when the DUKWs carrying them foundered on the reef, but the remaining divisional artillery of IIIAC was landed successfully. Even

[11] 3d Bn, 5th Mar SAR, 1-21Apr45, dtd 30Apr45, and 22Apr-22Jun45, dtd 10Jul45, hereafter *3/5 SAR*.

[12] *Okinawa Operations Record*, p. 69.

[13] In commenting on this section, Brigadier General Robert G. Ballance, the shore party officer of the 1st Marine Division, noted that the "bridge was a stone arch affair, dating back around 1200 A.D." *Ballance ltr.*

[14] *1st MarDiv SAR*, pt VII, pp. 1–2.

though the artillery arrived early, a combination of the rapid infantry advance and the resulting strain on communications made it difficult for forward observers to register their battalions. Corps artillery reconnaissance parties began landing at 1300, and found that "selection of position areas from map and photo study proved suitable in every case."[15]

The advance was halted between 1600 and 1700, and the attacking infantry dug in, established contact all along the IIIAC line, and carried out extensive patrolling to the front. To maintain the impetus of the attack of his division on L-Day, General Shepherd had committed his entire reserve early. The 6th Marine Division remained in good shape and was well disposed to resume the advance on the 2d. Both the 4th and 22d Marines still maintained a company in reserve, while the corps reserve (29th Marines, less 1/29) was located northwest of Yontan airfield, in the vicinity of Hanza, after its landing at 1535.

General del Valle's division was unable to close the gap on the corps boundary before dark and halted some 600 yards to the rear of the 7th Infantry Division on the right.[16] A company was taken from the reserve battalion of the 5th Marines and put into a blocking position to close the open flank. Two 1st Marines battalions, 1/1 and 2/1 (commanded by Lieutenant Colonels James C. Murray, Jr., and James C. Magee, Jr., respectively), landed at 1757. Lieutenant Colonel Stephen V. Sabol's 3d Battalion was on the transfer line at 1800 but, unable to obtain LVTs, remained in the boats all night. The 1st Battalion was attached to the 5th Marines for administrative control and moved inland to Furugen, while 2/1, similarly attached to the 7th Marines, dug in east of Sobe by 1845.[17]

Artillery support for the Marine infantry was readily available by nightfall. The 15th Marines had established its fire direction center (FDC) by 1700, and its batteries were registered by 1830. All of Colonel Brown's artillery battalions were prepared to fire night defensive missions, though two of the 11th Marines battalions did not complete their registration because of the late arrival of spotter planes. Since enemy action was confined to unsuccessful attempts at infiltration of the lines and to intermittent mortar and machine gun fire in the 4th Marines sector, there were relatively few requirements for artillery support that first night on Okinawa.

ICEBERG's L-Day had been successful beyond all expectations. In conjunction with the extended initial advance of IIIAC, XXIV Corps had captured Kadena airfield by 1000, driven inland to an average depth of 3,500 yards, and advanced south along the east coast to the vicinity of Chatan. On 1 April, the

[15] *Henderson ltr.* The same source stated: "The [Japanese] failure to fight for the beachhead never gave us a chance to see how well the artillery would have performed in the assault phase. I am convinced however that it would have been the best field artillery support of the war."

[16] LtCol John D. Muncie ltr to CMC, dtd 27Mar47.

[17] Col Russell E. Honsowetz ltr to CMC, dtd 9Oct54, hereafter *Honsowetz ltr.*

Tenth Army had landed an estimated 50,000 troops in less than eight hours and established a beachhead that was 15,000 yards wide and varied from 4,000-5,000 yards in depth. (See Map 7.) For the entire day's operations by four assault divisions, casualties were reported to Admiral Turner as 28 killed, 104 wounded, and 27 missing.

PROGRESS INLAND

Following a relatively quiet night, which was punctuated only by sporadic sniper, machine gun, and mortar fire, Tenth Army units resumed the attack on L plus 1 at 0730. Perfect weather again prevailed, as the early morning was cool and a bright sun soon dispelled the ground fog and haze to provide unlimited visibility. While no artillery preparation preceded the jump off on the second day, all guns were available on call for support fires, and registration of all battalions, including those of the 11th Marines, had been completed. Carrier planes were on station at 0600 before the attack began.

With the resumption of the advance, the 6th Marine Division continued to the east, while its left flank unit, 1/29, began clearing operations to Zampa Misaki. Admiral Turner wanted the point captured as a site for a radar station. He also wanted the Black Beaches uncovered so that unloading operations could begin. Throughout the day, the 22d Marines and 1/29 advanced rapidly against light resistance. By 1025, the latter unit had seized Zampa Misaki and found that the beaches there were unsuitable for use by IIIAC.

Major Anthony Walker's 6th Reconnaissance Company was then ordered to reconnoiter the beaches on the north coast of Zampa and the villages of Nagahama and Maeta Saki. Walker's scouts accomplished this mission before noon without opposition, and while doing so, encountered 50 civilians, who were taken into custody and transferred to stockades. The Nagahama beach was reported satisfactory for landing supplies.

On the 6th Division right, the 4th Marines advanced through rugged terrain, meeting intermittent enemy reaction. As the day wore on, however, pockets of stiff resistance were increasingly encountered, and at 1100, 3/4 came up against strong enemy positions, consisting of mutually supporting caves on both sides of a steep ravine. When the leading platoon entered the draw, it was met by a hail of small arms fire so heavy that the Marines could not bring out their 12 wounded until four hours later. "Every means of painlessly destroying the strongpoint was unsuccessfully tried and it was finally taken by a typical 'Banzai' charge with one platoon entering the mouth of the draw and one platoon coming down one side of the two noses that formed the pocket." [18]

The speed of the second day's advance again caused the assault units to become overextended. About midmorning, Colonel Shapley reported that there was a gap between his regiment and the 7th Marines, which he believed to be some 1,000 yards south of the division bound-

[18] *4th Mar SAR, Ph I&II*, p. 8.

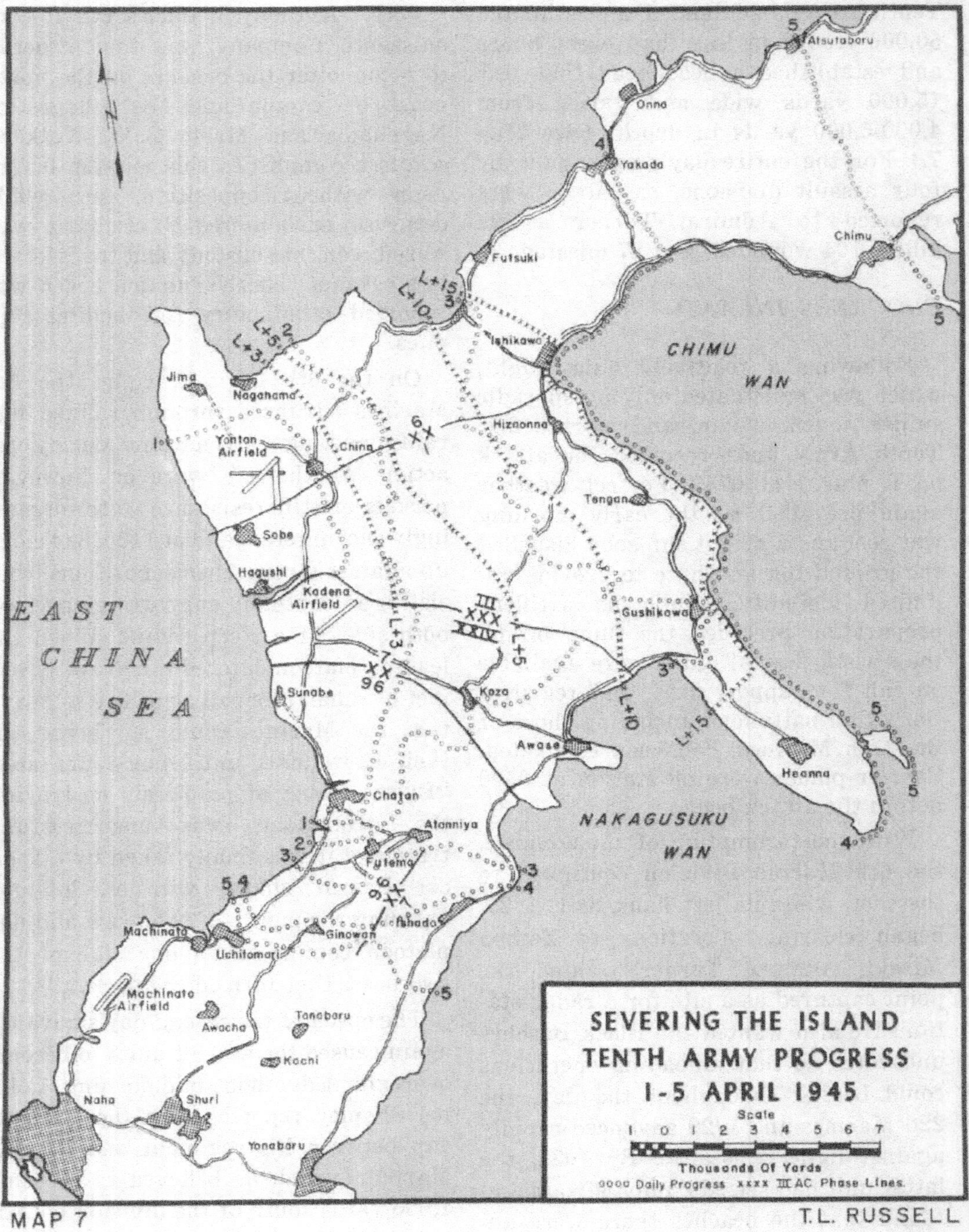

MAP 7 T.L. RUSSELL

ary,[19] and an adjustment was requested by Shepherd. The 7th was ordered to cover the gap, a movement that placed it ahead of 1/4. In sideslipping back to its own zone, 1/4 met with stiff opposition from strongly entrenched enemy forces similar to those that had held up 3/4. With the aid of a platoon of tanks, this position was finally reduced. The two battalions killed some 250 Japanese in the course of the day's operations before the 4th Marines attack ceased at 1830, some 1,000 yards ahead of the L plus 3 line.

During the day, Major Paul F. Sackett's 6th Engineer Battalion repaired the strips on Yontan and placed one taxiway in good enough condition to permit a VMO–6 spotter plane to land at 1500. By 4 April, all three Yontan runways were ready for emergency landings.

Spearheaded by extensive advance patrolling, the 1st Marine Division moved out on L plus 1 unopposed except for the slight interference presented by local defense units, which were part of a force officially designated the *1st Specially Established Regiment.* Activated by the *Thirty-second Army* on L minus 4, it was composed of 3,473 airfield service troops and *Boeitai,* less than half of whom were armed with rifles. In addition, the equipment of this regiment consisted of 55 light machine guns, and 18 grenade launchers. The heaviest weapons of this unit were 10 heavy machine guns and 5 20mm dual-purpose machine cannon. For the most part, the troops were completely untrained, and even the regular Army service troops had not been given such basic infantry training as firing a machine gun.

When Ushijima pulled his combat troops south, the *1st Specially Established Regiment* had been assigned the mission of servicing final air traffic on the Yontan and Kadena fields. The regiment was to destroy those fields on order after the Americans had landed and then to retire to positions from which it could deny their use to the invaders. The *1st Battalion* was located in the 6th Division zone; the *2d* and part of the *3d Battalion* in the 1st Marine Division area; the remainder of the *3d Battalion* faced the 7th Infantry Division; and the *5th Company* of the *12th Independent Infantry Battalion* was assembled in regimental reserve at Hanza. Uniquely enough, upon interrogation of some troops captured from these organ-

[19] "Lack of readily identifiable terrain features made it impossible to quickly identify the boundaries of zones of action in the rapid advance. Maps at this time were also poor and difficult to follow. Hill 165 [3,600 yards west of Yontan airfield] and certain towns, however, were unmistakable. [The] 7th Marines ZA was approximately 2,000 yards wide. While there was some slippage to the right all along the front, I know that 2/7 on the left and 1/7 on the right substantially covered their [zone of action]. My opinion is that the left flank of the 7th was not over 400 yards from the division boundary at any time. The slant distance from the actual right flank of the 4th Marines and the left flank of the 7th Marines may have been greater as the attacks of the two regiments were not at an even rate." BGen Edward W. Snedeker ltr to CMC, dtd 18Oct54, hereafter *Snedeker ltr I.* "The maps were not accurate. The 7th felt that they were on the boundary and so did the 4th. Who was right was never determined." *Honsowetz ltr.* "There was a gap, but it never was determined how much or in what area. In any case it was not a serious gap for a maneuvering assault." *Snedeker ltr 1965.*

izations, it was found that not one knew that he was in this paper organization, and only one had ever heard of it. Without exception, the prisoners gave as their unit the service or home guard element with which they had served at the airfields.

Despite their motley makeup, their commanding officer, Lieutenant Colonel Tokio Aoyanagi, determined to employ them in slowing the advance of the invaders. At 1400 on L-Day, he issued an order directing all of his battalions to hold every strongpoint, to carry out night raids, to destroy all bridges, and to construct tank obstacles. The colonel pleaded for "each and every one [to] carry out his duty with the conviction of certain victory." [20]

His men were poorly armed and mostly leaderless. Moreover, they lacked communications. When the escape routes to the organized forces in the south were cut off, therefore, this haphazardly organized group collapsed. The greater portion of these troops apparently fled to the northern hills, while a few, undoubtedly, escaped to the south; 26 were captured and 663 killed by the 1st Marine Division alone. Most of those who remained in the combat area quickly divorced themselves from the military, but other operated as snipers or guerrillas dressed in civilian attire.

Reliable intelligence was meager and hard to come by owing to the lack of contact with the enemy regular forces. As tactical operations developed rapidly against light opposition, hundreds of dazed civilians filtered through the lines and into the paths of the assault forces. When the Marines met these Okinawans, they interrogated, screened, and sent natives back to division stockades. Attempts to obtain information of the enemy from the older inhabitants were stymied because of the difficulties imposed in translating the Okinawan dialect. Younger natives of high school age, who had been forced to learn Japanese by the Nipponese overlords, proved to be a lucrative source of information, however.

Even though most civilians were cooperative with the Americans, they could provide very little information of immediate tactical importance. Nonetheless, the Okinawans confirmed the picture of the Japanese withdrawal to the south, clarified the presence or absence of units suspected of being in the area, aided in establishing an order of battle, and revealed the general and specific areas to which the rest of the civilian population had fled.

Military government personnel soon discovered that local inhabitants had moved with all their belongings to caves dug near their homes to escape from the path of war. Although interpreters roving the area in trucks mounted with loudspeakers assured the natives that they would be saved and induced them to leave their refuges voluntarily, other Okinawans continued to believe Japanese propaganda and viewed the American "devils" as barbarians and cutthroats. In many cases, particularly in isolated regions, it was necessary for language and civil affairs personnel "to enter the caves and verbally pry the

---

[20] "1st Specially Established Regiment OperO A2, dtd 1Apr45," in CinCPac-CinCPOA Bul 107–45, Translations and Interrogations No. 28, dtd 14May45, pp. 12–13.

dwellers loose." [21] Sometimes this resulted in troops coming upon a tragic scene of self-destruction, where a father, fearing for the lives of his family and himself at the hands of the invaders, had killed his wife and children and then had committed suicide. Fortunately, there were no instances of mass suicide as there had been on Saipan or in the Keramas.[22]

Specifically organized patrols were dispatched to round up civilians and transfer them to stockades in areas predesignated in military government plans. These patrols were often accompanied by language officers searching for documents, but most of the material found was of no military value. When a rewarding find was made, pertinent information was orally translated to the regimental S–2, who took down matters of local significance. The paper was then sent to the division G–2 language section.

The 1st Division learned from documents captured in its zone that the Japanese authorities had actively conscripted Okinawan males between the ages of 17 and 45 since the bombings of 10 October 1944. After their induction, these men were placed into three types of organizations; regular *Thirty-Second Army* units, specially organized engineer units, and labor forces. In order to neutralize the effects that might result from the presence of such a large hostile group in its midst, the Tenth Army decreed that all able-bodied males between the ages of 15 and 45 years were to be detained for further screening together with bona fide prisoners of war.[23]

Special agents from Army Counterintelligence Corps detachments assigned the division assisted the Marines in interrogating and screening each male Okinawan detainee. Eventually, after their clearance by the American agents, cooperative and intelligent natives were enlisted to aid in the interrogations. Specially qualified Okinawans were moved to liberated villages and districts to serve as informants for the Island Command.

Even more of a problem than the inability of the Marines to come to grips with the enemy or their difficulty in obtaining usable intelligence was the dislocation of the logistical plan and the subsequent strain on supporting units that resulted from the unopposed and rapid troop advance. The logistics annex of the operation plan had been based on the premise that the landings would be stubbornly contested, and unloading priorities were assigned accordingly. When the uncontested landing permitted the immediate debarkation of troops who had not been scheduled ashore until 2 or 3 April, landing craft originally allocated to move cargo were diverted for this troop movement. As a result, the unloading of supplies was delayed on L-Day.

Because the road net ranged in degree from primitive to nonexistent and in

[21] *1st MarDiv SAR*, pt VIII, p. 2.

[22] For examples of Japanese-induced mass hysteria among civilians see: Appleman, *et. al.*, *Okinawa Battle*, p. 58; Maj Carl W. Hoffman, *Saipan: The Beginning of the End* (Washington: HistDiv, HQMC, 1950), p. 245.

[23] "The wisdom of these precautions was illustrated by several incidents which confirmed our suspicions. Many a Kimono hid a uniform, and a number of civilians were found to be armed." *1st MarDiv SAR*, pt VIII, p. 7.

order to prevent traffic congestion, LVTs were not sent too far inland. The front lines were supplied, therefore, by individual jeeps, jeep trailers, weasels, and carrying parties, or a combination of all four. As forward assault elements moved farther inland, the motor transport requirement became critical and a realignment of unloading priorities was necessary. As a consequence, the unloading of trucks from AKAs and APAs was given the highest priority.

By the night of L plus 1, all of the units supported by the 11th Marines had moved beyond artillery range, and Colonel Brown's regiment had to displace forward. The movement of 1/11 (Lieutenant Colonel Richard W. Wallace), in particular, was long overdue. Artillery displacements were not yet possible, however, since organic regimental transportation had not landed and no other prime movers were available. Two battalions were moved forward on L plus 2 by shuttle movements and an increased transportation capability, which resulted when more trucks came ashore and were made available. The other two battalions moved up on the 4th. Although it had been planned to have the bulk of Corps Artillery on the island by the end of L plus 1, and all by the end of L plus 2, it was not until 10 April that General Nimmer's force was completely unloaded. "This melee resulted from the drastic change in unloading priorities." [24] Fortunately, unloading operations were improved by L plus 2.

Engineer battalions organic to the Marine divisions were generally relieved of mine removal tasks by the unfinished state of Japanese defenses, but there was no letup in their workload. Owing to the accelerated movement forward, the "narrow and impassable stretches of roads [and] lack of roads leading into areas in which operations against the enemy were being conducted, the engineers were called upon more than any other supporting unit." [25] A priority mission assigned to the engineers was the rehabilitation of the airfields after they had been captured. Their early seizure permitted work to begin almost immediately, and after the engineers had reconditioned Yontan airfield beyond merely emergency requirements, the first four-engine transports arrived from Guam on 8 April to begin evacuating the wounded.[26]

The only real problem facing General del Valle's units during the second day ashore, and one that tended to check a more rapid advance, was "the difficulty of supply created by the speed with which our units were moving and by lack of good roads into the increasing rough terrain." [27] In viewing the lack of any formidable resistance to either one of his assault divisions, General Geiger gave both of them permission to advance beyond the L plus 5 line without further orders from him.

[24] *Henderson ltr.*

[25] *4th Mar SAR,* Ph I&II, p. 29.

[26] *Blakelock ltr.* This source continues: "These planes were required to make the return flight to Guam without refueling at Okinawa due to the shortage of AvGas ashore to refuel the planes. Okinawa had five flights daily from Guam on a nonfueling basis until 12 April, when 500 gal/plane was furnished for the return flight."

[27] *1st MarDiv SAR,* chap VII, OpAnx, p. 4.

BEWILDERED CIVILIANS *wait to be taken to military government camps in the wake of the swift American advance across the island. (USMC 117288)*

TWO MARINES *of the 6th Division safeguard a young Okinawan until he can be reunited with his family. (USMC 118933)*

By 1500 of L plus 1, the progress of the 5th Marines had caused its zone of action to become wider, and in order to secure the division right flank and maintain contact with the 7th Infantry Division, the 1st and 2d Battalions of the 1st Marines were echeloned one behind the other on the corps boundary. Upon landing that morning, 3/1 moved inland to a point east of Sobe and remained there in division reserve.

As the 1st Marine Division had not yet located the center of the enemy defenses or determined his strength, "the weakness of the resistance . . . [remained] a source of astonishment" to General del Valle.[28] During the day, attempts at infiltration and the occasional ambushing of patrols by small hostile groups had little effect on the tactical situation. When the troops dug in at 1600, the division position was stabilized for the night on a line generally conforming to the L plus 5 line in the 7th Marines zone, while the 5th was slightly short of it.

To search out enemy positions, the 1st Reconnaissance Company (1st Lieutenant Robert J. Powell) was ordered to scout the division zone on 3 April, taking a route that followed along the corps boundary to the base of the Katchin Peninsula on the east coast. On this same day, the assault regiments were to continue the attack with an advance to the L plus 10 line. Because there had been only slight activity on the 1st Division front during L plus 1, the 11th Marines fired only five missions. While night defensive fires were planned for the second night ashore, they were not called for.

On this day in the XXIV Corps zone, the Army divisions also were able to exploit unexpectedly light resistance. The 7th Infantry Division had reached the east coast at Tobaru overlooking Katsuren Wan (Bay), effectively cutting the island in half and severing the enemy line of communications. Units of the 96th Division had advanced to the east and south, and succeeded in penetrating irregularly defended positions, some of which consisted merely of road mines and booby-trapped obstacles. At the end of L plus 1, General Bradley's front lines extended from the vicinity of Futema on the west coast to approximately one mile west of Unjo in the east.

By the close of 2 April, all assault division command posts had been established ashore, and the beachhead and the bulk of the high ground behind the landing beaches firmly secured. Enemy observation of Tenth Army movement and dispositions was thus limited, and any land-based threat to unloading operations removed.

Commenting on the conduct of Marine operations for the two days, General Buckner signalled General Geiger:

> I congratulate you and your command on a splendidly executed landing and substantial gains in enemy territory. I have full confidence that your fighting Marines will meet every requirement of this campaign with characteristic courage, spirit, and efficiency.[29]

During the night of 2–3 April, enemy activity was confined to sporadic sniper,

---

[28] LtGen Pedro A. del Valle ltr to CMC, dtd 29Sep54, hereafter *del Valle ltr*.

[29] *6th MarDiv Jnl, Ph I&II*, 3Apr45.

machine gun, mortar, and artillery fire, and intermittent infiltration attempts by individuals and small groups. Undeterred by this harassment, General Geiger's corps jumped off at 0730 on the 3d and again found slight opposition to the attack.

The 6th Marine Division resumed the offense in the same lineup with which it had ended the previous day. (See Map 7.) Both the 4th and 22d Marines advanced an average of more than 7,000 yards through difficult and heavily broken terrain to seize the dominating Yontan hill mass and the division objective beyond. While the 22d Marines moved forward, 1/29 patrols covered the entire Zampa Misaki without discovering any enemy troops on the peninsula. There was no repetition of the fierce clashes experienced by the 4th Marines the preceding two days. Throughout the day, the division was supported by tanks, which operated along the hazardous narrow trails existing on the precipitous ridge tops to the front.

By midmorning, the axis of advance of the 6th Division began to swing to the north as the towns of Kurawa and Terabaru were gained. Scattered rearguard action from withdrawing enemy troops was the only resistance encountered by the 22d Marines advancing on Nakadomori. In order to develop the situation on the division front and to determine the nature of Japanese defenses in the Ishikawa Isthmus, General Shepherd ordered his reconnaissance company to scout the coastal road from Kurawa to Nakadomori, and, at that point, to cross the isthmus to Ishikawa. Supported and transported by a reinforced platoon from the 6th Tank Battalion, the tank-infantry reconnaissance force completed the assignment and returned to its lines before nightfall. In the course of the patrol, the Marines made no enemy contacts and were fired upon only once from the vicinity of Ishikawa.

While 1/4, the right battalion, had relatively easy going most of the day, 3/4 on the left lagged behind because of the increasingly difficult terrain. When the division attack ceased at 1700, General Shepherd's troops were tied in with the 1st Division nearly a mile northeast of Kubo.

Following its sweep of Zampa Misaki, Lieutenant Colonel Moreau's 1/29 occupied new reserve positions east of Yontan airfield from which it could support either division assault regiment. At 2000, when IIIAC warned of an imminent enemy airborne attack, Moreau was reinforced with a tank company, which was deployed to defend the airfield by 2300; no Japanese paratroops or airborne infantry landed that night.

The only notable enemy activity experienced by the 1st Division during the hours of darkness, 2–3 April, occurred in the 7th Marines sector where the Japanese attempted extensive infiltration. In the fire fight that ensued, 7 Marines were killed and 7 wounded, while approximately 20–25 of the would-be infiltrators perished.

This brief flurry was not an indication of an imminent major engagement, for when del Valle's three combat teams pushed forward on 3 April they met only light opposition on the left and virtually none on the right. "Our ever-widening zone of action prohibited the 'hand-in-hand' advance of some small island operations and our units were able to maintain contact and clear their areas only by patrolling to the flanks and front." [30] With the resumption of the advance, motorized units of the 1st Reconnaissance Company began a series of patrols which were to encompass almost all of the division zone of action. In the morning, the Ikebaru-Napunja area was reconnoitered, after which the company was ordered to proceed down the Katchin Peninsula. Completing this mission by 1300, Lieutenant Powell's scouts were ordered up the east coast to Hizaonna and to return to division lines before dark. During the entire trip the only sign of enemy activity was a lightly held tank trap.

All units of the division were ordered to halt at 1700 on ground most favorable for defense. On the left, the 7th Marines had pushed forward against moderate opposition over increasingly difficult terrain. As the regimental commander later stated:

> The movement from the west coast landing beaches of Okinawa across the island to the east coast was most difficult because of the rugged terrain crossed. It was physically exhausting for personnel who had been on transports for a long time. It also presented initially an almost impossible supply problem in the Seventh's zone of action because of the lack of roads.[31]

Despite these hardships, Snedeker's troops gained 2,700 yards of enemy territory and dug in for the night after overrunning a strongpoint from which heavy mortar, 20mm, and small arms fire had been received. Shortly thereafter, Colonel Snedeker received permission to exploit what appeared to be an apparent enemy weakness and to continue the attack after the rest of the division had ended it for the day. He then ordered his reserve battalion to pass between 1/7 and 2/7 and advance towards the village of Hizaonna, on the high ground overlooking the east coast.[32]

The major fighting in this advance occurred when the 81mm mortar platoon was unable to keep up with the rest of 3/7. Company K, following the mortars, became separated from the main body upon reaching a road fork near Inubi after night had fallen. When he became aware of the situation, Lieutenant Colonel Hurst radioed the company to remain where it was and to dig in after its repeated attempts to rejoin the battalion were defeated by darkness and unfamiliar terrain. An estimated platoon-sized enemy group then engaged the Marines in a heavy fire fight, which continued through the night as

[30] *1st MarDiv SAR*, chap VII, OpAnx, p. 5.

[31] Col Edward W. Snedeker ltr to CMC, dtd 27Mar47, hereafter *Snedeker ltr 1947*.

[32] *1st MarDiv G–3 Jnl*, 3Apr45; Col Edward H. Hurst, interview with HistBr, G–3 Div, HQMC, dtd 3Mar55. "The forced march of over three miles of the Third Battalion, Seventh Marines to Hizaonna, late on 3 April is noteworthy. Orders for the march were not issued until about 1430 on that date. Information received from the Division Reconnaissance Company indicated no substantial enemy strength on the east coast in the Seventh's zone. What might be encountered from the front line to the east coast was unknown. The march was made over rugged unfamiliar terrain, with the probable expectancy of meeting enemy forces at any time, and it resulted in advance elements, including the battalion CO reaching Hizaonna at 1830, 3 April. Hizaonna was at that time well beyond our front lines. . . ." *Snedeker ltr 1947*.

the Japanese effectively employed mortars, machine guns, and grenades against the isolated unit. By noon on 4 April, a rescue team from 3/7 was able to bring the situation under control and the company was withdrawn after having sustained 3 killed and 24 wounded.[33]

With its right flank anchored on Nakagusuku Wan, 1/1 held a line sealing off two-thirds of the Katchin Peninsula. The 2d Battalion, 1st Marines, meeting negligible resistance from armed civilians, occupied the high ground immediately west of Gushikawa, where the eastern shore could be covered by fire. During the day's gains, "supply had been almost nonexistent and the troops were without water and still depending on the food they landed with." [34]

The advance of the 5th Marines gained momentum throughout the day, the troops having met only a four-man enemy patrol. The 1st Battalion reached Agina, where 3/5 was committed on the right to contact 2/1. One thousand yards away, on the left of the regiment 2/5 occupied the village of Tengan and then advanced to the east coast of Okinawa.[35]

[33] Hurst interview, *op. cit.;* 1st MarDiv Interviews, Co K, 7th Mar. This last source consists of a series of interviews conducted by Sergeants Kenneth A. Shutts and Paul Trilling, historians assigned to the 1st Marine Division for the Okinawa operation. These interviews form a valuable record of the important actions of the campaign at the small-unit level.

[34] *1st Mar SAR*, p. 6.

[35] "This was accomplished by 1700 on 3 April. 2/5, commanded by LtCol W. E. Benedict, had marched approximately eight miles over hilly country since 0800 when they left their 2 April position near Ishimine." Col John H. Griebel ltr to CMC, dtd 18Oct54.

By the end of L plus 2, the 1st Marine Division had driven to the coast, advancing 3,000–5,000 yards, and thus placed its lines 8–13 days ahead of the ICEBERG schedule. The 6th Division, meanwhile, had moved through difficult and heavily broken terrain honeycombed with numerous caves to gain 3,500–7,000 yards of enemy ground in its zone.

At dark on 3 April, the 6th Division left flank was anchored at the base of the Ishikawa Isthmus, thereby placing the Tenth Army 12 days ahead of schedule in this area. During this same day, the XXIV Corps had reached the eastern coast in force and its units had begun reorganizing and regrouping for the attack to the south. The 7th Infantry Division had secured the Awashi Peninsula, and pivoting southward in a coordinated move with the 96th Division, secured an additional 3,000 yards before the end of the day. In the vicinity of Kuba-saki, the 32d Infantry came up against its first real opposition on Okinawa, when it made contact with an enemy force estimated at 385 men. The regiment overran the enemy position and finally took Kuba-saki. After completing the wheel to the right, the 96th Division reorganized its front lines, putting its units in position for the southerly drive.

While observation planes, OYs (Consolidated-Vultee Sentinels), operated from Yontan airfield during the day, the 6th Engineer Battalion and the 58th Seabees continued working on the field. An F6F (Grumman Hellcat) from the carrier *Hancock* made an emergency landing at 1110, and the pilot reported that, in his opinion, the runway could satisfactorily accommodate all types of carrier planes. The other runways were expected to be operational for fighter-type aircraft by noon on the 4th.

Because of the very favorable situation that had developed during the day, General Buckner removed all restrictions he had placed on movement past the L plus 10 line and ordered IIIAC to seize the L plus 25 line at the earliest possible time. Geiger then ordered the 6th Marine Division to continue the attack on 4 April and to take the L plus 15 line, prepared to continue the advance to the L plus 20 line. General del Valle's division was ordered to advance to the L plus 20 line. (See Map 7.)

The continuing cool and clear weather on 4 April again served as a welcome change from the torrid humidity of the Philippines and the Solomons. Following a quiet night, broken only by the fighting in the Inubi area, the IIIAC jumped off on schedule at 0730 on L plus 3.

As the 6th Division pushed forward, no enemy hindered the 4th and 22d Marines advance to the L plus 15 line. General Geiger's reserve, the 29th Marines (Colonel Victor F. Bleasdale), less its 1st Battalion, had reverted to division control earlier, and Shepherd assigned it as his reserve. When it became apparent that the day's objectives would be reached by noon, the assault regiments were ordered to continue beyond the L plus 15 line to additionally assigned division objectives.

With all three battalions in the assault, the 22d Marines reached their Phase I objective at 1250. Organized as a fast tank-infantry column, Lieutenant Colonel Woodhouse's 2/22 sped up the west coastal road. All the while, he sent patrols inland along the route to maintain contact with 1/22 patrol columns in the interior of the regimental attack zone. On the right, the 3d Battalion reached Ishikawa before noon, having gradually pinched out the 4th Marines when that regiment reached the coast at about the same time. Colonel Shapley was then ordered to reconsolidate his unit in the Ishikawa area, and to prepare to support either division flank unit in the attack northward. In the course of the morning operations, exceedingly rough terrain, and the logistical support problems it posed, created greater obstacles to the advance than did the enemy.

At 1300, the attack up the Ishikawa Isthmus was resumed, with RCT–22 and 1/29 attached [36] taking over the entire division front from the west to the east coast. Advance was rapid in the afternoon as patrols met only scattered resistance until 1730, when a Japanese strongpoint, built around several heavy machine guns, fired upon a 3/22 patrol north of Yaka. Night defenses were not taken up until after this obstacle was reduced by units of Donohoo's 3/22 and Moreau's 1/29, the latter having assumed the 1/22 sector when Major Myers' unit was placed in regimental reserve.

When the 6th halted for the day, its Marines had advanced over 7,500 yards and held a line that stretched across the isthmus from a point just south of Yakada on the west coast to Yaka on

[36] After being relieved of its Yontan airfield defense mission, 1/29 moved to Nakadomari in the morning and was attached to the 22d Marines at 1155, when it moved up to the lines.

the east. In this day's fighting, the increasingly rugged terrain forecast the difficulties to be faced during the march northward. Supply lines were strained almost to their limit as they were extended across numerous ravines and steep valleys in the mountainous interior. Despite this, the troops were fed and logistical replenishment continued as the division prepared to continue the advance the next day.

With the exception of the few enemy positions encountered in its push to the east coast, the 1st Marine Division still did not have a clear picture of what Japanese defenses lay ahead on 4 April. As on the day before, the attack jumped off without artillery preparation. Rapid progress with little resistance was the general order, except on the left where the 7th Marines was still busy with the enemy in the vicinity of Inubi. The 2d Battalion reached the east coast by 1130 and, shortly thereafter, made firm contact with the 4th Marines. On the extreme right of the regiment, light but stubborn enemy fire enfiladed the 1/7 right flank, and delayed its arrival at the coast until 1700. Because of the rapid advance of the regiment over a roadless terrain, Colonel Snedeker requested supply airdrops during the day. The first drop was made at Hizaonna at 1400, about the same time that General del Valle's new CP was opening at a point between Ikebaru and Napunja.

After dark, when the 7th Marines was consolidating its positions on the L plus 15 line, the enemy began numerous attempts at infiltrating American lines. Although 45 Japanese were killed as they probed the regimental positions, it was difficult to obtain any information regarding the units represented by these men, who employed rifles, grenades, bayonets, and sharply pointed bamboo spears, which the Marines promptly dubbed "idiot sticks." [37]

In the center of the division line, the 5th Marines reached the shores of Kimmu Wan by early afternoon, when the battalions consolidated their positions and established firm contact with all flanking units. The same day, the 1st Marines occupied Katchin Peninsula in orderly fashion by noon and set up its defenses. Once these two regiments were in position on the L plus 15 line, they initiated patrolling to the rear to eliminate bypassed positions, a task in which the reconnaissance company and the division reserve (3/1) also participated.

That evening, 3/1 was ordered to take over the defense of Yontan airfield from the 29th Marines on 5 April. Tentative plans were formulated to release the 7th Marines to IIIAC in order to assist the 6th Division in its drive north. The next day, the 7th Marines (less Hurst's 3/7, which was attached to the 5th Marines) went into IIIAC reserve with orders to occupy and defend the village of Ishikawa, pending further tactical developments.[38]

In the course of its four-day drive across Okinawa, the 1st Marine Division found only negligible resistance, and this from Japanese units of undetermined strength employing delaying or rearguard tactics. The question remained: Where was the enemy? The division had killed 79 Japanese, cap-

[37] *1st MarDiv SAR*, chap. VII, p. 6.

[38] "A History of the 7th Marines on Okinawa Shima," p. 7, encl A to *Snedeker ltr 1947*, hereafter *7th Mar Hist*.

tured 2 prisoners of war (POWs), and encountered 500–600 civilians, who were quickly interned.

To the south of the 1st Division, the tactical situation in the XXIV Corps zone had been undergoing radical change. (See Map 7.) Army assault divisions had aggressively exploited the initial lack of enemy resistance. During the same time, they were hampered less by supply difficulties than the Marine divisions had been. Once General Hodge's divisions had wheeled to the right on 3 April for the drive southward, the lines were reorganized and preparations made for fresh assault units to effect passage of the lines the next day. The corps was now ready for the Phase I southern drive.

The 7th Division pushed forward on 4 April only to meet stiffening resistance from hostile artillery-supported infantry at Hill 165.[39] After a day's fighting, the division drove the Japanese from this dominant piece of terrain and continued forward to net approximately 1,000 yards for the day. Meanwhile, in the 96th Division zone, Army infantry battalions were held up by reinforced enemy company strongpoints several times during the day. Heavy Japanese machine gun, mortar, and artillery fire impeded the advance, but by the night of 4 April, the XXIV Corps had seized the L plus 10 line, which had been originally designated the southernmost limit of the Tenth Army beachhead.

[39] The reader should keep in mind the fact that elevation was expressed in meters on the basic 1:25,000 map used for Okinawa. When Hill 165's metric height is converted to feet, 544, it becomes a sizable hill.

310-224 O - 69 - 10

An increasing volume of enemy defensive fires was placed on the Army divisions as they moved out for their fifth day of ground action. On the XXIV Corps left flank, resistance came mainly from small, scattered enemy groups in the hills and ridges bordering the east coast. In the 96th Division zone to the right, both assault regiments became heavily engaged with Japanese outpost strongpoints during L plus 4. About noon, an enemy counterattack was broken up by tank-artillery supported infantry action just when the right flank regiment of the 96th, the 383d Infantry, drove unsuccessfully against the first of a series of prepared ridge positions guarding the approaches to Kakazu. Four tanks were lost during the day's fighting, one to a mine and the others to enemy antitank fire. Compared to the long advances of the previous four days of ground action, the 96th Division was able to take only 400 yards on L plus 4.

THE SWING NORTH [40]

Although the Japanese forces in the south offered an increasingly stiff defense as their positions were uncovered, the exact whereabouts of the main enemy strength in the northern part of the island remained as clouded as was his order of battle. Even though Phase I of the ICEBERG plan did not specify any action beyond isolation of the area

[40] Unless otherwise noted, the material in this section is derived from: *CTF 51 AR; Tenth Army AR; IIIAC AR; 1st MarDiv SAR; 1st MarDiv G–3 Jnl; 1st Mar SAR; 6th MarDiv SAR, Ph I&II; 6th MarDiv Jnl, Ph I&II; 4th Mar SAR, Ph I&II; 22d Mar SAR, Ph I&II; 29th Mar SAR, Ph I&II.*

above the Ishikawa Isthmus for the IIIAC, General Buckner believed that it would be profitable to exploit initial Marine successes. As a result, on 5 April he ordered IIIAC to reconnoiter Yabuchi Shima, to conduct a vigorous reconnaissance northwards to the Motobu Peninsula, and to initiate preparations for the early completion of Phase II. (See Map II, Map Section.)

At the same time that the 6th Division conducted its reconnaissance up the isthmus, the 1st Division entered a period devoted primarily to defensive activity. Supplies were brought up from the rear, positions were improved and camouflaged, and all units began heavy patrolling to the rear. At noon on L plus 4, a 1st Marines patrol waded across the reef to Yabuchi Shima from the Katchin Peninsula, captured five *Boeitai*, and reported the presence of some 350 civilians.

The nearly perfect weather which had prevailed since L-Day, deteriorated with light rain over scattered areas in the early evening of the 5th. Although there was no evidence that it was organized, enemy activity behind the lines increased during the day but only from small separated groups apparently operating independently of each other. Of this period, a regimental commander noted:

> There [were] almost daily patrol contacts with well-armed enemy groups. . . . Some of these groups were wandering aimlessly about while others occupied well defended, organized, and concealed positions. These patrol operations were extremely valuable in giving to the officers and men of the regiment added confidence in each other and helped all to reach a peak of physical perfection . . . independent patrols . . . often under fire, added greatly to the ability of the leaders of small units.[41]

Because it was necessary to move supplies forward to support the advance, the 6th Division delayed H-Hour on the 5th until 0900. At that time, armor-supported infantry columns were dispatched on deep reconnaissances up both sides of the isthmus, the 6th Reconnaissance Company on the left (west) flank and Company F, 4th Marines on the right. The latter advanced 14 miles before turning back in the late afternoon. During the day, the patrol had been delayed three times by undefended roadblocks but met no opposition until the tanks entered Chimu, where two of the enemy encountered were killed and a Japanese fuel truck was destroyed. The drive up the other side of the island was unopposed, but the tanks could not bypass a destroyed bridge at Onna. The reconnaissance company, forced to continue on to Nakama by foot, returned to the lines that evening.

While 6th Division mobile covering forces searched out routes of advance, the assault battalions rapidly moved forward, detaching companies as necessary to reduce bypassed enemy pockets of resistance inland. Although the terrain had become more difficult to negotiate and the enemy increasingly active, the division gained another 7,000 yards. The 22d Marines held the general line Atsutabaru-Chimu, with the 4th Marines (less 1/4 bivouacked at Ishikawa)[42] located in assembly areas just behind the front line, prepared to pass through early the next day. At 1000 on

[41] *Snedeker ltr 1947.*

[42] *1/4 SAR, Ph I&II,* p. 7.

HEAVY UNDERGROWTH *on the Ishikawa Isthmus hinders the progress of a 4th Marines patrol advancing to the north of Okinawa. (USMC 116527)*

GRINNING TROOPS *of the 29th Marines in M–7s heading for Chuta in their drive towards Motobu Peninsula. (USMC 117507)*

the 5th, the 29th Marines were released to parent control by IIIAC and moved to an assembly area in the vicinity of Onna.

As the 6th Division began its dash up the isthmus on 6 April, the 7th Marines in corps reserve patrolled the division zone south of the Nakadomari-Ishikawa line[43] while the 6th Reconnaissance Company mopped up enemy remnants from this boundary north to the Yakada-Yaka line. After its lines had been passed by the 29th Marines on the left and the 4th Marines on the right, the 22d reverted to division reserve and began patrolling back to the area of responsibility of the reconnaissance company.

Because there were only a few roads inland, Colonel Shapley planned to move rapidly up the main road along the shore, detaching patrols from the advance guard to reconnoiter to their source all roads and trails into the mountainous and generally uninhabited interior. In order to maintain control during the anticipated rapid advance, the regimental march CP moved out in a jeep convoy at the head of the main body. By 1300, 2/4 had been used up by the detachment of small patrols, and the 3d Battalion then passed through in accordance with the prearranged plan of leapfrogging the battalions. When the regiment halted for the day at 1600, it had advanced seven miles, encountering only scattered enemy stragglers. The supply operations of the 4th Marines on L plus 5 were hampered more than usual by the fact that three bridges along the route had been bombed out earlier by friendly air.

The 4th Marines resumed operations the next morning deployed in the same manner in which it had halted the night before—3/4, 1/4, and 2/4 in that order. The advance on L plus 6 was virtually a repeat of the previous day as the regiment continued the push up the east coast, the lead battalion dissipating its strength with the dispatch of patrols into the interior. As a result, the 1st Battalion (Major Bernard W. Green) passed the 3d at noon and led the way to the regimental objective, opposed only by the difficult terrain, poor roads, and fumbling enemy defense measures.

Nearly all such efforts failed, however, for in very few instances was the 6th Division drive slowed. Enemy defensive engineering efforts were almost amateurish, for abatis, with neither mines nor booby traps attached or wired in place, were pushed aside easily by tank-dozers or bulldozers. Even basic defensive combat engineering principles were violated by the Japanese, who did not distribute their mines in roads and defiles in depth. They even failed to cover with either infantry fire or wire what they had placed. On the whole, the mines were little more than a nuisance and caused but few casualties.[44] Bridges were often incompletely destroyed by Japanese demolitions, and Marine engineers were able to save valuable time by utilizing the remaining structural members as foundations for new spans in hasty bridge construction.

---

[43] Effective at 0600 on 7 April, the rear boundary of the 6th Division was readjusted to exclude the area south of this line. IIIAC OperO 2-45, dtd 6Apr45.

[44] *6th EngrBn SAR, Ph I&II,* chap VIII, p. 3.

When the 6th Division drive towards the north began, each assault regiment was assigned one company of the 6th Engineer Battalion in direct support. A platoon from each of these companies was attached to advance guards to clear roadblocks, remove mines, and build bypasses for combat vehicles around demolished bridges. The remainder of each company followed up the advance, repairing and replacing bridges and widening narrow thoroughfares wherever possible to accommodate two-way traffic. Following closely in the wake of the assault regiments, the third company of the engineer battalion further improved roads and bridges.

At the end of the Ishikawa Isthmus, where the mountains came down to the sea, engineer services were in even greater demand as they were required to widen roads that were little more than trails.[45] The infantry advance was slowed by the terrain as well as by the near-physical exhaustion of the patrolling Marines, who had been going up and down the thickly covered broken ground. Despite this tortuous journey, the 4th Marines had made another seven miles by the late afternoon of 7 April. Then, just north of Ora, the 1st Battalion set up a perimeter defense with its flanks secured on the coast. Colonel Shapley's CP and weapons company were located in the village itself, while 3/4 and 2/4 were deployed in defensive perimeters at 1,000-yard intervals down the road.

On the west coast, the 29th Marines had seized its next objective on 7 April, again with little difficulty. Advance armored reconnaissance elements reached Nago at noon to find the town leveled by naval gunfire, air, and artillery. Before dark, the regiment had cleared the ruins and organized positions on its outskirts.

As the advance northwards continued, the difficult road situation had made it imperative to locate forward unloading beaches from which the 6th Division could be supplied. (See Map II, Map Section.) When Nago was uncovered, it was found suitable for this purpose, and IIIAC requested the dispatch to this point of Marine maintenance shipping from the Hagushi anchorage. On 9 April, cargo was discharged for the first time at Nago, relieving the traffic congestion on the supply route up the coast from Hagushi.[46]

When planning for ICEBERG, General Shepherd had determined that Major Walker's company would be employed only in the reconnaissance mission for which it was best fitted and trained. In effect, the unit was intended to serve as the commanding general's mobile information agency. Pursuant to this decision, the reconnaissance company, supported and transported by tanks, was dispatched up the west coast road ahead of the 29th Marines in an effort to ascertain the character of Japanese strength on Motobu. After the company scouted Nago, it swung up the coastal road to Awa, and then, after retracing its steps to Nago, crossed the base of the peninsula in a northeasterly

[45] *Ibid.*, pp. 3–4.

[46] *Blakelock ltr 1965.*

direction to Nakaoshi.[47] Before returning to Nago for the night, the patrol uncovered much more enemy activity than had been previously revealed in the division zone of action and had met several enemy groups that were either destroyed or scattered.

From the very beginning of the drive to the base of the Motobu Peninsula, the 15th Marines was employed so that each assault regiment had one artillery battalion in direct support and one in general support. The rapidity of the 6th Division advance during this phase of the campaign forced the artillery regiment to displace frequently, averaging one move a day for each battalion and the regimental headquarters. To keep up with the fast moving infantry, the artillerymen were forced to strip their combat equipment to a bare minimum; they substituted radio for wire communications and by leapfrogging units, managed to keep at least one artillery battalion in direct support of each assault infantry regiment throughout the advance up the isthmus.[48]

Augmenting the 15th Marines, 6th Division artillery support was reinforced by the 2d Provisional Field Artillery Group (Lieutenant Colonel Custis Burton, Jr.) which displaced to positions north of Nakadomari on the eve of the drive up the Ishikawa Isthmus. Four days later, when resistance on Motobu Peninsula began to stiffen, the 15th Marines was reinforced further by the attachment of the 1st Armored Amphibian Battalion as artillery. The following day, the corps artillery supporting the advance was moved to Besena Misaki, a promontory at the southern extremity of Nago Wan, where it remained throughout the period of Marine operations in the north.[49]

MOTOBU UNCOVERED

Owing to the lack of intelligence about the location of the enemy, and a Tenth Army order to avoid unwarranted destruction of civilian installations unless there was a clear indication or confirmation of enemy presence, naval gunfire support was not used extensively in the drive up the Ishikawa Isthmus. After 5 April, however, all IIIAC naval fire support was diverted to the 6th Division zone of action. As the Marines moved north, these ships kept pace, firing up the numerous ravines leading down to the beach. Each assault battalion was furnished a call-fire ship during the day, and each regiment was furnished a ship to fire illumination at night.[50]

The Tenth Army gained land-based air support when TAF squadrons from MAG–31 and –33 arrived ashore on 7

[47] It was later revealed that, as the 6th Reconnaissance Company moved westward toward Awa, the Japanese were close on the company's northern flank, observing its movement, and holding fire. CMC [Gen Lemuel C. Shepherd, Jr.] memo to Head, HistBr, G–3 Div, HQMC, dtd 2Mar55, hereafter *Shepherd memo II;* LtCol Anthony Walker ltr to CMC, dtd 3Jun55.

[48] *15th Mar SAR, Ph I&II*, chap VII, n.p.

[49] *IIIAC Arty AR*, pp. 20–21. At this time, LtCol Burton's group consisted of only the Headquarters Battery and the 7th 155mm Gun Battalion, as the remainder of the IIIAC Artillery gun and howitzer battalions had passed to XXIV Corps control on 7 April. *Ibid.*, p. 19.

[50] *Ibid.*, p. 39.

JAPANESE 105MM GUN *captured in the heart of Mount Yae Take had previously commanded the entire coastal road along southern Motobu Peninsula. (USMC 122207)*

SUICIDE BOATS *found at Unten Ko on 10 April by Marines of 2/29. Note warning that boats had been booby trapped. (USMC 127905)*

and 9 April.[51] The 6th Division did not need them immediately during the first two weeks in the north, however, for the division advance had been rapid and suitable targets scarce. Daylight combat air patrols were flown almost as soon as the squadrons landed, but strikes in support of ground operations did not begin until L plus 12, and then they were directed at Japanese targets in the XXIV Corps zone in the south. As enemy resistance stiffened on Motobu Peninsula, Marine air was called upon to destroy emplacements, observation posts, and troop concentrations.[52]

After the division had gained the base of the Motobu Peninsula and had begun extending reconnaissance operations to the west on 8 April, aerial observation and photo studies confirmed the fact that the enemy had chosen to make his final stand in the rugged mountains of the peninsula. In order to reduce this Japanese bastion, and at the same time maintain flank security and continue the drive to the northernmost tip of Okinawa, General Shepherd needed to reorient the axis of operations and redeploy his forces. Consequently, the 22d Marines was taken out of division reserve and set up on a line across the island from Nakaoshi to Ora to cover the right and rear of the 29th Marines attacking to the west. Assembled near Ora, also, was the 4th Marines, which was positioned to support either the 29th Marines on Motobu or the 22d in the north.

[51] TAF WarD, Apr45.

[52] *6th MarDiv SAR, Ph I&II*, chap VII, p. 14. The 6th Division received its first TAF support missions of the campaign on L plus 16. TAF Periodic Rpt No. 2, dtd 20Apr45, p. 3.

During the next five days, the 4th and 22d Marines combed the interior and patrolled in the north, while the 29th probed westward to uncover the enemy defense.[53] On L plus 7, 2/29 moved northeast from Nago to occupy the small village of Gagusuku. The 1st Battalion, initially in reserve, was ordered to send one company to secure the village of Yamadadobaru, a mission accomplished by Company C at 0900. An hour later, the battalion as a whole was ordered to the aid of Company H, 3/29,[54] which had encountered heavy resistance in the vicinity of Narashido. By 1500, 1/29 had converged on this point and, despite heavy enemy machine gun and rifle fire, had reduced two strongpoints, after

[53] In this five-day sparring period "the 29th had a platoon of war dogs attached. These dogs gave an excellent account of themselves. Twenty-nine alerts were noted by the regimental S–3 section. All [alerts] enabled the Marine patrol involved to avoid a Japanese ambush. In one instance a patrol leader chose to ignore the dogs and was badly wounded." LtCol Angus M. Fraser ltr to CMC, dtd 24Mar55, hereafter *Fraser ltr. 2/29 SAR, Ph I&II*, p. 4.

[54] "Prior to a reorganization of Marine divisions in the spring of 1944, each infantry battalion had five companies: headquarters, weapons, and three rifle companies. When the weapons companies (D, H, and M) were absorbed [by their respective battalions], the rifle companies retained their original alphabetical designations, so that the three battalions of a regiment had companies lettered: 1st Bn, A, B, and C; 2d Bn, E, F, and G; 3d Bn, I, K, and L. The 2d and 3d Bns of the 29th Mar, formed after this reorganization took place, were lettered straight through in sequence after 1/29. Therefore, the rifle companies of the 29th Mar were A, B, and C in the 1st Bn, D, E, and F in the 2d Bn, and G, H, and I in the 3d Bn." Nichols and Shaw, *Okinawa Victory*, p. 94*n*.

which Lieutenant Colonel Moreau's men dug in for the night.

Intending to locate the main enemy force on Motobu, the 29th Marines moved out on 9 April in three columns; the 3d Battalion (Lieutenant Colonel Erma A. Wright) on the left flank, the 2d Battalion (Lieutenant Colonel William G. Robb) on the right flank, and Moreau's 1st Battalion up the mountainous center of the peninsula. All three columns encountered opposition almost immediately. This was an indication that the division may have at last hit the major enemy resistance in the north, and it was located in the area from Itomi west to Toguchi.

In the 3/29 zone on the left, roads were found to be virtually impassable as a result of effective enemy use of roadblocks, mines, and demolitions. The 6th Engineer Battalion reported that from Nago westward on the Motobu Peninsula the enemy had been even more destructive. They had demolished every bridge and blasted numerous tank traps in the roads. The Japanese had been careful to place these obstacles at points where no tank bypass could be constructed. Traps that had been made in the narrow coastal roads were put at the foot of cliffs where back fill was unavailable. Those in the valleys were always located where the road passed through rice paddies. When the crater was in a cliff road, trucks had to travel long distances to obtain fill for the hole.[55]

In the center, 1/29 was to occupy and defend Itomi before nightfall; about 600 yards short of the objective, however, the battalion was met by a strong enemy force and compelled to dig in for the night in place. The north coast was patrolled as far as Nakasoni by Robb's 2d Battalion, which destroyed supply dumps and vehicles and dispersed small enemy groups. The battalion also scouted Yagachi Shima with negative results.

The next day, L plus 9, Robb's men seized Unten and its harbor, where the Japanese had established a midget submarine and torpedo boat base. The base had been abandoned and large amounts of equipment and supplies were left behind by approximately 150 Japanese naval personnel, who were reported to have fled inland to the mountains. Toguchi, on the other side of the peninsula, was captured by 3/29, which sent patrols inland to Manna. On 10 April, 1/29 pushed forward through Itomi, and on the high ground north of the town it uncovered numerous well-prepared positions from which the enemy had fled.

During the first two days of the drive to clear Motobu Peninsula, frequent enemy contacts were made in the difficult terrain northwest and southwest of Itomi. Night counterattacks increased in intensity; one particularly strong attack, supported by artillery, mortars, machine guns, and 20mm dual-purpose cannon, struck the 1/29 defense perimeter on the night of 10-11 April and was not broken up until dawn.

Patrols from 2/29 were sent out on 11 April to make contact with 1/29 near Itomi. They met little opposition but substantiated previous intelligence estimates locating the main Japanese battle position in an area between Itomi and Toguchi. As a result of this verification, 2/29 (less Company F) was recalled from the north coast and ordered to set

[55] *6th EngrBn SAR, Ph I&II*, pt VII, p. 4.

up defensive installations and tie them in with 1/29 on the high ground near Itomi. Company F continued patrolling. During the day, 1/29 patrols scouted just to the north and northeast of Itomi and met only light resistance. On the other hand, 3/29, moving inland to contact the 1st Battalion, ran into heavily defended enemy positions at Manna and was forced to withdraw under fire to Toguchi.

In compliance with Admiral Turner's expressed desire that Bise Saki was to be captured early for use as a radar site, the 6th Reconnaissance Company was ordered to explore the cape area on 12 April, and to seize and hold the point unless opposed by overwhelming force. As anticipated, resistance was light, and the area was captured and held. That evening, Company F, 29th Marines, reinforced the division scouts. Overall command of this provisional force was then assumed by the reconnaissance company commander.[56]

In order to fix more definitely the hostile battle position, the 29th Marines continued probing operations. On the 12th, the 1st and 2d Battalions were disposed in positions near Itomi, and 3/29 was located in the vicinity of Toguchi. (See Map 8.) Company G was sent north to contact the reinforced division reconnaissance company and to meet 2/29 at Imadomari. Company H was ordered east to meet 1/29 at Manna, and Company I was ordered to patrol to the high ground south and east of Toguchi and to remain there overnight. As these last two companies proceeded on their missions, they came under intense fire that prevented the completion of their assignments unless they were to risk sustaining unacceptable casualties. Under cover of prompt call fires from the destroyer *Preston*, LVT(A) fire, and an 81mm mortar barrage, Company I was withdrawn while Company H served as rear guard. Both companies had organized a perimeter defense at Toguchi by midafternoon when the battalion CP received considerable artillery and mortar fire. The day's action cost the battalion 9 killed and 34 wounded.[57]

Because of this significant enemy reaction in the Toguchi area, Company G, upon its arrival at Imadomari at 1415, was recalled by the battalion. When Company H had been hit in the morning, 3/22 was alerted for possible commitment, and in the afternoon it was ordered to assemble in division reserve at Awa. Battalion headquarters and Companies I and K completed the motorized move after 1700, and L arrived at 0900 the following morning.

By the night of 12 April, General Shepherd's division was confronted with a fourfold task: to continue occupation and defense of the Bise area; to secure the line Kawada Wan-Shana Wan and prevent enemy movement through that area; to seize, occupy, and defend Hedo Misaki, the northernmost tip of Okinawa; and to destroy the Japanese forces on Motobu Peninsula.[58] On 10 April, 1/22 had established a perimeter defense at Shana Wan from which it

[56] *6th MarDiv Jnl, Ph I&II*, 12Apr45.

[57] *3/29 SAR, Ph I&II*, p. 4.

[58] IIIAC OperO 4-45, dtd 12 Apr45.

conducted vigorous patrolling eastward to the coast and north towards Hedo Misaki. By the 12th, battalion patrols had contacted the 4th Marines on the east coast; 3/4 was ordered to move to Kawada the next day.

During the period 8–12 April, the 4th Marines, located near Ora, patrolled all areas within a 3,000-yard radius of the regimental bivouac. On the 10th, Company K was sent north along the east coast on extended patrol after which it was to rejoin its battalion at Kawada. While in the field, the company relied on LVTs for daily support and evacuation. In a week's time, the patrol had travelled 28 miles up the coast.

Assigned the capture and defense of Hedo Misaki, Woodhouse's 2/22 moved rapidly up the west coast on 13 April in a tank- and truck-mounted infantry column, "beating down scattered and ineffective resistance." [59] At 2110, the 2/22 commander reported that a patrol had entered Hedo by way of the coastal road and that the entry had been opposed only by 10 *Boeitai*.[60] As soon as the rest of the battalion arrived, a base was set up and patrols were sent out to make contact with the 4th Marines advancing up the east coast.

At the end of the second week on Okinawa, on Friday, 13 April (12 April in the States), ICEBERG forces learned of the death that day of President Franklin D. Roosevelt. Memorial services were held on board American vessels and behind Tenth Army lines; those who could attend these services did so if the fighting permitted. One senior officer of the 1st Marine Division said later:

> It was amazing and very striking how the men reacted. We held services, but services did not seem enough. The men were peculiarly sober and quiet all that day and the next. Plainly each of them was carrying an intimate sorrow of the deepest kind, for they paid it their highest tribute, the tribute of being unwilling to talk about it, of leaving how they felt unsaid.[61]

THE BATTLE FOR YAE TAKE

While three of its four assigned missions in the north were being accomplished by extensive patrolling against little or no opposition, the 6th Marine Division found that destroying the firmly entrenched bulk of the enemy was becoming an increasingly difficult problem. Company I had apparently touched a sensitive nerve during its probings near Toguchi, judging by the immediate enemy reaction. This assumption was confirmed on the night of 12–13 April, when the 29th Marines encountered some English-speaking Okinawans, who had at one time lived in Hawaii.

The Marines were told that there was a concentration of 1,000 Japanese on the high ground overlooking the Manna-Toguchi road south of the Manna River. The civilians said further that the enemy force was commanded by a Colonel Udo, and that it contained an artillery unit under a Captain Kiruyama.[62] Previous

---

[59] Cass, *6th MarDivHist*, p. 57. General Buckner wanted this area captured early to gain a radar and fighter-director site. *CTF 51 AR*, pt III, p. 37.

[60] *6th MarDiv Jnl, Ph I&II*, 13Apr45.

[61] Quoted in McMillan, *The Old Breed*, p. 368.

[62] *6th MarDiv Jnl, Ph I&II*, 13Apr45.

reports of enemy order of battle were corroborated by the operations of strong combat patrols; the 6th Division now firmly fixed the Japanese defenses in an area some six by eight miles surrounding the rugged and dominating Mount Yae Take.[63] (See Map 8.)

The ground around this towering 1,200-foot-high peak prohibited extensive manuevering and completely favored the defense. Yae Take was the peninsula's key terrain feature and its heights commanded the nearby landscape, the outlying islands, and all of Nago Wan. The steep and broken approaches to the mountain would deny an attacker any armor support. Infantry was sure to find the going difficult over the nearly impassable terrain. The Japanese defenses had been intelligently selected and thoroughly organized over an obviously long period. All natural or likely avenues of approach were heavily mined and covered by fire.

It was soon concluded that approximately 1,500 men were defending the area and that the garrison, named the *Udo Force* after its commander, was built around elements of the *44th Independent Mixed Brigade.* Included in this group were infantry, machine gun units, light and medium artillery, Okinawan conscripts, and naval personnel from Unten Ko. In addition to 75mm and 150mm artillery pieces, there were two 6-inch naval guns capable of bearing on the coastal road for 10 miles south of Motobu, on Ie Shima, and all of Nago Wan.[64]

General Shepherd's estimate of the situation indicated that reduction of the Yae Take redoubt was beyond the capabilities of a single reinforced infantry regiment. In face of this conclusion, the 4th Marines (less 3/4) was ordered to move from the east coast to Yofuke. The 29th Marines was ordered to continue developing the enemy positions by vigorous patrolling on 13 April and to deploy for an early morning attack on the next day.[65]

Complying with General Shepherd's orders, Colonel Bleasdale again attempted to clear the Itomi-Toguchi road[66] and join his 1st and 3d Battalions. As elements of 1/29 moved out of Itomi towards Manna, they were ambushed and hit hard again by the 20mm cannon fire coming from the commanding heights. Probing north from Awa, 3/22 patrols also came under fire. Before these patrols could withdraw under the cover of their battalion 81mm mortars, an hour-long fire fight ensued. Adding to the general harassment from the enemy, artillery fire was placed on 3/22 positions in the afternoon.

At this same time, Japanese counterbattery fire was delivered against the

[63] Since "take" translated is mountain, *Mount* Yae Take is redundant, but was commonly used by participants of the battle.

[64] "The 77th Infantry Division on Ie Shima was pleased when 'Mt. Yae Take' was captured with its two 6-inch naval guns." LtGen Andrew D. Bruce ltr to Asst G-3, HQMC, dtd 28Oct65, hereafter *Bruce ltr 1965.* At the time that the battle for Yae Take was shaping up, the 77th was fully involved with landing operations on Ie Shima.

[65] *6th MarDiv Jnl, Ph I&II,* 13Apr45.

[66] MajGen Lemuel C. Shepherd, Jr., memo for the OIC, HistDiv, HQMC, dtd 30Oct47.

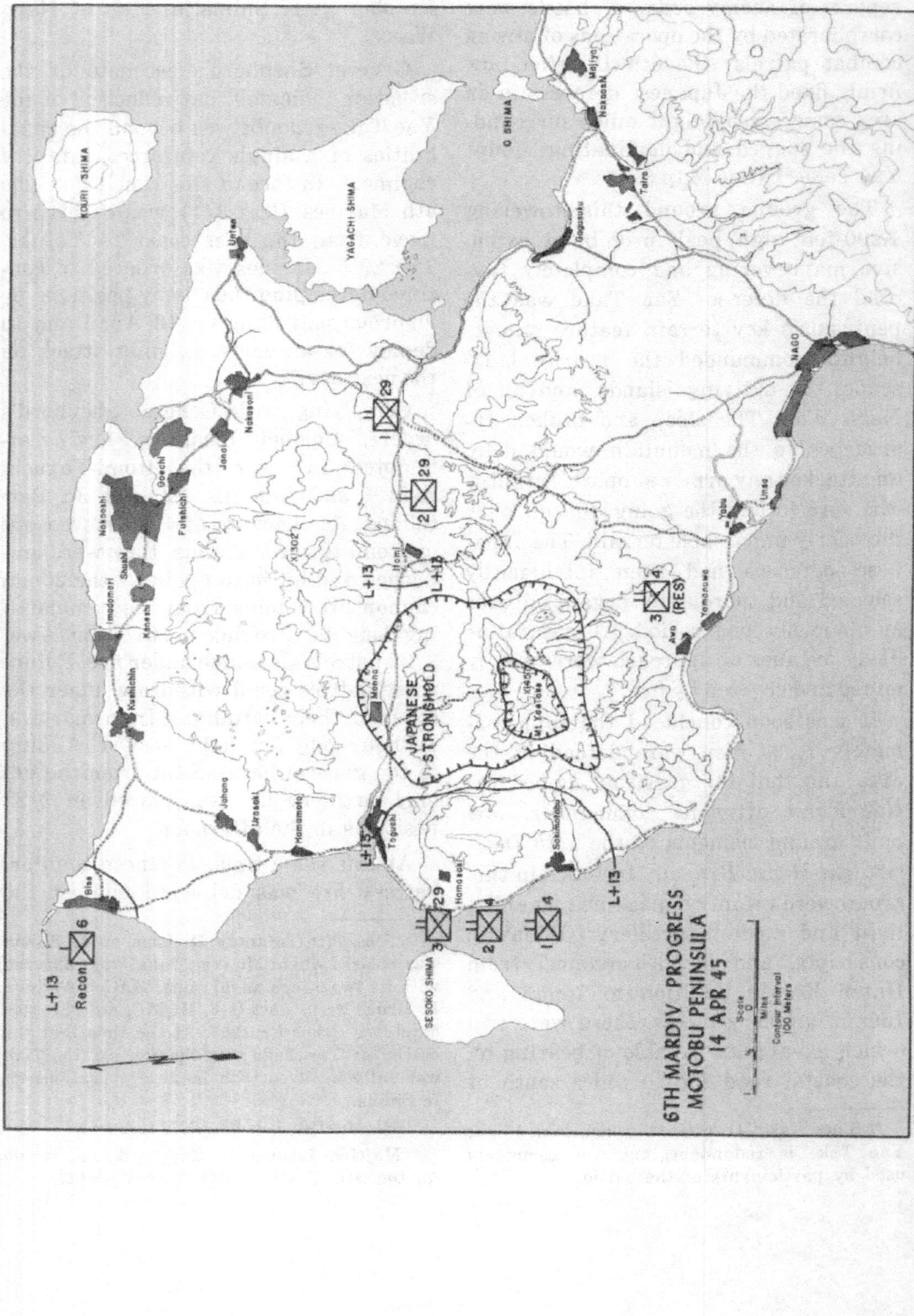

MAP B

T.L. RUSSELL

emplaced artillery of 2/15 (Major Nat M. Pace). This heavy bombardment inflicted 32 casualties, including two battery commanders and the executive officer of a third battery, and destroyed the battalion ammunition dump and two 105mm howitzers.[67] Air strikes were called in on the suspected sources of the fire and 3/22 dispatched patrols in an attempt to locate the enemy mortar batteries. Fires and exploding ammunition made the Marine artillery position untenable, so Pace's men withdrew to alternate positions.[68]

Earlier in the day, the 4th Marines (less 3/4) began its move to Yofuke with Hayden's 2d Battalion in the lead. The west coast was gained after a difficult hike over primitive roads, but Hayden was ordered to continue the march to a point on the southwest corner of the peninsula just below Toguchi, and the battalion arrived there at 1700. Green's 1st Battalion arrived at Yofuke at 1630 and, while digging in for the night, was ordered to move to a position just west of Awa. This displacement was accomplished just prior to darkness by shuttling the battalion by truck, a company at a time. When nightfall came, the 4th Marines was disposed with the 1st and 2d Battalions in perimeter defense, a little less than three miles apart on the southwest coast of Motobu; the 3d Battalion was 20 miles away on the east coast; and regimental headquarters was set up at Yofuke with the Weapons Company.[69]

[67] *15th Mar SAR, Ph I&II*, chap X, n.p.; LtCol Nat M. Pace ltr to CMC, dtd 22Mar55.

[68] *Ibid.*

[69] Maj Orville V. Bergren ltr to CMC, dtd 6Feb48, hereafter *Bergren ltr.*

Based on his original estimate of the situation, General Shepherd planned a coordinated attack for 14 April when the 4th Marines, with 3/29 attached, would advance inland to the east. At the same time, the 29th Marines (less 3/29) would drive to the west and southwest from the center of the peninsula. In effect, this was a situation where two assault regiments attacked a target from directly opposing positions. The danger of overlapping supporting fires was lessened, in this case, by the intervention of the high Yae Take mass. Nevertheless, success of this rare maneuver required close and careful coordination of all supporting arms. (See Map 8.)

In the 4th Marines zone of action, Colonel Shapley's troops were ordered initially to seize a 700-foot-high ridge about 1,200 yards inland and dominating the west coast and its road. It was immediately behind this ridge that Company I of 3/29 had been mauled on the 12th. Intermittent machine gun fire had been received from this area since that time.

The attack jumped off at 0830 on the 14th with 3/29 on the left, 2/4 on the right, and 1/4 initially in regimental reserve. Preceded by an intense artillery, aerial, and naval bombardment, the Marines advanced against surprisingly light resistance. Disregarding scattered Japanese machine gun, mortar, and light artillery fire, the Marines gained the ridge before noon with the left flank of 3/29 anchored to a very steep slope.[70]

[70] *4th Mar SAR, Ph I&II*, p. 12; *2/4 SAR, Ph I&II*, p. 7; *3/29 SAR, Ph I&II*, p. 4.

In order to protect his open right flank, Colonel Shapley moved 1/4 up to an assembly area to the right rear of 2/4. Company C was ordered to take a dominating ridge 1,000 yards to the right front of the 2d Battalion. By noon the company made contact with small enemy groups and soon thereafter began receiving mortar and machine gun fire. Company A was then committed on the left of Company C and the advance was continued.

At the same time, 2/4 and 3/29 resumed the attack to seize the next objective, another ridge 1,000 yards to the front. As the troops headed into the low ground approaching the height, enemy resistance began to stiffen appreciably even though the advance was again preceded by heavy naval gunfire and artillery barrages, and two air strikes. The ground, ideally suited for defense, consisted of broken terrain covered with scrub conifers and tangled underbrush, and the Japanese exploited this advantage to the utmost.

The enemy defense was comprised of small, concealed groups which formed covering screens to the main positions. The Japanese employed every possible stratagem to delay and disorganize the advance, and to mislead the attackers as to the location of the main battle position. Enemy soldiers would lie in a concealed position with their weapons zeroed in on a portion of the trail over which the Marines would have to pass. After allowing a sizable force to pass without interference, the enemy would open up on what they considered a choice target. When a company commander passed the ambush point with his headquarters section, the machine guns opened up, killing him and several other nearby Marines. There were many officer casualties. It was in this manner that the commander of 1/4 was killed in an area where there had been no firing for over half an hour. No one else was hurt, though Major Green's operations and intelligence officers were standing on either side of him. Lieutenant Colonel Fred D. Beans, regimental executive officer, assumed command of the battalion.

Although the hills and ravines were apparently swarming with Japanese, it was difficult to close with them. "It was like fighting a phantom enemy," stated one Marine officer.[71] The small enemy groups, usually armed with a heavy Hotchkiss machine gun and several light Nambu machine guns, frequently changed positions in the dense undergrowth. When fired upon, furious Marines raked the area from where the volleys had come. After laboriously working their way to the suspected enemy position, the Marines came upon only an occasional bloodstain on the ground; they found neither live nor dead Japanese.

Company G of 2/4 made the first strong contact with the enemy at 1350 when it came under rifle, machine gun, mortar, and artillery fire. Less than five minutes later, Company E began receiving similar treatment. After being spotted, a Japanese artillery piece was silenced by naval gunfire and artillery

[71] Maj Orville V. Bergren, "School Solutions on Motobu," *Marine Corps Gazette*, v. 29, no. 12 (Dec45), p. 3, hereafter Bergren, "School Solutions."

brought to bear on it.[72] Despite heavy casualties in Company G and stubborn enemy delaying tactics, Hayden's battalion drove the covering forces back and took the ridge with a frontal attack combined with an envelopment from the right. By 1630, the attack had halted with both 3/29 and 2/4 on the regimental objective and 1/4 on the high ground to the right. Contact was then established all along the line.

East of Yae Take, the 29th Marines jumped off from Itomi in a column of battalions to clear the Itomi-Toguchi road and to eliminate the strongpoints that patrols had discovered the previous four days. (See Map 8.) As the attack developed, it became apparent that an advance in a westerly direction would be both difficult and costly. The axis of the attack was reoriented, therefore, to the southwest in order to take advantage of the high ground. With Lieutenant Colonel Moreau's 1st Battalion leading, the 29th Marines advanced 800 yards up steep slopes against determined enemy resistance. By late afternoon, 1/29 had become pinned down by overwhelming fire from the high ground to its front. The 2d Battalion was committed on the left flank to strengthen the defense and the troops dug in for the night.

When it was relieved during the day by 1/22, 3/4 made a motor march from its east coast position to relieve 3/22 in division reserve. The latter then returned to its patrol base at Majiya.

The following day, L plus 14, Colonel William J. Whaling assumed command of the 29th Marines from Colonel Bleasdale, and the regimental CP displaced to Itomi.[73] During the day, the regiment consolidated its position and organized defensive positions on the high ground. (See Map 9.) Constant pressure in the rear of Yae Take was maintained by vigorous patrolling which assisted the 4th Marines on the other side of the mountain. At 1600, heavy 20mm cannon fire began raining down on the battalion command posts [74] and, about the same time, enemy forces unsuccessfully attempted to infiltrate 2/29 lines under the cover of grenade, rifle, and mortar fire. By 1700, 2/29 had tied in with the 1st Battalion, and shortly after was able to stem the forces of the attack, but not before 35 Marines had become casualties.[75] Artillery and mortar fire, and naval gunfire from the main and secondary batteries of the *Colorado* were placed on the suspected 20mm cannon emplacements and silenced them for a time.[76]

When the 4th Marines began its attack at 0700 on the 15th, it was in the same formation in which it had halted the previous night. The advance was resisted by small scattered groups such as those that opposed the Marines the day before. At noon, as the regiment approached the half-way mark to that day's objective, Japanese resistance be-

[72] The piece, however, was not destroyed as the Japanese had employed their familiar tactic of firing three or four rounds from a position at the mouth of a cave, and then withdrawing the weapon back into the tunnel, where counterbattery fire could not reach it. *2/4 SAR, Ph I&II*, p. 5.

[73] Col William J. Whaling ltr to CMC, dtd 16Dec47, hereafter *Whaling ltr.*

[74] *29th Mar SAR, Ph I&II*, p. 2.

[75] *Ibid.*

[76] CO, USS *Colorado* (BB-45) Rpt, ser 002, dtd 6Jun45, p. 29 (OAB, NHD).

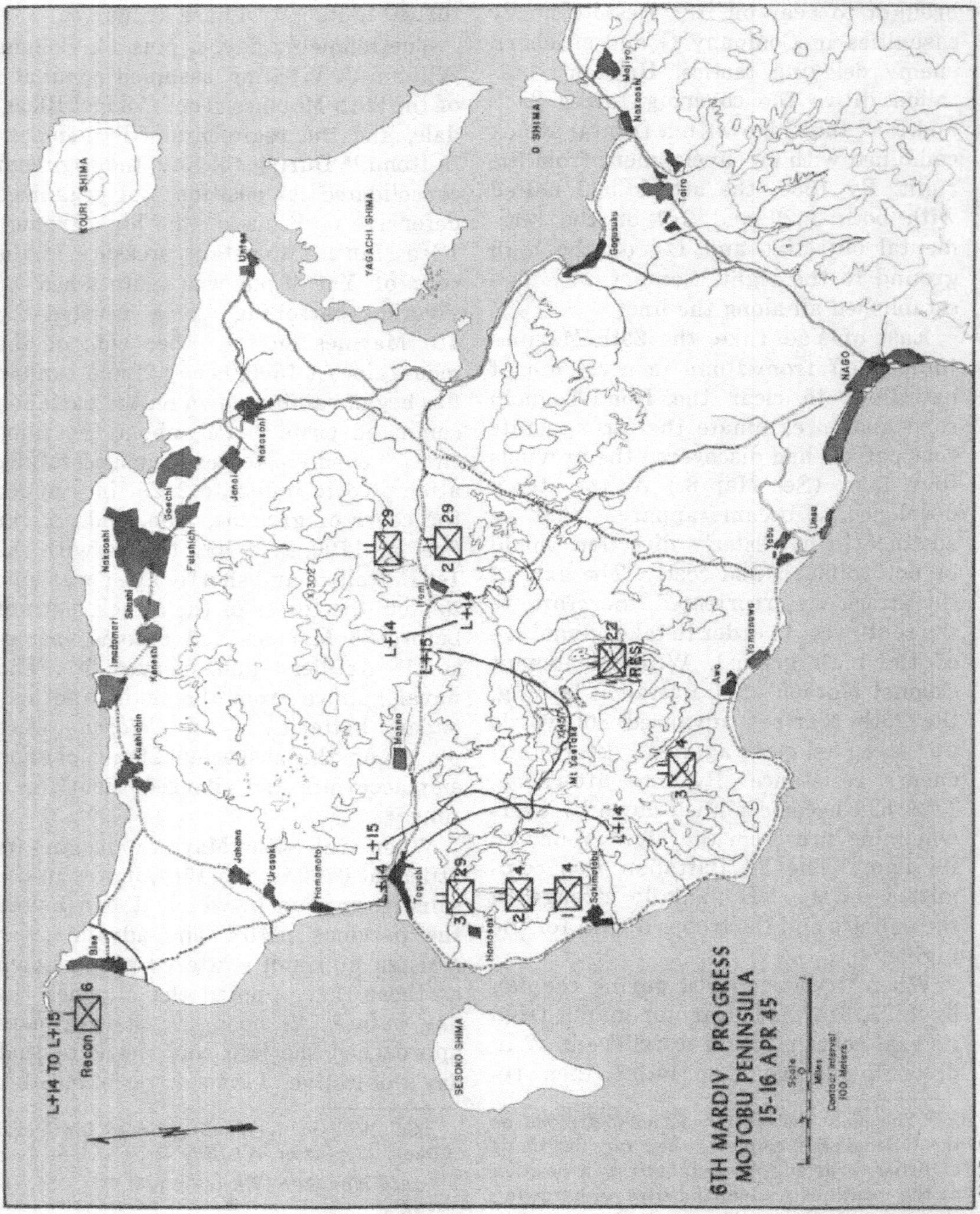
6TH MARDIV PROGRESS
MOTOBU PENINSULA
15-16 APR 45
MAP 9
T.L. RUSSELL
SESOKO SHIMA
YAGACHI SHIMA
KOURI SHIMA
O SHIMA
NAGO
Mt Yaetake
L+14 TO L+15
Recon

came markedly stiffer. From caves and pillboxes emplaced in dominating terrain, the enemy poured down effective fire as the assaulting units climbed the steep mountainside. (See Map 9.)

As 3/29 pushed forward some 900 yards to the east and south, it engaged in numerous fire fights while it received intense machine gun, mortar, and artillery fire. An enemy strongpoint on Hill 210, 500 yards to the battalion right front, held up the advance. In addition to well-dug-in machine guns and mortars, the position also contained the mountain gun that had been pinpointed the day before. For the second day, attempts were made to destroy this devastating weapon with naval gunfire and artillery, as well as air strikes which employed 500-pound bombs and napalm. Despite these efforts, the piece continued functioning and causing considerable damage.

All along the line, bitter fighting ensued as 2/4 again bore the brunt of the rugged going in attempting to capture the high ground dominating the right flank. Although it jumped off with three companies abreast (less one platoon in battalion reserve), 2/4 was able to make only small gains against intense small arms fire. After a day's fighting, the battalion managed to place two companies on Hill 200, while the third one, despite severe casualties (65, including 3 company commanders), eventually advanced three-fourths of the way up a hill to the right of 200. In order to establish a better position, Company G withdrew partway down the hill where it tied in with Company F. On the right of the regimental line, a 200-yard gap between the 2d and 1st Battalions was covered by fire. In the late afternoon, in the area immediately southwest of Yae Take, 1/4 finally seized a key hill mass from which it had been driven back earlier in the day.

When the attack ceased at 1630, the center and right battalions were on their objectives and 3/29 was slightly behind them, organizing ground favorable for defense. During the day, resupply operations and the evacuation of the mounting number of casualties over the tortuous terrain became more and more difficult, and the troops had become very tired. Nonetheless, many caves had been sealed and there were 1,120 enemy dead counted. Colonel Udo apparently foresaw defeat; that night he decided to resort to guerrilla operations and also to move his command to the mountain strongholds of northern Okinawa by way of Itomi.[77]

The 4th Marines knew by this time that it was attacking a force of at least two companies which had organized the terrain to their best possible advantage. Moreover, it became apparent that the Japanese had oriented their defenses to face the anticipated direction of the attack. Owing to these circumstances, and since the advance was still toward friendly troops and artillery, it was decided to contain Udo's mountain force and envelop his defenses by a flanking

[77] *Okinawa Operations Record*, p. 138. "During the period 16–19 April while pressure was being applied to the Udo force on Mt. Yae Take a considerable number of the enemy either were directed to disband and directed to infiltrate through our lines or were cut off by our columns. They followed the natural line of drift, were engaged by our CP at Itomi, particularly at night. Many were killed at our perimeter defenses." *Whaling ltr.*

action from the south; this shifted the direction of the main Marine effort to the north. Implementing these decisions, 3/4 reverted to regimental control and was to be committed in the attack the next day, and 1/22 was ordered into division reserve at Awa.

On 16 April, the 6th Marine Division was deployed to wage a full-scale attack on the enemy from three sides. (See Map 9.) As the 29th Marines continued pressuring in from the east, the 4th Marines with 3/29 would complete the squeeze play from the west and southwest. A juncture between the 4th and 29th Marines would be effected when 1/22 sent strong patrols north into the gap between the two regiments. Each of these three principal assault elements was assigned an artillery battalion in direct support. The artillery was so deployed that the fires of two battalions of the 15th Marines, one company of the 1st Armored Amphibian Battalion, and a battery of the 7th 155mm Gun Battalion could be placed in any of the three zones of action.[78]

In the 4th Marines zone, 3/29 was to seize the high ground 500 yards to its front, including the redoubtable Hill 210. To the right of this battalion, 2/4 was to remain in position and support the attacks of Wright's 3/29 and Beans' 1/4 by fire, while units on the right flank of 1/4 wheeled to the north. The 3d Battalion, 4th Marines, was to attack to seize the division objective, establish contact with 1/4 on the left, and to protect the right flank until 1/22 drew abreast of the line. Weapons Company, 4th Marines, organized as an infantry company,[79] was ordered to patrol thoroughly the right rear of Beans' and Wright's battalions, since 1/22 was not scheduled to start from Majiya until first light.

Because of resupply difficulties,[80] the attack did not resume until 0900 on the 16th. By 1200, 3/29 had seized its objective with a perfectly executed basic maneuver, a single envelopment. As the attack began, Company H, on the 3/29 right flank, faced Hill 210 frontally. Company G in the center was ordered to break contact with Company I on its left and to make an end-around play assaulting the enemy from the south. A Company H support platoon moved into the gap left by G and supported that company by fire, as did 2/4 from its commanding position on the right.

Supporting fires effectively neutralized the Japanese defenders and kept their heads down until Company G Marines had gained the top of 210 and swarmed over the forward slope. Grenades and demolitions blasted the shocked enemy from their caves and

[78] *15th Mar SAR, Ph I&II*, chap VII, n.p.

[79] *4th Mar SAR, Ph I&II*, p. 14.

[80] "Supply and evacuation soon became a difficult problem. The road net was far from adequate, and engineers were working feverishly to build new roads where necessary. However, the rugged terrain prevented them from catching up with the infantry, and usually the last 500 to 1,500 yards of the trip of the chow, ammunition, and water from the forward dumps to the front lines could be negotiated by no other means than manpower. For the next three days this was the case. Division sent up as many replacements from the division pool as could be spared. Battalion headquarters companies were used. Support platoons were used. Evacuation of wounded men was equally difficult if not more so." *4th Mar SAR, Ph I&II*, p. 13.

they retreated hastily, pursued all the while by effective fire from both the assault and support units. In capturing this objective, the Marines had silenced the troublesome mountain gun and killed 147 of the enemy. The positions of Companies H and G were now inverted, with Company H in the center of the line and G on the right flank firmly holding Hill 210.

While 3/29 was securing its objective, 1/4 completed its pivot northwards and had established contact with 3/4. Well to the rear, Myers' 1/22 advanced to cover the open 4th Marines flank. With 3/29 and 2/4 solidly established as landlords of the high ground facing east, 1/4 and 3/4 looked north in positions at a right angle to the other two battalions. When the attack resumed a half-hour later, 3/29 and 2/4 remained in position providing fire support to the advancing 1/4 and 3/4. At this time, the formidable Mount Yae Take was in the 1/4 zone.

The 1st Battalion moved out with Company A on the left attacking frontally up one nose, and Company C working up a draw on the right. Progress up the steep slope was arduous and not helped by enemy small arms fire, light and scattered though it was. As Company A reached the crest, the Japanese met it with withering fire at very close range. In the face of the rifle, machine gun, grenade, and knee mortar [81] drumfire, the Marines withdrew below the summit, and in turn, employed their own 60mm mortars and grenades against the enemy entrenched on the reverse (north) slope. The battle waged fiercely at close quarters as neither side was able to hold the height for long. At last the tide turned in the Marines' favor, helped mainly by supporting fires of 2/4 coming from the high ground overlooking the enemy.

The victory was not bought cheaply; even though the two companies possessed Yae Take, the situation was critical. Over 50 Marine casualties had been sustained in the assault and the ammunition supply was nearly spent. It also appeared that the Japanese were regrouping for a counterattack. Fortunately, effective 15th Marines artillery fire and the excellent mortar and machine gun support of 2/4 held the enemy in check until ammunition could be brought up.

Recalling this phase of the battle for Mount Yae Take, the operations officer of the 4th Marines wrote:

> If the supply problem was difficult before, it was a killer now. That 1.200-foot hill looked like Pike's Peak to the tired, sweaty men who started packing up ammunition and water on their backs. Practically everyone in the 1st Bn headquarters company grabbed as much ammunition as he could carry. A man would walk by carrying a five-gallon water can on his shoulder and the battalion commander would throw a couple of bandoleers of ammunition over the other! . . . The Battalion commander, on his way up to the front lines to get a closer look at the situation, packed a water can on his way up. Stretchers also had to be carried up, and all hands coming down the hill were employed as stretcher bearers.[82]

Additional assistance in resupply and evacuation was afforded the 1st Bat-

[81] "Knee mortar" was a misnomer Americans commonly applied to the Japanese grenade launcher.

[82] Bergren, "School Solutions," p. 6.

talion when Company K, coming up from the rear in late afternoon to revert to the control of 3/4, took out the 1/4 wounded and returned with water and ammunition. The resupply of the 1st Battalion occurred just in time, for at 1830, an hour after Yae Take had been seized, the enemy reacted with a fanatic *Banzai* charge across the battalion front. An estimated 75 Japanese made up the wildly attacking group, but again the supporting fires of artillery and 2/4 stemmed the rush and virtually annihilated the force. As the Marines dug in for the night, Mount Yae Take was held securely.

Lieutenant Colonel Beans' battalion consolidated its holdings in the afternoon while receiving small arms and mortar fire. On the left, Company B was committed to tie in 3/4 with the 1st Battalion. Because its progress was slowed more by the terrain than the enemy, 1/22 on the right was unable to gain contact with either the 4th or the 29th and established a defense perimeter for the night.

While the 4th Marines was storming Yae Take, Colonel Whaling's regiment maintained unrelenting pressure against the enemy's rear positions. As the attack rolled forward, the Japanese resisted stubbornly from log-revetted bunkers and occasional concrete emplacements, and from machine gun, mortar, and artillery positions concealed in ravines and in caves on the heights.

In this phase of the 6th Division's northern campaign, the Japanese exhibited their well-known ability to exploit the terrain and gain maximum benefit from weapons emplaced in caves and pits and concealed by natural camouflage. Of all the weapons that the enemy employed effectively, his use of the 20mm dual-purpose cannon was most noteworthy. Marine battalion CPs received a daily ration of fire from these weapons, and all roads and natural avenues of approach were covered. Any Marine attempt to move over these easier routes often proved disastrous.

Since there was no alternative, "the method of reducing the enemy positions followed a pattern of 'ridge-hopping'," [83] in which all supporting arms covered the attacking force as it enveloped hostile defenses and reduced them in detail. In some cases, the 29th discovered abandoned positions and weapons, suggesting that the Japanese determination to resist was considerably diminished when attacked on the flank. The action in the 29th Marines zone was characterized by simultaneous attacks which, in effect, consisted of a series of local patrol actions to seize critical positions, followed by mopping-up activity within the area.

A heavier-than-usual artillery preparation was laid down before the jump off on the morning of 17 April. (See Map 10.) At 0800, the 29th Marines began an advance to join up with the 4th Marines along the Itomi-Toguchi road. From here, the two regiments would then sweep northward abreast of each other. Moving out over difficult terrain against light resistance, 1/29 made slow progress, but by 1300 had secured its objective, the highest hill in its area.

The enemy positions which confronted 1/29 were on the crest and face of this

---

[83] *29th Mar SAR, Ph I&II*, pt VII, p. 5.

hill and presented a problem in precision naval gunnery to the *Tennessee*, whose line of supporting fire was almost parallel to the target. As troops rapidly advanced, the ship's main and secondary batteries delivered such an intense bombardment that the hill was taken without Marine casualties. On the way to the top, the infantry killed 8 Japanese and 32 more on the crest itself, but the huge craters produced by the *Tennessee's* guns contained in excess of 100 more enemy dead.[84]

Within an hour after 2/29 had resumed its attack, some 50 enemy troops had been flushed out and were observed fleeing to the northwest. Shortly after, the battalion was able to move forward against negligible opposition, stopping only to destroy large enemy stores of equipment, ammunition, and supplies. Before noon, physical contact had been established with 1/22, which had reduced the positions met in its zone and had captured a considerable amount of enemy clothing and ammunition. After having made contact with the 4th Marines on its left, 1/22 was pinched out of the line and withdrew to Awa, where it set up defenses for the night.

The first missions flown by TAF squadrons in support of Marine ground forces during the Motobu campaign struck enemy targets early in the morning of the 17th. At 1000, eight VMF-322 aircraft attacked and destroyed Manna. Upon completion of that mission, the flight was radioed by the ground commander that "the town was wiped out. One hundred per cent of bombs and rockets hit target area." [85] VMF-312, -322, and -323 flew a total of 47 sorties during the day. In the afternoon, one mission of eight planes was cancelled when the assigned target was overrun by Marine infantry.[86]

As it still faced a critical supply shortage on the 17th, the 4th Marines did not launch its attack until 1200, after replenishment. Then the advance toward the Itomi-Toguchi road was resumed with the 1st and 3d Battalions on the right. In reverse of the previous day's situation, 2/4 and 3/29 on the left faced east at a right angle to the front of the other two battalions in assault. They were, therefore, ordered to remain in place and to support the assault from present positions until the attacking units masked their fires.

The attacking element made rapid progress as their downhill path was blocked only by isolated enemy stragglers. Without too great an effort, the Marines overran elaborately fortified positions, intricate communications systems, and bivouac areas. The hastily departing enemy left behind a scene of an undisciplined retreat—for dead bodies and military paraphernalia were strewn all over the area. Large stores of equipment, food, weapons, and clothing were either captured or destroyed. As 1/4 swept across the 3/29 front, 2 8-inch naval guns, 5 artillery

[84] In appreciation of this naval gunfire support, General Shepherd sent to the Commander, Northern Attack Force, a message, which stated in part: ". . . the effectiveness of your gunfire support was measured by the large number of Japanese encountered. Dead ones." Quoted in VAdm Lawrence F. Reifsnider ltr to CMC, dtd 21Mar55.

[85] ADC Daily Intel Sum, 161800 to 171800, dtd 17Apr45, hereafter *ADC IntelSum* with date.

[86] *Ibid.*

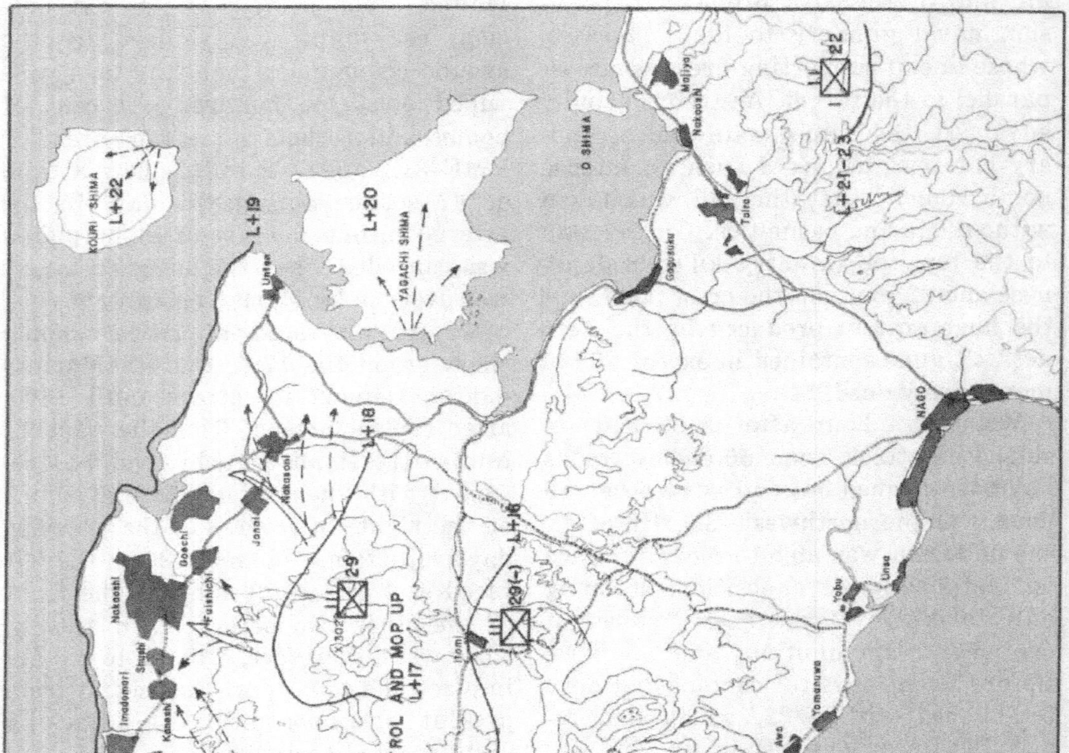
6TH MARDIV PROGRESS
MOTOBU PENINSULA
17-23 APR 45
L+16 TO L+22
PATROL AND MOP UP
L+17
KOURI SHIMA
TAGACHI SHIMA
O SHIMA
SESOKO SHIMA
NAGO
MAP 10
T.L. RUSSELL

pieces, 8 caves full of ammunition, and over 300 dead Japanese were found before the Company G position on Hill 210.[87] Although the 1st Battalion met but few of the enemy during the day, Hochmuth's 3d Battalion killed 56 without losing a Marine.[88]

After their attack axes had shifted northward, the 4th and 29th Marines made contact with each other in late afternoon on the high ground overlooking the Itomi-Toguchi road. At that time, 2/29 was withdrawn from the line to clear out any bypassed enemy pockets in the regimental zone. By the end of 17 April, a review of that day's operations indicated that the enemy was unable to maintain his position and was, in fact, attempting to retreat in order to escape annihilation. There was little doubt that the 6th Marine Division had broken the back of enemy resistance on the peninsula, an assumption that was confirmed when an enemy map captured by the 4th Marines showed that the Yae Take position was the only organized Japanese defense on Motobu.

After four days of vigorous fighting, activities on the 18th were confined to reorganizing, resupplying, and consolidating the gains of the previous day, and patrolling the Itomi-Toguchi road. (See Map 10.) In an attempt to prevent the further escape of any of the enemy and to destroy his trenches and camouflaged emplacements in front of the lines, at 0750, four VMF–312 planes attacked targets with general purpose (GP) and napalm bombs, rockets, and then strafed the smoking positions. The ground troops later reported that all of the hits were in the target area and the enemy trenches were completely destroyed. During the rest of the day, VMF–312 and –322 flew 12 additional sorties in support of General Shepherd's troops.[89]

The now-bypassed 3/29 was detached from the 4th Marines and moved around the base of the peninsula by truck to rejoin its parent organization at Itomi. In the same way, 1/22 rejoined its regiment at Majiya. In the 4th Marines area, the 1st Battalion went into reserve, bivouacking near Manna. Upon its reversion to regimental control, 3/29 took up blocking positions on the right flank, north of Itomi, to prevent any enemy escaping to the east. The 29th Marines left flank was pushed northward to straighten out the division lines. As 3/4 conducted local patrols, the 2d Battalion patrolled the area through which the 1st and 3d Battalions had attacked the previous day. Resupplying the assault regiments continued to be difficult for the enemy had thoroughly mined the area now held by the division and had denied the Marines use of the Itomi-Toguchi road by digging tank traps there. In addition, many trees had been felled across the road, which was pockmarked with numerous shell craters.

On L plus 18, the final drive to the northern coast of the peninsula began with the 4th and 29th Marines abreast. (See Map 10.) Preceding the 0800 jump off, four Corsairs from VMF–312 struck at a hillside containing gun emplacements and strongpoints that opposed the ground attack. Again napalm, GP

[87] *3/29 SAR, Ph I&II*, p. 4.

[88] *1/4 SAR, Ph I&II*, p. 6; *3/4 SAR, Ph I&II*, p. 7.

[89] *ADC IntelSum*, 18Apr45.

bombs, rockets, and strafing attacks were employed to ease the infantry advance. When the Marines pushed forward against negligible resistance, they came across elaborate cave and trench systems filled with numerous enemy dead, undoubtedly the victims of the artillery, naval gunfire, and air bombardments. All organized resistance ended on Motobu Peninsula when the 4th and 29th Marines gained the north coast on 20 April. General Shepherd assigned garrison and patrol sectors to his units on Motobu;[90] at the same time, mopping-up operations continued in the rest of the IIIAC zone.

In the course of the fighting for the peninsula, the 6th Marine Division had sustained casualties amounting to 207 killed, 757 wounded, and 6 missing in action. The Marines counted over 2,000 Japanese dead, men who had forfeited their lives while defending their positions with a tenacity that was characteristic.

Of the 6th Division drive up the isthmus and into the peninsula, Brigadier General Oliver P. Smith noted:

> The campaign in the north should dispel the belief held by some that Marines are beach-bound and are not capable of rapid movement. Troops moved rapidly over rugged terrain, repaired roads and blown bridges, successively opened new unloading points, and reached the northern tip of the island, some 55 miles from the original landing beaches, in 14 days. This was followed by a mountain campaign of 7 days' duration to clear the Motobu Peninsula.[91]

[90] 6th MarDiv OperO 41–45, dtd 20Apr45.

[91] Smith, *Personal Narrative*, p. 82.

USS ENTERPRISE *is hit on 14 May 1945 by a* Kamikaze *which dove out of low cloud cover. (USN 80-G-331011)*

TRACERS *fill the sky as AAA gunners repulse a Japanese raid over Yontan Airfield. (USMC 118775)*

Phase 1 Continued

PROGRESS OF LOGISTICAL SUPPORT [1]

Shortly after the L-Day landings, Radio Tokyo predicted that the beachhead on Okinawa would be wiped out.[2] From L-Day on, the impressive flow of troops and supplies ashore gave little support to this optimistic enemy forecast, however, as the Tenth Army hold on the island rapidly tightened. While the assault units fanned out to gain assigned initial objectives, battalion shore party commanders assumed control of their beach sectors. During L-Day, successively higher command echelons landed, and, by nightfall, divisions had assumed control of shore party operations.

A coral reef extending the length of the beaches was the only real obstacle to early unloading operations. During floodtide, a steady procession of DUKWs and LVTs shuttled cargo across the reef, and only within this 4-to-5 hour period of high tide could ships' landing craft make runs directly to a few scattered places on the beaches. Low tide, however, exposed the coral outcroppings, and necessitated the establishment of offshore transfer points to maintain the flow of supplies to the beach. Barge cranes required at the transfer points to transship cargo were not available in appreciable numbers until L plus 2.

Increasingly intense *Kamikaze* raids posed a threat to the transport groups and caused delays in the buildup of supplies ashore. Additionally, the unexpected rapid infantry advances disrupted the unloading schedule. Meanwhile, shore party officers faced such other problems as the lack of suitable beach exits and the scarcity of engineering equipment to prepare them. Another critical matter of note was the shortage of transportation to clear the beaches of supplies. As the volume of cargo being landed increased, the number of trucks available for hauling to inland dumps decreased. According to the operation plans, organic assault division motor transport was to have supported the efforts of the shore parties initially. When frontline troops began to outdistance their support elements, the divisions were forced to withdraw their trucks

[1] Unless otherwise noted, the material in this section is derived from: *CNO Record; CTF 51 AR; Tenth Army AR.*

[2] IIIAC G–2 PeriodicRpt No. 2, dtd 3Apr45.

from the beaches to resupply forward assault units.[3]

The effort beginning on L-Day to bridge the reef barrier off the Hagushi beaches bore fruit by 4 April. In place opposite Yontan airfield on Red Beach 1 were ponton causeways that had been side-lifted to the target by LSTs. Earth fill ramps were constructed across the reef to Purple Beach 1 and the Orange Beaches near Kadena. Within the mouth of the Bishi Gawa, close to Yellow Beach 3, a small sand bar had been cleared of surface obstructions and enlarged. A loop access road was then cut through the beach cliff to the bar by engineers with Seabee assistance. As soon as these facilities were ready, cargo from landing craft as large as and including LCTs, could be unloaded directly over the two causeways and the improved sand bar.[4]

A total of 80 self-propelled barges, also side-carried to Okinawa, was in use constantly from the beginning of unloading operations. The barges were employed in various ways, essentially at the discretion of the division commanders. One barge was assigned to each LST(H) as a landing float onto which the bow ramps of the landing ships were dropped to ease the transfer of casualties from small boats or amphibians. The majority of the barges served as floating supply dumps. These were particularly valuable for supplying critical items to the units ashore at night when cargo ships carrying needed supplies retired from the transport areas.

IIIAC mounted cranes on 12 of these self-propelled units and positioned them at the reef where netted cargo was transferred from boats to LVTs or DUKWs for the final run to inshore dumps. Referring to the demonstrated success of this method, one Marine shore party commander commented:

> This was the [1st Marine Division's] innovation, first practiced successfully at Peleliu. Two of these barge-mounted cranes were loaned to [the] 6thMarDiv on [L plus 1] to facilitate their cargo handling, and XXIV Corps took up the method. That method accounted for the comparative lack of clutter on the 1st MarDiv beaches. That [the 1st Marine] Division had no beach dumps is a fact of prophetic import for future operations, for I believe establishment of such will invite their destruction in an assault landing.[5]

Encouraged by the satisfactory tactical picture, Admiral Turner authorized the use of floodlights and night unloading on all beaches starting 2 April, and directed that ships' holds be cleared of all assault cargo immediately. On the same day, he ordered that the personnel

[3] During logistics planning, it was assumed that, because of the immensity of the ICEBERG operation and the number of forces and equipment involved, there would be a commensurately huge consumption of fuels and lubricants (Class III supplies) by both ground and aviation units. Therefore, supplies for the construction of bulk storage facilities, and offshore pipelines to connect with tankers, were loaded for early delivery. Because the initial advance was so rapid and the resultant consumption of motor fuel unexpectedly high, the construction of these bulk handling facilities was expedited. Beginning 19 April, gasoline and diesel fuel was pumped to dumps from beached large gasoline barges. Four days later, tankers began pumping Class III supplies ashore; by 26 April, bulk storage facilities for aviation fuel had been established at Yontan and Kadena airfields. *Blakelock ltr 1965.*

[4] *Ballance ltr.*

[5] *Ibid.*

and equipment of the aviation engineer battalions and the MAGs be expeditiously unloaded. On 3 April, General Geiger recommended to Turner that all priorities established for LSTs unloading over IIIAC beaches be suspended until every member of the airfield headquarters, service, construction, and maintenance units had arrived ashore.

Planned unloading priorities were upset, however, by Tenth Army insistence on getting Yontan and Kadena airfields operational at the earliest possible time, and by General Buckner's authorization on L plus 2 for corps commanders to bring garrison troops ashore at their discretion. Those on board control vessels and shore party personnel soon viewed many situations wherein low priority units and equipment intermingled with the shoreward flow of essential assault materiel. This interruption of supposedly firm unloading schedules was due, in part, to the natural desires of ships' captains to unload their vessels and to clear the vulnerable Hagushi anchorage as quickly as possible. The inadequacy of the motor transport available to the shore parties and the radical change in the unloading priorities, however, forced many ships to stand off shore with half-empty holds while awaiting the return of boats which were, meanwhile, stacked up at the control vessels.

Further complicating the critical control problem were the efforts of individual landing boat coxswains who, disregarding their instructions, attempted to "get to the beaches at all costs." Commenting on this matter, one transport group commander said:

> There seemed to exist on the part of most coxswains an almost fierce determination to be first ashore with their individual boats, regardless of the orderly assignment to unloading points, which it is the function of the control vessel to carry out. Coxswains simply would not follow orders to form and remain in cargo circles, but jockeyed for positions of advantage from which to come along side the control vessel. Many even attempted to ignore the control vessel and bypass it, proceeding directly to whatever beach they had a preference for.[6]

Despite this, the control of ship-to-shore traffic was probably handled better at Okinawa than in previous Pacific operations, except those at Peleliu and Iwo Jima. After observing the assault landings in the Marianas, Admiral Turner was convinced that only "the most experienced personnel obtainable should be used in the Control Parties for assault landings." [7] Consequently, the key members of the control groups which operated in the Palaus and Bonins served on board the control vessels at Okinawa, where their collective experience helped make ICEBERG a more efficient operation.

Although the ship-to-shore cargo transfer procedures were soon ironed out, problems at the beaches still existed. Organization of the northern landing beaches, for example, progressed slowly. In a critical but friendly evaluation of Marine shore party operations, experienced British observers stated that:

> There seemed to be little or no traffic control, no sign posting of roads or dumps, and no orderly lay-out of the beach areas. It has been said already that the speed of advance inland outran the landing of

---

[6] *CNO Record*, chap 7, p. 55.

[7] CTF 51 OpRpt FORAGER, dtd 25Aug44, Recommendations, p. 2.

> vehicles. The rapid landing of [motor transport] therefore became an imperative need and there is no doubt . . . that the rate of landing could have been greatly accelerated by proper organization. For instance, although vehicles were able to wade ashore at low tide on Yellow 2, they were only using one exit. This had no beach roadway on it, although its gradient and surface were such that a tractor was frequently required to pull vehicles through it. This considerably retarded progress. (It was noted both here and at other beaches that where beach matting had been laid down, it had usually been cut up by tractors. Separate exits for wheels and tracks is not one of the Marine Shore Party rules!)
>
> It is easy to be critical, but the general impression remains that unloading organization in this sector was insufficiently flexible to cope with the unexpected military situation. However, the Shore Party work in this Corps [IIIAC] must be judged by results, and the fact is that after L-plus 1 day, no serious criticism of the unloading progress was made by the Corps Commander.[8]

The planned and orderly transition of shore party control to progressively higher troop echelons continued as the beachhead expanded. On 3 April, the XXIV Corps commander took charge of the southern beaches, and, three days later, the commander of the III Amphibious Corps Service Group assumed control for the unloading of the Marine divisions.[9] After a conference of responsible fleet and troop logistics officers on board Admiral Turner's flagship on 8 April, arrangements were made for Tenth Army to take over all shore party activities on the Hagushi beaches the following morning. Major General Fred C. Wallace, the Island Commander, was placed in charge and his 1st Engineer Special Brigade was directed to assume control of all beaches, with the exception of the one which had recently been opened at Nago. In order to operate a much-needed forward supply dump for the far-ranging infantry units of the 6th Marine Division, the IIIAC Service Group retained control of this northern landing point.

Many of the shore party troops in the IIIAC zone of action were from replacement drafts. They had trained with the divisions as infantrymen and accompanied the assault echelon to the target. Until needed to replace casualties in the combat units, these Marines fulfilled a vital function while assigned to shore party and ships' working parties.

Although the weather remained perfect until the afternoon of L plus 3, heavy rain and winds during that night and most of the following day hampered unloading activities. With the abatement of high winds on 6 April, a stepped-up

[8] British Combined Operations Observers POA Rpt to Chief of Combined Operations Representative, British Joint Staff Mission, Subj: Operations for the Capture of the Okinawa Gunto, dtd 18Apr45, p. 16 (OAB, NHD), hereafter *British Observers Rpt.*

[9] Based on the post-operation recommendations of Lieutenant Colonel Francis M. McAlister, the commander of the Marine corps service group especially established for the landing on Guam, a like unit was organized for ICEBERG. Its mission was to operate all of the IIIAC shore installations and facilities required for debarkation, supply, and evacuation activities, and to provide local security in the service area. About 1 February 1945, the staffs of this service group and the Corps Shore Party assembled at Guadalcanal. Officers for these staffs came from IIIAC headquarters, FMFPac, and Tenth Army, and, on a temporary duty status, from the 1st and 6th Marine Divisions. *IIIAC AR*, chap 3, p. 21.

YELLOW BEACH 3 *on L plus 2. As soon as LCTs and LCVPs are unloaded, others arrive to take their place. (USMC 118214)*

CAUSEWAYS *relieve logistical problems as tons of supplies are transported inland. (USMC 118304)*

unloading pace resulted in the emptying of 13 APAs and AKAs, and 60 LSTs. The day before, in the midst of the storm, 32 empty cargo and transport vessels left the target area. Between L-Day and 11 April, when the first substantial increment of garrison shipping arrived, unloading over the Hagushi beaches was confined primarily to assault shipping. By noon of 11 April, 532,291 measurement tons of cargo had been unloaded, an amount greater than had been put ashore during the entire course of the Marianas campaign.[10]

SECURING THE EASTERN ISLANDS AND IE SHIMA [11]

Since the rapid sweep of the Tenth Army had cleared the shoreline of Chimu Wan and a large section of the upper portion of Nagagusuku Wan by 5 April, Admiral Turner was anxious to utilize the beaches and berths on the east coast as soon as possible. Although minesweepers were clearing the extensive reaches of both anchorages, before unloading operations could be safely started the Japanese strength on the six small islands guarding the mouths of the two bays had to be determined. To acquire this information, the FMFPac Amphibious Reconnaissance Battalion was attached to the Eastern Islands Attack and Fire Support Group and assigned the mission of scouting the islands. (See Map 11.)

Tsugen Shima, the only island suspected of being heavily defended, was the first target of the battalion. Although Tsugen is relatively small, its position southeast of the Katchin Peninsula effectively controls the entrances to Nakagusuku Wan. Aerial observers reported that the village of Tsugen and the high ridge overlooking it contained extensively developed strongpoints. After midnight, early on 6 April, high-speed APDs carrying the battalion arrived off the objective, and Companies A and B embarked in rubber boats to land on the western coast of the island at 0200. Just a short way inland from the landing point, four civilians were encountered; two were made prisoner, but the other two escaped to alert the garrison.[12]

Enemy reaction came almost immediately. Company A began receiving machine gun fire from the vicinity of Tsugen, while Company B was similarly taken under fire from a trench system in the northwest part of the island. Japanese mortars soon found the range of the landing party, whereupon the Marines withdrew to the beach under an unceasing shower of shells. Since the battalion assignment was to uncover enemy opposition and not engage it Major Jones reembarked his unit at

[10] "The first garrison shipping to arrive carried assault cargo that could not be lifted in the assault shipping" and, by 15 April, 577,-040 measurement tons of cargo, mostly assault supplies, had been unloaded against an estimated beach capacity of 529,070 measurement tons. *Blakelock ltr.*

[11] Unless otherwise noted, the material in this section is derived from: *CTF 52 AR; CTF 53 AR; 27th InfDiv OpRpt; 77th InfDivOpRpt,* Ie Shima; *PhibReconBn AR;* LtCol Max Myers (ed.), *Ours to Hold It High: The History of the 77th Infantry Division in World War II* (Washington: Infantry Journal Press, 1947), hereafter Myers, *77th InfDiv Hist.*

[12] Company B had rejoined the battalion on 3 April after its release at Iwo Jima by V Amphibious Corps.

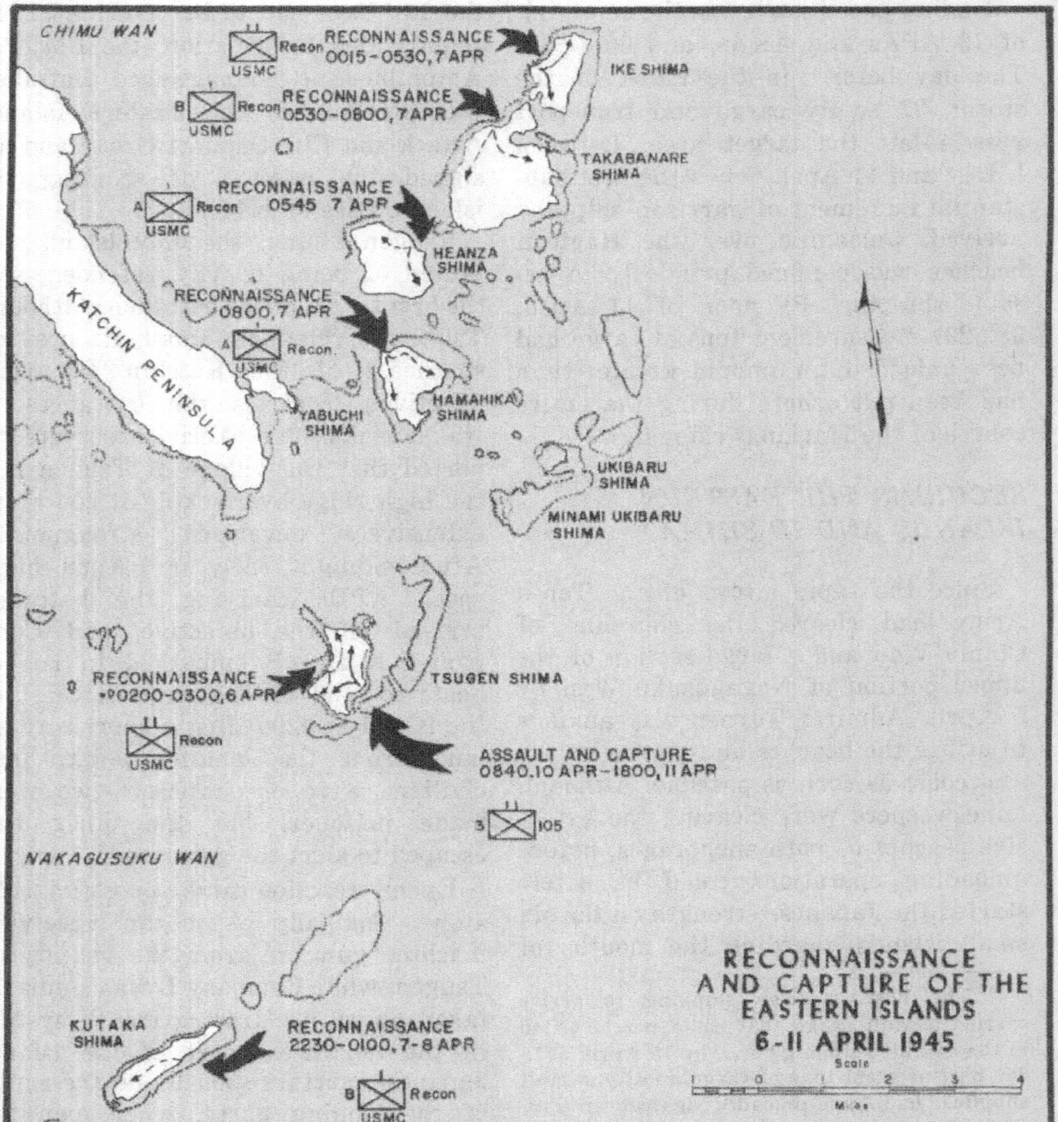

MAP 11 T.L. RUSSELL

0300. Although the Japanese claimed an easy victory over an "inferior" force,[13] the scouts had accomplished their mission. Company A lost two Marines killed and eight wounded.

On the evening of 6 April, Major Jones' men resumed their investigation of the rest of the islands in the offshore group. At 0015 on 7 April, the entire battalion landed on Ike Shima, the northernmost island. When no sign of enemy troops or installations and only one civilian was discovered there, Company B went on to Takabanare Shima. Landing at 0530, it discovered that 200 thoroughly frightened Okinawan civilians were the island's only inhabitants. At about the same time, two platoons of Company A went to Heanza Shima and, using their rubber boats, crossed over to Hamahika Shima. Daylight patrols confirmed the absence of enemy soldiers, but 1,500 more civilians were added to those already counted. These islands were occupied later in April by 3/5.[14]

After nightfall on 7 April, Company B reboarded its APD, which then circled Tsugen Shima to land the Marines on Kutaka Shima, opposite enemy-held Chinen Peninsula. As the company paddled in to shore, the heavy surf capsized three of the boats and one man drowned. The island had neither enemy troops, installations, nor civilians, and the scouts withdrew shortly after midnight.

While the reconnaissance battalion was searching the rest of the Eastern Islands on 7 April, UDT swimmers checked the proposed landing beach on the east coast of Tsugen Shima preparatory to the assault there. The capture of the Eastern Islands had been assigned to the 27th Infantry Division as its part in Phase I of the Tenth Army preferred invasion plan. The information gained from the 6–7 April reconnaissance indicated that commitment of an entire division was not warranted, and only one regiment was assigned for the operation.[15]

As the main body of the Army division was landing over the Orange beaches near Kadena on 9 April, the ships of the 105th RCT were rendezvousing at Kerama Retto with the command ship of the Eastern Islands Attack and Fire Support Group. The assault unit selected for the landing on Tsugen was 3/105, while the other two battalions of the RCT were designated floating reserve to be called up from Kerama if needed. Although Tsugen had been pounded intermittently by air and naval gunfire since L-Day, the ships' guns again blasted the island on 10 April, the day of the landing. Initial resistance was light when the soldiers landed at 0839, but the enemy, strongly entrenched in the stone and rubble of Tsugen, soon engaged the invaders in a day-long fire fight. The battle continued throughout the night, during which time the Army battalion sustained many casualties from the incessant enemy mortar fire coming from the heights above the village.

At daylight on the 11th, the rifle companies of 3/105 made a concerted

[13] *Okinawa Operations Record*, pp. 72–73.

[14] *1st MarDiv G–3 Jnl*, 17-23Apr45.

[15] In response to General Hodge's request, on 8 April General Buckner ordered the remainder of the 27th Division landed on Okinawa to reinforce the XXIV Corps offensive.

MARINE RECONNAISSANCE *personnel prepare rubber boats for landings on Eastern Islands. (USMC 120002)*

DOMINATING *Ie Shima is Iegusugu Yama, beyond which are the airfields—the primary objectives of the operation. (USN 80–G–315059)*

attack against stubborn opposition which gradually died out. Organized resistance was eliminated by 1530, and the battalion was ordered to embark shortly thereafter to join the rest of the regiment at Kerama Retto. In a day and a half of fighting, the battalion lost 11 men, had 80 wounded, and 3 missing.[16] An estimated 234 Japanese were killed and no prisoners were taken. The seizure of Tsugen Shima opened the approaches to Nakagusuku Wan, and ensured that XXIV Corps would receive supply shipments over the eastern as well as the western beaches. This operation also uncovered beaches in Chimu Wan which were developed by the Seabees and used for unloading the LSTs which brought construction supplies and equipment from the Marianas. This action relieved the load which had been placed on the Hagushi beaches, expedited base development, and hastened the building of additional unloading facilities.[17]

In its rapid advance leading to the capture of the Motobu Peninsula, the 6th Marine Division demonstrated that Okinawa north of the Ishikawa Isthmus could be taken by an attack overland. ICEBERG commanders were forced in turn to reappraise the original plans for Phases I and II. They found that naval requirements were now reduced to resupply and fire support operations, and that the ships which might have been needed for an amphibious assault of Motobu Peninsula—a possibility considered in all advance planning—were now available for the capture of Ie Shima. Losing no time, Admiral Turner issued the attack order directing the seizure of the island and its vital airfield, and designated the Northern Attack Force commander, Admiral Reifsnider, as Commander, Ie Shima Attack Group.

Ie Shima was important because its size and physical features permitted extensive airfield development. Three and a half miles northwest of Motobu Peninsula, the island plateau was mostly flat land, broken only by low hills and scattered clumps of trees. Located in the middle of the eastern part of the island was a rugged and extremely steep 600-foot-high limestone mountain, Iegusugu Yama. There were few obstacles to widespread construction of airdromes besides this prominent terrain feature. This factor escaped the attention of neither Japanese nor American planners. The enemy had already laid out three runways, each a mile in length, on the central plateau, and the ICEBERG plan called for the expansion of these existing strips as well as the addition of others which would eventually accommodate an entire wing of very-long-range fighter aircraft.

The landing force selected for the invasion was General Bruce's 77th Infantry Division. After the Keramas landing, this unit spent two weeks on board ship in a convoy which steamed in circles approximately 300 miles southeast of Okinawa. Without warning, on 2 April enemy aircraft dove out of clouds which had hidden their approach and crash-dived four ships (three of which were command ships), before antiaircraft fire could open up on the intruders. The entire regimental

[16] *27th InfDiv OpRpt*, p. 30.

[17] *Blakelock ltr 1965.*

staff of the 305th Infantry was killed and wounded, and the total number of casualties listed in this one attack was 17 soldiers killed, 38 wounded, and 10 missing.[18] Ten days after this disaster, the division was committed to land on 16 April, its second assault landing in less than a month. (See Map 12.)

Major Jones' Amphibious Reconnaissance Battalion was assigned to execute the first mission of the operation. His unit was directed to seize and occupy Minna Shima, a small crescent-shaped island lying 6,500 yards southeast of the main target. Two 105mm and one 155mm howitzer battalions from 77th Division artillery were to be emplaced there to provide supporting fires during the Ie Shima battle.

The Marine scouts landed at 0445 on 13 April and within two hours had swept the island. They discovered 30 civilians but found no enemy soldiers. The battalion remained on the island the rest of the 13th and, on the morning of the 14th, occupied positions from which it covered UDT preparations of the reef and beach for the landing of artillery. By noon of 14 April, Major Jones had reembarked his men on board the APDs. Three days later, the battalion was released from attachment to the 77th Division and attached to IIIAC.

As scheduled, the preliminary bombardment of Ie Shima began at dawn on 16 April and was stepped up at 0725 when missions in direct support of the landing were fired. Five minutes before S-Hour (as the landing time was designated for this operation), 16 fighter planes made a strafing and napalm attack on the beaches while other fighters and bombers orbited over the island, ready to protect the attack group and support the ground assault.

Although there was little opposition to the landing, the troops experienced stiffening resistance by afternoon when enemy delaying groups, concealed in caves and fortified tombs, started to contest every yard of advance. For a period of six days, 77th Division ground forces struggled. Initially making only slight gains, in many cases, they fought hand-to-hand with defenders who contested every inch of ground. As the battle unfolded, it was found that Japanese defenses were centered about Iegusugu Yama and the small village of Ie, which lay at the foot of the southern slope of the mountain. A masterful camouflage job had been performed by the Ie Shima garrison, for nearly 7,000 people were concealed on the island. The mountain contained a maze of hidden firing positions; Ie itself had been converted into a veritable fortress. The ground approaching the mountain and the town was honeycombed with caves, tunnels, bunkers, and spider holes on which the Japanese had expended their great industry and defensive skills. The advance route to the core of enemy defenses was open land and uphill all the way, flanked by Japanese positions in the village and dominated by emplacements located in a reinforced concrete building on a steep rise facing the attacking troops. The infantry soon named this structure "Government House" and the terrain on which it stood "Bloody Ridge."

[18] *77th InfDiv OpRpt,* Ie Shima, p. 7.

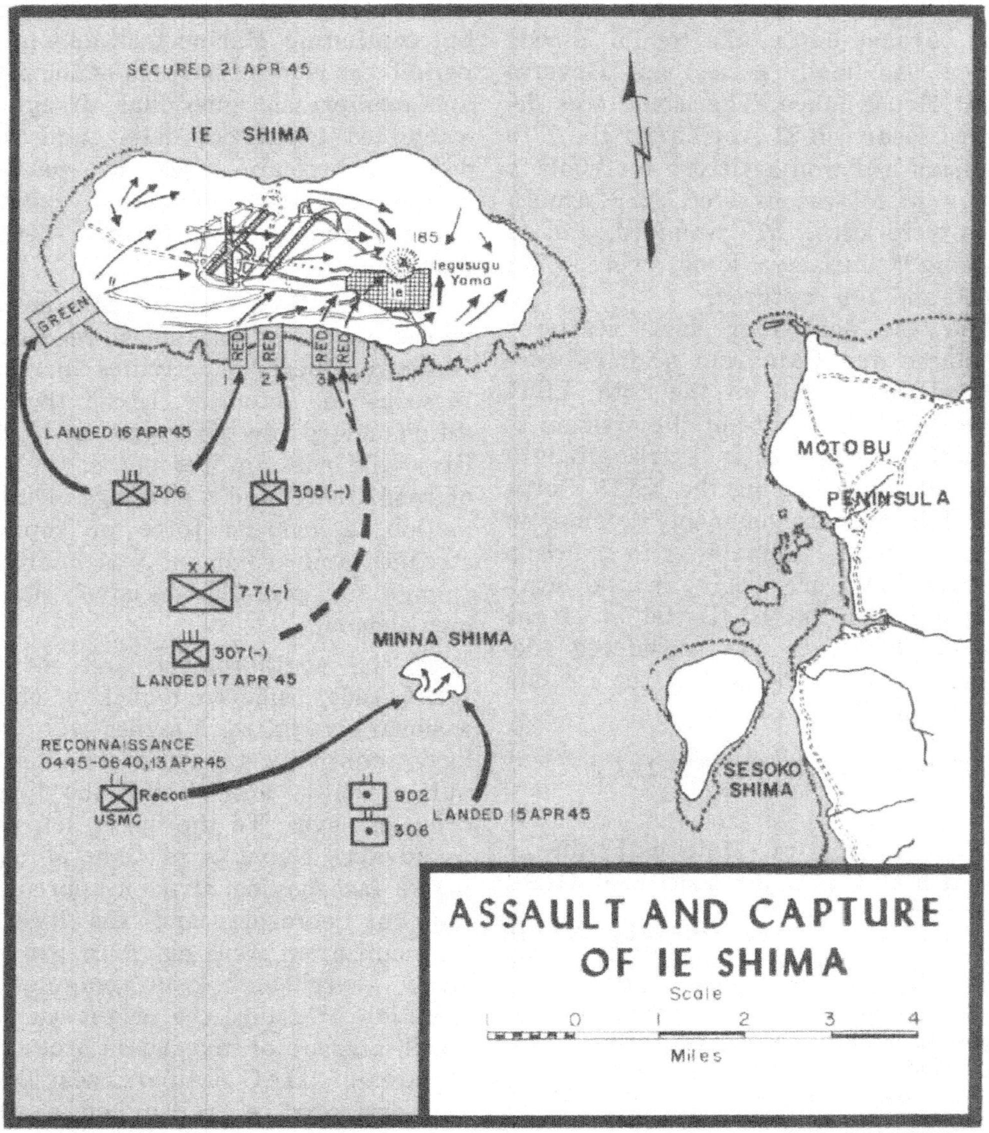

MAP 12 T. L. RUSSELL

On 20 April, after a grim grenade and bayonet battle, the top of Bloody Ridge was finally gained and Government House taken. The island was declared secure on 21 April after the 77th Division had won a victory for which a heavy price was exacted; 239 Americans were killed, 879 wounded, and 19 missing.[19] Japanese losses were 4,706 killed and 149 captured.

For the next four days, scattered Japanese and Okinawan soldiers were hunted down and, on the 25th, LSTs began shuttling units of the division to Okinawa, where their extra strength was needed in helping the XXIV Corps maintain pressure on enemy defenses in front of Shuri. Remaining in garrison on Ie Shima were the regimental headquarters and the 1st Battalion of the 305th. This force was considered adequate to handle the rest of the cleanup operations in the island.

THE MARINES' "GUERRILLA WAR" [20]

The capture of Motobu Peninsula constituted the major portion of IIIAC offensive operations in April. A lesser but continuing Marine task during the period was ridding the area of the pesky and omnipresent guerrillas. Irregulars attempted to harass, delay, and wear down American units by partisan tactics classically employed against patrols, convoys, or isolated detachments.

Once Yae Take fell and Marines advanced to the northernmost reaches of Okinawa, guerrilla activities increased in scope and intensity. Under the conditions offered by the rugged and primitive wilderness of the north, the lack of roads there, and a shortage of information, a modern force of superior strength and armament was unable to engage the guerrilla decisively in his own element.

In the southernmost area of the IIIAC zone, aside from picking off occasional stragglers, Marines were kept busy improving the road net, sealing burial vaults, and closing the honeycomb of caves. To the north, however, as advance elements of General Shepherd's fast-moving division approached Motobu Peninsula, and the lines of communication were extended progressively, guerrillas took advantage of the situation. During the night of 8–9 April, a group of marauders broke into the area of IIIAC Artillery, near Onna, and destroyed a trailer and a small power plant. Following this attack at dawn, other enemy groups attempted to disrupt north-south traffic passing through Onna by rolling crudely devised demolition charges down upon passing vehicles from the cliffs above.[21]

---

[19] *Tenth Army AR*, Corrections by CG, 77th InfDiv, dtd 11Oct45. Amongst those killed during the 18 April fighting was Ernie Pyle, the renowned war correspondent. He was buried on Ie Shima in the 77th Division cemetery. The division erected a marker in his memory near the spot where he was shot by a Japanese machine gunner. Inscribed simply, the marker reads: "On this spot the 77th Infantry Division lost a buddy, Ernie Pyle, 18 April 1945." Myers, *77th InfDiv Hist*, p. 265.

[20] Unless otherwise noted, the material contained in this section is derived from: *IIIAC* AR; IIIAC G–2 PeriodicRpts Nos 1–30, 1Apr–1May45; *1st MarDiv SAR; 1st MarDiv G–3 Jnl; 6th MarDiv SAR, Ph I&II; 6th MarDiv Jnl, Ph I&II.*

[21] Tenth Army G–2 Rpt No. 15, 10Apr45.

MARINE ROCKET LAUNCHERS *in support of the drive south. (USMC 121342)*

AWACHA POCKET, *showing the gorge which was the scene of hard fighting by the 5th Marines. (USMC 121104)*

In the south of the 6th Division zone, on 7 April the 7th Marines (less 3/7), in corps reserve at Ishikawa, was assigned to patrol tasks.[22] The northern half of the regimental patrol sector was covered by Lieutenant Colonel John J. Gormley's 1/7, which had moved to Chimu, while Berger's 2d Battalion and certain designated regimental troops, in a perimeter defense around the bombed-out ruins of Ishikawa, had a related mission of patrolling north and inland from the village.[23]

One 7th Marines task was warding off nightly infiltration attempts by individual or small groups of Japanese and Okinawan irregulars in search of food. Most of them were killed or wounded either entering the village or leaving it. The initial patrols in the region were without incident, but, as pressure was applied to Colonel Udo's force in the mountain fastnesses of Motobu Peninsula, the quiet that had prevailed in the supposed-rear zone was dispelled. On 12 April, a 2/7 patrol fell victim to a well-planned ambush on Ishikawa Take, the highest point on the isthmus. By the time that the entrapped Marines were able to pull out under cover of the fires of the regimental weapons company, 5 men had been killed and 30 wounded.

The next day, Lieutenant Colonel Berger sent two companies into the ambush zone and occupied it against only token resistance. In customary partisan fashion, the elusive guerrillas had departed the area, seemingly swallowed up by the heavy vegetation, deep gorges, and spiny ridges of the complex terrain.

After spending a quiet night on the twin peaks of the heights, the two companies, E and F, were withdrawn to approach the guerrilla lair from a different direction. While retiring, the Marines were fired upon from above by the reappearing enemy, and a number of men were hit. After circling to the far (west) side of the island and establishing a skirmish line, the two companies moved in on the commanding ground where the guerrillas were well dug-in and concealed. The irregulars were engaged, but "did not appear to be well organized."[24] Those of the enemy who escaped were hunted down by patrols.

This task proved to be painstaking and time-consuming, for the vegetation on the western slopes of Ishikawa Isthmus seriously hampered effective patrolling despite the fact that this section was the least precipitous in the neck of the island. Visibility off the trails frequently was limited to five feet, at most, by dense stands of bamboo and scrub conifer. Since flank security was impractical in this terrain, the war dogs accompanying the Marines proved a valuable asset in alerting their masters to enemy hidden in the undergrowth.[25] Lack of roads and the difficult terrain here raised resupply problems which were solved by the organization of supply pack trains[26] to support 2/7

[22] BGen Edward W. Snedeker ltr to CMC, dtd 10Mar55, hereafter *Snedeker ltr II*.

[23] *7th Mar SAR*, pp. 2–3.

[24] IIIAC G–2 PeriodicRpt No. 15, dtd 16Apr45.

[25] Capt Verle E. Ludwig memo for HistBr, G–3 Div, HQMC, dtd 28Oct54.

[26] Because of the difficulty of resupply under these conditions, a previous Tenth Army ban on the use of captured native horses was lifted.

patrols. Enemy resistance continued here for nearly two weeks, during which time Berger's Marines killed about 110 of the guerrilla force.[27]

As the 6th Marine Division closed in on the main Japanese position in the Motobu heights, the tempo of guerrilla activity on the fringe of the battle increased proportionately. A daily occurrence at dusk was the harassing of artillery positions by irregulars, who caused the registration of night defensive fires to be delayed.[28] When Major Pace's 1/15, in direct support of the 22d Marines, displaced to cover the infantry drive to the northernmost limit of the island, its perimeter was hit almost nightly by sporadic sniping and knee mortar fire. In addition, grenades, demolition charges, and even antipersonnel land mines were thrown into the defensive installations encircling the battalion area. The hills in the rear of the 1/15 position afforded the enemy excellent observation and apparently permitted him to coordinate his attacks on the Marines.[29]

From 14 through 16 April, as the battle for Yae Take was coming to a climax, fires mysteriously broke out in various west coast villages from the southern extremity of Nago Wan to the northern tip of the island. On 17 April at dawn, Nakaoshi was struck by an enemy hit-and-run attack that simultaneously swept over the 6th Engineer Battalion command post (CP), water point, and supply installations nearby. Civilian collaboration with Japanese military forces appeared to be a factor in these incidents, when evidence of native sabotage was uncovered during an investigation of the series of fires on the west coast.

The security threat presented by Okinawan civilians appeared to be pervasive, for it arose within the 1st Marine Division zone also. As early as 9 April, Lieutenant Colonel Miller, the 3/5 commander, reported that many civilians were destroying their passes and appeared to be roaming about freely at night. It was reasonable to assume that they were contacting the Japanese at this time.[30]

For better zonal security control, the 1st Marine Division began rounding up all civilians on 11 April and herding them into stockades built on Katchin Peninsula. The following day, all able-bodied Okinawan males were taken into custody in order to determine their military status. The prevailing tactical situation in the north, at this time, required that organized resistance be broken before Marine control over civilians could be established and combat troops spared for this duty.

From the beginning of the 6th Division drive north, an increasing number of Okinawans was encountered on the roads. Only a few men were of obvious military age and were detained. The others, stopped and questioned, were allowed to continue on with their affairs. At the height of operations in the north, 12–16 April, the division was unable to collect able-bodied males methodically in the manner of the 1st Division in central Okinawa. Civilians

[27] *7th Mar SAR*, p. 3.

[28] *4/15 SAR, Ph I&II*, chap VII, n.p.

[29] *1/15 SAR*, Ph I&II, chap III, n.p.

[30] *5th Mar S–3 Jnl*, 9Apr45.

of doubtful character and background, however, were seized. When hostilities on Motobu ceased, the 6th Division organized a civilian control center at Taira where, beginning 16 April, from 500 to 1,500 natives were interned daily until operations in the north were ended.

On 15 April, Hurst's 3/7 (attached earlier to the 5th Marines) reverted to parent control and began active patrolling from its base at Chuda on the west coast. General del Valle regained the 7th Marines the next day, and, as the 6th Division began meeting increased resistance, the boundary between the Marine divisions was readjusted along the Chuda-Madaira road.[31]

From 17 to 19 April, it appeared that, parallel to the steady reduction of their positions on Yae Take, the Japanese were shifting from a tactical policy of defense to one based on partisan warfare. After the 6th Division took the mountain redoubt, and following a reorganization of Marine units, General Shepherd's command moved to assigned garrison areas. Here it began patrolling vigorously to fix and destroy remaining pockets of enemy resistance. To assist the division in securing northern Okinawa, the Amphibious Reconnaissance Battalion, part of IIIAC since 17 April, was attached with a mission of seizing and occupying the small islands lying off Motobu Peninsula.

In a period of two days, 21–22 April, the battalion reconnoitered the islands of Yagachi and Sesoko with negative results. Though no enemy forces were encountered, the Marines found a leper colony containing some 800 adults and 50 children on Yagachi Shima. Before they landed on Sesoko, the scouts met more than 100 natives moving by canoe from islands to the west in search of food, and "considerable difficulty was involved in coralling and controlling" them.[32] On the 23d, Walker's 6th Reconnaissance Company scouted Kouri Shima and found no enemy.

While the battle for Yae Take raged, and even after it had ended, 6th Division rear area patrols began making contacts with enemy troops attempting to escape from the fighting on Motobu. On 22 April, near Nakaoshi, 1/22 patrols killed 35 enemy in a fire fight. On the next day, this battalion met a strong force, estimated at three rifle squads, three light machine gun squads, and one mortar squad, firmly entrenched in previously prepared positions, including caves and pillboxes, in the mountainous area east of Nago. Two Marine companies assaulted the Japanese killing 52, before an ammunition shortage forced the battalion to break off the action. It returned to the battle scene on 24 April, this time with 4/15 (Lieutenant Colonel Bruce T. Hemphill) in direct support, and the strongpoint was

[31] Lieutenant Colonel Sabol's 3d Battalion, 1st Marines, passed to General Shepherd's control this same day. After moving north, 3/1 was attached to the 22d Marines and ordered to Kawada. From this point, the battalion assisted the 22d Marines in patrolling the vast regimental zone of responsibility, an area which covered 140 square miles and included a 95-mile coast line. On 16 April, 1/22 moved to the vicinity of Awa, where it became 6th Division reserve on Motobu Peninsula. *22d Mar SAR, Ph I&II*, pp. 12, 14.

[32] *PhibReconBn AR*, p. 12.

reduced.[33] The engagement ended towards evening with the deaths of a Japanese officer and two NCOs and the remainder of the group fleeing. The battalion continued patrolling the region on the next day and cleaned out the enemy pocket.[34]

Intensified patrolling of the Ishikawa Isthmus began on 23 April after a small IIIAC military police group was extricated from an ambush by a 7th Marines detachment. The 2d Battalion, 1st Marines, reinforced the 7th, and all available 1st Division war dogs were attached to that regiment. At the same time, stricter travel regulations within the IIIAC area were enforced, and the movement of a single vehicles in the corps zone during hours of darkness was forbidden.

In the 6th Division zone, while the 29th Marines remained on Motobu Peninsula, the 4th Marines moved to its assigned area in the northern part of the island. At Kawada, 3/1 was relieved by Hochmuth's 3/4 and returned to parent control on the 23d. During the next two days, the rest of Colonel Shapley's regiment was disposed with Hayden's 2/4 at Ora, and Beans' 1/4, regimental troops, and the headquarters complement bivouacked in the vicinity of Genka, a small west coast village located about five miles north of the juncture between Motobu Peninsula and the rest of the island. From this point, Colonel Shapley's mission was to seek and exterminate stragglers in the southern half of what had been the 22d Marines area. Upon being relieved, 1/22 prepared to move to the west coast to a point just south of Ichi, which had been the 3/22 patrol base since 16 April.

The mountainous interior of the north was combed continually by Marine patrols for *Udo Force* survivors and semi-independent guerrilla bands. The 6th Division learned from civilians in the area that small groups of Okinawan home guardsmen were in the hills of the northern part of the island and had been preparing to wage partisan warfare for nearly a year. As part of the preparations, they had reportedly established stockpiles of supplies in the interior. The civilians further stated that some of the guardsmen had returned to their homes and civilian pursuits. They also said that home defense units were being trained in the villages by Okinawan veterans who had served previously in China with Japanese forces.[35]

Until the afternoon of 27 April, however, patrol results were negative with the exception of an occasional flushing out of individuals or small groups. At this time, a 3/4 reconnaissance patrol sighted a 200-man enemy column moving through the northeastern corner of the Marine regimental zone toward the east coast. It was believed that these Japanese had survived the Motobu Peninsula fighting by infiltrating in groups of 20 to 40 from the combat area by way of Taira and that they were going to try to join up with the main enemy force in the south.

Steps were taken immediately to destroy the group. Two battalions of the 22d Marines were ordered to the south to block the column, while 3/4 moved

[33] *22d Mar SAR, Ph I&II*, p. 14.

[34] *Ibid.*

[35] IIIAC G-2 PeriodicRpt No. 28, dtd 29Apr45.

inland from Kawada. Further ringing the escape-minded enemy was Donohoo's 3/22, which proceeded toward the interior on a cross-island trail 1,000 yards north of, and parallel to, the 1/22 advance from Hentona. Since it was anticipated that the fugitives would be apprehended in the 22d Marines zone, 3/4 was attached to that regiment, Additionally, two artillery battalions were to support the pursuers.

The first contact was made just prior to noon on 28 April, when one of 3/4's companies engaged the escaping Japanese in a fire fight. At the end of the three-hour contest, 109 enemy soldiers were dead; 1 Marine was killed and 8 wounded.[36] The other pursuing units were unable to reach the scene of the action because of the difficult terrain; 1/22 encountered small scattered groups as it advanced, while 3/22 was still underway when 3/4 radioed that it had destroyed the enemy. Thereupon, Colonel Schneider ordered his 3d Battalion to continue on to the east coast, and Colonel Shapley's 3/4 returned to Kawada and parent control.[37]

Even though the guerrillas in the IIIAC area had forced the Marines to remain constantly on the alert, General Geiger was able to declare the end of organized resistance in the north on 20 April. Continuous patrolling remained the general order, however. As usual in counterguerrilla operations, the number of combat troops employed was out of proportion to the size and number of guerrillas hunted. In most cases, it was a one-sided fight, for a substantial percentage of the partisan ranks were filled with the poorly trained and equipped *Boeitai*. The primary contribution of native Okinawans to the guerrilla effort was a knowledge of the land over which they fought; their offensive efforts were limited mainly to night forays against supply installations, disrupting communications systems and centrals, and attacking water points and hospitals. Although these destructive attempts usually ended in failure, they forced friendly units to maintain extensive security detachments, sometimes in platoon or company strength.

As of the 20th, when Motobu Peninsula was reportedly cleared of enemy troops, the Tenth Army began to pay greater attention to the native population in occupied sections of the island. All civilians, irrespective of age or sex, found in the areas of combat units were to be interned. Furthermore, Okinawans were prohibited from moving about freely unless accompanied by an armed guard.[38] General Geiger established eight internment camps in the IIIAC zone, but the number of collection points in the Marine area was later re-

[36] *3/4 SAR, Ph I&II*, p. 8.

[37] Regarding this small operation, General Smith commented that it "was an excellent example of alert patrolling; [a] rapid decision by the commander; and very effective execution of a difficult approach." Smith, *Personal Narrative*, p. 91.

[38] This order could not be complied with immediately by the 6th Marine Division because of the large numbers of civilians in its zone, and emphasis continued to be directed towards the detention of able-bodied men. Adding to the difficulties in this area was the fact that many of the civilians "were already in the category of displaced individuals, having fled to the north from Naha and other southern areas well in advance of the assault." *IsCom AR*, chap 8, sec XXIV, pp. 2–3.

duced to three; Katchin Peninsula, Chimu, and Taira.

Although tighter security controls prevailed in the corps zone, isolated incidents behind the battleline still occurred. In the last week of April, a 7th Marines patrol killed a Japanese corporal who was wearing a kimono over his uniform. Intelligence agencies found evidence of a Japanese-planned and -sponsored program of espionage and sabotage for the rear areas. In the XXIV Corps zone, the following document was recovered:

> Permit
>
> Army line probational officer Inoye Kuchi and two others: The above mentioned are permitted to wear plain clothes for the purpose of penetrating and raiding enemy territory from April 25, 1945, until the accomplishment of their mission.[39]

TAF OPERATIONS IN APRIL AND THE KAMIKAZE THREAT[40]

Owing to the early and unopposed capture of Yontan and Kadena airfields, Tactical Air Force, Tenth Army, began land-based operations sooner than expected. On 2 April, General Mulcahy and his staff went ashore and selected a CP site midway between the two fields. General Wallace's ADC headquarters was dug in nearby.

While TAF personnel were kept busy constructing camp and repair facilities, Marine engineers and Seabees began repairing the runways on Yontan and Kadena. The airfields were found to be lightly surfaced and badly damaged by naval gunfire and bombings. Hurried grading permitted the use of Yontan by 7 April, but the problems at Kadena were more extensive. Damage here was greater, and the source of coral for surfacing was at some distance from the field. Nevertheless, the strips on Kadena were ready for dry-weather use two days after those on Yontan and, by 1 May, they were all-weather operational.

Three weeks earlier, the ADC Air Defense Control Center (ADCC) had come ashore and, on 7 April, begun operating from three LVTs specially rigged to serve as the defense command CP and to function as both an ADCC and the Air Defense/Fighter Command operations center. On 19 April, the center moved to more spacious quarters in an abandoned farmhouse nearby.

[39] Translated in IIIAC G-2 PeriodicRpt No. 28, dtd 29Apr45.

[40] Unless otherwise noted, the material contained in this section is derived from: CinCPac WarD, Apr45; Fifth Flt AR, 1Apr-17May45, Ryukyus Op, dtd 10Jul45, hereafter *Fifth Flt AR; CTF 51 AR; Tenth Army AR;* 2d MAW WarD, Apr45; *TAF AR;* TAF Periodic Rpts, Apr45; ADC (MAG-43) WarD, Apr45; *ADC IntelSums,* Apr45; G-2 Sec, ADC, TAF, Historical Summary of the Organization and Activities of the ADC, Okinawa, n.d., hereafter *ADC Hist;* MAG-31 WarD, Apr45; MAG-33 WarD, Apr45; MilAnalysisDiv, USSBS, *Air Campaigns of the Pacific War* (GPO, Jul47), hereafter USSBS, *Air Campaigns;* MilAnalysisDiv, USSBS, *Japanese Air Power* (GPO, Jul46), hereafter USSBS, *Japanese Air Power;* Craven and Cate, *The Pacific;* Hayashi and Coox, *Kōgun;* Capt Rikihei Inoguchi and Cdr Tadashi Nakajima, former IJN, with Roger Pineau, *The Divine Wind: Japan's Kamikaze Force in World War II* (Annapolis: USNI, 1958), hereafter Inoguchi, Nakajima, and Pineau, *Divine Wind;* Morison, *Victory in the Pacific;* Masatake Okumiya and Jiro Horikoshi, with Martin Caidin, *Zero!* (New York: E. P. Dutton & Co., Inc., 1956), hereafter Okumiya, Horikoshi, and Caidin, *Zero!;* Sherrod, *Marine Air Hist.*

When General Wallace opened his CP on Okinawa, the air defense commander became the land-based agent of CASCU, which continued operating on board Admiral Turner's flagship. Under ADC operational control were land-based aircraft, radar air warning and control installations, and antiaircraft artillery units. It was the air defense commander's primary mission to coordinate the combined efforts of these three disparate support activities so that they meshed with the operations of the overall air defense system of the expeditionary force. ICEBERG plans had stipulated that TAF would assume full responsibility for the air defense of Okinawa when the amphibious landings were completed, but, because of "the all-out efforts of Japanese aircraft and the success of their *kamikaze* suicide attacks directed against naval units, operational control of aircraft in the Ryukyus remained with the Navy until the area was secured." [41]

General Wallace believed that the major tactical task of ADC was to meet the *Kamikaze* threat. From 7 April, when VMF–311 pilots scored the first TAF kill [42] of a suicider as they flew in to Yontan from their CVE lift, ADC efforts were directed toward confronting and stopping the destructive enemy air attacks. The fighter squadrons of MAG–31 and –33 mounted combat air patrols from Yontan and Kadena fields on the first days that they arrived at these bases.

As the battle was joined on Okinawa by the Tenth Army and General Ushijima's forces, the American fleet in surrounding waters was engaged in a desperate battle of its own. The Japanese air attacks on the Kerama Retto invasion group merely heralded even greater enemy attempts to destroy the radar pickets and support vessels safeguarding the troops on Okinawa. Many of these enemy aircraft were on either conventional bombing or reconnaissance missions; others in the aerial attacks were part of the *Special Attack Force*, the *Kamikazes*.

As the success of American operations in the Philippines became apparent and MacArthur's air strength reigned supreme, enemy naval air commanders saw that there was no prospect of any advantage to be gained in the sky while Japanese squadrons continued employing orthodox tactics. The *Kamikaze* effort evolved as a result of these considerations. Appearing first in the Philippines,[43] this was an organized and

[41] *Tenth Army AR*, chap 11, sec VII, p. 5.

[42] This event marked another "first" also, for it was the first time that Marine pilots had shot down an enemy plane with the package of four 20mm cannon newly installed in Corsairs in lieu of the six .50 caliber machine guns heretofore mounted in the planes. "All of the pilots expressed enthusiasm over the great destructive power of the new and heavier armament." IntelSec, DivAvn, HQMC (OpNav-37), *Marine Air Intel Bul*, May45, p. 2, hereafter *AirIntelBul*, with date.

[43] The authors of the official AAF history suggest that the first *Kamikaze* attack took place in April 1944 in the area of the Andaman Islands, when a Japanese pilot dived his plane into an American torpedo aimed at shipping in an enemy convoy. In official recognition of this act, the hapless aviator was posthumously decorated by Field Marshal Count Hisaichi Terauchi, the commander in chief of the *Southern Area Army*. Craven and Cate, *The Pacific*, p. 352. The point here is that the *Kamikaze* attacks in the Philippines were the first so organized of their type.

desperate attempt by suicide-bent Japanese naval aviators to deprive American shipping at Leyte of aerial protection by crashing the flattops of the covering carrier force. The enemy anticipated that the success of their tactics would then guarantee a Japanese surface victory in the event of an all-out engagement with United States naval forces. Although Japanese commanders felt that suicide missions were a "temporary expedient" only, used "because we were incapable of combatting you by other means . . .,"[44] initial success gave added impetus to their fuller employment.

Correctly anticipating that the next invasion attempts would be at Iwo Jima and, after that, Okinawa, *Imperial General Headquarters* withdrew the remnants of some Army and Navy air units from the Philippines in early January 1945 to strengthen the defense of the Home Islands and the Ryukyus. Upon completion of this transfer, designed to "produce a more unified [defense] strategy,"[45] brigades and regiments of the *Sixth Air Army* and naval squadrons of the *Fifth Air Fleet* were combined into a single tactical command on 19 March under Admiral Soemu Toyoda, Commander in Chief, *Combined Fleet*.[46] At the outset, it was determined that operations of this combined force of about 1,815 planes[47] were to be well planned and organized—a definite contrast to the sporadic, albeit somewhat successful, *Kamikaze* attacks at Leyte.

One of the first opportunities for the Japanese to mount coordinated suicide and conventional air attacks occurred during the TF 58 raids of 18–19 March 1945 on Japan. Although the carriers were damaged and there were some American casualties, the enemy lost 161 aircraft. Most of this damage, strangely enough, was not caused by *Kamikazes*. An important result of this raid was the destruction, while still on the ground on Kyushu, of many of the Japanese planes scheduled to be employed in the defense of the Ryukyus. This disaster forced the *Fifth Air Fleet* to reevaluate its plans. Moreover, a Tenth Army landing relatively unharassed by enemy air raids was guaranteed, for Toyoda's squadrons were unable to mount a major air offensive until after the beginning of April.[48]

Scattered conventional and *Kamikaze* flights from Japan and Formosa carried the attack to the Western Islands Attack Group of the ICEBERG force first; later these planes began swarming all

[44] LtGen Torashiro Kawabe, *IJA*, quoted in USSBS, *Japanese Air Power*, p. 60.

[45] LtGen Michio Sugawara, *IJA*, quoted in *Ibid.*, p. 66.

[46] MilHistSec, Japanese Research Div., HQ-USAFFE, Japanese Monograph No. 123, Homeland DefNavOps, pt II, Mar43-Aug45, dtd Jun49, p. 73, hereafter *Homeland DefNavOps*.

[47] RAdm Toshiyuki Yokoi, IJN, "Kamikazes and the Okinawa Campaign," *USNI Proceedings*, v. 80, no. 5 (May54), p. 508, hereafter Yokoi, *Kamikazes*.

[48] HistSec, G–2, GHQ, FEC, Japanese Studies in the World War II, Monograph No. 86, History of the Fifth Air Fleet, Operational Record 10Feb-19Aug45, n.d., p. 35 (OCMH, DA), hereafter *Fifth AirFlt Hist*.

over the transports and picket line off Okinawa. During the first few days of April, the toll of ships damaged and sunk grew at a steady rate while naval casualties mounted in consequence. By 6 April, Admiral Toyoda was prepared to launch from Kyushu the first of ten carefully planned *Kamikaze* attacks, which were to be flown over a period ending 22 June. A total of 1,465 sorties emanated from Kyushu to sink 26 American ships and damage 164 others.[49] Not included in these loss figures are the victims of small-scale *Kamikaze* efforts by another 250 planes which rose from Formosa air bases, and the 185 additional sorties flown from Kyushu, independent of the mass attacks.[50]

The Japanese decision to turn to large-scale air operations was arrived at after Toyoda had studied both his and the *Thirty-second Army* situations and had found that "it would be futile to turn the tide of battle with present tactics." [51] He therefore dispatched the first and largest coordinated suicide attack—*Kikusui* Operation No. 1—[52] against ICEBERG forces on 6 April.

[49] USSBS, *Campaigns*, p. 328.

[50] USSBS, *Japanese Air Power*, p. 23.

[51] *Fifth AirFlt Hist*, p. 41.

[52] The attacks "were given the cover name Kikusui . . ." which ". . . literally means 'chrysanthemum water' and the characters making up the word were used in the crest of a 15th Century Japanese hero who took the side of the Emperor in a prolonged civil war against heavy odds." USSBS, *Japanese Air Power*, p. 66.

Spearheading the *Kamikazes* were 14 planes sent to bomb and strafe Okinawa airfields before dawn in order to destroy Allied aircraft suspected of being there. Apart from their nuisance value, the raids did little damage to the runways and none to TAF planes, for the squadrons had not yet flown ashore. Following the first group of enemy hecklers were more than 100 fighters and bombers sent to engage TF 58 off Amami-O-Shima in order to draw American carrier-based planes away from the suiciders heading for Okinawa.

For a 36-hour period, 6–7 April, the Japanese flew 355 suicide sorties, which were accompanied by nearly an equal amount of conventional cover, reconnaissance, and bombing planes. As these aircraft bore in to crash, torpedo, and bomb the ships at anchor in Hagushi transport area, crewmen in exposed positions and troops on the beaches were subjected to a deadly rain of antiaircraft artillery shell fragments. Friendly fighters were not immune from the effect of the hundreds of guns firing from the beaches and ships; three American pilots were shot down when they followed Japanese planes too closely into the murderous barrage.

The main attack, which began about 1500 on 6 April, spread out all over the combat zone with the outer ring of radar pickets and patrol craft—lacking a protective smoke-screen cover—catching the full fury of the battle. Ships of all types, however, were fair game for the *Kamikazes.*[53] Before the Okinawa landing, the Japanese confined the direction of their suicide attack efforts to American carrier task forces. After 1 April, the attacks were mounted against convoys, and, just prior to the first *Kikusui*, the enemy began hitting all surface forces. After the time of the 6–7 April attack, the Japanese reserved the carrier forces for *Kamikaze* attention while their conventional bombers and fighter craft were directed to hit other American vessels and transports around Okinawa.[54] In this first mass suicide attack, Admiral Turner's forces claimed to have shot down at least 135 Japanese planes, while the pilots from the Fast Carrier Task Force reported splashing approximately 245 more, bringing the total American claims of enemy losses to nearly 400 pilots and planes. Contemporary Japanese sources place the losses in *Kikusui* No. 1 at 335.[55]

As a sidelight to the air battle over and the land fighting on Okinawa, the Japanese mounted their only real surface threat to the success of the American invasion. Intending to attack Allied shipping at Okinawa, the 69,100-ton battleship *Yamato* and a covering group steamed out of the Tokuyama Naval Base, on Honshu, at 1500 on 6 April. Less than two hours later, the enemy vessels were sighted by two U. S. submarines in the screen lying off the east coast of Kyushu. Within 24 hours, TF 58 pilots had administered death blows to the *Yamato* and a part of her group, and had forced the remainder to scurry home.

Since TAF pilots had not yet begun operations from Okinawa when *Kikusui* No. 1 struck, the four Marine squadrons on board the carriers *Bennington* (VMF–112 and –123) and *Bunker Hill* (VMF–221 and –451) carried the ball for Marine aviation during the time that General Mulcahy's planes and pilots were still on board their carrier transports. Until late in April, as much as 60 percent of the ground support missions flown for Tenth Army units were carried out by Navy and Marine carrier pilots, while the primary concern of TAF flyers was to blunt the

[53] In the Kerama anchorage, an LST and two merchant cargo ships, the *Logan Victory* and the *Hobbs Victory*, were hit and eventually sank. The sinking of these two Victory ships was a serious matter and could have affected the course of the campaign critically. Because they had been loaded on the west coast with most of the 81mm mortar ammunition then available in the United States, these sinkings created a shortage of this type of ammunition in subsequent resupply shipments. In order to make up for the loss, at least partially, Army and Navy transport planes airlifted 117 tons of 81mm ammunition to Okinawa before the end of the campaign. To satisfy immediate needs, LSTs were loaded and dispatched to the target with stock taken from Marianas and South Pacific supply reserves. *Blakelock ltr; Tenth Army AR*, chap 11, sec IV, p. 12.

[54] *Homeland DefNavOps*, p. 113.

[55] Inoguchi, Nakajima, and Pineau, *Divine Wind*, pp. 225–226; Hattori, *War History*, v. IV, (table facing p. 132). Despite the extensive American and Japanese material available relating to their respective air losses in World War II, it is still difficult to reconcile this numbers conflict. For a discussion of this matter, see Shaw and Kane, *Isolation of Rabaul*, pt V, "Marine Air Against Rabaul," *passim*.

Kamikaze menace. To at least one TAF air group commander, "it seemed strange for planes off the carriers to come in for close-support missions, passing [Okinawa-based] Marine pilots flying out for CAP duty. . . ." [56]

Almost as soon as Colonel Munn's MAG–31 squadrons touched down at Yontan, a 12-plane combat air patrol was organized and launched to remain airborne until dark. Prior to the time that TAF joined the fighting, CAPs had been flown by planes from both the Support Carrier Group and TF 58. Originally, a large CAP, varying from 48 planes in relatively quiet periods to 120 or more during critical times, was flown to protect the surface forces from air attacks. Basically, the aircraft were deployed "in a circle in depth" over the invasion and picket craft.[57]

Generally, TAF planes were airborne from dawn to dusk on CAP flights, and they flew special early morning and twilight CAPs as well. On 14 April, the commander of the ICEBERG operation transferred the responsibility for flying night CAPs from TF 58 to TAF. In addition, TAF was to maintain another four planes constantly on patrol during the hours of darkness. This last mission was assigned alternately to the night fighters of VMF(N)–542 and –543 commanded by Majors William C. Kellum and Clair "C" Chamberlain, respectively. In order to guard the radar picket ships—special objects of the *Kamikaze* attacks—General Wallace's fighter command was ordered on 14 April to maintain a continuous two-plane daylight CAP over each of the three picket ships that were stationed offshore northeast of TAF airfields. Each flight leader was to report directly to the captain of the ship he was guarding. In turn, the naval officer would control the flight and ensure that its planes were kept out of range of the ship's antiaircraft guns. Two days after this mission was first initiated, the number of ships protected by this CAP was increased by two.

By the time that TAF had been established ashore, the three Marine Landing Force Air Support Control Units, commanded overall by Colonel Vernon E. Megee, had landed also. Although they were shore-based representatives of CASCU and outside of the TAF chain of command, by the very nature of their functions the LFASCUs worked closely with the Marine aviation units. Once air support operations began, coordinating agencies relayed all orders concerning aircraft missions directly to General Mulcahy's command in a smoothly functioning system. At Tenth Army headquarters, Megee's LFASCU–3 screened all requests for air support received from LFASCU–1 (Colonel Kenneth H. Weir) and –2 (Colonel Kenneth D. Kerby) which were working with IIIAC and XXIV Corps respectively. If a review of TAF and carrier aircraft commitments indicated that an air support request was consistent with priority requirements, the mission was approved. At that time, if Marine planes were assigned, LFASCU–3 relayed the

[56] A fuller story of the operations of Marine carrier squadrons is found in pt III, chap 2, *infra.*, and Sherrod *Marine Air Hist*, pp. 357–368.

[57] *CNO Record*, chap 1, p. 47.

order for the mission directly to the TAF operations section.[58]

Frontline control of the ground support missions flown by both land- and carrier-based aircraft was provided by Air Liaison Parties (ALPs) from the Joint Assault Signal Companies attached to each division. Ground unit requests for air support were reviewed first with respect to the capabilities and availability of the other supporting arms to fulfill a specific mission, and then passed on to the LFASCU at corps headquarters. If the request was approved here, the LFASCU would requisition the necessary number and types of planes, and stipulate the armament they needed for successful completion of the mission. In addition, the LFASCU provided strike direction and supervised the scheduling of all air support in unit fire support plans.[59]

Not all close air support missions were ground controlled in this campaign. Employed at Okinawa was an air coordinator, or airborne traffic director, who spotted and marked the ground target for the planes flying the mission. The coordinator would direct the flight to the best target heading, observe attack results, and correct subsequent runs if he decided that they were needed.[60] At times when smoke and weather conditions over the target denied the airborne controller suitable visibility, the support mission would be run nevertheless, but directed by the ALP.

It took time to establish land-based radar reporting, control, and homing stations on Okinawa and the outlying islands. In addition to the problems involved in getting the Air Warning Squadrons (AWSs) and their equipment ashore rapidly, initially it proved difficult to net the ground-to-ground communications systems with the overall ship-to-shore warning system. Prior to the establishment of the ADCC, the individual radar stations had reported directly to CASCU aboard the *Eldorado*. After 8 April, the day on which the control center first began to provide shore-based operational homing facilities, AWS early warning teams began reporting directly to the ADCC which, in turn, passed on to Navy control the reported enemy and friendly plots.

Early warning teams were also assigned temporarily to each assault division and corps headquarters. They then operated in coordination with the AAA units already assigned to the defense of corps and division sectors. Here, the teams monitored ships' radar telling circuits and local air warning and inter-fighter director nets, from which air raid warning information was obtained and passed on to the ground units. In addition to radar coverage, the air warning squadrons provided radio monitoring services, the results of which figured prominently in and assisted the operation of the Air Defense Command.

The AWSs also worked very closely with and were, in fact, supervised by

[58] *TAF AR*, chap 6, sec III, p. 1.

[59] IIIAC Rpt, Air Support, encl H to CG, AirFMFPac ltr to CMC, Subj: Air Support, dtd 28Aug45.

[60] *ADC Hist*, p. 16.

the senior Marine AAA officer, Colonel Kenneth W. Benner, commanding the 1st Provisional Antiaircraft Group. He was responsible for coordinating the disposition and operation of his organic radar with that of the AWSs in order to ensure maximum surface and low-angle electronic surveillance for defense against enemy air attacks.[61]

Because theirs was a vital role in the overall air defense of the ICEBERG forces, land-based AAA units, although attached to the assault corps, were directly under the operational control of General Wallace's ADCC. On 20 April, the antiaircraft units reverted to the Tenth Army which then assigned the 53d Antiaircraft Artillery Brigade the mission of coordinating all AAA activities. At the same time, the brigade became the TAF agency for providing the ground forces with early air raid warning services while continuing to fulfill its AAA defense mission.

Initially, the 1st Provisional Antiaircraft Artillery Group was assigned to support IIIAC during Phases I and II of the operation by providing AAA defense for corps units, installations, and beaches, and the captured airfields in the corps zone.[62] Additionally, the group was to provide anti-boat defense of corps beaches, supplement field artillery units in both direct and general support mission, and be prepared to fire seacoast artillery missions.[63]

The assault elements of the group were the 2d (Lieutenant Colonel Max C. Chapman) and 16th (Lieutenant Colonel August F. Penzoll, Jr.) Antiaircraft Artillery Battalions, which supported the 6th and 1st Marine Divisions respectively. Scheduled to land on order at later dates, the 5th (Lieutenant Colonel Harry O. Smith, Jr.) and 8th (Lieutenant Colonel James S. O'Halloran) AAA Battalions were to reinforce the group and extend antiaircraft defenses already existing.

Because of the rapid progress of the infantry and the assignment of higher priority to items needed ashore immediately, the landing of the Marine AAA battalions was delayed. Group and battalion reconnaissance parties landed on L plus 2 to select sites, and beginning on 5 April, the units themselves were given an unloading priority. By 12 April, the battalions were in position ashore.

Initially, one heavy and two light AAA batteries of the 2d Battalion were assigned a defense sector on 6th Marine Division beaches; the 16th Battalion supported the 1st Division with two heavy and two light batteries. The remaining five 90mm gun batteries of the group defended Yontan airfield.[64]

When the 53d Antiaircraft Artillery Brigade assumed control of Tenth Army AAA units, it found that the defenses in the IIIAC and XXIV Corps zones were unbalanced. On 27 April, the brigade adjusted the dispositions and,

---

[61] *Tenth Army TntvOplan 1-45*, anx 9, p. 4.

[62] 1st Prov AAA Gru Gen AR, Okinawa Shima Op, Ph I&II, 1-22Apr45, dtd 18May45, p. 1, hereafter *1st Prov AAA Gru AR.*

[63] *Ibid.*

[64] LtCol Peter J. Speckman ltr to CG, FMFPac, Subj: AAA Observer's Rpt on Okinawa Shima Op, n.d., encl A to CG, FMFPac ltr to DistrList, dtd 28Jun45, p. 4, hereafter *Speckman rpt.*

in addition, extended AAA defenses across the island to the east coast in order to break up enemy raids coming from that direction. The 1st Provisional Group continued the Yontan area defense, but was made responsible for defense of the entire Yontan-Kadena sector also. To aid in this last mission, Colonel Benner's group was augmented by two Army AAA-Automatic Weapons battalions.[65]

Lieutenant Colonel O'Halloran's 8th Antiaircraft Battalion landed at Nago Wan on 17 April to defend IIIAC units and supply dumps in that area. Its most immediate problem, in view of the tactical situation on the Motobu Peninsula, was achieving ground security.[66] For that reason, battalion .50-caliber heavy machine guns and some .30-caliber light machine guns obtained locally were assigned a primary mission of ground defense.

Enemy air attacks on shore installations were directed at Yontan and Kadena airfields mainly, and usually took place at night. The only firing opportunities afforded shore-based antiaircraft artillery during daylight occurred when Japanese aircraft, with the obvious intention of attacking the transport area, made their approaches from the landward side of the anchorage.[67] Usually, the illumination or visual sighting of an enemy plane, and sometimes even an American one, was the signal for a wave of wild uncontrolled firing both from shipboard and the island. "Carbines, rifles, and even .45 caliber pistols enthusiastically joined the fun on occasion." [68] Under these conditions, casualties and materiel damage resulted from falling shell fragments and wild shots until the Tenth Army insisted upon the enforcement of greater fire discipline by all unit commanders.

From the beginning of ADCC operations, there was no satisfactory communications and control system linking the fighter command and the antiaircraft artillery units. When ADC was heavily engaged with enemy air attacks, liaison with AAA units weakened or broke down completely. At times, permission for the guns to fire on unidentified or enemy planes was withheld, even when the area was definitely under attack. On several occasions, air raid warning flashes were not relayed to the AAA command until after the infantry and shore party units had been informed.[69] Of necessity, an efficient control system was soon initiated. By the end of the month, Marine AAA units were credited with the destruction of 15 planes and 8 assists, 5 probably destroyed, and 6 damaged.[70]

---

[65] It should be noted that Marine AAA battalions were a composite type of unit, containing both heavy and light AA weapons. The Army groups consisted of heavy and light AAA battalions, the former containing only the 90mm gun batteries and the latter being comprised entirely of automatic weapons units.

[66] 8th AAA Bn AR, dtd 22Apr45, encl E to *1st Prov AAA Gru AR*, p. 6, hereafter *8th AAA Bn AR*.

[67] *Speckman rpt*, p. 1.

[68] *1st Prov AAA Gru AR*, p. 14.

[69] 53d AA Brig After-Action Rpt, dtd 21Jun45, p. 48, hereafter *53d AA Brig AAR*.

[70] 1st Prov AAA Gru AAA ARs for 6-16Apr45, dtd 23Apr45, and 17-30Apr45, dtd 10May45.

It was noted that, during enemy air attacks, Japanese aviators showed an increasing knowledge of radar evasion measures, and frequently used "window" [71] in both conventional and *Kamikaze* attacks. In commenting on the enemy failure to mount air attacks on the ground forces, one observer stated that "it was difficult to understand why they had not resorted to formation bombing from low altitudes," [72] since low-angle radar detection of approaching aircraft was almost impossible. He concluded that the concentration on suicide attacks was too great; he might have added that the Japanese just did not have enough planes by this time to divert their air strength to missions other than the *Kamikaze* attacks.

While TAF fighter pilots added to the expanding bag of downed enemy planes, other types of air missions in support of the Tenth Army were performed at the same time by General Mulcahy's command. Upon its arrival, the Army Air Forces' 28th Photo Reconnaissance Squadron rephotographed the entire Okinawa Gunto area to obtain more accurate and complete coverage than had been available for the maps used on L-Day. The squadron also provided infantry commanders with enlarged aerial photographs of masked terrain features to their zones. As soon as Major Allan L. Feldmeier's VMTB-232 arrived on 22 April, it was given tasks other than its original mission of antisubmarine warfare.[73] During the remainder of the month, the squadron flew numerous artillery observation missions daily, bombed and strafed enemy lines and installations in southern Okinawa, and conducted heckling raids in the same areas almost nightly.

The second mass *Kamikaze* attack took place during 12–13 April. Although as frenzied and almost as destructive as the first attack, it was mounted by only 392 planes, on both conventional and suicide missions,[74] as opposed to the 699 total in the first attack. As in *Kikusui* No. 1, TF 58 pilots downed most of the enemy, but carrier-based Marine flyers were active also. Flying Leathernecks from the *Bennington* shot down 26, and *Bunker Hill* Marines downed 25. Okinawa-based TAF pilots accounted for 16 more.

During the interval between the first and second mass raids, the Japanese command had recognized the threat presented to their air attacks by American land-based aircraft, so Kadena was bombed early on the 15th by planes that preceded the *Kamikazes.* TAF personnel and airplanes were endangered further when both of the fields occupied by Tenth Army squadrons were fired upon by an artillery piece, or pieces, nicknamed "Kadena Pete" in not-too-fond memory of

[71] "Window" is the name given to strips of cut metal foil, wire, or bars usually dropped from aircraft as a radar counter-measure.

[72] *Speckman rpt,* p. 1.

[73] On their arrival, the squadron pilots were pleased to discover that their primary mission had been cancelled since their planes were not equipped with sound-ranging equipment and they themselves had not been trained in its use. This mission was then given to Kerama-based seaplanes. VMTB-232 WarD, Apr45; *CTF 51 AR,* pt V, sec E, p. 5.

[74] Hattori, *loc. cit.*

" 'Pistol Pete' at Henderson Field on Guadalcanal." [75]

TAF reports evaluating *Kikusui* No. 2 noted that the evasive tactics employed by the enemy "do not tend to indicate that the flyers were top-flight fighter pilots," and that "a definite lack of aggressiveness" seemed "to confirm the belief that the pilots were green." [76]

A third mass raid of 498 aircraft (196 suiciders) occurred 15–16 April. As the furious air battle carried over into the second day, TAF planes began to score heavily. The largest bag made by land-based aircraft to that date was accomplished by VMF–441 (Major Robert O. White) pilots, who had shot down 17 of the 270 Japanese pilots and planes allegedly splashed on these two days.[77]

In this attack, a TAF pilot made the first sighting of the so-called "Baka" bomb [78] in its maiden appearance over Okinawa. This small, single-engined, wooden craft, powered by rockets, carried a one-man crew and over a ton and a half of explosives. Carried by a twin-engined bomber to a point near the target, the *Baka* was released when its pilot had verified the weapon's target and position, oriented his own position, and started the rocket motors. Although the destructive powers of the *Baka* were real, its employment was erratic and it appeared too late in the war to be influential.

TAF operations for the rest of April tended to fall into a routine of CAPs and support missions. On 22 April, the dusk air patrol was vectored to a point over part of the radar picket line then being attacked by enemy aircraft. When the half-hour battle had ended, Marine pilots claimed 33¾ Japanese planes. Five days later, during the fourth mass *Kamikaze* attack (27–28 April), 115 suicide-bent Japanese pilots were launched against friendly shipping and the steadfast radar pickets. On the second day of the attack, at about 1600, the airborne TAF CAP and an additional 36 Corsairs were vectored out 40 miles northwest of Okinawa to intercept an approaching *Kamikaze* formation. After dark, when the two-hour fight was over, the Marine fighters were credited with downing 35½ enemy planes. Upon

---

[75] *Henderson ltr.* General Henderson continues with a narration of the steps taken to discover and neutralize the artillery menace, which turned out to be not one but six 15cm guns cleverly hidden from aerial observation. "As I recall our attack, we first pounded all FA and AAA positions heavily with massed 155mm fire to immobilize the guns, so they couldn't be moved that night and also [to] cause maximum casualties and incidental damage. Then we went to work with the 8-inch howitzers and 155mm guns to get positive destruction of each piece. When darkness came, we put heavy interdiction fire on the positions to keep [the guns from being moved] during the night. The next morning at first light we found them still there and went to work polishing them off. Kadena Pete didn't bother us any more and the flyers and brass were all happy."

[76] G–2 Sec, ADC, Analysis of Ops of Okinawa Based Fighter Aircraft, 1-30Apr45, n.d., hereafter *ADC AcftOpAnalysis* with month.

[77] Hattori, *op. cit.*, notes 182 planes were downed.

[78] This was a derisive term, meaning "foolish," given the weapon by Americans. The Japanese named it *Ohka* (cherry blossom), and put it into full production late in 1944. Inoguchi, Nakajima, and Pineau, *Divine Wind*, pp. 140, 141. See also USSBS, *Japanese Air Weapons And Tactics* (GPO, Jan47), pp. 20–22.

being congratulated by the Tenth Army commander on the accomplishments of his pilots, General Mulcahy sent a message to the ADC: "Not only brilliant work by fighter pilots but excellent command control and most efficient reservicing by ground personnel were admiration of and inspiration to all." [79] By the end of April, TAF pilots had flown 3,521 CAP sorties and shot down or assisted in the downing of 143¾ enemy aircraft.

[79] MAG–31 WarD, Apr45.

The Defense Stiffens

APPROACH TO SHURI [1]

In the days immediately following the facing movement of the XXIV Corps and the beginning of its drive to the south, increasingly stiff and bitter resistance gave proof that the prepared enemy defenses were being uncovered. The nature of the contacts with the Japanese also heralded the end of the relatively easy and fast-moving XXIV Corps advance. By the morning of 6 April, it was evident that the Japanese "lines were drawn for a full-scale battle." [2] (See Map III, Map Section.)

What the 7th and 96th Divisions had encountered was a strong enemy position that extended the width of the island and roughly followed the line through Machinato, Kakazu, Kaniku, Minami-Uebaru, and Tsuwa. With flanks anchored on the East China Sea and the Pacific Ocean, the Japanese barrier was the outermost of a series of defense rings centering about Shuri, headquarters of the *Thirty-second Army*. The veteran troops of the *62d Division* were entrenched in this outpost sector, which was composed of well-prepared positions on high ground that was liberally studded with machine guns and mortars, and surrounded by barbed wire, antitank ditches, and minefields. Unknown to the Americans, the enemy was prepared to fight a "prolonged holding action" here.[3]

Limited gains through highly developed defenses in the Nakama-Kakazu-Ouki area were made on 7 April. In the 96th Division zone, Army troops advanced over broken ground and wooded ridges to reach the approaches of Kakazu. By 1600, after a furious struggle, one infantry battalion—supported by three air strikes, four artillery battalions, and the 14-inch rifles of the *New York*—managed to penetrate to a point within 500 yards of the northern limits of Kakazu.

To break through the increased resistance, General Hodge had concluded that additional artillery support was essential, and on 5 April he had requested that Tenth Army give him whatever battalions were available. Because III Amphibious Corps Artillery could not be employed with maximum effect in the north, General Buckner ordered most of the 155mm units of IIIAC Artillery to be attached to XXIV Corps. On L plus 6 and 7, the 8th and 9th 155mm Gun Battalions and the 1st, 3d, and 6th 155mm Howitzer Battalions were detached from IIIAC and displaced south to support the attack there.

The howitzer battalions were assigned to the 419th Field Artillery Group and paired off with Army artillery battalions to form three firing

[1] Unless otherwise noted, the material in this section is derived from: *Tenth Army AR; XXIV Corps AR; XXIV Corps Arty AR; 7th InfDiv AR; 27th InfDiv OpRpt; 96th InfDiv AR.*

[2] *Okinawa Operations Record*, p. 72.

[3] *Ibid.*

groupments, which were controlled by the Army battalion commanders. The Marine gun battalions, a IIIAC Artillery headquarters detachment, and the 749th Field Artillery Battalion (8-inch howitzers) were formed into a provisional group, named The Henderson Group after its commander, Lieutenant Colonel Frederick P. Henderson, IIIAC Artillery Operations Officer.[4] Brigadier General Josef R. Sheetz' XXIV Corps Artillery could now support the drive against the Shuri defenses with four 155mm gun battalions, one 8-inch and six 155mm howitzer battalions, and two 155mm gun battalions from the 420th Field Artillery Group.

During the night of 7–8 April, XXIV Corps units repulsed minor enemy infiltration attempts. The Japanese had planned that their first major counterattack against Tenth Army troops would coincide with the *Kikusui* attack on 6 April, but when aerial reconnaissance reported the presence of a more lucrative target for aircraft, a large American convoy steaming south of Okinawa, the *Kamikaze* and ground attacks were rescheduled for the night of 8 April. This attack was cancelled indefinitely when another large convoy was spotted off the west coast of Okinawa just prior to the jumpoff. Because the situation was not favorable in either case, cooler heads amongst the *Thirty-second Army* staff prevailed and were able to stave off the launching of an unsupported Army counterattack. It was only a question of time, however, before the advocates of an all-out offensive would have their day.[5]

When it could no longer be employed profitably in the north, the 11th Marines was also sent south to provide additional Marine artillery in answer to Hodge's request of 5 April. The three 105mm howitzer battalions of Colonel Brown's regiment displaced southward on 9 April to reinforce the direct support battalions of the 7th and 96th Divisions. This reinforcement was in addition to the IIIAC artillery dispatched earlier. The 2d and 3d Battalions of the 11th Marines were attached to the 96th Division, and the 4th assigned to the 7th Division. On 12 April, the remaining battalion of the 11th Marines, 1/11, (75mm pack howitzers) was also attached to the 96th Division.[6]

Moving south at the same time as the Marine artillery were elements of the 27th Infantry Division (less RCT 105), which had landed at noon that day.

[4] *IIIAC Arty AR*, pp. 20–21.

[5] *Okinawa Operations Record*, pp. 70–76.

[6] *11th Mar SAR*, n.p. The 11th Marines commander noted later that when his 105mm battalions joined the Army divisions: "The two Division Artillery Commanding Generals . . . did not want the 75mm Pack Howitzer Battalion. The Army had very little respect for 75s then. . . ." Shortly after the 11th displaced to the south, Colonel Brown "again asked Brigadier General [Robert G.] Gard of the 96th Division Artillery to use 1/11. Then, as happened throughout the campaign, there was a surplus of 75mm ammunition, and Brigadier General Gard was finally convinced that the Battalion would be valuable for harassing and interdiction fires, thus saving on his scarce 105mm ammunition." After 1/11 finally went south, Colonel Brown joined the staff of General Sheetz as liaison officer and remained there until 20 April, when his battalions reverted to his command. MajGen Wilburt S. Brown ltr to CMC, dtd 26Feb55, hereafter *Brown ltr II*.

Released from Tenth Army reserve, the division moved to a bivouac area just east of Kadena airfield, where it awaited a combat assignment.

Heavy enemy opposition and torrential rains driven by strong winds hampered the efforts of XXIV Corps when it resumed the attack on the 10th. On the corps front overall, the 7th Division was able to advance approximately 400 yards in its zone, but an antitank ditch and a minefield near Ouki, and mutually supporting caves and pillboxes on the right flank, seriously limited the division attack.

The 96th Division, which had begun the battle for Kakazu Ridge on the previous day, continued its attack against this key feature in the enemy's Shuri defense system. On 9 April, the division had attempted to take the position with two battalions in a predawn surprise attack. Frequent Japanese counterattacks and withering fire caused heavy casualties and forced the soldiers to relinquish their gains at 1630 and withdraw to positions from which the attack was launched. Nothing was left to chance on L plus 9, as all three regiments attacked after an intense artillery and naval gunfire bombardment lasting 30 minutes was placed on previously located positions. Air cover was not available because of the continuing bad weather, which turned the ground into a quagmire and bogged down the tanks scheduled to support the advancing infantry. When the day's fighting ended, the division had made an average gain of 300 yards along the entire front.

As night fell and the fighting died down all along the XXIV Corps lines, the *Thirty-second Army* issued orders for a counterattack to be mounted on 12 April. Encouraged by the overly optimistic reports of the success of *Kikusui* No. 1 during the 6–7 April raids, the Japanese planned the counterattack to coincide with the second mass *Kamikaze* raid. Although the suicide flights began to pour into the skies above Okinawa at 1300 on 12 April, it was not until more than nine hours later that the ground effort was launched. The mission of the enemy assault units was to inflict as much damage as possible in rear areas, where their close proximity to Tenth Army troops would protect them from the devastating fire of American naval guns and artillery.

Under the cover of a mortar barrage, Japanese troops attempted the penetration of American lines. They were thrown back as artillery and small arms fire caught them fixed in the light of star shells thrown up by gunfire support ships. The enemy made several more attempts, but XXIV Corps units repulsed each one. On the night of 13–14 April, two lesser attacks occurred, but these also were driven off. In the two days' action, XXIV Corps reported 1,584 Japanese troops killed and four captured.[7]

A partial explanation for the failure of the counterattacks is found in the strength of American reaction to them. In addition, Japanese sources offer another approach:

> When the Army chief of staff, after the opening of the offensive, visited the headquarters of the 62d Division, he learned that the senior staff officer [of the *Thirty-second Army*], Colonel Yahara, after the

[7] Tenth Army G–2 Rpt Nos. 18–20, dtd 13-15Apr45.

> issuance of the Army order for the attack, personally communicated to the responsible operational officers of both the 24th and 62d Divisions that commitment of a few shock troops would suffice for the attack instead of employing a major force, since the attack was bound to fail.[8]

Colonel Yahara's opposition to the attack sprung from his belief that it was not in keeping with the defensive mission of the *Thirty-second Army* and that it would result in a sheer waste of manpower. He was right, for, in effect, the attack was very costly to the Japanese, who concluded that "the night assault resulted in a complete failure." [9]

Although XXIV Corps estimated that its troops had destroyed 6,883 of the enemy by 14 April,[10] its order of battle maps still indicated that the *12th, 13th,* and *14th Independent Infantry Battalions* of the *62d Division's 63d Brigade* opposed the corps advance. Although Tenth Army intelligence agencies knew that elements of four new battalions had been added to the enemy line after the 12–13 April counterattack, the Americans were unable to explain the continued identification of those infantry units that had received enough casualties to be considered destroyed. Actually, the inability of the Tenth Army to maintain a current order of battle file stemmed directly from the replacement system of the *Thirty-second Army*. The Japanese gradually fed individuals and small groups coming from service and support assignments into forward units. At the same time, entire companies and battalions—as yet uncommitted—were absorbed temporarily, or permanently in some cases, into the existing defensive lineup, and were given the designation of the unit into which they had been incorporated.

The first reorganization of the *12th Independent Infantry Battalion* on 23 March serves as an excellent example of this practice. At that time, its organic strength was 1,043; attached special guard, labor, and naval elements raised the total to 1,333. The battalion was armed with 49 light and 9 heavy machine guns, 42 grenade launchers, and 2 75mm guns. On 12 April, after more than a week of continuous fighting against the XXIV Corps, *12th Independent Infantry Battalion* strength was listed by the Japanese as 1,257. Only 414 men remained of the original battalion and 61 from the unit attached originally, but the battalion had been strengthened by the addition of the *2d Battalion, 22d Regiment* (less one rifle company) and the entire *1st Light Mortar Battalion*.[11] Surprisingly enough, the battalion was more heavily armed than it had been before L-Day, for it now had 45 light and 13 heavy machine guns, 45 grenade launchers, 19 90mm mortars, and 3 75mm guns.[12]

By the end of the second week of April, Tenth Army intelligence officers had obtained a fairly accurate picture of Japanese defense plans from captured enemy maps and documents. The Americans were forced to revise their L-Day estimate of enemy strength upward by 7,000 to a total of 72,000,

[8] *Okinawa Operations Record*, p. 82.

[9] *Ibid.*

[10] Tenth Army G–2 Rpt No. 10, dtd 14Apr45.

[11] IIIAC G–2 Rpt No. 38, JOB Suppl, dtd 9May45.

[12] "Organization of the 12th IIB, dtd 12Apr45," in 1st MarDiv G–2 PeriodicRpt No. 37, dtd 8May45, Trans No. 46.

which was "deemed a conservative minimum." [13] It was apparent that the bulk of the *Thirty-second Army* had not yet been met.

As the Tenth Army prepared for this encounter, it became evident that the ammunition supply chain could not keep up with the demand, and it was necessary to apply command restrictions on ammunition expenditure as early as 9 April.[14] Concerning this shortage, the Marine Deputy Chief of Staff of the Tenth Army noted:

> The artillery, in fact, was used too freely. For a considerable period, artillery ammunition was being unloaded over the beaches at the rate of 3,000 tons per day. . . . It was considered normal to fire a concentration of four or five battalions. A good bit of TOT [time on target] firing was done.[15]

The nature of the Shuri defenses demanded the fullest employment possible of all available weapons. Artillery, especially, was needed to reduce prepared positions and denude them of their skillfully prepared camouflage, to seal off the firing ports, and to collapse the labyrinth of interconnecting tunnels that housed and protected the defending troops. Since their operations were not subject in the same degree to the restrictions of inclement weather and enemy air attacks, as were air and naval gunfire, corps and divisional artillery, of necessity, served as the support workhorses for assaults.

Because General Hodge knew that a maximum effort would be needed if *Thirty-second Army* lines were to be penetrated, he scheduled a corps attack, three divisions abreast, for 19 April. Beginning 15 April, four days were spent in preparation for the attack. While guns and howitzers steadily hammered at enemy forward positions and troop concentration areas, artillery ammunition reserves were stockpiled both at the batteries and distribution points. In the pre-attack period, planes from TAF, and Task Forces 51, 52, and 58 flew a total of 905 sorties in direct support missions for XXIV Corps. The pilots dropped 482 tons of bombs and expended 3,400 rockets and over 700,000 rounds of .50 caliber and 20mm ammunition on Japanese installations.[16] Added to this firepower was that coming from the strong force of TF 51 battleships, cruisers, and destroyers that remained offshore both day and night.[17]

Prior to the attack, the frontline units attempted to improve their positions with small local attacks, while patrols were sent forward in order to pinpoint enemy positions and weapons emplacements. When the 27th Division entered the line on 15 April, a general

[13] JOB Sum for 8-14Apr45, in Tenth Army G-2 Rpt No. 22, dtd 17Apr45.

[14] For amplification of this artillery ammunition shortage, see section entitled "Logistical Progress" in chap 7, *infra*.

[15] Smith, *Personal Narrative*, p. 95. Time on target is an artillery technique in which several units fire on the same target and so time their fire that all projectiles hit the target simultaneously. The point here is not that five battalions firing one volley will expend more ammunition than one battalion firing five volleys, but that the five-battalion TOT is much more effective. BGen Frederick P. Henderson ltr to Asst G-3, HQMC, dtd 28Oct65, hereafter *Henderson ltr 1965*.

[16] *XXIV Corps Arty AR*, Anx C, enc 2, p. 4.

[17] *CTF 51 AR*, pt V, sec C, pp. 39-42.

reshuffling of the XXIV Corps front took place. (See Map III, Map Section.) On that date, General Griner assumed responsibility for the corps right flank and, on the following day, regained his 105th Infantry, which had been released from army reserve following the capture of Tsugen Shima. All initial XXIV Corps assault deployments were completed two days before the jump-off.

The support provided by air, naval gunfire, and artillery prior to the 19 April attack might seem pallid in comparison with the destructive potential of the nuclear weapons of a later era. To the assault force leaders and their troops, however, the immensity of the preparatory and supporting fires was awesome. The firepower of 6 battleships, 6 cruisers, and 9 destroyers was assigned to direct support of the attacking corps, and 650 Navy and Marine aircraft were directed to hit enemy defenses, assembly areas, and supply points.

Beginning at 0600 on 19 April, 27 battalions of artillery, covering the five-mile front with a density greater than one weapon to every 30 yards, fired in their pre-attack bombardment everything from 75mm to 8-inch howitzers. Regarding this massing of battalions, one observer remarked:

> Not many people realize that the . . . artillery in Tenth Army, plus the LVT(A)s [mounting 75mm howitzers] and NGF equivalent gave us a guns/mile of front ratio on Okinawa that was probably higher than any U. S. effort in World War II. We look with awe on the Russian doctrine of 300 guns/mile of front for an attack. But if you take our Okinawa figures, and apply a reasonable multiplication factor for our flexible fire direction system that rapidly enabled us to mass all guns within range on a target, we equalled or exceeded the Russians in effective available fire support.[18]

Equally impressive was the air support provided the ground troops during this offensive. At one time alone during 19 April, "we had 375 aircraft on station, and . . . LFASCU–2, controlling seven simultaneous air strikes on a ten-mile front, had literally reached the point of saturation." Commenting on this, the commander of the LFASCUs stated that "I do not believe that we have ever exceeded, or since equalled, this magnitude of close air support on any given day."[19]

To the troops poised for the attack, it seemed incredible that anyone could survive in that terrible downpour of steel, yet it soon became apparent that almost all of the enemy did. The Japanese were hidden in caves and protected by solid limestone walls deep within the hill-ridge complex astride the XXIV Corps route of advance.

Initially, the assault infantry made moderate gains, but when the enemy remanned his positions, the attack slowed and then halted under the resumption of intense mortar, machine gun, and artillery fire. Generally, all along the line, advances were negligible to nonexistent as enemy resistance stiffened. Kakazu Ridge, the formidable bastion opposing the 27th Division, proved to be as difficult to take at this time as it had been when the 96th Division made the attempt. The 27th mounted a battalion-size infantry attack, supported by a re-

---

[18] *Henderson ltr.*

[19] Gen Vernon E. Megee ltr to Asst G–3, HQMC, dtd 19Oct65.

inforced tank company, in an attempt to bypass the ridge through a cut between Kakazu and Nishibaru. Anticipating the probable use of this route, on the night of 18 April the Japanese had emplaced mortars, machine guns, antitank guns, and antiaircraft cannon to cover the Ginowan-Shuri road, which crossed through the cut. The enemy cut off the tank company from its covering infantry by planned protective fire.[20]

The tanks were able to get behind the ridge to shoot up the village of Kakazu, but without infantry support, they were forced to withdraw to their own lines. Only 8 of the original 30 tanks in the foray made it back through the cut; the remaining 22 fell victim to the fire of antitank and antiaircraft guns, mines in Kakazu village, and satchel charges borne to the tanks by suicide-bent enemy soldiers.

By the end of the day, on the corps right flank, the 27th Division was halted at the western end of the Urasoe-Mura escarpment; the 96th Division, in the center, had pushed through Kaniku to gain positions on the forward slopes of Nishibaru Ridge; and the 7th Division, on the left, was held up by fanatic opposition and heavy fire, with the net result that it made no progress at all.

As the XXIV Corps ground out the second day of its offensive, the pattern of future fighting emerged—little yardage gained at a high cost in lives to both sides. Heavy casualties were sustained by all the attacking divisions, but the Japanese frontline units also were punished and considerably reduced in size. Only the sheer courage and fanatic determination of the enemy and the strength of his natural defenses kept the XXIV Corps at bay. Action during the period 20–23 April consisted of heavily supported local attacks against key strongpoints.

When General Hodge renewed the XXIV Corps attack on 24 April, he was ready to throw the full weight of its power against the forces holding Shuri's outer defense ring. During the night of 23–24 April, however, unknown to the Americans and under cover of "the most intensive artillery fire yet experienced on the XXIV Corps front," [21] General Ushijima had withdrawn his defending units from the line that had held up the 7th and 96th Divisions for two weeks. All along the front, American forces now made sweeping and significant gains, and the heretofore-difficult Kakazu Ridge was taken with little effort.

After the 27th Division had entered the lines on the corps right flank in mid-April, 2/11 and 3/11 were reassigned from support of the 96th Division to reinforce the fires of 27th Division artillery. This change was made because "General Sheetz thought, even then, that the 1st Marine Division would be needed in the south, on the coast." [22] At this time, the 11th Marines commander, Colonel Brown, heavily reinforced his three 105mm battalions with regimental headquarters personnel so that as many men as possible could gain battle experience.[23]

---

[20] "62d Div Battle Lessons, Priority Dispatch No 19, dtd 20Apr45," in CinCPac-CinCPOA Bul 212–45, Translations and Interrogations No. 39, dtd 30Aug45, pp. 47–49.

[21] Tenth Army G–2 Rpt No. 30, dtd 25Apr45.

[22] *Brown ltr II.*

[23] *Ibid.*

Firing battery crewmen were not the only Marines in the 11th to gain on-the-job training in the south, for regimental communications personnel were kept exceptionally busy. Although radio was depended upon primarily, wiremen laid telephone wires from Army fire direction centers to the Marine units supporting XXIV Corps after frequent interference in 11th Marines radio circuits had made reliance on wire communications necessary.

Owing to a shortage of trained wiremen in the Army battalions, these same Marines in addition had to lay and maintain all lateral wire communications for three Army divisional artillery headquarters. This communication system permitted the rapid massing of all XXIV Corps and attached artillery fire whenever all other means of communication broke down.[24] Forward observer teams of the 11th Marines also gained valuable experience when they went forward to the XXIV Corps infantry units their artillery battalions were supporting. The knowledge gained by the teams supporting the 27th Division was especially useful later when the 1st Marine Division relieved the 27th in the same general area.

Indications that greater Marine participation in the Shuri battle would be forthcoming occurred on 21 April when Tenth Army ordered General Geiger to make the 1st Tank Battalion available for attachment to the 27th Division. Although the IIIAC commander had no compunction about his Marines fighting in the south, he was not happy at the prospect of their being committed piecemeal.[25] If Marine assistance was needed in the south, it was Geiger's opinion that the entire 1st Marine Division should be committed.[26] Although a warning order for the tank battalion displacement had been dispatched to the 1st Division, the actual movement orders were never issued and the matter was apparently dropped by Tenth Army.[27]

General Buckner acknowledged the need for a substantial infusion of fresh troops into the main battleline, and directed General Geiger, on 24 April, to designate one IIIAC division as Tenth Army reserve. One regiment of that division was to be ready to assemble and move south on 12 hours' notice. General Geiger selected the 1st Marine Division, and General del Valle placed the 1st Marines on alert status.[28]

At this point, the question arises why the 2d Marine Division, in Tenth Army reserve, was not committed in action on Okinawa when it was apparent that it was needed. On 9 April, Admiral Nimitz authorized General Watson's division to return and debark at Saipan; on the 14th, the division was released from Tenth Army reserve and reverted to IIIAC control, although it remained on Saipan. Both at this time, and in later critiques of the fighting on Okinawa, there was a strong body of senior offi-

[24] *Ibid.* General Brown added that "the 11th Marines used well over 1,000 miles of wire from April 29th to June 22d, having as a peak as much as 262 miles in operation at one time."

[25] Smith, *Personal Narrative*, p. 88.

[26] *Ibid.*

[27] *Ibid.*; IIIAC G–3 Jnl, 1Apr–30Jun45, 22Apr45 entry, hereafter *IIIAC G–3 Jnl* with date of entry cited.

[28] *IIIAC AR*, p. 4.

310-224 O - 69 - 14

cers who felt that there was no sound reason why the 2d Marine Division could not have been employed to make an amphibious assault on the southeastern coast of Okinawa. Possibly, a second landing could have succeeded in cracking the Shuri barrier where the attack of the XXIV Corps in mid-April failed.

General Vandegrift suggested that the 2d Division be employed when he visited Okinawa on 21 April with Brigadier General Gerald C. Thomas, Admiral Nimitz, and Nimitz' chief of staff, Rear Admiral Forrest P. Sherman. When, during a meeting at Geiger's CP, Buckner stated that he was going to commit IIIAC divisions in the south shortly, Vandegrift:

> did not object to the Marines being committed to the main fight—they were on Okinawa for this purpose. But I did question Buckner's tactical plan. Instead of trying to slug it out with the enemy, Geiger, Thomas, and I argued for an amphibious landing in the rear or anyway on the flank of the enemy by Buckner's reserve, the 2d Marine Division on Saipan.
>
> Forrest Sherman, among others, objected to a landing on the far east coast as impractical. We replied that the bay of our choice was the alternate landing area for the original operation, so apparently Buckner had thought it quite practical. Having been shot down on this point, Sherman claimed it would take too long to load out the 2d Division from Saipan. We promised him it could be underway in six hours.
>
> Despite these and other arguments Sherman refused to back us, nor did Buckner seem impressed. I learned later that General Bruce, commanding the 77th Army Division which had fought so well on Guam, proposed a similar plan as did Kelly Turner, whose transports were being hurt by the kamikaze tactic.[29]

Although General Bruce had also pressed for a second landing, for his troops had all but captured Ie Shima, Buckner refused because his G–4 had told him that food but not ammunition could be supplied for this project. In addition, the site of the proposed 77th Division landing was so far south of the main Tenth Army line at that time, neither XXIV Corps artillery nor troops could support it. Besides, at the time that the 77th was available, it was needed in the line as the 7th, 27th, and 96th Divisions were in bad shape because of casualties and fatigue. Nor did Buckner want to use the 2d Marine Division for a second landing, for it was scheduled to invade Kikai Shima, north of Okinawa, in July.

The Tenth Army commander was evidently convinced that the greater need was for fresh troops on the Shuri front and that a landing on the southeastern beaches was logistically infeasible. Despite the arguments presented by General Vandegrift in favor of such a tactic, he was not supported by either Nimitz or Sherman, and Buckner remained unmoved in his decision.

He faced the basic alternatives of a two-corps frontal attack against Shuri or an envelopment of the enemy forces facing his troops. Having decided against the landing in the enemy rear, his next step was to commit IIIAC in the south.

---

[29] Alexander A. Vandegrift and Robert B. Asprey, *Once a Marine* (New York: W. W. Norton and Company, Inc., 1964), p. 290, hereafter Vandegrift and Asprey, *Once A Marine*.

The Marines were available to the Tenth Army commander as a result of change in the ICEBERG plan of operations. On 26 April, General Buckner was informed that Phase III of the plan, the projected invasion of Miyako Shima in the Sakashima Group east of Taiwan, was cancelled. This high-level decision by the JCS freed IIIAC from the Miyako operation and permitted Buckner to insert that corps into the southern Okinawa line.[30] The next day, the army commander declared his intent to attach the 1st Division to XXIV Corps at an early date in order to relieve the understrength and badly battered 27th Division for garrison duty under Island Command.[31] Also on 27 April, the 77th Infantry Division completed its move from Ie Shima to Okinawa, and its leading elements moved into position to relieve the 96th the next day.

Matters concerning the future employment of IIIAC units were discussed at a conference held at Tenth Army headquarters on 28 April. Colonel Walter A. Wachtler, General Geiger's G–3, was informed that the 1st Marine Division was to be attached to XXIV Corps on the last day of April. General del Valle's troops would begin moving south to relieve the 27th Division on that same day. Upon its relief, the 27th was to move north to relieve the 6th Marine Division, which would then move to an assembly area near Chibana to await further orders for the movement south. It was planned that, on or about 7 May, the IIIAC was to take over the zone held at that time by the 1st Marine Division and, simultaneously, Tenth Army would then assume tactical control of the two-corps front. A coordinated army attack would be made soon thereafter.[32]

1ST MARINE DIVISION JOINS XXIV CORPS[33]

While the XXIV Corps made preparations to relieve two of its frontline divisions, the attack to the south continued. The enemy reacted savagely to the grinding advance of the 96th Division, throwing counterattacks, repeated artillery and mortar barrages, and never-ending infiltration attempts in the soldiers' path. The division objective was the Maeda Escarpment. Retention of this position was vital to *Thirty-second Army* defense plans, because the terrain offered a commanding view of all of the Japanese positions as far as the Shuri foothills, and at the same time guaranteed continued enemy ob-

[30] Of this assault, the FMFPac commander informed the Commandant of the Marine Corps that it was his "opinion that the target is unnecessary—practically in a [Japanese] rear area and its capture will cost more than Iwo Jima." LtGen Holland M. Smith ltr to CMC, dtd 27Mar45 (Vandegrift Personal Correspondence File, HistBr, HQMC).

[31] *1st MarDiv G–3 Jnl*, 27Apr45.

[32] *IIIAC AR*, p. 44.

[33] Unless otherwise indicated, the material in this section is derived from: *Tenth Army AR; XXIV Corps AR; TAF AR; TAF PeriodicRpts*, Apr-May45; *ADC Daily IntelSums*, Apr-May45; *1st MarDiv SAR; 7th InfDiv AR; 27th InfDiv OpRpt; 77th InfDiv OpRpt*, Okinawa; *96th InfDiv AR; 1st Mar SAR;* 5th Mar SAR, 22Apr-23Jun45, n.d., hereafter *5th Mar SAR; 7th Mar SAR.*

servation into American lines.[34] The region surrounding Maeda, therefore, became the focus of ferocious fighting when the enemy attempted to retain the dominating ground of the second Shuri defensive ring. On 29 April, units of the 77th Division began relief of the 96th Division and immediately took up the attack. The soldiers of the 77th were very tired from the fighting they had just experienced on Ie Shima, and the division was far understrength because of casualty losses. As a result, it could make but slight gains against the highly developed defenses.

The 27th Division lines on the 29th had been pushed through Kuwan and Miyagusuku during a daylong drive, which exposed deeply dug-in, heavily mined Japanese positions. These extended throughout the rugged hills and ridges that bordered the east and southeast sides of Machinato airfield. At 0600 on the following day, the 1st Marine Division was attached to XXIV Corps. Immediately thereafter, march serials of the 1st Marines and 1st Tank Battalion began moving to the 27th Division area in the south. The Army division, meanwhile, continued its attack south of Machinato airfield. At 1000, the first of the Marine units began moving into 27th Division lines even as it halted its forward progress and disengaged its advance patrols, which had been caught in a heavy fire fight.

On the extreme right of the corps line, beginning at Kuwan, 1/1 took up positions, which made a half-circle around the south of Machinato and joined up with 3/1 just to the northeast of Nakanishi. No orders were issued for the resumption of the attack in this area, so the Marines spent the rest of 30 April digging in, improving existing defenses, and registering defensive fires.[35] (See Map III, Map Section.)

The commander of 3/1 had been informed by the commander of the relieved Army battalion that some Japanese were still holed-up in Miyagusuku. Marines, dispatched to mop up, moved towards the village, whereupon enemy artillery and mortar fire began falling on them. Under this cover, Japanese troops began infiltrating back into Miyagusuku in some strength. After being pinned down in the village ruins by the concentrated fire, the Marines were forced to withdraw; they set up north of the village for the night. At dusk, the 3/1 reserve was committed on the left of the line, where it tied in with the one yet-unrelieved unit of the 27th Division.[36]

The 77th Division completed the relief of the 96th at the same time the 1st Marines took over the right of the 27th Division lines. At noon, General Bruce assumed command of the zone from General Bradley and, throughout the day, 77th Division troops attempted to improve their positions on the escarpment. Despite heavy supporting fires, the sheer fury and fanatic determination of the defenders forced the attackers back to defensive positions of the previous night.

[34] *Okinawa Operations Record*, p. 91.

[35] 1st Bn, 1st Mar SAR, 23Feb-21Jun45, n.d., p. 3, hereafter *1/1 SAR*.

[36] 3d Bn, 1st Mar SAR, Phase I and II of the Nansei Shoto Operations, dtd 10Jul45, p. 9, hereafter *3/1 SAR*.

By midafternoon of 1 May, two of the 1st Division assault regiments, the 1st Marines on the right and the 5th Marines on the left, had relieved the 27th Division; General del Valle assumed command of the former Army zone at 1400. (See Map 13.) Although this action marked the official entry of the 1st Division into the southern front, the 11th Marines had been in the vanguard when it supported the 7th, 27th, and 96th Divisions throughout most of the fighting in April.[37] The artillery battalions of the 27th Division remained in position to continue supporting the 1st and 77th Divisions in their attack to the south.[38]

Even before completing relief of the 27th Division on 1 May, General del Valle's Marines saw clearly the results of combat in southern Okinawa, and soon learned of the tenacity of the defenders. When the 5th Marines relieved the remaining Army regiments during the afternoon of the 1st, they learned how hard these units had been hit. Each 2/5 infantry company replaced one depleted battalion of the 105th, and 3/5 took over the area held by the 106th. At 1400, while consolidating their positions south of Awacha, 2/5 Marines observed about a platoon of Army tanks moving south in the town. As soon as the tanks had emerged from the town, they were hit by 47mm AT fire within 20 yards of the Marine line.[39] Perhaps even more disturbing to the Marines was the news that their unit identification and the location of their front had already been noted on an enemy map captured just that day.[40]

General del Valle's men continued their defensive activities on their first day in the south. By 1700, all lines had been tied in, and 1/5 had taken up positions in a reserve area. As the frontline Marines dug in, preparations were made for a fully supported division attack scheduled for the following day, with the north bank of the Asa Kawa (River) as the objective. The corps commander instructed General del Valle to exert constant pressure against the enemy and to support the 77th Division attack with fire and maneuver.[41] Because it flanked the Army division, the 5th Marines was assigned this task.

In order to prevent an American penetration after the 27th Division had taken Gusukuma, the Japanese had been forced to reform their west coast battleline. It now was held by a major portion of the *62d Division*, which was positioned along a line that ran generally from Jichaku and Uchima through the ridges north of Dakeshi to Awacha. In addition, there were "powerful elements of the Division scattered and re-

[37] "On 29 April, Headquarters, 11th Marines assumed command of its scattered battalions in the zone of and reinforcing the 27th Division, the two battalions with the 7th and 96th Divisions displacing into that area. All four battalions of the 11th Marines were commended by the Army divisions with which they served." *Brown ltr II*, copies of commendations appended.

[38] One battalion, the 249th Field Artillery, accompanied the 27th Division when it went north and furnished support during infantry mop-up activities. This unit returned south on 29 May to reinforce the fires of IIIAC Artillery.

[39] 2d Bn, 5th Mar SAR, Okinawa, 1Apr-22Jun45, n.d., pp. 2–3, hereafter *2/5 SAR*.

[40] *1st MarDiv G–3 Jnl*, 1May45.

[41] *Ibid.*

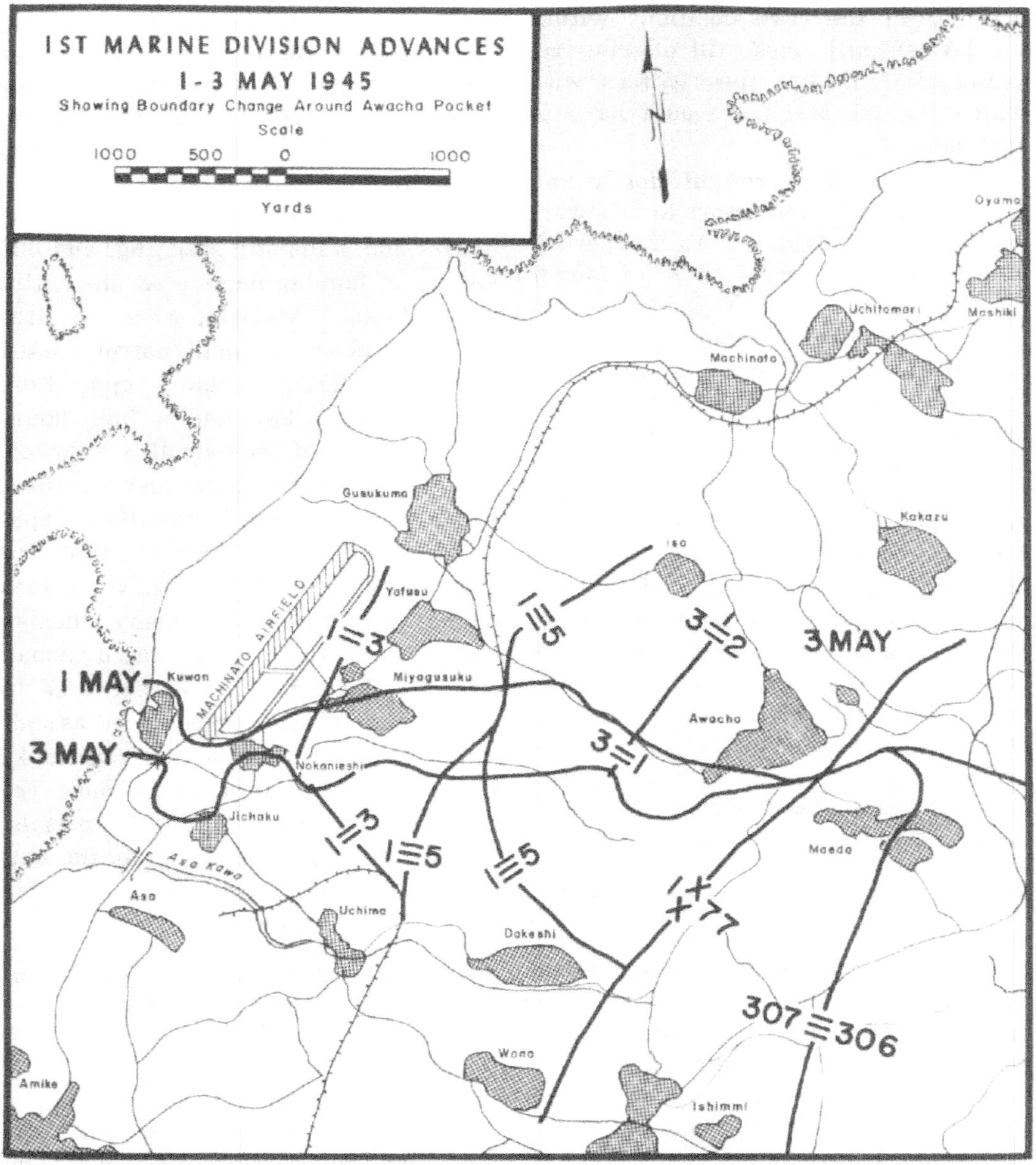

MAP 13

T. L. RUSSELL

maining in the cave positions within the [American] lines still offering resistance." [42] It was these forces with which the 1st Marines would have to contend.

The 1st of May brought cloudy and cooler weather, and sporadic showers heralded the Okinawa rainy period, which, in itself, serves as a harbinger of the approaching typhoon season (July–November). During the previous night, all 1st Marines battalions had received intermittent mortar and artillery fire. The day was devoted by 1/1 to patrolling its front and attempting to readjust its lines. A reconnaissance patrol discovered that a deep L-shaped ravine cut across the entire battalion front and that it formed a natural barrier to the next logical objective. The Marines also found that the retreating enemy had blown out the fill where the main north-south highway crossed this chasm. Added to the enemy artillery and antitank guns registered on the area, this obstacle obviously would prohibit an armor-supported infantry penetration.

A patrol sent from one of the Marine rifle companies to the west of the ravine to scout out other approaches was taken under extremely heavy fire that came from the steep cliffs along the far side of the declivity. It was apparent that the enemy, from positions on the high ground to the south and southeast of the Asa Kawa, had excellent observation of the battalion approach route and that the Marines were going to have a difficult time reaching and crossing the river.

[42] *Okinawa Operations Record*, p. 97.

On the left of the regiment, meanwhile, 3/1 prepared for a second attempt to secure Miyagusuku, this time with the support of seven tanks from the 1st Tank Battalion, four of them mounting flamethrowers. The flat trajectory tank cannon fire blasted the houses and walls still standing, and 300 gallons of flaming napalm set the entire pyre aflame.[43] At 1045, when the fires had died down, a small patrol passed through without enemy opposition. Approximately two and a half hours later, the rest of the battalion followed, and, as it cleared Miyagusuku, there was a step-up in the intermittent Japanese mortar and artillery fire that had been falling since before dawn. Added to this fire was that of enemy riflemen and machine gunners. Increased casualties and the difficulties encountered in evacuating them forced the assault companies to withdraw under a smoke screen and mortar barrage once the battalion commander had given permission for such a move. By 1900, the battalion returned to the positions it had occupied on 30 April.

[43] 1st TkBn Summaries of Tank Action, 15Apr-23Jun45, hereafter *1st TkBn Sum*, with dates of action. According to the *3/1 SAR*, p. 10, the tanks supporting this action were Army mediums, but the 1st Tank Battalion lists them as belonging to its own Company A. The flame tanks came from Company B, 713th Tank Battalion, which was attached first to the 1st Marine Division and later to IIIAC. This battalion was one of the first Army armored units to be equipped with and trained in the use of an improved tank-mounted flame thrower, which replaced the 75mm gun in the medium tank. Appleman, *et. al.*, *Okinawa Battle*, p. 256.

The 1st Division attack to the Asa on 2 May began in driving rain which seriously limited visibility and reduced the amount of effective air support supplied that day. The two frontline Army divisions in XXIV Corps attacked enemy-held ridge positions containing pillboxes and mutually supporting small arms and automatic weapons emplacements. Added to these barriers and the destructiveness of Japanese artillery fire was the foe's determined and ferocious refusal to yield any ground.

When the 5th Marines jumped off at 0900, following artillery, naval gunfire, and some air preparation, 2/5 on the extreme left came under flanking fire from positions in front of the 307th's lines. Within an hour, the 2d Battalion was pinned down, and, by 1100, heavy casualties forced its withdrawal under a smoke cover to its original positions. As soon as the 3d Battalion crossed its line of departure, it too came under the frontal and flanking fire that had driven 2/5 back. "The advance was untenable and had to be withdrawn to initial positions." [44] Because the 5th Marines was unable to advance, Company L of 3/1 (on the battalion left) was stopped in its attempt to move beyond Miyagusuku with Company K. The latter, however, was not pinned down and was able to progress beyond the edge of the ruins that were once a village. At 1446, the 1st Marines commander, Colonel Chappell, was ordered to change the axis of the regimental attack from due south to the southeast. General del Valle had reasoned that the new attack direction, which would hit the flank of many of the positions holding up the 5th Marines, would enable both Company L and his left regiment to continue the advance.

After new attack orders had been issued, battalion boundaries adjusted, and a 10-minute artillery barrage laid, the 1st Marines attack was resumed at 1630 against very heavy fire. By dark, 3/1 had fought its way to a small series of hills approximately 300 yards south of Miyagusuku. Pouring rain, machine gun fire, and grenades began falling on the leading elements of the assault as the troops gained this ground. Here they began digging in at 2000 for the night. Almost immediately, the enemy began the first in a series of infiltration attempts which were to mark the hours of darkness. The violence of the hand-to-hand clashes on the hill held by Company K was reflected in a comment the next morning by one of the Marines, who said that this had been the grimmest night he had spent so far on Okinawa.[45]

On the extreme right of the division, 1/1 ran into equally heavy resistance in its effort to cross the ravine facing it. Although one company had already passed through Nakanishi, and was in position by 1000 to plunge into the ravine where it angled north towards Miyagusuku, the slowness of the attack to the left of the 1st Marines presaged caution, and the 1/1 commander was told to be wary of exposing his troops to enfilade fire.

[44] LtCol Martin C. Roth ltr to CMC, dtd 18Mar55.

[45] *3/1 SAR*, p. 13.

Despite all precautionary measures, including the blasting of enemy positions by self-propelled assault guns at the beginning of the attack, Japanese fire continued to pour from the caves and heights overlooking the defile to catch the advancing troops. Disregarding this fusilade and the casualties resulting from it, Company B managed to gain a defiladed position just short of the initial objective. At this point, however, the Marines were cut off both front and rear by enemy fire. At 1300, orders for a general withdrawal were issued, and the company disengaged and pulled back to high ground under the cover of smoke.

At 1630, when the attack was resumed in the new direction, the 1st Battalion attacked straight across the ravine in order to ensure flank protection before making the southeasterly move. The impetus of the drive carried Company A to the outskirts of Jichaku, where it dug in. By the end of a quiet but wet night, the assault units had established a firm line where the division awaited the joining of RCT–7.

The 7th Marines had displaced south on the morning of the attack, its battalions moving to the vicinity of Uchitomari. On the following day, 3 May, the regimental CP displaced forward to a point about 200 yards north of Gusukuma, while the 1st and 2d Battalions took up beach defense positions in the vicinity of Machinato airfield. The 3d Battalion was attached to the 5th Marines to assist the advance of that regiment.

Continuing the attack on 3 May, the division assigned intermediate and final objectives to the 1st Marines. The first began at the railroad spur bridge crossing the Asa River between Asa and Uchima, and extended northeast, generally following the spur initially and then the main line itself, to a point just east of Miyagusuku. The second began at the same bridge, but ran generally east along the high ground between Dakeshi and Wana to the division boundary. The line between the attacking regiments, bent back to reflect the assignment given the 1st Marines, gave to the 5th the thankless task of clearing out the confused terrain that soon would be called Awacha Pocket.

Although the 5th Marines gained about 500–600 yards in its zone, the 1st became heavily engaged in fire fights all along the line and was restricted to limited gains. Forward elements of 1/1 ran into difficulty in every direction; Company F,[46] attacking Jichaku, was held up by a stubborn defense, and Company A was cut up in its attempt to reduce the ravine position that had stymied the battalion on the previous day. Neither company was able to move forward and both were forced to withdraw under the cover of smoke late in the morning, carrying with them the large number of casualties they had sustained. Because it appeared that the attack would not succeed without armor support, plans were made to use tanks after the road south of Kuwan had been cleared of mines. After dark, an engi-

[46] Because both Companies B and C were badly shot up on the previous day, Company F was attached to 1/1 prior to the redirected attack that had begun at 1630 on 2 May.

neer mine-clearing team, protected by infantry began reconnoitering the proposed armor attack route.[47]

Company L, on the left of the 1st Marines line, was unable to move until 3/5 had seized a high hill to the left front of the company. After it took the hill, the battalion was driven off at 1555 by a heavy enemy artillery concentration. Company L was ordered to retake the objective, and, following a 10-minute 81mm mortar barrage, gained the hilltop 20 minutes later. Here, the company was pinned down by Japanese fire coming from high ground to the front and on its flanks. When Company K was unable to close in on L and had to fall back, Company I was committed on the right of K to close a gap that had developed along the battalion boundary. The regiment then assigned Company G of 2/1 to back-up the 1,200-yard front of 3/1.

To escape the furious machine gun and mortar fire that had followed it after it was driven off the hill on the right of 3/1, 3/5 was forced to fall back another 100 yards; its advances on 3 May were limited to 200-300 yards. Passing through 2/5 that morning to begin its attack with two companies in the assault, the 1st Battalion, 5th Marines, was able to gain some 500–600 yards, but was forced in late afternoon to bend its lines back to tie in with 3/5. Immediately upon being relieved, 2/5 swung over to its left to take over part of the 307th Infantry lines on the outskirts of Awacha. (See Map 13.)

Once its 2d Battalion was relieved by 2/5, the 307th Infantry moved it to the left and, with all three of its battalions on line, mopped up the top of the escarpment—and the upper part of its reverse slope—during the day. By nightfall, the Army regiment held positions commanding the Japanese defensive alignment all the way back to the Shuri foothills. Despite having been pushed back, the enemy still determinedly refused the Americans further gains and fanatically resisted from reverse-slope caves, sometimes counterattacking in company and platoon strength to regain critical terrain.

The ferocity of Japanese resistance continued unabated all along the XXIV Corps line, for as veteran units were annihilated, they were quickly rebuilt with fresh rear area troops, or replaced with new infantry elements. General Hodge's dire prediction at the beginning of the 19 April attack that "it is going to be really tough . . . and that I see no way to get [the Japanese] out except blast them out yard by yard. . . ."[48] was being all too grimly substantiated.

[47] This team removed all mines in the road until it came to a spot where a shell crater, flanked by rice paddies, would effectively block tank traffic. Upon returning to its lines, the troops were nearly cut off by a large enemy unit that had attempted to work its way around the rear of the patrol. After a brisk fire fight in which there were some Marine casualties, the patrol regained its lines. *1st Mar SAR*, p. 10.

[48] CG, XXIV Corps ltr to CGAFPOA, dtd 17Apr45, quoted in Appleman *et. al.*, *Okinawa Battle*, p. 185.

THE JAPANESE COUNTERATTACK[49]

During the grueling see-saw battle in the south, both sides suffered heavily. The slow but perceptible American gains were costly, but the Japanese paid the higher price. The *62d Division* bore the brunt of the April attack and by the end of the month its combat strength was less than half of what it had been originally. Although many *Thirty-second Army* officers viewed the Japanese cause on Okinawa as hopeless, they were buoyed up by the fact "that after thirty consecutive days of systematic fighting the main body of [our] fighting forces should remain intact. . . ."[50] Not yet bloodied in the fight for Shuri were most of the units of the *24th Division*, *44th IMB*, and *5th Artillery Command.* An attitude favoring the offense permeated General Ushijima's command, whose members considered that commitment of these fresh troops in one major effort would effectively blunt the American drive.

Prior to the landings on the west coast, the expectation of an American amphibious assault at Minatoga had caused the *Thirty-second Army* commander to deploy a considerable portion of his strength in that area. But, by the end of April, the steady attrition of the forces manning the Shuri outer defense ring caused General Ushijima to reappraise his situation and reexamine his mission. Since he had been ordered to prolong the battle as long as possible and inflict heavy casualties on the invaders, Ushijima decided to utilize the units immobilized in the southeast to reinforce the Shuri positions.

Implementing this decision, the *24th Division* and the *44th IMB* were ordered to begin a movement north on 22 April. The *24th*, recovering control of its *22d Regiment* from the *62d Division*, was to occupy defensive positions in a line from Gaja on the east coast to Maeda at the eastern end of the Urasoe-Mura escarpment. The depleted battalions of the hard-hit *62d Division* were to concentrate in the area from Maeda to the west coast near Gusukuma. Taking up blocking positions behind the *62d* on the high ground to the south and east of the Asa River, the *44th IMB* was to cover Naha and the ridges and draws flanking to the west of Shuri.

To protect the area south of the Naha-Yonabaru valley, and to forestall further American landings on the west coast, both Admiral Ota's force and a provisional guard group, formed to guard the Chinen Peninsula, were kept in place. The Chinen units were not to make a last-ditch stand, but were to make a fighting withdrawal to Shuri if the southeastern beaches were invaded.

In less than a week, by 27 April, the new enemy defensive setup had been established. But even small local Japanese counterattacks failed, despite the reconstitution of the frontlines and the infusion of fresh troops. Steadily,

[49] Unless otherwise noted, the material in this section is derived from: *Tenth Army AR; XXIV Corps AR; 1st MarDiv SAR; 7th InfDiv AR; 77th InfDiv OpRpt*, Okinawa; *1st Mar SAR; 5th Mar SAR; Okinawa Operations Record; IntelMono; Shimada Interrogation; Yahara Interrogation;* Hayashi and Coox, *Kōgun.*

[50] *Okinawa Operations Record*, 92–93.

XXIV Corps units encroached upon enemy positions and forced their defenders back. In the *Thirty-second Army* headquarters deep below Shuri Castle, General Cho led other firebrands in an attempt to convince the army commander that conditions were favorable for an all-out, army-sized counterattack, employing the relatively intact *24th Division* as the spearhead.

Colonel Yahara was a lone dissenter to the plan. His belief that the Japanese attack would end in abject failure and certain defeat was based on several factors. He noted that the Americans, positioned on commanding ground, were materially and numerically superior. Fatalistically prophesying an inevitable Japanese defeat no matter what, the colonel reasoned that the army should "maintain to the bitter end the principle of a strategic holding action." [51] Any other course of action would doom the army, be detrimental to its mission, and open the way for an otherwise earlier invasion of the Japanese homeland.

Despite Colonel Yahara's impassioned and reasoned arguments, General Cho, backed by other proponents of the offensive—the division and brigade commanders—swayed Ushijima to their way of thinking, and, in the end, prevailed. In scope and desired objectives, the attack plan was exceedingly ambitious; it called for nothing less than the destruction of XXIV Corps and capture of Futema and its environs.[52] (See Map IV, Map Section.)

The counterattack was to begin at 0500 (Y-Hour) on 4 May (X-Day). The Japanese believed it would be successful because they knew a relief of the American lines was then taking place.[53] At Y-Hour, the *89th Regiment* (on the right) would begin a penetration of the 7th Division front to gain its objective, the Minami-Uebaru foothills, by sunset. In the center, the *22d Regiment* was to hold its positions near Kochi and Onaga, where it would support the assaulting units with fire. When the *89th Regiment* formed an east-west line at Tanabaru, the initial objective, the *22d* would move out, destroying any American unit remaining to its front, and follow up in rear center of the division main effort to be made by the *32d Regiment*. At Y-Hour, the *32d* would drive forward to seize 77th Division positions southeast of Maeda, and then continue on to gain the heights west of Tanabaru by sunset also.

Armored support of the attack was to be supplied by the *27th Tank Regiment*, after it had moved from positions near Ishimmi to penetrate the 77th Division lines west of Kochi. Here, the tanks would take up new positions to assist the *22d* and *32d Regiments*. The day before the attack, the *44th IMB* was to move to the area northwest of Shuri, where the brigade would provide left-flank security until the initial objective was taken. Immediately thereafter, the *44th* would swing north to Oyama and the coast just beyond, to isolate the 1st Marine Division from the battle. This task would be supported by the hereto-

[51] *Okinawa Operations Record*, pp. 76-77.

[52] Mistakenly, the Japanese believed that Tenth Army headquarters was located at Futema; actually, this is where the 96th Division CP was sited.

[53] Hayashi, *Kōgun*, p. 142.

fore-uncommitted *62d Division.* To make certain that the Marines would be cut off, the Japanese planners had reinforced the *44th IMB* with a considerable number of armored, artillery, and antitank elements. On the night of 3–4 May, the guns, mortars, and howitzers of the *5th Artillery Command* were to move out of their hidden positions into the open to provide the Japanese attack with full gunfire support. The *Thirty-second Army* also called upon Admiral Ota to participate in the massive counterattack, for he was directed to form from his naval command four infantry battalions to be used as army reserve in exploiting the breakthrough.

The Japanese attack plan provided also for hitting the open flanks of the XXIV Corps. Embarking from Naha on the night of 3–4 May, a makeshift navy of landing barges, small craft, and native canoes was to land a major portion of the *26th Shipping Engineer Regiment* behind 1st Marine Division lines at Oyama. Concurrently, elements of the *26th, 28th,* and *29th Sea Raiding Squadrons* were to wade the reef on the Marine flank, go ashore in the vicinity of Kuwan, and more inland to support the counterlanding of the *26th.* Committed to the west coast attack was a total of approximately 700 men.

Another envelopment was to be attempted on the east coast where about 500 men from the *23d Shipping Engineer Regiment* and the *27th Sea Raiding Squadron* would land behind the 7th Division at Tsuwa. The mission of both regiments was to infiltrate American rear areas in small groups and to destroy equipment and harass CPs with grenades and demolitions. No concerted attacks were to be made unless assault groups numbered more than 100 men. If all went according to plan, the two counterlanding elements would join up near the center of the island to assist the *24th Division* advance.[54]

A never-changing assumption in ICEBERG intelligence estimates was an enemy capability to mount a large-scale counterattack. As of the evening of 3 May, however, an analysis of recent enemy tactics indicated that he was more likely to continue fighting a series of delaying actions from successive positions, defending each one "until the troops on the position are nearly annihilated." [55] Since the American order of battle of enemy elements facing XXIV Corps was then current, and each enemy move and countermove had been viewed with respect to the related tactical situation, indications of an imminent major attack were not perceptible. Local counterattacks and stiffened resistance were merely attributed to the infusion of new strength into Japanese lines. XXIV Corps troops were not caught offguard, however, when the attack was finally mounted.

Preceding the two-day struggle—called by Colonel Yahara "the decisive action of the campaign" [56]—the fifth mass *Kamikaze* attack struck at dusk on 3 May. Tokyo had notified the *Fifth Air Fleet* on 30 April of the impending

---

[54] "POW Interrogation No. 38, Superior Pvt, Hq, 2d Co., 26th ShpgEngrRegt (CO's orderly)," in 1st MarDiv G–2 Periodic Rpt No. 39, dtd 10May45.

[55] Tenth Army G–2 Rpt Nos. 33 and 39, dtd 28Apr and 4May45 respectively.

[56] *Yahara Interrogation.*

Thirty-second Army attack.[57] The Japanese air command then issued orders for a mass suicide raid to be launched on 3 May, prior to the beginning of the ground assault. *Kikusui* No. 5 targets were to be American supply areas, airfields, and the ever-suffering radar pickets.[58]

Although they were to have played only a secondary role in the overall attack, the *Kamikaze* pilots were more destructive and successful in what they did than was the Japanese infantry. In two hours, however, 36 of the 125 suiciders in this raid were shot down according to the claims of antiaircraft gunners ashore and afloat, and those of carrier- and land-based American pilots.[59] Japanese sources note that a total of 159 planes of all types participated in the 3–4 May raid.[60]

Although a barrier of antiaircraft fire kept the conventional bombers at such heights over the airfields and the anchorage that they could cause only superficial damage, the suiciders bore in to inflict wide-spread havoc on the radar pickets. A destroyer and an LSM were sunk; two minelayers and a support landing craft (LCS) were damaged.[61] Enemy bombers again appeared over the island shortly after midnight to hit Tenth Army rear area installations, but as before, accurate AAA fire kept them high over their potential targets and caused the bombing to be erratic. A string of bombs fell near Sobe, however, and crashed through the overheads of IIIAC Evacuation Hospital No. 3, destroying two dug-in surgery wards, killing 13 and wounding 36 patients and medical personnel.[62] Radar-directed TAF night fighters were unable to close with the enemy bombers because American electronic early-warning equipment was disrupted by "window" that had been dropped by four Japanese reconnaissance aircraft.[63]

Beginning on 4 May at 0600, and for four hours thereafter, the *Kamikazes* pushed a murderous onslaught against the radar pickets to coincide with the *Thirty-second Army* ground effort. By the time that the morning forays and the one later at dusk against the escort carrier group were over, the number of naval casualties and ships damaged and sunk was sobering. There were 91 Americans killed, 280 wounded, and 283 missing on the 4th,[64] and on the picket line, two destroyers and two LSMs were sunk; two other destroyers, a minesweeper, a light minelayer, and an LCS were damaged.[65] A turret on the cruiser *Birmingham* was hit by a suicider in the morning attack, and another enemy pilot succeeded in crashing the flight deck of the carrier *Sangamon* in the afternoon, causing an explosion which damaged both elevators and destroyed 21 planes.

[57] *Fifth AirFlt Hist*, p. 68.

[58] *Ibid.*

[59] *CTF 51 AR*, pt III, pp. 75–76; TAF pilots downed 3 planes to bring their claimed total to 145¼ kills. TAF Periodic Rpt No. 4, 3May45.

[60] Hattori, *op. cit.*, table facing p. 132.

[61] *CTF 51 AR*, pt III, pp. 75–76.

[62] Tenth Army G–2 PeriodicRpt No. 40, dtd 5May45.

[63] *Fifth AirFlt Hist*, p. 72. Japanese use of "window" enabled many of their night sorties to reach Okinawa relatively unscathed.

[64] *CTF 51 AR*, pt III, pp. 77–79; pt V, sec H, pp. 7–10.

[65] For a closer view of the radar pickets' sufferings, see Morison, *Victory in the Pacific*, pp. 251–256.

Enemy air did not go unpunished on the 4th, for American pilots claimed to have destroyed a total of 95 Japanese aircraft. ADC flyers had their second most successful day, next to 12 April, for their claims totalled 60¾ kills, bringing TAF claims overall to 206 in less than a month of operations.[66]

The Navy was not concerned solely with helping to beat off the aerial attacks, for Admiral Turner alerted his surface force to the possible threat accruing from suiciders in the enemy *Sea Raiding Squadrons.* He cautioned his "flycatcher" [67] screen of cruisers, destroyers, and gunboats on both coasts to be especially watchful. It was this screen that discovered the shipping engineer regiments attempting to slip behind American lines and assisted the ground forces in combating the counterlandings by illuminating and shelling them. When daylight of 4 May revealed the extent of the Japanese ground effort, the two battleships, five cruisers, and eight destroyers assigned as daytime gunfire support for XXIV Corps joined with artillery and air to blunt the Japanese infantry advance and silence its weapons support.

The Japanese ground offensive began shortly after dark on 3 May with a steadily accelerating rate of artillery fire placed mainly on the frontlines of the 7th and 77th Divisions. As American guns replied in kind, the normal battlefield sounds became an almost unbearable cacophony. In a comparatively less noisier sector near Machinato airfield, LVT(A) crews on guard opened up on unidentified individuals they heard on the beach, and shortly thereafter, support chaft were seen firing at targets in the water just offshore.[68] Less than an hour after this outbreak of firing, the 1st Marines reported enemy barges heading in for shore at Kuwan.[69]

The landing took place here, instead of at Oyama as originally planned, because the landing craft carrying the bulk of the attack force had trouble negotiating the route through the reefs and lost their way.[70] This error was further compounded by the fact that the troops went ashore at the exact point where Company B, 1/1, had anchored its night defense position.

The stealthy enemy approach went undetected by beach sentries and became known only when a clamorous babble signalled the opening of the Japanese attack. This alert resulted in an immediate response from the Marines; they opened up immediately at the overcrowded barges with fire from machine guns and mortars, previously sited to cover the reef. A combination of burning barges, flares, and tracers soon gave the battle scene an infernal glow. This illumination over the reef

[66] TAF PeriodicRpt No. 5, 4May45.

[67] The supporting force assigned craft nightly to a anti-small boat screen. The heavier fire support ships were designated to control the screen, to illuminate points of activity on request, and to harass suspected boat locations. These ships were nicknamed "flycatchers" as the result of their success one night following Admiral Turner's admonition to "be particularly alert as this looks like a fine night to catch flies." *CTF 51 AR,* pt V, sec C, pp. 6, 16.

[68] 3d ArmdAmphBn SAR, Nansei Shoto Operation, dtd 1Jul45, pp. 12–13.

[69] *1st MarDiv G-3 Jnl,* 4May45.

[70] "POW Interrogation No. 38, Superior Pvt, Hq, 2d Co, 26th ShpgEngRegt," *op. cit.*

revealed Japanese heads bobbing in the water and provided the Marine riflemen and machine gunners with targets which they raked unmercifully, blunting the raid.[71] The 1st Marines commander immediately reinforced the threatened area, and LVT(A)s from the 3d Armored Amphibian Battalion took up blocking positions on the reef above Kuwan.

By 0245, those survivors of the ill-fated landing attempts who had gained the beach were being pounded steadily by all available weapons. Despite the immediate Marine reaction to the attempted Japanese envelopment from the sea, some enemy troops managed to infiltrate to the rear of 1/1 before the fighting began on the beach. These raiders were engaged by Company F, 2/1, in an intense fire fight, which ended with 75 enemy dead lying where they had fallen around the Marine positions.[72]

Because he was left with but one rifle company as his regimental reserve, Colonel Chappell requested the attachment of a 7th Marines battalion to his regiment. Division approved the request and ordered 2/7 to move south to report to 1/1 for orders. Preceding the rest of his battalion, Lieutenant Colonel Berger and his staff arrived at the 1/1 CP at 0500 to find all in order and the Japanese threat contained. With the exception of scattered enemy remnants holed up in Kuwan, most of the 300–400 Japanese who had attempted the landing were dead, and were seen either lying on the beach or floating aimlessly in the water amidst the flotsam of the early-morning battle. Lieutenant Colonel Berger's battalion, assigned to mop up the counterlanding area, began relieving the right flank elements of 1/1 at 0645 so that the 1st Battalion could continue the attack to the south.[73]

Other enemy landings were attempted before dawn behind 1st Division lines farther up the west coast. Most of these Japanese efforts were doomed to failure either when the combined fire of naval vessels, LVT(A)s, infantry, and service troops caught the boats in the water or when, by light of day, the few Japanese able to reach shore were hunted down and killed. An estimated 65 enemy landed near Isa in the vicinity of the division CP; some who hid in the cane fields survived until dawn, only to be tracked down by 1st Reconnaissance Company scouts accompanied by war dogs and their handlers.

On the east coast, the counterlanding met with the same lack of success, for the "flycatchers" and 7th Division troops cut the shipping engineers to pieces, killing an estimated 400.[74] Thus, the *Thirty-second Army* gambit failed; there was little indication that the rest of Ushijima's counterattack plan could be fulfilled.

Japanese artillery fire continued through the night of 3–4 May, reaching a deafening thunder at 0430, when a half-hour cannonade was fired in preparation for the *24th Division* attack. Added to the bursting fragments of the high-caliber shells were those of many

---

[71] *1/1 SAR*, p. 8.

[72] 2d Bn, 1st Mar Narrative Rpt-ICEBERG, 23Feb-22Jun45, n.d., p. 4, hereafter *2/1 SAR*.

[73] 2d Bn, 7th Mar SAR, 2May-22Jun45, dtd 2Jul45, p. 4, hereafter *2/7 SAR*.

[74] Tenth Army POW Interrogation Rpt No. 10, Sgt Hiroshi Tamae, Hq, 23d ShpgEngRegt.

thousands of mortar projectiles which fell on the frontlines when the attackers attempted to breach XXIV Corps defenses. The Japanese assault units suffered heavily as they moved through their own fire to gain the American lines. The attack was blunted, however, under a blanket of steel laid down by naval gunfire, air, and 16 battalions of divisional artillery, backed up by 12 battalions of 155mm guns and 155mm and 8-inch howitzers from XXIV Corps artillery.

Beginning at daybreak, the first of 134 planes to fly support for XXIV Corps made its initial bombing run. By 1900, 77 tons of bombs, 450 rockets, and 22,000 rounds of machine gun and cannon ammunition had been expended on Japanese troop concentrations and artillery positions.[75] Even in the face of the *Kamikaze* attacks, gunfire support vessels, from battleships to patrol and landing craft, ranged the coastal waters delivering observed and called fires on enemy targets.[76]

The heavy smoke that *Thirty-second Army* had ordered laid on American lines obscured from the Shuri heights the enemy's view of the progress of the battle. Despite the fact that it was a bald-faced lie, good news, telling of "the success of the offensive carried out by the 24th Division," [77] poured into the army command post at the opening of the attack.

The initial impetus of the attack on the Tenth Army left flank by the *89th Regiment* was blunted by 7th Division troops, who had begun mopping up isolated pockets by noon. In the center, the *22d Regiment*, unable to maintain attack momentum by following up what were to have been "successful" advances by right flank units, spent the day locked in a violent fire fight with 7th Division infantry in the Kochi-Onaga region.

The major drive of the *24th Division,* mounted by the *32d Regiment,* was towards the Urasoe-Mura escarpment, where the *44th IMB* was to exploit and pour through the break it made to hit the rear of the 1st Marine Division. A day-long series of enemy attacks in strength all along the line fell far short of General Ushijima's goals, and darkness found Tenth Army units still in firm control of the escarpment. An inescapable conclusion was that the Japanese push had failed. Not only had XXIV Corps troops securely retained their original positions, but in some cases, even in the face of withering enemy fire and stubborn Japanese resistance, the Americans had attacked and captured some enemy territory.

The 1st Marine Division attack on 4 May was delayed twice, from 0800 to 0900 and then to 1000, owing to the need for its units to be reorganized and resupplied. As soon as the assault battalions of the 1st and 5th Marines resumed their advance, heavy and well-placed fire from the *62d Division* pinned down the left flank company of 1/1 east

[75] *XXIV Corps Arty AR*, p. 8; Anx C, encl No. 2, p. 8.

[76] *CTF 51 AR*, pt V, sec C, p. 53.

[77] *Okinawa Operations Record*, p. 98. Students of the Pacific War will note that, in general, Japanese battle commanders consistently were unable or refused to report their failures or the prospect of an imminent defeat until long after these were all but accomplished facts.

310-224 O - 69 - 15

of Jichaku, and caused heavy casualties. The left flank of 3/1 received machine gun fire from both its front and from its flanks, and was unable to advance. By 1700, however, except for a short stretch of enemy territory extending from the gap in the left center of 3/1 lines to the eastern edge of Jichaku, the leading elements of the 1st Marines were only a few hundred yards away from the final regimental objective, the north bank of the Asa Kawa. In mid-afternoon, division attached 3/7 to the 5th Marines, and Colonel Griebel moved the battalion into blocking positions behind his 3d Battalion.

As darkness fell on the evening of 4 May, its gloom was no greater than that already pervading *Thirty-second Army* headquarters. As the shambles of the thwarted counterattack were surveyed by the staff, it was quite apparent that the effort was a failure. The commander of the *24th Division*, General Amamiya, nonetheless, ordered the *32d Regiment* to try again after dark what it had failed to accomplish earlier that day. Following an extremely heavy artillery and mortar barrage, the regiment hit the frontlines of the 77th Division at 0200 on 5 May in an attempt to penetrate the 306th Infantry positions. Despite the blunting of its initial attack by American artillery, the *32d* returned at dawn, this time with armored support. The assaulting force received the same reception it had been given earlier; six tanks were destroyed and the remnants of the regiment forced to withdraw. In the course of these attacks, *3/32* had suffered crippling casualties and the *1st* and *2d Battalions* of the regiment had been wiped out.[78]

The survivors of the several counterattacks were hunted down by Tenth Army troops at the same time the frontline divisions consolidated their positions and prepared to resume the advance. Only in the 1st Marine Division zone was the pattern of enemy opposition consistent with that occurring before the counterattack of the 4th. Desperation arising from the failure of the major *24th Division* effort further spurred the shaken troops of the *62d Division* to make a more steadfast stand against Marine advances. Japanese strength was concentrated on the left of the 1st Division zone, where last-ditch attempts were made to guard the vital western approaches to Shuri. Attacking platoons were hit from all sides by fire emanating from caves, pillboxes, and fortified tombs.

Overcoming this opposition with difficulty, General del Valle's troops made substantial gains during the day. On the left flank, the 5th Marines registered encouraging progress, advancing up to 600 yards in some parts of its zone. Following close behind the 1st Marines, the 7th filled the gap on the right flank which resulted from the eastward swing of the division. In the course of the day's action, the 1st Marine Division succeeded in reaching the Asa Kawa; and by evening, frontline units began digging in on the commanding ground overlooking the river line,

[78] *Okinawa Operations Record*, "Record of the 24th Division," pp. 174, 177, states: "The 2d Battalion was completely enveloped by the enemy and its escape became impossible."

awaiting new enemy counterattacks which never materialized.

Casualty figures following these two days of battle revealed that the 7th and 77th Divisions, which had felt the full fury of the counterattack, lost a total of 714 soldiers killed, wounded, and missing in action.[79] The 1st Marine Division, which had continued its southerly drive in the same period, suffered corresponding losses totalling 649 Marines.[80] Reflecting the fury with which the enemy had fought and the punishment that he had sustained, the Japanese losses were at least 6,227 men,[81] all dead and almost all of them irreplaceable veteran infantry troops.

Checked by the tremendous firepower of the Tenth Army, each Japanese division in the attack had been chopped down to approximately 20 or 25 percent of its original strength, and enemy artillery strength was halved. In addition to these losses, Ushijima lost 59 artillery pieces destroyed in American air-naval gunfire-artillery bombardments. As a result, never again in the Okinawa campaign did Tenth Army troops receive such intensively destructive Japanese artillery fire as that which had preceded the doomed enemy counterattacks. The net result of this two-day Japanese effort was that the *Thirty-second Army* was compelled to abandon the offensive on the evening of the 5th and to return to its old positions.[82]

In the end, the decisive defeat of the Japanese counterattack bore out the dire predictions of Colonel Yahara. The senior operations officer also won a tearful promise from his army commander that his, Yahara's, counsel would be followed in the future. The defensive pattern then in effect in the *62d Division* zone was to be duplicated across the entire army front. Additionally, the *24th Division* and the *5th Artillery Command* were to reorganize; their tactics would be revised to consist of holding actions in previously prepared and strongly fortified positions. This revision would force the Americans to advance in the face of withering fire, gaining little. The final judgment on the worth of the Japanese counterattack was given by its strongest proponent, General Cho. "After this ill-starred action," Ushijima's chief of staff was reported by a reliable observer as having "abandoned all hope of a successful outcome of the operation and declared that only time intervened between defeat and the 32d Army." [83]

[79] Appleman, *et. al.*, *Okinawa Battle*, p. 302.

[80] 1st MarDiv WarD, May45.

[81] This is the number of enemy dead that was counted in front of XXIV Corps lines. It may be considered a minimum figure, since the Japanese undoubtedly evacuated some of their dead and some of the wounded who died later.

[82] Hayashi and Coox, *Kōgun*, pp. 142–143.

[83] *Shimada Interrogation.*

Forging Ahead

IIIAC ON THE LINES [1]

On 5 May, Tenth Army ordered the attack to the south continued on the 7th with two corps abreast; IIIAC on the right, XXIV Corps on the left.[2] (See Map IV, Map Section.) With two corps now poised for the assault, Tenth Army assumed direct command of the southern front. The day before the attack, 6 May, General Geiger's CP opened at a new location near Futema, where the operation order was received. Effective at 0600 on the 7th, the 1st Division would revert to IIIAC control and the latter would then take over the zone of action for which General del Valle had been responsible previously.[3] The 7 May attack was a prelude to a second and major assault to be launched on 11 May. (See Map 14.) The objective of the first attack was to gain favorable jump-off positions for the second one, which was to be directed against the Shuri defenses. Before IIIAC units could get into position for the scheduled 7 May attack, the Marines had to get across the Asa River estuary.

General Geiger ordered the 1st Division to attack south on the 7th, with the main effort on its left. The 6th Division was to relieve the 1st on the right of the corps zone with one regimental combat team before 1600 on 8 May. Both Marine divisions spent 6 May readying for the next day's work; General Shepherd's division moved to assembly areas preparatory to its commitment, and the 1st maintained pressure on the enemy. During the course of this day, as 2/7 [4] stood fast along the Asa River estuary line, other elements of the division were unable to budge virtually a stonewall defense. Despite accurate counterbattery fire, enemy artillery activity increased noticeably. Two tanks working in front of the 5th Marines lines were knocked out by well-placed Japanese antitank fire as they blasted enemy cave positions at close range.

The 1st Marines attack zone was narrowed considerably when new regimental boundaries were established. A concentration of the force of the Marine attack against enemy positions in the western approaches to Dakeshi hill defenses was necessary because the 1st Marines line cut back sharply from the Asa Gawa to the 5th Marines positions north of the Awacha Pocket. This situation subjected attacking units of the 1st Marines to punishing frontal and

---

[1] Unless otherwise noted, the material in this section is derived from: *Tenth Army AR; XXIV Corps AR; IIIAC AR; 1st MarDiv SAR; 6th MarDiv SAR, Ph III; 7th InfDiv AR; 77th InfDiv OpRpt*, Okinawa; *96th InfDiv AR; 1st Mar SAR; 5th Mar SAR; 7th Mar SAR; 22d Mar SAR, Ph III.*

[2] Tenth Army OpOrd 7–45, dtd 5May45, in *Tenth Army AR.*

[3] *IIIAC AR*, p. 46.

[4] The 7th Marines took over the division right (coastal) flank at 0730 on the 6th, at which time 2/7 relieved 1/1.

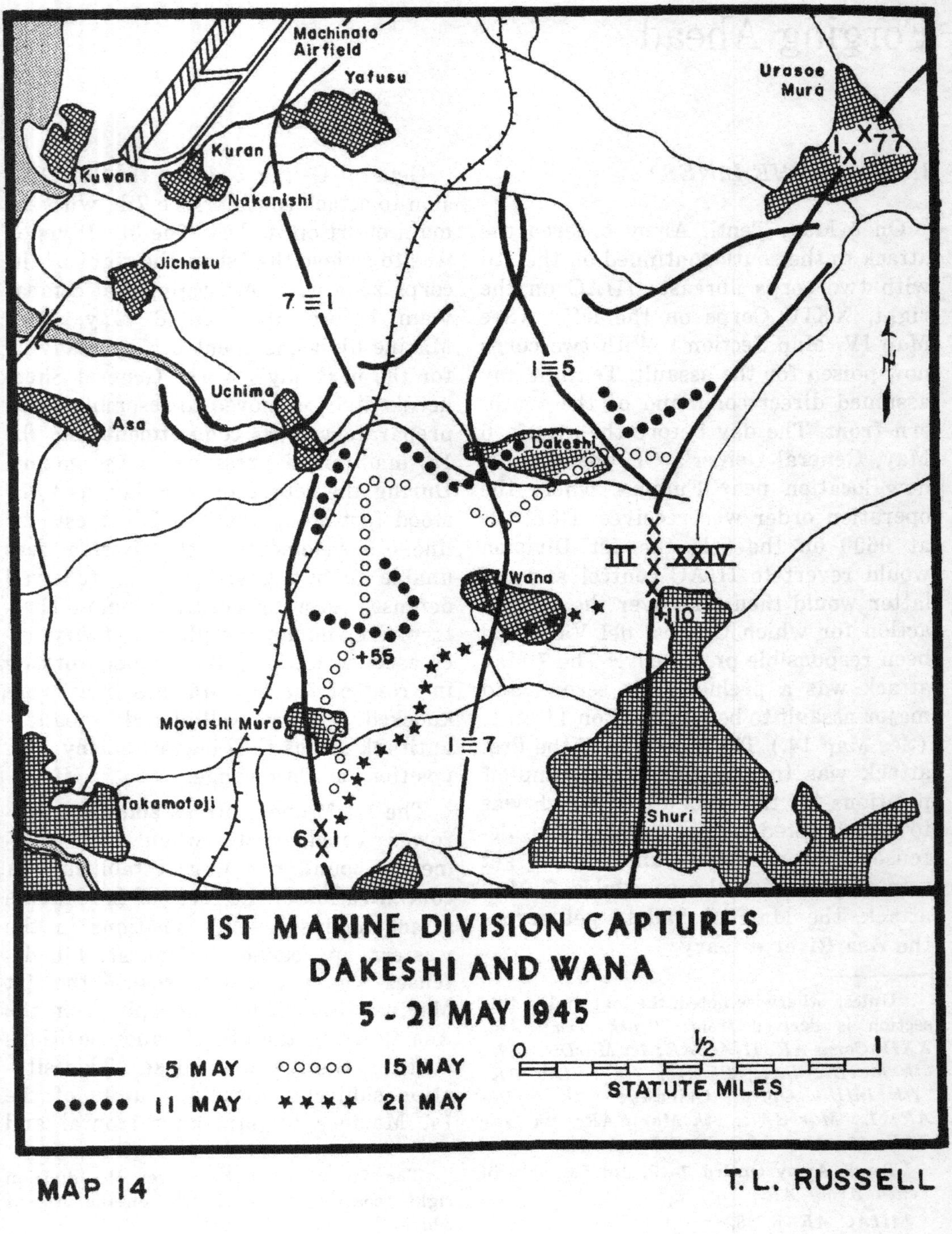

1ST MARINE DIVISION CAPTURES DAKESHI AND WANA 5 - 21 MAY 1945

MAP 14 T. L. RUSSELL

flanking fire from a well-organized maze of hills and ridges protecting Dakeshi.

In a downpour which lasted two days, 3/1 attacked on 6 May. Its attempts to breach the Dakeshi defenses were unavailing and easily fended off by *62d Division* troops. Similarly, the efforts of 2/1 were stymied. Fierce enemy fire held up the attacking Marines and forced elements that had been able to gain even a little ground to withdraw.

Despite an intense four-battalion artillery preparation,[5] and air and naval gunfire bombardment, the 5th Marines could penetrate only slightly into the Awacha Pocket, then held by *23d Independent Infantry* and *14th Independent Machine Gun Battalions.*[6]

On the left, 2/5 moved its lines forward about 200 yards to tie in with the 307th Infantry, after which the Marines coordinated their advance with that of the soldiers. When 2/5 resumed the assault, it called for mortar and artillery fire on enemy reverse slope positions impending the advance of the units on the regimental right flank.

At 0600 on 7 May, General Geiger assumed IIIAC command responsibility for the 1st Marine Division zone and regained control of his corps artillery battalions which had been attached to XXIV Corps up to this time.[7] General Nimmer then reorganized his command into three groups: the IIIAC artillery battalions comprised the first; XXIV Corps Artillery made up the second; and the third group consisted of the 27th Infantry Division Artillery (104th, 105th, and 106th Field Artillery Battalions), which had remained in position when the rest of the division headed north to relieve the 6th Marine Division. The purpose behind Nimmer's action was to provide IIIAC divisions with maximum effective tactical support.

Commanded by Lieutenant Colonel Custis Burton, Jr., the 2d Provisional Field Artillery Group (3d and 6th 155mm Howitzer Battalions and XXIV Corps 145th Field Artillery Battalion) was given the mission of providing general support to the 1st Division and reinforcing the fires of the 11th Marines. Similarly, the 27th Division Artillery under Brigadier General William B. Bradford, USA, was to support the 6th Marine Division and the 15th Marines.[8] Lieutenant Colonel Ernest P. Foley commanded the third group—named after him—consisting of the 7th, 8th, and 9th 155mm Gun Battalions, which were to deliver long-range reinforcing, counterbattery, interdiction, and harassing fires in support of IIIAC generally.

The Marine 1st 155mm Howitzer Battalion remained under XXIV Corps

[5] This was just one of the many examples of interservice artillery missions fired in support of the southern drive. In this particular case, the fires of 2/11 (1st Marine Division) were reinforced by those of the 104th Field Artillery Battalion (27th Infantry Division), 3d 155mm Howitzer Battalion (IIIAC Artillery), and the 145th Field Artillery Battalion (XXIV Corps Artillery). *11th Mar SAR*, S-3 Periodic Rpt for 7May45.

[6] *Okinawa Operations Record*, p. 107.

[7] *1st MarDiv G-3 Jnl*, 7May45.

[8] "If the wording of the orders didn't spell this out—this is what happened in practice. . . . The heavier calibers (2d Provisional Field Artillery Group) were put in support of the 1st MarDiv because [the latter's] 105s and 75s couldn't blast out the Jap caves and bunkers on the Shuri Hill mass." *Henderson ltr 1965.*

Artillery control until 23 May, when it reverted to its parent unit, IIIAC Corps Artillery. During the period that Marine artillery units had supported XXIV Corps (7 April to 6 May), they fired 53,988 rounds in a total of 2,344 missions.[9]

As the 1st Marine Division once again came under IIIAC control, it could look back on the six difficult days of combat it had been with XXIV Corps. During this period, the division sustained 1,409 battle casualties, including 199 men who were either killed or subsequently died of wounds in the fighting to gain the northern bank of Asa Kawa and the outer reaches of the Dakeshi defense system.[10]

The attack of the 1st Marines, scheduled to begin at 0900 on 7 May, was held up because muddy terrain prevented supporting armor from arriving on time. In the meanwhile, Colonel Arthur T. Mason, the 1st Marines new commander,[11] ordered his 3d Battalion to support the attack of 2/1 on Hill 60—a height commanding the battalion front—by bringing all available fire to bear on the reverse slope positions of this enemy-held hill. The supporting fires continued until 1400, when tanks arrived at the front to assist in the attack.

As mortars and assault guns pounded the top and reverse slope of the hill, and artillery fire covered the foot of the objective, the coordinated tank-infantry assault was launched against determined, fanatic, and well dug-in Japanese troops. In less than half an hour, Company E, spearheading the attack, swept to the hilltop in a practical application of "the effect of properly massed supporting fires in front of the assault troops." [12] A hand-grenade duel ensued when the enemy defenders emerged from their caves after the fires supporting the attack were lifted. Almost immediately, the volume of Japanese fire of all types "grew noticeably stronger and progressively more intense so that it was evident that the enemy was receiving large reinforcements." [13] In view of this potential threat, the position was adjudged untenable by the battalion commander, who withdrew his company to their lines of the previous night.

It had become apparent by the morning of 7 May that the deep draw cutting across the front of 1/5 and to the right of 2/5 positions contained the bulk of the enemy's Awacha defenses. At 0900, General del Valle and Colonel Griebel conferred with the commanders and staffs of the two assault battalions, and discussed the methods by which the Japanese positions rimming the draw and studding its steep slopes were to be reduced. It was decided that an extensive air, artillery, and rocket preparation would precede the infantry jump-off scheduled for 1200; a reinforced tank company was moved up in time to support the assault.

---

[9] *IIIAC Arty AR*, pp. 22–26.

[10] *XXIV Corps AR*, p. 29.

[11] Colonel Mason relieved Colonel Chappell on 6 May.

[12] *2/1 SAR*, p. 5.

[13] *Ibid.* It was at this time that General Ushijima attached the *3d Battalion, 2d Infantry Unit* to the *62d Division* "in order to relieve the crisis of the troops fighting in the vicinity of Dakeshi." *Okinawa Operations Record*, p. 106.

The fighting that afternoon was marked by tactics which "General Buckner, with an apt sense of metaphor, called . . . the 'blowtorch and corkscrew' method. Liquid flame was the blowtorch; explosives, the corkscrew." [14] Marine flamethrower and demolition teams burnt out and sealed many of the enemy cave installations in their zone. By 1700, the time the battalions dug in for the night, 1/5 had gained 300–400 yards in the center, but 2/5 and 3/5 could do little more than attempt to straighten their lines. Even though *62d Division* troops holding this area gave way slightly during the day, it was obvious that Awacha Pocket was not going to be taken quickly or easily.

News of the collapse of Nazi Germany and the announcement of V–E Day on 8 May, the day of the Allied victory in Europe, drew little response of any sort from either side on Okinawa. Most of the cold, rain-soaked Americans and Japanese in the frontlines were concerned only with that very small but vital part of the war where their own lives were at stake. Still, V–E Day did not go unnoticed. The Americans conducted Thanksgiving services on board many of the ships off Okinawa. In addition, the voices of naval guns and artillery pieces helped in the celebration. At exactly noon, every available fire support ship directed a full-gun salvo at the enemy; [15] in addition, three battalions from IIIAC Artillery massed fires on a suspected Japanese CP.[16] The results of the noontime shoot were not ascertained, but, in the words of one observer, "It made one hell of a big noise." [17]

Heavy, driving, and cold rains on the 8th continued to immobilize Tenth Army troops. The attack was bogged down in the 1st Marine Division zone; but in the area directly in front of the lines, numerous caves and pillboxes were destroyed in a general mop-up. The 1st and 5th Marines each received a battery of 75mm pack howitzers that were manhandled up to the front in an unsuccessful attempt to place direct fire on enemy dispositions.[18]

General Shepherd's 22d Marines, selected to lead his 6th Division drive in the south, moved out from Chibana on 8 May, and by 1530 the same day its 1st and 3d Battalions had relieved the 7th Marines along the Asa Kawa.[19] At 1600, the 6th Division commander assumed

[14] Appleman, *et. al.*, *Okinawa Battle*, p. 256.

[15] *CTF 51 AR*, pt III, p. 84.

[16] *IIIAC Arty AR*, p. 23.

[17] *CTF 51 AR*, pt V, sec C, p. 21.

[18] 1st MarDiv SAR, pt VII, p. 3. "One battery was sent to the 1st Marines, another to the 5th Marines (over artillery protest). Since neither the organic M–7's of the infantry nor the tanks would function satisfactorily [in the mud], it was obvious that the truck-drawn pack howitzers could not. Neither battalion accomplished anything. The battery with the 5th Marines was never used. Both batteries were returned on 11 May, but their parent battalion was emasculated throughout the period of their absence. 1/11 was then built up to a strength of 24 guns by adding LVT(A)'s. 75mm ammunition was plentiful, as contrasted with the heavier calibers, so 1/11 (reinforced) was used to fire interdiction, harassing, and 'appeasement' missions across the front. Later, two more batteries of LVT(A)'s were formed as artillery to thicken the fires of 2/11 and 3/11." *Brown ltr II*.

[19] The 7th Marines moved to an assembly area just east of Nakama where it became 1st Division reserve.

responsibility for his zone in the corps front. Once in position opposite the Asa Kawa estuary, the 6th spent 9 May in patrolling and reconnoitering before the crossing was attempted. A patrol from 3/22 (on the division right) inspected a ruined bridge crossing the river and later reported that it was not passable to either foot or vehicular traffic, that the water in the estuary was four feet deep at high tide in its most shallow portion, and that the river bed had a thick mud cover. Other patrols, also from 3/22, were sent out at noon to discover possible crossing sites and to determine enemy strength and dispositions. These patrols drew fire from positions across the river, but noted that other caves and pillboxes farther south appeared to be unmanned. The Marines also reported that the soft stream bed could not support a tank ford.

To the left of 3/22 was the 2d Battalion in positions near Uchima. Its only enemy contact during the day took place when a patrol that crossed the river drew heavy enemy fire. Before withdrawing, the Marines rescued both the pilot and observer of an artillery spotting plane that had been shot down in enemy territory. As a result of the information collected by the patrols, both the division and regiment were better able to plan for an effective exploitation of a beachhead following the crossing of the Asa River.

The 6th Engineer Battalion moved light bridging material up to the 22d Marines line in daylight, and under the cover of darkness began constructing a footbridge near the site of a ruined bridge. At 0300, 10 May, 3/22 was to cross the river with 1/22 in support, and then attack the high ground overlooking the south bank of the Asa at dawn. The 2d Battalion was to provide fire support from a strongpoint set up on high ground southwest of Uchima. This river crossing was only a part of the all-out army attack scheduled for 11 May. Envisioned in the objectives of the plan were the envelopment and destruction of enemy forces occupying the Shuri bastion, and finally, the total annihilation of General Ushijima's command.[20]

General del Valle's division made fairly substantial gains on 9 May, even in the face of miserable weather conditions which prevented the attack from being launched until 1200. When armor support became available and was able to move forward over the muddy terrain, the troops advanced 200–300 yards and generally straightened the division line. Until this attack, Colonel Mason had kept 1/1 in reserve, where the battalion took in 116 replacements for the 259 casualties it had sustained in the period between 30 April and 6 May.

On 9 May, the immediate objective of 1/1 was to penetrate and destroy enemy defenses on Hill 60. Just before the attack began, the assault battalion moved behind 2/1, which was to exploit the successful penetration by seizing and consolidating the captured positions; 3/1 was to support all of this action with fire.

Because there was no contact with the 5th Marines on the left, a general shift was made into its zone to wipe out the

[20] Tenth Army OpO 8–45 was placed into effect on 9 May. *Tenth Army AR*, chap 7, sec III, p. 19.

sources of enemy fire that were taking a heavy toll of troops in 1/1. As a result of a combination of heavy casualties and exhaustion occurring in the ranks of 1/1, its attack almost bogged down, but was revitalized at 1600, when the Marines swept forward over the difficult final 150 yards to gain the initial objective line.[21]

Following a thorough air, artillery, naval gunfire, and mortar preparation, the 5th Marines attacked the mouth of Awacha Draw at noon on the 9th, with 3/5 and 3/7 in the assault; 1/5 and 2/5 furnished fire support from positions facing the draw. Initially, the attack moved rapidly and the first objective—the same ridgeline that had faced 1/1—was soon reached, but fire in large volume on 3/7 from the left of its exposed flank held up the attack. At 1515, 1/7, which had moved up from Gusukuma that morning, was committed in the line to fill the gap that had appeared between 3/7 and 1/5.[22]

To overcome the determined stand of Awacha's defenders, and spurred on by the need to continue the division attack to gain the objectives assigned by Tenth Army, General del Valle issued a new operation order late in the afternoon of the 9th. Assigned a limited zone on the left of the division front, the 5th Marines (less 3/5) was to reduce the Awacha Pocket beginning with an attack the next morning. New boundaries were given the 1st and 7th Marines (with 3/5 attached) which placed them in jump-off positions across the division front for the planned 11 May attack.[23] Colonel Snedeker assumed responsibility for the new 7th Marines zone at 1855, relieving the 5th, and placed 3/5, 3/7, and 1/7 on line for the 10 May attack. (See Map 14.)

As an aftermath to the unsuccessful 4–5 May counterattack, the Japanese attempted to readjust and reinforce their lines against expected American reprisals and a continuation of the Tenth Army onslaught. To gain the time and breathing space needed to rebuild the *62d Division* somewhat, General Ushijima gradually withdrew the division from the tangled Maeda–Asa Kawa complex toward Shuri where replacements from the *Naval Base Force*, service and supply troops, and *Boeitai* could join. Remaining in front of the 96th Infantry Division in positions extending from Dakeshi to Gaja were the *22d* and *89th Regiments* of the *24th Division*, which had been reinforced in the second week of May by division troops and the now-defunct sea raiding base battalions.

IIIAC forces were opposed by the *44th Independent Mixed Brigade*, which

[21] In late afternoon, the battalion commander, Lieutenant Colonel James C. Murray, Jr., was inspecting his night defenses when he was hit by a sniper. Before being evacuated to the rear, he appointed the Company B commander, Captain Francis D. Rineer, temporary battalion commander; Rineer directed the 1/1 night defense until relieved the next morning (10 May) by the 1st Marines executive officer, Lieutenant Colonel Richard F. Ross, Jr. The latter, in turn, was relieved by Lieutenant Colonel Austin C. Shofner, formerly Division Provost Marshal, on 13 May, and then returned to his duties as regimental executive officer, *1/1 SAR*, p. 14; *1st Mar SAR*, p. 14.

[22] 1st Bn, 7th Mar SAR, 22Apr-22Jun45, dtd 29Jun45, p. 5, hereafter *1/7 SAR*. On 9 May, both 1/7 and 3/7 were attached to the 5th Marines. The CP of the 7th Marines (–) remained near Nakama. Late that afternoon, 3/5 was attached to the 7th.

[23] 1st MarDiv WarD, May45, p. 8.

was built around a nucleus composed of the *15th IMR*, now at full strength, and fresh replacements from the *3d Battalion, 2d Infantry Unit, 7th Independent Antitank Battalion, 1st* and *2d Independent Battalions*, and the *26th Shipping Engineer Regiment*. Although American forces were in close contact with the enemy and fully engaged by them at Dakeshi, it was difficult to determine Japanese troop strength in the area because the core of this strength was screened by outposts and scattered strong defensive positions still held by remnants of the *62d Division*.

During the night of 9–10 May, the enemy was particularly active in the 1st Marine Division zone, and he made numerous attempts to infiltrate the 5th Marines area. Between 0200 and 0300, 1/5 fought off two counterattacks in which the Japanese had closed to bayonet range; enemy troops were driven off only after an extended hand-to-hand battle. After daybreak, upwards of 60 enemy bodies were found in front of the battalion lines. Despite this early-morning action, the division continued the attack at 0800, following a heavy artillery and smoke preparation, with three infantry regiments abreast.

Although General del Valle's troops encountered stiff opposition all along the line, the 5th Marines, on the corps left, received the most violent enemy reaction from positions centered around the Dakeshi Ridge and the high ground running generally along the corps flank. Armor assigned to support the 5th Marines attack did not arrive in time for the jumpoff; poor roads had again bogged down the tanks and prevented them from aiding the infantry.

Less than an hour after the attack began, 1/5 was pinned down by heavy enemy machine gun and mortar fire that skyrocketed casualty figures. At 1700, the battalion was withdrawn under the cover of smoke. Although 1/5 was unable to move forward, the rest of the regiment made inroads into enemy positions. Supported by artillery and flamethrower tanks, 2/5 overran all enemy resistance in that portion of Awacha Draw which lay in its zone. This action placed the regiment in the heart of the Awacha defenses; it did not account for the many other remaining Japanese pockets which the 5th Marines was to meet in the next few days.

As the 5th fought its battle, the 7th Marines attacked with the 1st and 3d Battalions in the assault, 2/7 and 3/5 in reserve. On the right, 3/7 was immobilized at its line of departure by accurate mortar and artillery shelling, and heavy small arms fire from pillboxes and caves to its front. The 1st Battalion, however, attacked on time; by noon, its forward elements were on the low ground north of the Dakeshi Ridge, where visual contact was reestablished with 1/5. Japanese fire on this advanced postion increased as the morning wore on, and shortly before noon, machine gun fire from a draw in the 5th Marines zone began hitting the 1/7 assault company. This fire continued unabated and finally halted the attack. When 1/5 was forced to withdraw, 1/7, exposed on both flanks, found its position untenable. The 7th Marines assault units were pulled back to their original lines at 1754.

For the 10 May attack, Colonel Mason's 1st Marines was given the task of gaining the road leading west out of Dakeshi. Although 3/1 jumped off and reached its planned intermediate objective on time, it was forced to hold up and wait for the 1st Battalion, which did not begin its scheduled advance because of the late arrival of its supporting armor.[24] At 1020, both battalions resumed the attack, now supported by armor, and reached a low ridge overlooking the Dakeshi road at 1600.

All attempts by assault companies to move beyond this point were met by extremely heavy machine gun fire from Dakeshi ridge, driving back the combat patrols attempting to bypass the nose of the ridge. It soon became painfully evident that no further advance would be possible until the ridge was taken.

While the 1st Division set in its defenses for the night, the 6th Division remained active. As soon as it became dark on 9 May, the 6th Engineer Battalion began building a footbridge for the planned infantry crossing over the Asa Kawa estuary. At 0530, two and a half hours after 22d Marines assault elements had crossed over to the south bank of the river, a Japanese two-man suicide team rushed out of hiding to throw themselves and their satchel charges onto the south end of the footbridge; both the bridge and the enemy soldiers were destroyed. Prior to this destructive act, however, 1/22[25] and 3/22 each had succeeded in moving two assault companies across the river under the cover of darkness, and to positions for a continuation of the attack to the south. The loss of the bridge, therefore, posed no great hardship; included in the attack plans were contingency provisions that were to go into effect if this, in fact, took place. Therefore, when the bridge was blown up, engineer demolition teams with the assault elements breached the seawalls on the south bank of the river to permit immediate access to the frontlines to supply- and troop-laden LVT(A)s.

At 0520, under the cover of a protective smoke screen and an artillery preparation, the attack south of the Asa began. The assault companies were at first hampered by fog and the smoke of battle. During the early morning hours, enemy resistance was moderate and limited to small arms and machine gun fire. Soon, however, Japanese artillery shells began falling on the bridgehead area.

By noon, the Marines had succeeded in driving only 150 yards into enemy defenses, while, at the same time, the volume of both small arms and artillery fire increased steadily. Under the cover of heavy supporting fire, each assault battalion brought its reserve company across to join the attack. Even with continuous artillery and naval gunfire support, the ground troops could not crack the Japanese line.

The only weapons capable of breaching these defenses—tank-mounted flat trajectory cannon—were not available because supporting armor was unable to ford the mud- and silt-bottomed stream despite its numerous attempts to do so. The tanks were then forced to withdraw to the northern bank of the

[24] The 1st Battalion was held up when a tank from the 6th Tank Battalion hit a mine and blocked the only tank road into the 1/1 area.

[25] Company A of 1/22 crossed over by wading the shallow eastern portion of the stream.

Asa, there to await the construction of a Bailey bridge, which the 6th Engineers were to begin building after dark on the night of 10–11 May.

The 22d Marines advanced along their entire front during the afternoon; the 1st Battalion made the greatest gain, 350 yards. As darkness fell, the division halted and forward companies dug in for the night. The engineers began work on the Bailey bridge at 2200, stopping only when the crossing site was shelled. This intermittent shelling successfully delayed completion of the bridge by six hours. The first Marine tanks did not cross the river until 1103 on 11 May, some four hours after the coordinated, two-corps Tenth Army attack had begun.[26]

AIR OPERATIONS IN MAY [27]

A mass *Kamikaze* raid on 10–11 May, the fifth of the campaign, unintentionally served to preface the 11 May ground attack of the Tenth Army. There was no indication that the Japanese had prior knowledge of the impending attack or that their air assault had been planned to forestall the American push. An enemy air raid should have been anticipated, however, because the first relatively clear weather since the previous *Kikusui* attack appeared at this time.

Between midnight of 10 May and 0420 on the following day, the Air Defense Control Center plotted 19 enemy raids approaching Okinawa, each one ranging in size from one to nine aircraft. Most of the planes orbited over the water about 40 miles northwest of the island, where they formed up for a furtive pass at American targets. None, however, approached any closer than 10 miles to Okinawa.

This situation soon changed on the 11th, when, at 0630, TAF pilots intercepted the first in a series of suiciders attempting to crash targets in the Ie Shima and Hagushi anchorages. By the time of this mass raid, enemy air tactics generally followed the pattern previously observed during the major landings in the Philippines, but with an increased emphasis on the use of the *Kamikaze*. Like the tactics employed at Leyte, at Okinawa attacking groups approached at altitudes ranging from 9,000 feet to sea level; the low-level approach was usually made during periods of limited visibility. Japanese pilots also would approach a target at low altitudes if their attack was covered by clouds and poor visibility, or when they felt that American radar units could not detect their planes. While this indicated

[26] 6th MarDiv Unit Jnl, Okinawa Operation, Ph III, 23Apr-30Jun45, dtd 11May45, hereafter *6th MarDiv Jnl, Ph III*. In commenting later on the 6th Engineers' effort, the 6th Division Commander noted that "the task of constructing the Bailey bridge across the Asa Kawa Estuary was a splendid feat of combat engineering. It was not only built under observed fire, both artillery and small arms, but it was built in a difficult locality where the actual engineering constituted a considerable problem. Even the dump site where the bridging material was stacked was occasionally under fire." *Shepherd memo II*.

[27] Unless otherwise noted, the material in this section is derived from: *CNO Record; CTF 51 AR; TAF AR; ADC AcftOperAnalysis*, 1-31May45; *ADC IntelSums*, May45; Morison, *Victory in the Pacific;* Sherrod, *Marine Air Hist*. Also, May45 WarDs of following organizations: TAF, 2d MAW, ADC, and MAG–14, –22, –31, and –33.

some training in low-level evasive tactics, it did not show that enemy pilots had a proper appreciation of the range and coverage of American radar.[28]

At Okinawa, enemy air activity usually began when the final night CAP had withdrawn to home fields, and ended when the early morning patrol flight approached. As successful as the alert American air patrols had been in protecting assault shipping and radar picket vessels, it was impossible to prevent losses as long as even a single Japanese plane penetrated the ICEBERG air screen. In at least five instances in May, *Kamikazes* that had been so seriously damaged by fighter aircraft that they could not have possibly returned to home bases—and conceivably could not have even recovered level flight—managed to remain on course, break through the American screen, and hit their targets.[29]

On-station planes from TF 51 and TAF fended off the raids of 11 May, but not before a Dutch merchantman, two American destroyers, and an LCS had been hit. A total of 217 Japanese planes were employed in *Kikusui* No. 6; 104 of these were suiciders.[30] The claims of opposing sides regarding the number of planes their pilots had downed conflict similarly in this particular engagement as they did in others. Uniquely enough, the number the Japanese admitted losing in this air battle was in excess of the number that the Americans claimed that they had shot down. A Japanese source lists 109 of their planes shot down or missing,[31] while ICEBERG forces claimed only 93. Of this number, the two destroyers that were under attack have been credited with blasting 34 enemy aircraft out of the sky; ships' AAA and defending air patrols claimed the remainder. TAF pilots downed 19 planes of this last portion in slightly more than two hours of fighting in the morning, and increased the score of the Tenth Army air arm to 234.[32]

A sidelight of the 11 May raid occurred when USS *Hugh W. Hadley*, a radar picket, was under direct attack from *Kamikazes*. Protecting the destroyer overhead was a two-plane CAP maintained by VMF-323 pilots.[33] The conduct of these Marine flyers is best described by the ship's action report: "One very outstanding feat by one of these two planes . . . was that, though out of ammunition, he twice forced a suicide plane out of his dive on the ship, and the third time forced him into such a poor position that the plane crashed through the rigging but missed the ship, going into the water close aboard. This was done while all guns on the ship were firing at the enemy plane . . . His wingman also stayed at masthead height in the flak and assisted in driving planes away from the ship." [34]

[28] *CNO Record*, chap 3, p. 1.

[29] *ADC AcftOpAnalysis*, 1-31May45, p. 7.

[30] Hattori, *op. cit.*, table facing p. 132.

[31] *Ibid.*

[32] TAF PeriodicRpt No. 6, 11-17May45.

[33] VMF-323 WarD, May45; Sherrod, *Marine Air Hist*, p. 394. Although the Aircraft Action Report for this particular sortie, included in the war diary, did not specifically describe the feat noted above, Sherrod deduced that these were VMF-323 planes because he later identified the call letters, which appeared in the 11 May ship's log entry relating to this action, as theirs.

[34] Quoted in *ADC AcftOpAnalysis*, 1-31May-45, p. 8.

On the day before, Captain Kenneth L. Reusser and First Lieutenant Robert R. Klingman of VMF-312 destroyed an enemy plane in a manner described as "one of the most remarkable achievements of the war." [35] Several times earlier in the month, extremely fast Japanese reconnaissance craft—apparently on photographic missions—had been encountered at high altitudes, usually 30,000-38,000 feet. Klingman was flying wing on Reusser, division leader of a four-plane CAP, which was then at an altitude of 10,000 feet. Reusser noticed the presence of vapor trails at about 25,000 feet, and obtained permission to investigate. He led his division in a climb to 36,000 feet, where two of the planes were forced to disengage after reaching their maximum altitude. Klingman and Reusser continued to climb and close with the Japanese intruder only after they had fired most of their ammunition to lighten their aircraft.

At 38,000 feet, they intercepted the enemy and Reusser opened fire first. Expanding all of his remaining ammunition in one burst, he scored hits in the left wing and tail of the enemy plane. Klingman then closed in, but was unable to fire because his guns had frozen at this extreme altitude. After a two-hour air chase, he finally downed the Japanese plane by cutting off its tail control surfaces with his Corsair's propeller. Although Klingman's plane had holes in the wing and engine, and the propeller and engine cowling were damaged, he managed to land the plane intact and without injury to himself.[36]

An almost tragic aftermath to this encounter occurred two days later, when Klingman was flying another mission. His plane's hydraulic system failed and he chose to bail out over the water rather than attempt a crash landing on one wheel. A destroyer escort recovered the lieutenant from the water and carried him to the *Eldorado*, where he had dinner with the Expeditionary Force commander, Admiral Turner.

Marine and Navy night fighter aircraft came into their own during May, especially with the arrival at Okinawa on the 10th of Lieutenant Colonel Marion M. Magruder's VMF(N)-533 following its long over-water flight from Engebi, Eniwetok Atoll, in 15 F6F-5Ns (radar-equipped Grumman Hellcats) and 5 transport planes.[37]

---

[35] Unknown source quoted in MAG-33 WarD, May45.

[36] *Ibid.; ADC AcftOpAnalysis*, 1-31May45, p. 8.

[37] VMF(N)-533 WarD, May45. Before Lieutenant Colonel Magruder had received orders to take his squadron to Okinawa, the pilots who had been with the unit since arrival overseas in May 1944 were scheduled to be replaced. The relief pilots arrived at Engebi on the same day that VMF(N)-533 was ordered to the Ryukyus. According to Lieutenant General Louis E. Woods, who commanded the 4th MAW, of which the squadron was a part: "The higher echelon of command did not feel that [Magruder's planes] had the range to fly from Engebi, but [the squadron] had very accurate fuel consumption records that proved beyond a doubt that they could." After Magruder received permission to fly the squadron to Okinawa, Woods "went to Engebi personally and talked to all the pilots, telling them they were badly needed in the war zone even though their tour of duty was up and their replacements were present. All concerned agreed that the original members of the squadron should go." LtGen Louis E. Woods ltr to Asst G-3, HQMC, dtd 30Oct65, hereafter *Woods ltr I*.

Prior to the Okinawa operations, the quality of the direction and conduct of night CAPs was poor and the results of most operations negligible. As an example, in the first year of night fighter operations—November 1943 through October 1944—Navy and Marine pilots accounted for only 39 enemy aircraft; in less than two months of the Okinawa campaign, night fighters shot down 35 Japanese planes,[38] and VMF(N)-533 fliers claimed 30 of them. Noteworthy is the fact that, with six enemy planes to his credit, Captain Robert Baird of 533 was the only Marine night fighter ace in the war.[39]

The drastic change for the better in night fighter squadron operations resulted from improved electronic equipment, techniques, and performances of both pilots and the ground director crews. The new Hellcats also were a large factor in this improvement. With the arrival of additional air warning squadrons and their radar equipment on Okinawa, and their establishment on outlying islands as they were captured, the intricacies of guiding night fighters to targets were overcome. Within a short time after their appearance in the Okinawa battle zone, fighter directors could bring a pilot to within 500 feet of an enemy plane, at which point the flyer could establish visual contact with the intruder aircraft and down it.

One of the most spectacular, unique, and perhaps the only air-to-air rocket kill in the war occurred in the early-morning darkness of 17 May, when VMTB-232 pilot First Lieutenant Fred C. Folino spotted an unidentified plane while flying his TBM (Avenger) on a night heckling mission. He radioed an ICEBERG control ship for information and identification of the stranger, all the while climbing to gain altitude to get into attack position. Assured by the controller that there were no friendly planes in the area, and having requested and received permission to attack, the Marine pilot dove on the now-fleeing enemy. Folino expended all of his ammunition as the torpedo bomber strained to close the gap. He then began firing his rockets. The first was short of the target, the next one struck the plane, and a third tore off a large portion of the wing. "Momentarily lost to the TBM, the plane next appeared on the beach below, a blazing wreck."[40] Acknowledging this act, Admiral Spruance sent his personal congratulations to Lieutenant Folino.

After dark on 17 May, TAF pilots extended the range of their operations to Japan for the first time. The arrival of AAF fighter squadrons and their Thunderbolts (P-47s) on Ie Shima in mid-May provided General Wallace's ADC with a long-range strike capability. This was demonstrated when a pair of Thunderbolts rocketed and strafed three airfields on southern Kyushu on this first extended mission, and then added to the insult by strafing the brightly lit streets of Kanoya before

[38] Naval Aviation Confidential Bul No. 4-45, pp. 11-13, quoted in *ADC AcftOpAnalysis*, 1-31May45, pp. 10-11.

[39] For a more complete story of the early development and organization of Marine night fighter squadrons, see Sherrod, *Marine Air Hist*, chap 11, pp. 159-169.

[40] TAF WarD, May 45.

returning to home base unchallenged by enemy pilots. Following this night run, AAF pilots began making daylight runs over southern Japan. Judging by the quality of the opposition they received, it became apparent that the Japanese had been holding back their more skillful pilots and newer and faster aircraft for the close-in defense of the Empire. The American pilots reported that they were encountering "pilots who were . . . skillful and aggressive. . . ." [41]

The majority of the enemy planes rising to meet the Army Air Force's flyers were Zekes, but some were the newer single-seat fighters, the faster Jacks and Franks,[42] which had just begun to appear in the air war over Okinawa and Japan. In the final analysis, the AAF Thunderbolts outperformed the Japanese aircraft without exception. They outclimbed and outturned the enemy planes and especially excelled in the high altitudes, where Japanese aircraft performance had been superior earlier in the war. If nothing else, these performance factors and the raid on Kyushu further dramatized the complete ascendancy of American air power. This evidence, however, did not convince the Japanese that continuation of the mass *Kamikaze* raids was merely an exercise in futility.

[41] TAF, Tenth Army, AirOpNotes, No. 2, dtd 15Jun45.

[42] For a fuller discussion of American nomenclature for Japanese aircraft, see Shaw and Kane, *Isolation of Rabaul*, p. 450; also, Sherrod, *Marine Air Hist*, pp. xiv, 135*n*. A primary source of information for both of the books noted above was Vern Haugland, *The AAF Against Japan* (New York: Harper & Bros., 1948), pp. 367–371.

310-224 O - 69 - 16

American planes rising from crowded fields on Yontan and Ie Shima successfully blunted the suicide attacks, and as a result, *Special Attack Force* aircraft and pilot losses mounted all out of proportion to the results achieved. *IGHQ* then decided that the only way to reverse the situation was by destroying the U. S. planes at their Okinawa fields. A surprise ground attack mission was therefore assigned to the *Giretsu* (Act of Heroism) *Airborne Raiding Force.* Armed with demolition charges, grenades, and light arms, the commandos of this unit were to land on Kadena and Yontan fields, where they would make one desperate effort to cripple American air operations—even temporarily—by destroying or damaging planes and airfield facilities. The men undertaking this raid were to be flown to Okinawa on the night of 24 May [43] in planes that would accompany those in the formation of *Kikusui* No. 7.[44]

The *Giretsu,* consisting of 120 men, was divided into five platoons and a command section, and was transported to the assigned target in 12 twin-engine

[43] While intercepting an enemy flight in the early evening twilight of 24 May, a VMF–312 pilot reported sighting and downing a Tony (single engine fighter). Interestingly enough, the Marine flyer reported that the Japanese plane was dark gray and had a black cross painted on the underside of each wing, very much like the Maltese Cross markings on German military aircraft. VMF–312 WarD, May–45, Although the configuration of the Tony and the Messerschmitt ME–109 were similar, a search of available records does not indicate the presence of either German pilots or planes in the Pacific, nor is any reason given for a Japanese plane to be painted in this fashion.

[44] Hattori, *War History*, p. 129.

bombers.[45] The general attack began at about 2000, when Yontan and Kadena fields were bombed as a prelude to the airborne raid. Approximately two and a half hours later, antiaircraft artillerymen and aviation personnel based at Yontan were surprised to see several Japanese bombers purposely but rashly attempting to land. With one exception, the planes that were shot down over the field either attempted to crash ground facilities and parked aircraft or went plummeting down in flames, carrying entrapped troops with them.

The plane that was the one exception made a safe wheels-up landing, and troops poured out even before it had come to a halt. As soon as the raiders deplaned, they began to throw grenades and explosive charges at the nearest parked aircraft, and sprayed the area with small arms fire. The confusion which followed this weird gambit is difficult to imagine. Uncontrolled American rifle and machine gun fire laced the airfield and vicinity, and probably caused most of the ICEBERG casualties. TAF pilots and ground personnel, as well as the men in the units assigned to airfield defense, took part in the general affray, which saw the death of 2 Marines and the wounding of 18 others.[46]

When the attack was over, no prisoners had been taken and 69 Japanese bodies were counted. Despite his losses, the enemy accomplished one part of his mission: he had destroyed 8 planes (including the personal transport of Major General James T. Moore, Commanding General, AirFMFPac, who had arrived that morning), damaged 24 others, and set fire to fuel dumps, causing the loss of some 70,000 gallons of precious aviation gasoline.[47]

Meanwhile, approximately 445 aircraft, of which nearly one-third were suiciders, struck at the American naval forces, concentrating on the radar pickets. The first phase of the attack was broken off about 0300 on the 25th, only to resume at dawn with a renewed fury that continued during the day. At the end, the enemy planes had damaged an APD and a LSM, both so severely that the former capsized later and the latter had to be beached and abandoned. Eight other vessels, generally destroyer types, were also damaged, but in varying degrees. In this action, the Japanese pilots exacted a toll of 38 Americans

[45] *Ibid.* There is some discrepancy regarding the number of men and planes involved in this abortive foray. *ADC IntelSums* states that there were three planes, 2d MAW WarD, May45, counts four, and *Tenth Army AR*, chap 7, sec III, pp. 23–24, indicates five planes attacked the field. Hattori, *op. cit.*, notes that four of the planes in the airborne mission either made forced landings elsewhere—Japanese territory assumed—or turned back, and that the remaining eight succeeded in landing their passengers. The last part of this statement is contradicted by the numbers cited above in the American sources, which all agree that only one plane landed, a crash landing at that, to disgorge its shaken passengers, and that the other aircraft—whatever their numbers, two, three, or four—were shot down while approaching Yontan field.

[46] 2d MAW WarD, May45.

[47] *Ibid.*

killed, 183 wounded, and 60 missing in action.[48]

The raiders suffered also, for friendly air claimed the shooting down of over 150 enemy planes. Of this number, ADC planes claimed an all-time high to date, 75 destroyed in this 24-hour period, to bring the total of TAF claims to 370. High scorer during the 24–25 May raid was the 318th Fighter Group with 34 kills listed, followed by MAG–31.[49] Postwar Japanese sources dispute these statistics, stating that only 88 planes failed to return to base.[50] Regardless of this conflict in numbers, enemy air continued to suffer.

The last mass *Kamikaze* attack in May began just two days later, 27 May, and lasted until the evening of the 28th. The raid caused TAF to establish the longest single enemy air alert of the campaign thus far—9 hours and 16 minutes. Japan sent up 292 aircraft, of which nearly one-third again were suiciders.[51] Heavy antiaircraft artillery fire and combat air patrols fought off the invaders, but not before a destroyer had been sunk and 11 other ships damaged in varying degrees. As before, personnel losses to the fleet were great: 52 men killed, 288 wounded, and 290 missing.[52] The enemy did not escape unscathed, for ICEBERG forces claimed to have splashed more than 100 intruders, and of these, TAF fliers claimed 40. Japanese sources again show figures that differ from those in American records, and show losses of only 80 planes for *Kikusui* No. 8.[53]

By the end of May, ADC fighter pilots had added 279¼ claimed kills to April figures. This gave TAF a total of 423 enemy aircraft destroyed in the air in 56 days of operations.[54] In this same period, 7 April through 31 May, only three American planes were shot down by Japanese pilots out of the 109 aircraft lost to such other causes as pilot error, aircraft malfunctions, and cases of mistaken identity by friendly AAA units.

Prevailing bad weather during most of May had limited air activity, although both sides flew a number of missions even under minimal flying conditions. TAF records for this period indicate that its planes were grounded nine days in May, while cloud cover of varying degree existed during the other 22 days.[55] In addition, the continuing rain that bogged down wheeled and tracked vehicles also turned fighter strips into quagmires. On Ie Shima, for example, a total of 20.82 inches of rain fell from 16 May to the close of the month, causing one Marine air unit to note that "the resultant mixture of water and Ie

---

[48] ComFifthPhibFor AR, Capture of Okinawa Gunto, Phases I and III, 17May-21Jun45, pt III, pp. 20–26, hereafter *CTF 31 AR*. On 17 May, Vice Admiral Harry W. Hill, who had earlier relieved Admiral Turner as Commander, Fifth Amphibious Force, in addition became Commander Task Force 51. This numerical designation changed 10 days later to Task Force 31, when Admiral Halsey took over the Fifth Fleet and it became the Third Fleet.

[49] *ADC IntelSums*, May45.

[50] Hattori, *War History*, table facing p. 132.

[51] *Ibid.*

[52] *CTF 31 AR*, pt III, pp. 30–34.

[53] Hattori, *op. cit.*

[54] *ADC AcftOpAnalysis*, 1-31May45, p. 1.

[55] *Ibid.*

Shima soil produces mud of a character that surpasses description." [56]

Towards the middle of May, TAF strength was increased when pilots and planes of Marine Aircraft Group 22 and the Army Air Forces' 318th Fighter Group arrived and began operations.[57]

Although the ground troops showed a partisan interest in the aerial dogfights high above them and the vivid pyrotechnical displays occurring during the air raids at night, the infantry was more vitally concerned with winning its own battle and with the assistance the air units could give in the drive southward. Out of a total of 7,685 sorties flown by TAF pilots in May, 716 (against an April figure of 510) were in support of the ground forces. Included in the May figure were night heckler and intruder missions flown away from Okinawa. The majority of the ground support sorties were directed against enemy troop concentrations, caves, and truck parks.[58] As the campaign progressed, and as the pilots gained experience with their planes and improved ordnance and a greater familiarity with the area, the ground support effort became increasingly effective. By the middle of May, TAF had reached the state where it was fully prepared to assist the forces of the Tenth Army, then poised to strike the heart of the Shuri defenses.

[56] MAG-22 WarD, May45.

[57] Both groups were based on Ie Shima. The flight echelon of the AAF group (19th, 73d, and 333d Fighter Squadrons) began arriving on the island on 13 May; Colonel Daniel W. Torrey, Jr.'s Marine squadrons flew in over a 10-day period, 21-30 May, from Engebi where the group had been part of the 4th MAW. MAG-22 consisted of HqSqn-22, SMS-22, VMF-113, -314, and -422, and VMTB-131—all of which were to operate from fields on Ie—and VMF(N)-533, which was based at Yontan airfield on its arrival. *ADC AcftOpAnalysis*, 1-31May45, p. 1.

[58] *ADC AcftOpAnalysis*, 1-31May45, p. 15.

ON SHURI'S THRESHOLD [59]

Although the enemy chose 11 May to mount a mass *Kamikaze* attack and many Japanese planes were, in fact, then diving on American surface forces off Okinawa, both corps of the Tenth Army launched a coordinated assault at 0700. Two and a half hours earlier, enemy infantry units had attempted a counterattack following a heavy mortar and artillery barrage on the center of the 1st Marine Division line. Unfortunately for the attackers, the barrage lifted too soon and they were caught by American prearranged defensive fires while still forming. Though the enemy force sustained heavy casualties, the remnants attempted to reform and continue the assault, only to be wiped out by Marine close-range small arms fire.

The 6th Engineers had not yet bridged the Asa Kawa when the attack was to begin; nevertheless, the 6th Division jumped off on time before this vital support route was completed. With the 22d Marines in the lead, the assault troops advanced slowly against a stubborn and well-organized defense built

[59] Unless otherwise noted, the material in this section is derived from: *CTF 31 AR; CTF 51 AR; Tenth Army AR; IIIAC AR; 1st MarDiv SAR; 6th MarDiv SAR, Ph III; 77th InfDiv OpRpt*, Okinawa; *96th InfDiv AR; 1st Mar SAR; 5th Mar SAR; 7th Mar SAR; 22d Mar SAR, Ph III; 29th Mar SAR, Ph III.*

around machine guns and mortars concealed in cave mouths. In early afternoon, enemy troops hidden in a particularly formidable coral hill formation held up the movement of 1/22 on the left. All attempts to envelop the position from either flank failed. The battalion then halted to permit a heavy naval gunfire shelling of the Japanese defenses, after which the Marines resumed the advance under the cover of armor support which had crossed the now-completed bridge shortly before noon.

Flamethrowers, demolitions, and direct tank fire were employed when the attack again began; the position was reduced after a bitter close-in fight. Upon inspection, this hill proved to be a key feature of the Asa Kawa defense system, and contained a vast network of headquarters and supply installations within a large tunnel and cave complex.

Continuing on, the 22d Marines took nearly 1,000 yards of strongly defended enemy territory by 1800, after which mopping up operations continued well into the night. All during the day, the Bailey bridge had remained under continuous enemy artillery and sniper fire, in the face of which reinforcements and supplies poured over the crossing to support forward elements and maintain the momentum of the advance.

To the left of the 6th Division, the 1st Marine Division attacked following an intense air, artillery, and naval gunfire preparation. Substantial gains were made all along the line against a defiant enemy who contested every inch of the advance. Behind the continuing bombardment, the 1st Marines pushed forward along the railroad near the division right boundary, while the 7th Marines made slower progress in the center and left of the line as it reached positions west of the high ground protecting Shuri.

As 2/1 attacked towards its objective, the high ground west of Wana, it began receiving some of the heaviest enemy resistance experienced in the division zone that day. When the battalion passed the nose west of Dakeshi, troops on the left came under heavy flanking machine gun fire from the village. Unable to continue the advance in the direction of the objective, the battalion attacked in column down the west side of the railroad, taking advantage of the cover furnished by the high embankment. At 1600, 2/1 had advanced about 900 yards and was partially on its goal, but held up so that the 22d Marines could come up on the right. Here, the 2d Battalion became subject to accurate long-range flanking artillery fire which soon took a heavy toll in casualties. The situation became more difficult when supply and evacuation were prevented because all possible routes of approach were covered. It finally became necessary for the companies to dig in for the night where they stood.

The 3d Battalion jumped off at about the same time as 2/1, and moved out to cover a gap that had occurred between the latter and the 7th Marines, which was still fighting in the middle of Dakeshi. As the foremost elements of 3/1 reached the point where 2/1 was first fired upon, they were likewise hit and their supporting tanks were unable to get past the draw because of heavy and accurate 47mm antitank fire. Finally, the 3d Battalion negotiated the gap and took up positions just east of the railroad embankment, where it was subjected to frequent artillery and mortar shelling for the rest of the afternoon and all through the night.

Meanwhile, the 7th Marines continued the attack on Dakeshi, where enemy reaction to all forward movement proved costly to both sides. Because 3/7 had been held up throughout most of 10 May, the regiment jumped off the next day with 1/7 and 2/7 in the assault to envelop the strong positions in front of the 3d Battalion. While 3/7 contained the enemy to its front and the regimental reserve, 3/5, protected the rear, attacking forces pushed forward to gain troublesome Dakeshi Ridge.

On the right of the 7th's zone, 2/7 advanced in the face of Japanese mortars, grenades, and automatic weapons fire—the latter coming from pillboxes and coral caves—to gain approximately 800 yards and seize the ridge overlooking and running through Dakeshi. At 1800, the battalion attack was halted on the positions then occupied. As the men dug in for the night, some of the veterans of the Peleliu campaign were reminded of how much the fight for the ridge that day resembled the action at Bloody Nose Ridge.[60] Throughout the night of 11–12 May, the new defenders of Dakeshi Ridge fought off numerous Japanese attempts to infiltrate under cover the constant artillery and mortar barrages coming from enemy emplacements on Wana Ridge.

The fall of aggressively defended and vital Dakeshi Ridge, and its occupation by Marines, meant that one more barrier to the heart of the Shuri defenses had been raised. In addition, the Japanese were now denied the use of commanding ground from which the terrain from Shuri and Naha to Machinato Ridge, and the entire coastal area in between, could be covered by observation and fire. The taking of Dakeshi Ridge effectively and decisively breached the enemy's Naha-Shuri-Yonabaru line, and raised some question as to how much longer he could hold it before Shuri itself was threatened.

Dakeshi was further endangered by the manuever of 1/7, which swung towards the town from the northeast and placed the fanatically defended village in between a rapidly closing pincers. Although Dakeshi was now ripe for capture, positions on the reverse slope of the ridge, in the village itself, and a pocket of resistance to its north continued to be held by soldiers who were determined to defend to the death.

In the 1st Division rear, 2/5 eliminated the last remaining organized resistance in the Awacha Pocket, and 1/5 moved up behind 1/7 to wipe out scattered enemy remnants bypassed during the day. At 1800, 3/5 reverted to parent control.

Tanks providing close-in fire support to the 7th Marines on 11 May had been pressed into service to evacuate casualties. Some wounded were taken up into the tanks through the escape hatches; others rode on the rear deck of the tracked vehicles, which backed out of the battle area in order to provide an armored shield between the stretcher cases and enemy fire.

---

[60] *2/7 SAR*, p. 4; also, see that part dealing with the Peleliu campaign in Garand and Strobridge, "Western Pacific Operations."

By nightfall, 1/7 was positioned and linked on its left with the 305th Infantry of the 77th Infantry Division to form a solid line at the division boundary. The major effort of XXIV Corps was made in the left center of its zone by the 96th Infantry Division, which had completed the relief of the 7th the previous day. While the 77th pressed the enemy through central Okinawa towards Shuri, the 96th approached a hill mass directly northwest of Yonabaru. This terrain feature controlled the eastern reaches to Shuri, completely dominated the east-central coastal plain, and was the easternmost anchor of the enemy's main battle position. All natural routes to the hill were constantly under observation and thoroughly covered by Japanese fire.

Conical Hill, as this bastion was soon named, commanded a series of ridges and other lesser hills, whose capture was to be costly and time-consuming. Murderous fire during the 11 May attack forced the frontline units of the 96th Division to relinquish whatever gains were made that day, although the 383d Infantry on the division left had battled forward 600 yards to establish a foothold on the northwest slopes of Conical. On the same day, assault battalions of the 77th Division gained but 400–500 yards. Strongly entrenched enemy took advantage of the broken terrain to take the flanks, and at times the rear, of the advancing soldiers under fire. At nightfall, the division halted, consolidated its gains, and dug in to the accompaniment of sporadic mortar and artillery fire.

Enemy small-boat activity during the night of 11–12 May increased noticeably over that of previous nights. American patrol boats reported making many radar and visual contacts off Naha, and some enemy craft were spotted apparently heading for the Hagushi transport area. The "flycatchers" remained vigilant, however, and efficiently thwarted these Japanese surface ventures. Seven enemy boats were sunk between midnight of the 11th and 0400 the next day. Several other Japanese craft were fired upon with unknown results.[61]

These coastal skirmishes were tame in comparison to the bloody land battle which continued with unabated violence. When the Tenth Army attack resumed on the 12th, Marine assault elements found Japanese resistance undiminished. On the right of the 6th Division zone, 3/22 moved out in the face of small arms fire pouring down from positions in rocky cliffs overlooking its route of advance, and from the mouths of Okinawan tombs dug in the hillsides that lined it. By 0920, the battalion reached its objective, the high ground commanding a view of Naha below, and sent out patrols through the suburbs of the city to the banks of the Asato Gawa. Here, the Marines found the bridge demolished and the river bottom muddy and unfordable. Patrols from 1/22 also were sent down to the river bank after the battalion had reached the heights in its zone at 1400. Both battalions dug in for the night in firm control of the terrain on the northern outskirts of Naha.

The 2d Battalion was unable to keep pace with or match the advances of the

---

[61] *CTF 51 AR*, pt III, pp. 91–92.

troops coming down the west coast; for, in addition to fighting the enemy in its path, 2/22 was forced to contend with the telling effect of Japanese fire coming out of the 1st Division zone from positions on the dominating terrain standing between the division boundary and the Shuri hill mass. Nevertheless, at 1400 Company G reached the battalion objective, the high ground overlooking Naha. Because the left flank of 2/22 was overextended, at 1350 General Shepherd attached 3/29 to the 22d Marines and alerted the rest of the regiment for commitment into the lines. By the end of the day, 6th Division troops occupied positions from which they were to fight for pretty much the rest of the month.

After first having repulsed a counterattack, at 0730 on 12 May, 1/7 together with 2/7 launched a converging attack aimed at closing a 400-yard gap existing between the two battalions. As this assault force moved into the ruins of Dakeshi village, the enemy mortar and artillery fire that had been falling steadily since the Tenth Army advance had begun increased sharply. At 1522, however, the adjacent flanks of the two units made contact, and the battalions consolidated their positions for the night along the northern outskirts of the village and on the high ground to its east and west.

The 1st Marines attack to improve positions west of Wana was held up for three hours, while 2/1 was given an airdrop of rations, water, ammunition, and medical supplies. During the interim, this battalion came under extremely heavy and accurate mortar and small arms fire, which caused many casualties. At 1030, the battalion jumped off, but all companies reported that they had run into a swarm of sniper and heavy machine gun fire coming from positions in the vicinity of Wana. Casualty evacuation and resupply soon became increasingly difficult because all routes were exposed to enemy observers located on the heights to the left of the 2/1 advance. As the day wore on, the assault companies were forced to dig night defenses on ground then held, not too many yards ahead of their 11 May positions.

Attacking to the southeast on the left of 2/1, the 3d Battalion was partially protected by the overhanging bank of an Asa Kawa tributary, and penetrated 300 yards towards the mouth of Wana Draw. Forward movement ceased at 1630 and the 1st Division dug in for the night, 2/1 tying in with 3/29 on the right. All the while, 2/1 remained in an isolated forward position. An undetermined number of the enemy counterattacked the Marines at 2230, causing General del Valle to alert the 5th Marines for possible commitment in support of the 1st, but 3/1 contained the attack without need of reinforcement.

Enemy small boats were again active on the night of 12–13 May. An attempted counterlanding on the coast between the Asa and Asato Rivers was broken up by American patrol craft. The approximately 40 surviving Japanese were eliminated by 3/22 at the edge of the reef.

When the 6th Division attack resumed at 0730 on the 13th, the task of 3/22 was to reconnoiter the northern outskirts of Naha. As one patrol approached a village that another patrol had passed through safely on 12 May, it was turned back by enemy fire. Battalion 81mm mortars were laid on the settlement's houses, and an infantry platoon accompanied by a tank platoon was sent in at 1400 to overcome all resistance. Well-concealed and determined defenders, however, stymied this attack. One tank was disabled by a satchel charge placed by a suicidally inclined Japanese soldier and the rest of the Shermans were forced to turn back.

Another infantry and tank platoon teamed up, this time attacking from the north of the village, but this effort, too, was thwarted by the combination of heavy machine gun fire, an enemy determination to hold, and the narrow village streets which restricted tank movement. Regiment then ordered the enemy blasted out and the village burned. After levelling the buildings, and killing approximately 75 defenders, the Marine tanks and troops withdrew at 1630. In its zone, 1st Battalion Marines met resistance from enemy outposts holed up in houses on the north bank of the river.

The main division effort was made on the left by 2/22, with 3/29 assigned to clear high ground overlooking the Asato River from which the enemy fired into the left flank elements of 2/22. Because of the difficulty in getting essential supplies and the rocket trucks scheduled for preparation fires forward to the front, the attack was delayed until 1115. Despite the heavy rocket and artillery preparation, intense enemy resistance grew yet more determined as the day wore on, making the tank-infantry assault teams' way difficult. By the end of the day, the two assault battalions had gained no more than 200–300 yards. Just before dark, Company H, 3/29, rushed and seized the troublesome hill on the left, quieting the heavy flanking machine gun fire that had been coming from that sector.

At the close of the day, it was clear that the 22d Marines had been worn out and its battle efficiency sapped in the fighting that brought the division down to the outskirts of Naha. During the 2,000-yard advance south from the Asa Kawa, the regiment had suffered approximately 800 Marines killed and wounded. Therefore, General Shepherd ordered the attack resumed on 14 May with the 29th Marines making the main effort on the left, supported by the 22d Marines on the right. The 3d Battalion, 29th Marines, reverted to parent control at 1800, at which time the regiment officially assumed responsibility for its new lines. To take over the positions vacated by the 29th, the 4th Marines—IIIAC reserve—moved south, where it would guard the division rear and back up the LVT(A)s guarding the open seaward flank of the 6th Division.

In the 1st Marine Division zone on 13 May, the 1st Marines was forced to repel two predawn counterattacks in platoon to company strength before launching its own attack. The Marine assault was first delayed until supporting tanks got into positions, and then held up again until the 7th Marines had cleared Dakeshi. Organic crew-served weapons of the 1st Marines fired upon observed enemy positions in the village in support of the 7th. The 1st Marines attack finally began at 1230, when 3/1 jumped off to extend the battalion line to the right into the 2/1 sector and to clean up bypassed enemy positions. Primarily, the 3d Battalion objective was the high ground at the mouth of Wana Draw. Heavy machine gun fire from three sides and a deadly hail of mortar, grenade, and rifle fire greeted the tank-infantry assault teams as they gained the hill. Finding the position untenable, the attackers were forced to withdraw under the cover of smoke and fire furnished by the tanks, which also evacuated casualties. The same formidable obstacle of flying steel that met 3/1 forced 2/1 back and prevented the latter from moving its left flank up to extend its hold on the high ground west of Wana.

After first blunting a predawn enemy attack on 13 May, Colonel Snedeker's 7th Marines jumped off at 0730 with 2/7 in the assault, 1/7 and 3/7 in reserve. The 2d Battalion cleaned out Dakeshi, the 1st Battalion eliminated snipers and sealed caves on the ridge overlooking the village, and the 3d Battalion protected the rear of the regiment. Despite the employment of tanks, self-propelled 75mm guns, and 37mm antitank guns, the enemy was not subdued until late in the afternoon.

Opposing 2/7 on the reverse slope of Dakeshi Ridge was a honeycomb of caves centering around one which was later found to be the command post of the *64th Brigade*. These positions were discovered late in the afternoon and taken under close assault; so close, in fact, that Japanese postwar records note that even the brigade commander and his CP personnel took part in the fighting. After dark, the enemy ordered survivors of the last-ditch stand to attempt to infiltrate American lines in order to reach Shuri and reform.[62]

In the XXIV Corps zone, assault elements of the 96th Infantry Division executed a flanking maneuver west of Conical Hill and gained a foothold from which the stronghold could be reduced. On 13 May, the division captured the western and northern slopes of Conical, thus opening the way for the capture of Yonabaru and the unlocking of another door to Shuri's inner defenses.

Both corps of the Tenth Army attacked at 0730 on 14 May to clear the eastern and western approaches to Shuri and to envelop the flanks of that bastion. Fighting was especially bitter in the IIIAC zone, where the Marine divisions were unsuccessful in their attempts to break through the enemy line west of Wana and northwest of Naha. It soon became apparent that the Marines had run into the Japanese main line of resistance. This assumption was borne out by the heavy losses sustained by attacking infantry units and the number of tanks, 18, in the two Marine tank battalions destroyed, disabled, or damaged by enemy antitank, mortar, and artillery fire, mines, and suicide attacks.

General Shepherd's troops had jumped off at 0730 on the 14th to seize the high ground running generally along the north bank of the Asato Gawa. From the very beginning of this attack, the assaulting forces met strong, well-coordinated, and unremitting opposition. Attacking in conjunction with the

[62] *Okinawa Operations Record*, p. 107.

29th Marines, the 22d succeeded in seizing approximately 1,100 yards of the bank of the Asato, despite the presence of numerous machine gun and sniper positions in the path of the advance.

It was on the regimental left, however, that the going was roughest and the fighting most savage. In the face of mounting casualties during the day, the attack of 2/22 finally ground to a halt at 1500, when the battalion ran into a system of strongly defended and thoroughly organized defenses. These guarded a rectangularly-shaped hill, dominating and precipitous, that was quickly dubbed "Sugar Loaf." (See Map V, Map Section.) This hill itself was at the apex of an area of triangularly shaped high ground that pointed north. A concentration of Japanese power here had turned back 2/22 in the two previous days. Enemy dispositions on Sugar Loaf were so organized that the defenders could cover the front, rear, and flanks of any portion of the position with interlocking bands of automatic weapons fire and devastating barrages from mortar, artillery, and grenade launchers.

Although the intensity of Japanese resistance increased proportionally as assault troops approached this bastion—already recognized as a key defensive position—it was not realized at first that this bristling terrain feature and its environs constituted the western anchor of the Shuri defenses. At the time that the 22d Marines reached Sugar Loaf, the regimental line was spread thinly and excessive casualties had reduced combat efficiency to approximately 62 percent.

Nonetheless, despite the factors which forced the halt, 2/22 received direct orders from division at 1515 to seize, occupy, and defend the battalion objective—including Sugar Loaf Hill—this day at any cost.[63] In answer to the battalion commander's earlier request for reinforcements, Company K, 3/22, was attached to back up the attack. Moving out at 1722 behind a line of tanks and an artillery-laid smoke screen, Company F attacked Sugar Loaf for the second time on the 14th. In a little more than two hours later, some 40 survivors of Companies F and G were in position at the foot of the hill under the command of the battalion executive officer, Major Henry A. Courtney, Jr.

Snipers were everywhere, and the group also came under fire from mortars on the flanks as well as the reverse slopes of Sugar Loaf. To carry supplies and much-needed ammunition up to the exposed Marines, and to reinforce Major Courtney's pitifully small force, Lieutenant Colonel Horatio C. Woodhouse, Jr., 2/22 battalion commander, sent 26 newly arrived replacements forward. All during this time, the Japanese were rolling grenades down on the Marine position from the heights above, and Courtney saw no other alternative to remaining where he was than to attack up the hill to seize its crest. All American illumination of the area was stopped when Courtney and his 40-odd Marines stormed the hill at 2300, throwing grenades as they scrambled up the slopes. As soon as they carried the crest, they dug in to wait out a night of ex-

[63] *2/22 SAR, Ph III*, pp. 5–6.

pected counterattacks and the enemy's customarily heavy mortar fire.

On the left of 2/22, enfilade fire from flanking hills in the zone of the 29th Marines undoubtedly contributed to the battalion's hard going during this day. After 3/29, the regimental assault battalion, had jumped off at 0730 on the 14th, it tried to bypass Japanese strongpoints on its left to draw abreast of 2/22 on its right. The 3d Battalion was forced to halt and fight around this center of resistance the rest of the morning and part of the afternoon, when Japanese fire on its rear proved troublesome. At about 1630, the 29th Marines' commander regrouped his assault elements and moved 1/29 into the line on the left of 3/29. The attack was renewed with Companies A and H working over the flanks of the enemy position, slowly compressing and neutralizing it. Company G, in the meanwhile, continued the attack southward, and, after fighting its way 200 yards across open ground, gained the forward slopes of a hill northwest of Sugar Loaf, where it tied in with the lines of 2/22.

In the 1st Division zone, the objective of the 2d Battalion, 7th Marines, on 14 May was Wana Ridge. The battalion jumped off at 0730, with 1/7 prepared to pass through and continue the attack if 2/7 was unable to continue. As soon as the left element of 2/7 cleared past Dakeshi village and entered open terrain, it was pinned down by Japanese fire. While Company E was held up, Company G, followed by F, swung through the zone of 1/1 to approach a point within 100 yards of Wana Ridge. Before they could take cover, concentrated enemy machine gun and mortar fire inflicted heavy casualties upon these leading elements of 2/7, and they were ordered to hold their positions until relieved by 1/7.

At 1107, Colonel Snedeker ordered the relief of the 2d Battalion by the 1st, which was ready to effect the relief at 1252. At that time, however, the commander of 1/7 requested that all supporting arms under the control of 2/7 be transferred to him, and before the transfer had been completed the renewed attack was delayed until 1615. When this designated H-Hour arrived, and following an intensive naval gunfire, artillery, rocket, and 4.2-inch mortar preparation, 1st Battalion assault units moved out behind tanks and under the cover of a protective smoke screen. The main effort was made by Company B, which advanced through Dakeshi to the south into open terrain. Immediately upon coming into this clearing, these Marines were taken under the same fire that had pinned down Company E earlier.

When enemy fire from Wana and Shuri prevented the company from advancing further, it was ordered to withdraw to Dakeshi to set up a night defense. In the meanwhile, moving south in unfamiliar territory to take over the positions of Company G, Company A ran into numerous enemy groups attempting to penetrate Marine lines. The relief was finally effected at 1900, but not before the commander and executive officer of Company A had become casualties.[64]

Coordinated with the attack of the 7th Marines against Wana Ridge was the one launched that same day by the 1st Marines. The regiment's major effort was made by 1/1, with the western tip of the ridge as the initial objective; the 2d and 3d Battalions supported the assault by fire. By noon, Company C secured the objective and began digging in and consolidating the newly won position despite heavy enemy fire. There was no contact on the left with the 7th Marines, which was moving up slowly against bitter opposition. Meanwhile, the portion of the ridge to have been occupied by the 7th soon was swarming with Japanese soldiers forming for a counterattack. Because he could not be reinforced in time, the commander of Company C requested and received permission to withdraw. After doing so in good order, the company set up a strong line for night defense on the battalion left, where contact was made with the 7th Marines. Units of the 5th Marines began relieving assault companies of the 1st at 2200, so that the 1st Division could renew its attack against Wana the next morning with a relatively fresh regiment.

With the coming of darkness on 14 May, Tenth Army assault troops were probing deeply into the Japanese main line of resistance all along the island. Almost flying in the face of indisputable evidence indicating that nearly half of the enemy garrison had been killed—the heaviest losses consisting of first-rate infantrymen—was the undeniable fact that there were no signs of Japanese weakness anywhere along the Tenth Army front. Conversely, the nature of operations in the south promised that enemy defenses were not going to be breached without grinding, gruelling, and unrelenting tank-infantry combat.

LOGISTICAL PROGRESS[65]

All ICEBERG assault and first echelon transports and landing ships had been unloaded by the end of April and released for other assignments. Compared to that provided in Central Pacific operations earlier, the magnitude of the logistical support furnished the Okinawa invasion force can be best seen in the following table:

[64] In commenting upon the Japanese machine gun fire that prevented Company C from coming up to the line to fill in the gap and tie in with Companies A and B, the S–3 of 1/7 stated that this was "a standard Japanese tactic that gave our forces trouble throughout the war. There was always a reluctance on our part to withhold [hold back] an attack that was continuing to gain ground. Also, there was a general tendency to tie in the defenses and establish ourselves in a defensive posture prior to dark. These two factors often overlapped and we found ourselves in the position of frantically establishing a night defense in the few remaining minutes of twilight. The Japanese, of course, had long since reacted to this procedure and were wont to send infiltration groups and LMG teams against us in the awkward period between BENT [Beginning of Evening Nautical Twilight] and EENT [End of Evening Nautical Twilight]. It was in this period that the relief of G/2/7 by A/1/7 took place." Maj Don P. Wyckoff ltr to CMC, dtd 25Mar55.

[65] Unless otherwise noted, the material in this section is derived from: *CTF 51 AR; Tenth Army AR.*

Operation	No. of ships	Personnel	Measurement tons
Gilberts	63	35,214	148,782
Marshalls	122	85,201	293,792
Marianas	210	141,519	437,753
Leyte [66]	110	57,411	214,552
Palaus	109	55,887	199,963
Iwo Jima	174	86,516	280,447
Okinawa	*458*	*193,852*	*824,567*

As stated earlier here, the complexities of logistical support operations were compounded by several factors, not the least of which was the disrupted unloading schedule. Additionally, the jammed condition of the beach dumps and the shortage of shore party personnel and transportation gave the Tenth Army supply problems, also. The primary concern, however, was with the inability of the Tenth Army to maintain an adequate artillery ammunition reserve on the island. This situation arose because of the rapid expenditure of shells of all calibers and types needed in the drive to reduce the positions protecting Shuri. Beginning with the major XXIV Corps attack on 19 April, the initial Tenth Army ammunition support was quickly expended and replenishment shipments were gobbled up as soon as they arrived.

Although "ammunition resupply had been based on an estimated 40 days of combat," [67] it was necessary to revise shipping schedules upwards drastically in order to meet the increased demands. On 17 April, General Buckner made the first of many special requests for ammunition in short supply. Specifically, he asked CinCPOA to load five LSTs at Saipan with 155mm howitzer and gun ammunition for arrival at Okinawa by 27 April. The fulfillment of this request was a stop-gap measure and in no way guaranteed that the critical artillery ammunition shortage would be alleviated for the rest of the campaign.

Kamikazes played a large part in creating this shortage by sinking a total of three ammunition ships in April with a loss estimated at being well in excess of 22,000 tons of vitally needed cargo. Even after the release of a considerable amount of ammunition late in April, when contingent operations for Phase III of ICEBERG were cancelled, the shortage remained critical throughout the fighting. Artillery commands were never able to maintain more than a minimally satisfactory reserve level of shells in their ammunition points.

Although initial shipments consisted of "balanced loads" of ammunition, as the supply requirements of the Tenth Army became clearer it also became apparent that there would be a greater

[66] "The number of ships and cargo tonnage listed under 'Leyte' represents that loaded for the Palau-Anguar-Yap Operation which was diverted to the Leyte Operation but which was unloaded under direction of Com3dPhibFor." *CTF 51 AR*, pt V, sec I, p. 33.

[67] *Tenth Army AR*, chap 11, sec IV, p. 21. "In the preliminary planning the Tenth Army Ordnance Officer envisaged heavy ammunition expenditures. Original requests far exceeded the supplies made available initially and had to be drastically modified downward to fit availability of ammunition stocks and shipping to transport same." BGen David H. Blakelock, USA, ltr to CMC dtd 22Mar55, hereafter *Blakelock ltr 1955*.

demand for artillery shells than for small arms ammunition. Accordingly, logistics officers were able to schedule resupply shipments that more suitably filled the needs of the ground forces on Okinawa.

But even as L-Day ended, unloading facilities on the Hagushi beaches were already overtaxed. Because some of the assault beaches were not capable of sustaining heavy and continuous shore party operations, and other sites selected for eventual use were either not suitable or not uncovered on schedule, the program of beach unloading as set up in the logistics plan proved totally unrealistic. The most satisfactory tonnage unloading figure that could be attained under the then-present conditions was reached on 5 May, and the figures never equalled the planned goals thereafter.[68]

Four new beaches were opened up between 17 April and 17 May on the east coast of the island in Chimu and Nakagusuku Wan to support the southern drive of XXIV Corps and base development activities. Unfortunately, the gap between actual and planned unloading tonnage was never closed, even with the addition of these new points.

[68] "One of the most important factors in preventing the unloading from reaching planned goals after 5 May was the weather. Frequent interruptions were also caused by Red alerts and enemy air raids. During May there were 17 days of heavy seas and torrential rains (14.68 inches fell in May with over 11 inches falling between 20 and 30 May) which materially interfered with unloading operations. During June there were 15 days when typhoon warnings and heavy weather materially interfered with unloading." *Ibid.*

Because Phase III was cancelled and the mission of IIIAC changed, General Buckner could mount a four-division attack on Shuri. This increased ground activity vastly accelerated the consumption of all classes of supplies, and caused more supply ships to be called up than could be handled efficiently.[69] In essence, this move was a calculated risk in the face of numerous mass and individual *Kamikaze* attacks on the transport areas. Nonetheless, the risk had to be taken if an adequate reserve of essential supplies was to be maintained in the immediate area.

As the insatiable appetites of the ground units for supplies increased in late April and early May, quartermaster and shore party units made extensive efforts to speed the unloading and processing of all goods. To help ease the situation, Rear Admiral John L. Hall, Jr., Senior Officer Present Afloat at Hagushi, and General Wallace, the Island Commander, recommended to General Buckner that more cranes, transportation, and personnel be employed to empty beached landing ships and craft; that more LVTs and crane barges be used at the reef transfer line; that intermediate transfer dumps be established to prevent excessively long hauls by shore party vehicles; and that the requirements for ammunition and fuel oil dispersion be modified somewhat to conserve personnel and transport. The admiral made one other recommendation: that the beach at Nago Wan be transferred from the control of IIIAC

[69] According to the resupply schedule established before the landing, a six-day delay occurred between call-up and arrival of the replenishment echelons.

to that of the Island Command. General Buckner approved these recommendations almost immediately, but any gains made by these improvements were quickly minimized by the increasing size and variety of logistics tasks.[70]

At no time after the landing was there any prospect that the Tenth Army had not come to Okinawa to stay, but the problem of sustaining the momentum of the ground offensive became quite acute in late April and early May. In addition to his tactical responsibilities as commander of the Joint Expeditionary Force, during this early phase of the campaign Admiral Turner also had a logistic responsibility for maintaining adequate levels of all classes of air, ground, and naval replenishment stocks needed to support a successful Tenth Army operation.

Admiral Spruance, acting in accordance with the ICEBERG operation plan, announced that the amphibious phase of the Okinawa landing was ended on 17 May. At 0900 on that day, Vice Admiral Harry W. Hill, Commander, V Amphibious Force, relieved Admiral Turner as Commander, Task Force 51, and took over the control of his naval activities, and of air defense. In the shift of command responsibilities, Admiral Hill was directed to report to General Buckner, who took command of all forces ashore and assumed Turner's former responsibility to Admiral Spruance for the defense and development of captured objectives. At this time, all of Turner's former logistics duties were taken over by a representative of Commander, Service Squadron Ten, the Navy logistical support force in forward areas.

Admiral Turner's successful period of command responsibility at Okinawa was marked by his direction of the largest amphibious operation of the Pacific War. Forces under his command had killed 55,551 and captured 853 Japanese troops in ground action, and had claimed the destruction of 1,184 enemy aircraft. During the first 46 days of the campaign, *i.e.*, until 16 May, 1,256,286 measurement tons of assault, garrison, maintenance, and ground ammunition cargo had been unloaded over island beaches. Gunfire support force guns, from 5- to 16-inch in caliber, had fired over 25,000 tons of ammunition while covering Tenth Army ground troops and protecting the ships of TF 51.[71]

In the course of six weeks of incessant fighting, the enemy had exacted a terrible price for every inch of ground he

[70] "Another factor contributing to the lower rate of unloading was the increased need of service troops, particularly engineers, to improve roads and to endeavor to make existing roads passable as a result of the extremely heavy rains. These additional troops came from airfield construction projects and the beach areas, the latter causing [a] corresponding slow down of unloading. The arrival of engineer and service troops on the island never kept pace with the requirements for unloading." *Blakelock ltr 1955.*

[71] The immensity of this support may be seen in comparing the Okinawa figure with the tons of naval ammunition expended in Central Pacific operations earlier: Tarawa (1,833); Makin (1,800); Kwajalein (3,937); Roi-Namur (2,251); Eniwetok (2,020); Saipan (10,965); Tinian (3,627); Guam (8,000); Peleliu (5,579); and Iwo Jima (14,190). ComPhibsPac Amphibious Gunnery Bul No. 2, Assault on Okinawa, dtd 24May45.

yielded. On 17 May, Tenth Army casualty figures included 3,964 men killed, 18,258 wounded, 302 missing, and 9,295 non-battle casualties. Of these casualties, hospital ships had evacuated 10,188; APAs 4,887; and air transport, 5,093. The hard-hit naval forces had lost 1,002 men killed, 2,727 wounded, and 1,054 missing. Air defense units, both TAF and carrier-based squadrons, lost 82 planes to all causes, while TF 51 had 156 ships sunk or damaged in action with the enemy.

Despite the fact that a major portion of Okinawa had been taken, and Tenth Army ground units had punished the enemy unmercifully, all evidence pointed to a continuation of the hard fighting. Nevertheless, the tactical situation on the fringes of Shuri almost imperceptibly showed signs that the Japanese defenses were slowly giving way.

310-224 O - 69 - 17

Reduction of the Shuri Bastion

As conceived in Tenth Army plans, the object of the full-scale attack beginning on 11 May was to destroy the defenses guarding Shuri. In the end, this massive assault took the lives of thousands of men in two weeks of the bloodiest fighting experienced during the entire Okinawa campaign. For each frontline division, the struggle to overcome enemy troops on the major terrain feature in the path of its advance determined the nature of its battle. Facing the front of the 96th Infantry Division was Conical Hill; the 77th Division fought for Shuri itself. Marines of the 1st Division had to overcome Wana Draw, while Sugar Loaf Hill was the objective of the 6th Marine Division. (See Map V, Map Section.)

BATTLE FOR SUGAR LOAF HILL [1]

Sugar Loaf Hill was but one of three enemy positions in a triangularly shaped group of hills which made up the western anchor of the Japanese Shuri defense system. Sugar Loaf was the apex of the triangle, which faced north, its flanks and rear well covered by extensive cave and tunnel positions in Half Moon Hill to the southeast and the Horseshoe to the southwest. The three elements of this system were mutually supporting. In analyzing these defenses, the 6th Marine Division pointed out that:

> . . . the sharp depression included within the Horseshoe afforded mortar positions that were almost inaccessible to any arm short of direct, aimed rifle fire and hand grenades. Any attempt to capture Sugar Loaf by flanking action from east or west is immediately exposed to flat trajectory fire from both of the supporting terrain features. Likewise, an attempt to reduce either the Horseshoe or the Half Moon would be exposed to destructive well-aimed fire from the Sugar Loaf itself. In addition, the three localities are connected by a network of tunnels and galleries, facilitating the covered movement of reserves. As a final factor in the strength of the position it will be seen that all sides of Sugar Loaf Hill are precipitous, and there are no evident avenues of approach into the hill mass. For strategic location and tactical strength it is hard to conceive of a more powerful position than the Sugar Loaf terrain afforded. Added to all the foregoing was the bitter fact that troops assaulting this position presented a clear target to enemy machine guns, mortars, and artillery emplaced on the Shuri heights to their left and left rear.[2]

Following its successful charge to seize the crest of Sugar Loaf, Major Courtney's small group had dug in. An unceasing enemy bombardment of the newly won position, as well as the first in a series of Japanese counterattacks to regain it, began almost immediately. At midnight, 14-15 May, there were sounds of enemy activity coming from the other

[1] Unless otherwise noted, the material in this section is derived from: *Tenth Army AR; IIIAC AR; 6th MarDiv SAR, Ph III; 6th MarDiv Jnl, Ph III; 4th Mar SAR, Ph III; 22d Mar SAR, Ph III; 29th Mar SAR, Ph III.*

[2] *6th MarDiv SAR, Ph III*, p. 5.

side of the crest, signifying an impending *banzai* charge to Courtney. He forestalled the charge by leading a grenade-throwing attack against the reverse slope defenders, in the course of which he was killed.

At 0230, only a handful of tired and wounded Marines remained on the top of Sugar Loaf, and Lieutenant Colonel Woodhouse ordered his reserve, Company K, to reinforce the depleted group. With the coming of dawn, the forces on Sugar Loaf had been reduced again by enemy action and fire, while 2/22 itself had been hit by numerous Japanese counterattacks and attempts at infiltration all along the battalion lines. At 0630, Company D of 2/29 was attached to the 22d Marines to help mop up the enemy in the rear of 2/22.

There were less than 25 Marines of Courtney's group and Company K remaining in the 2/22 position on Sugar Loaf when daylight came; at 0800, the seven survivors of the Courtney group were ordered off the hill by the battalion commander. Within a short time thereafter, the enemy launched another attack against the battered position. During the height of this attack, a reinforced platoon of Company D arrived on the hilltop and was thrown into the battle. Suffering heavy casualties while en route to the position, the Company D platoon was hit even harder by the charging Japanese as soon as it arrived at the top of the hill. At 1136, the few survivors of Company K and the 11 Marines remaining of the Company D platoon were withdrawn from Sugar Loaf. The Company D men rejoined their parent unit, which was manning a hastily constructed defensive line organized on the high ground just in front of Sugar Loaf.

The enemy counterattack was the beginning of a series which soon reached battalion-sized proportions, and which, by 0900, had spread over a 900-yard front extending into the zones of 1/22 and 3/29. An intensive naval gunfire, air, and artillery preparation for the division assault that morning temporarily halted the enemy attack, but it soon regained momentum. By 1315, however, the Japanese effort was spent, though not before the 22d Marines in the center of the division line had taken a terrific pounding. In an incessant mortar and artillery bombardment supporting the enemy counterattack, the battalion commander of 1/22, Major Thomas J. Myers, was killed, and all of his infantry company commanders—and the commander and executive officer of the tank company supporting the battalion—were wounded when the battalion observation post was hit.[3]

Major Earl J. Cook, 1/22 executive officer, immediately took over and reorganized the battalion. He sent Companies A and B to seize a hill forward of the battalion left flank. When in blocking positions on their objective—northwest of Sugar Loaf—the Marines could effectively blunt counterattacks expected to be mounted in this area. Because the possibility existed of a breakthrough in the zone of 2/22, the regimental commander moved Company

[3] Upon learning of the death of Major Myers, General Shepherd commented: "It's the greatest single loss the Division has sustained. Myers was an outstanding leader. Whenever I called on him for a job he never failed me.'" Cass, *6th MarDiv Hist*, p. 117.

SUGAR LOAF HILL, *western anchor of the Shuri defenses, seen from the north. (USMC 124983)*

TANKS *evacuate the wounded as the 29th Marines continue the attack on Sugar Loaf. (USMC 122421)*

I of 3/22 into position to back up the 2d Battalion. At 1220, Lieutenant Colonel Woodhouse was notified that his exhausted battalion would be relieved by 3/22 as soon as possible, and would in turn take up the old 3d Battalion positions on the west coast along the banks of the Asato. The relief was effected at 1700 with Companies I and L placed on the front line, and Company K positioned slightly to the right rear of the other two. Company D, 2/29, reverted to parent control at this time.

During the ground fighting on the night of 14–15 May, naval support craft smashed an attempted Japanese landing in the 6th Division zone on the coast just north of the Asato Gawa. Foreseeing the possibility of future raids here, General Shepherd decided to strengthen his beach defenses. In addition to a 50-man augmentation from the regiment, 2/22 was also reinforced by the 6th Reconnaissance Company to bolster its night defenses. To further strengthen Lieutenant Colonel Woodhouse's command, he was given operational control of 2/4, which was still in corps reserve.

The objective of the 29th Marines on 15 May was the seizure of Half Moon Hill. The 1st and 3d Battalions encountered the same bitter and costly resistance in the fight throughout the day that marked the experience of the 22d Marines. A slow-paced advance was made under constant harassing fire from the Shuri Heights area. By late afternoon, 1/29 had reached the valley north of Half Moon and became engaged in a grenade duel with enemy defenders in reverse slope positions. Tanks supporting the Marine assault elements came under direct 150mm howitzer fire at this point. Several of the tanks were hit, but little damage resulted. At the end of the day, the lines of the 29th Marines were firmly linked with the 22d Marines on the right and the 1st Division on the left.

Facing the 6th Marine Division was the *15th Independent Mixed Regiment*, whose ranks were now sadly depleted as a result of its unsuccessful counterattack and because of the advances of 1/22 and the 29th Marines. More than 585 Japanese dead were counted in the division zone, and it was estimated that an additional 446 of the enemy had been killed in the bombardments of supporting arms or sealed in caves during mopping-up operations.[4] Expecting that the Americans would make an intensive effort to destroy his Sugar Loaf defenses, General Ushijima reinforced the *15th IMR* with a makeshift infantry battalion comprised of service and support units from the *1st Specially Established Brigade*.[5]

The success of the 6th Division attack plan for 16 May depended upon the seizure of Half Moon Hill by the 29th Marines. (See Map VI, Map Section.) Once 3/29 had seized the high ground east of Sugar Loaf, 3/22 was to make the major division effort and capture the hill fortress. Immediately after the attack was launched, assault elements on the regimental left flank encountered heavy fire and bitter opposition from enemy strongpoints guarding the objective. The 1st Battalion was spearheaded by a Company B platoon and its supporting armor. After the tank-infantry

[4] 6th MarDiv G-2 PeriodicRpt No. 45, dtd 16May45.

[5] *Okinawa Operations Record*, p. 109.

teams had passed through the right flank to clear the reverse slope of the ridge held by Company C, devastating small arms, artillery, mortar, and anti-tank fire forced them to withdraw. The fury of this fire prevented Company C from advancing over the crest of the ridge and the other two platoons of Company B from moving more than 300 yards along the division boundary before they too were stopped by savage frontal and flanking fire.

The night defenses of the battalion remained virtually the same as the night before; however, the units were reorganized somewhat and their dispositions readjusted. At 1400 that afternoon, Lieutenant Colonel Jean W. Moreau, commander of 1/29, was evacuated after he was seriously wounded by an artillery shell which hit his battalion OP; Major Robert P. Neuffer assumed command.

Continuously exposed to heavy enemy artillery and mortar bombardment, 3/29 spent most of the morning moving into favorable positions for the attack on Half Moon. Following an intensive artillery and mortar preparation, tanks from Companies A and B of the 6th Tank Battalion emerged from the railroad cut northeast of Sugar Loaf and lumbered into the broad valley leading to Half Moon. While Company A tanks provided Company B with direct fire support from the slopes of hills just north of Sugar Loaf, the latter fired into reverse slope positions in the ridge opposite 1/29, and then directly supported the assault elements of the 3d Battalion.

At about the same time that their armor support appeared on the scene, Companies G and I attacked and quickly raced to and occupied the northern slope of Half Moon Hill against slight resistance. The picture changed drastically at 1500, however, when the Japanese launched a violent counteroffensive to push the Marines off these advanced positions even while they were attempting to dig in. The enemy poured machine gun, rifle, and mortar fire into the exposed flanks and rear of the Americans, who also were hit by a flurry of grenades thrown from caves and emplacements on the south, or reverse, slope of the hill. As evening approached, increasing intense enemy fire penetrated the smoke screen covering the digging-in operations of the troops and they were ordered to withdraw to their earlier jump-off positions to set in a night defense.

On the right of the division, when the 22d Marines attack was launched at 0830 on the 16th, assault elements of the 1st Battalion were immediately taken under continuous automatic weapons fire coming from the northern edge of the ruins of the town of Takamotoji, just as they were attempting to get into position to support the attack of 3/22. The fact that this previously quiet area now presented a bristling defense indicated that the Japanese had reinforced this sector to confound any American attempt to outflank Sugar Loaf from the direction of Naha. In the end, because of the criss-crossing fires coming from the village, Half Moon Hill, and the objective itself, the 3d Battalion was unable to fulfill its assignment.

The battalion commander, Lieutenant Colonel Malcolm "O" Donohoo, had planned to attack Sugar Loaf from the east once the flank of the attacking unit, Company I, was safeguarded by a successful 3/29 advance. Company L, 3/22, was to support the attack by covering the south and east slopes of Sugar Loaf with fire, while 1/22, in turn, would take the high ground west of Sugar Loaf, where it would support the Company L movement by fire. The success or failure of the attack on the hill hinged on the success or failure of 3/29.

At 1500, despite the fact that 3/29 had not fully occupied the high ground, Company I moved out with its tank support and reached Sugar Loaf without serious opposition. Once the troops in the van attempted to gain the crest, however, they began receiving heavy enemy mortar and machine gun fire. In an effort to suppress this fire, the tanks began flanking the hill, but ran into a minefield where one tank was lost. Company I, nevertheless, gained the top of the hill at 1710 and began digging in. The situation was in doubt now, because both 1/22 and Company L were pinned down and 3/29 was forced to withdraw from Half Moon. Company I, therefore, was in an exposed position and its precarious hold on Sugar Loaf had become untenable. With both flanks exposed and its ranks depleted by numerous casualties, the company had to be pulled back from the hill under the cover of fire of both division and corps artillery. As 3/22 reorganized for night defense, enemy batteries bombarding the Marine lines wounded Lieutenant Colonel Donohoo, who was replaced by Major George B. Kantner, the battalion executive officer.

This day was categorized by the 6th Division as the "bitterest" of the Okinawa campaign, a day when "the regiments had attacked with all the effort at their command and had been unsuccessful." [6] One infantry regiment, the 22d, had been so sorely punished that, in assessing his losses for the day, Colonel Schneider reported that the combat efficiency of his unit was down to 40 percent.[7] Because the fighting of the preceding eight days had sapped the offensive capabilities of the 22d Marines and reduced the regiment to a point where its continued employment was inadvisable, it became apparent that the 29th Marines would have to assume the burden of taking Sugar Loaf. On 17 May, the regimental boundary was shifted west to include the redoubt in its zone and thereby lessen control problems in the attacks on both it and Half Moon.

In an effort to neutralize the seemingly impregnable Japanese defenses here, the attack of 17 May was preceded by an intensive bombardment of 29th Marines objectives by all available supporting arms. In this massive preparation were the destructive fires of 16-inch naval guns, 8-inch howitzers, and 1,000-pound bombs. Following this softening up, and spearheaded by a heavy and continual artillery barrage, the 29th Marines launched a tank-infantry attack with three battalions abreast. The 1st and 3d Battalions on the left had the mission of taking Half Moon, while 2/29, with Company E in assault, was to take Sugar Loaf.

Company E made three attempts to take its objective, and each proved costly and unsuccessful. The first effort, involving a wide flanking movement in which the railroad cut was utilized for cover, was stymied almost immediately when the troops surged onto open

[6] *6th MarDiv SAR, Ph III*, p. 7.

[7] *22d Mar SAR, Ph III*, p. 5.

ground. A close flanking attack around the left of the hill characterized the second effort, but the steep southeastern face of the height precluded a successful climb to the top. The axis of the attack was then reoriented to the northeast slope of Sugar Loaf, and the lead platoon began a difficult trek to the top, all the while under heavy mortar fire coming from covered positions on Half Moon. Three times the assaulting Marines reached the crest, only to be driven off by a combination of grenades and bayonet charges. Almost all fighting was at close range and hand-to-hand.

After quickly reorganizing for a fourth try, the now-fatigued and depleted company drove to the hilltop at 1830, when it was met again by a determined Japanese counterattack. This time, however, the Marines held, but heavy casualties and depleted ammunition supply forced the battalion commander to withdraw the survivors of the company from Sugar Loaf. Thus, the prize for which 160 men of Company E had been killed and wounded on that day fell forfeit to the Japanese. Some small sense of just retribution was felt by Company E Marines when the enemy foolishly and boldly attempted to reinforce Sugar Loaf at dusk by moving his troops to the hill along an uncovered route. Artillery observers immediately called down the fire of 12 battalions on the unprotected Japanese, decisively ending their reinforcement threat.

So well integrated were the enemy defenses on Half Moon and Sugar Loaf, capture of only one portion was meaningless; 6th Division Marines had to take them all simultaneously. If only one hill was seized without the others being neutralized or likewise captured, effective Japanese fire from the uncaptured position would force the Marines to withdraw from all. This, in effect, was why Sugar Loaf had not been breached before this, and why it was not taken on the 17th.

A combination of tank fire, flame, and demolitions had temporarily subdued the Japanese opposing the 1/29 approach on the 17th and enabled Companies A and C to advance swiftly across the valley and up the forward slopes of Half Moon. While Company C mopped up remaining enemy defenders, Company A renewed its attack across the valley floor and raced to the forward slopes of Half Moon. When Company B attempted to cross open ground to extend the battalion lines on the left, it was stopped cold by accurate fire coming from the hill, Sugar Loaf, and Shuri. At this time, the positions held by the exposed platoons of Company A became untenable. The battalion commander authorized their withdrawal to a defiladed area approximately 150 yards forward of their line of departure that morning.

By 1600, 3d Battalion companies had fought their way to Half Moon under continuous fire and begun digging in on the forward slope of the hill. They were not able to tie in with 1/29 until 1840, two hours after Company F had been ordered forward to fill in the gap between the battalions. Following a crushing bombardment of these hastily established positions on Half Moon and the exposure of the right flank of 3/29 to direct and accurate fire from enemy-held Sugar Loaf, the entire battalion was pulled back when Company A was withdrawn from its left. Strong positions were established for night defense —only 150 yards short of Half Moon. The gaps on either side of 3/29 were protected by interlocking lanes of fire established in coordination with 1/29 on its left flank and 2/29 on its right.

On 18 May at 0946, less than an hour after the 29th Marines attacked, Sugar Loaf was again occupied by 6th Division troops. (See Map VII, Map Section.) The assault began with tanks attempting a double envelopment of this key position with little initial success. A combination of deadly AT fire and well-placed minefields quickly disabled six tanks. Despite this setback and increasingly accurate artillery fire, a company of medium tanks split up and managed to reach and occupy positions on either flank of Sugar Loaf, from which they could cover the reverse slopes of the hill.

In a tank-infantry assault, Company D, 2/29, gained the top of the heretofore-untenable position, and held it during a fierce grenade and mortar duel with the defenders. Almost immediately after subduing the enemy, the company charged over the crest of the hill and down its south slope to mop up and destroy emplacements there. Disregarding lethal mortar fire from Half Moon that blanketed Sugar Loaf, Company D dug in at 1300 as well as it could to consolidate and organize its newly won conquest.

All during the attempts to take Sugar Loaf and Half Moon, the enemy on Horseshoe Hill had poured down never-ending mortar and machine gun fire on the attacking Marines below. To destroy these positions, Company F was committed on the battalion right. Supported by fire from 1/22 on its right and Marines on Sugar Loaf, the company pressed forward to the ridge marking the lip of the Horseshoe ravine. Here it was stopped by a vicious grenade and mortar barrage coming from the deeply entrenched enemy. Because of this intense resistance, the company was forced to withdraw slightly to the forward slope of the ridge, where it established a strong night defense.

Implicit in the 6th Marine Division drive towards the Asato Gawa was a threatened breakthrough at Naha. To forestall this, General Ushijima moved four naval battalions to back up the *44th Independent Mixed Brigade.* Few men in the rag-tag naval units were trained for land combat, much less combat at all, since the battalions were comprised of inexperienced service troops, civilian workers, and Okinawans who had been attached to Admiral Ota's *Naval Base Force.* The commander of the *Thirty-second Army* thought that the lack of training could be compensated in part by strongly arming the men with a generous allotment of automatic weapons taken from supply dumps on Oroku and the wrecked aircraft that dotted the peninsula's airfield.[8]

---

[8] *IntelMono*, pt I, sec B, pp. 15–16; *CICAS Trans* No. 202, Naval Attack Force T/O&E, dtd 4May45. No standards for the organization or equipping of these units had been established prior to their organization; but the makeup of one of them, found in the translation of the above-noted T/O&E, fairly well reflects the composition of the others. As of 4 May 1945, the *3d Battalion* of the *Iwao Force*, a three-battalion group organized to reinforce the *Thirty-second Army*, was composed of 415 men in two companies armed with a total of 28 machine guns, 258 rifles, 27 grenade dischargers, and 1,744 grenades.

Despite their lack of combat experience, the naval force was to perform a three-fold mission with these weapons: back up the Sugar Loaf defense system, hold the hills northwest of the Kokuba River, and maintain the security of Shuri's western flank in the event that the defenses of the *44th IMB* collapsed. The furious Japanese defense of the buffer zone stretching from the Naha estuary of the Kokuba to the western outskirts of the town of Shuri indicated their concern with the threat to the left flank of the Shuri positions.[9]

The coming of darkness on 18 May was not accompanied by any noticeable waning in the furious contest for possession of Sugar Loaf, a battle in which the combat efficiency of the 29th Marines had been so severely tested and drained. In the nearly nine days since the Tenth Army had first begun its major push, the 6th Marine Division had sustained 2,662 battle and 1,289 non-battle casualties,[10] almost all in the ranks of the 22d and 29th Marines. It was patently obvious that an infusion of fresh blood into the division lines was a prerequisite for the attack to be continued with undiminished fervor. Accepting this fact, General Geiger released the 4th Marines to parent control effective at 0800 on 19 May, at which time General Shepherd placed the 29th Marines in division reserve, but subject to IIIAC control.

At 0300 on the morning of the scheduled relief, a strong Japanese counterattack hit the open right flank of Company F, 2/29, poised just below the lip of the Horseshoe depression. The fury of the enemy attack, combined with an excellently employed and heavy bombardment of white phosphorous shells, eventually forced the advance elements of Company F to withdraw to the northern slope of Sugar Loaf.[11] At first light, relief of the three exhausted battalions of the 29th began, with 2/4 taking up positions on the left, 3/4 on the right.

Despite the difficult terrain, constant bombardment of the lines, and opposition from isolated enemy groups which had infiltrated the positions during the night, the relief was effected at 1430 at a cost to the 4th Marines of over 70 casualties—primarily from mortar and artillery fire. At approximately 1530, a counterattack was launched against 2/4, which then was in a precarious position on Half Moon Hill, on the division left flank. After nearly two hours of fighting, the attack was broken up. The advance Marine company was then withdrawn from its exposed point to an area about 150 yards to the rear, where the battalion could reinforce the regimental line after tying in with 3/5 and 3/4.

The area from which the attack had been launched against Company F, 2/29, was partially neutralized during the day by the 22d Marines. Under its new com-

[9] *Okinawa Operations Record*, pp. 110–111.

[10] *6th MarDiv SAR, Ph III*, p. 9. The division defined battle casualties as killed and wounded; non-battle casualties included the loss of men due to exhaustion, sickness, and combat fatigue.

[11] "Close examination of the aerial photo available was the cause of some concern since it appeared that the long ridge occupied by Co F was honeycombed with caves at, and slightly above, the level of the valley floor. . . . The counterattack which dislodged Co F apparently was initiated by Japanese from those caves." *Fraser ltr.*

mander, Colonel Harold C. Roberts,[12] the regiment pushed its left flank forward 100–150 yards to the high ground on the left of Horseshoe. Disregarding heavy artillery and mortar fire as well as they could, the Marines dug in new positions which materially strengthened the division line.

After a night of this heavy and accurate enemy bombardment, the two assault battalions of the 4th Marines jumped off at 0800 on 20 May. Preceded by a thorough artillery preparation and supported by the 6th Tank Battalion, the 5th Provisional Rocket Detachment, and the Army 91st Chemical Mortar Company, the Marines moved rapidly ahead for 200 yards before they were slowed and then halted. The determined refusal of the Japanese infantry entrenched on Half Moon and Horseshoe Hills to yield, and fierce machine gun and artillery fire from hidden positions in the Shuri Hill mass, where enemy gunners could directly observe the Marine attack, blocked the advance.

It soon appeared as though the fight for Half Moon was going to duplicate the struggle for Sugar Loaf. To reinforce the 2/4 assault forces and to maintain contact with the 5th Marines, Lieutenant Colonel Reynolds H. Hayden, commander of 2/4, committed his reserve rifle company on the left at 1000. In face of a mounting casualty toll, at 1130 he decided to reorient the axis of the battalion attack to hit the flanks of the objective rather than its front. While the company in the center of the battalion line remained in position and supported the attack by fire, the flank companies were to attempt an armor-supported double envelopment. At 1245, when coordination for this maneuver was completed, the attack was renewed.

Company G, on the right, moved out smartly, and, following closely behind the neutralizing fires of its supporting tanks, it seized and held the western end of Half Moon. While traversing more exposed terrain and receiving fire from three sides, the left wing of the envelopment—Company E—progressed slowly and suffered heavy casualties. Although subjected to a constant barrage of mortars and hand grenades, the company reached the forward slope of its portion of the objective, where it eventually dug in for the night. The night positions of 2/4 were uncomfortably close to those of the Japanese, and separated only by a killing zone along a hill crest swept by both enemy and friendly fire. Nonetheless, the battalion had made fairly substantial gains during the day and it was set in solidly.

Earlier that day, as 3/4 attacked enemy positions on the high ground forming the western end of Horseshoe, it had received fire support from the 22d Marines. The 4th Marines battalion employed demolitions, flamethrowers, and tanks to burn and blast the honeycomb of Japanese-occupied caves in the forward (north) slope of Horseshoe Hill. When the regiment halted the attack for the day at 1600, 3/4 had gained its objective. Here, the battalion was on high ground overlooking the

[12] Colonel Roberts and Lieutenant Colonel August C. Larson had relieved Colonel Schneider and Lieutenant Colonel Karl K. Louther, as commander and executive officer, respectively, of the 22d Marines at 1430 on 17 May.

Horseshoe depression where the Japanese mortars, which had caused so many casualties that day, were dug in.

To maintain contact with 2/4 and to strengthen his line, Lieutenant Colonel Bruno A. Hochmuth, 3/4 commander, had committed elements of his reserve, Company I, shortly after noon. Anticipating that a counterattack might possibly be mounted against 3/4 later that evening, Colonel Shapley ordered 1/4 to detail a company to back up the newly won positions on Horseshoe. Company B was designated and immediately briefed on the situation of 3/4, routes of approach, and courses of action to be followed if the Japanese attack was launched.

The sporadic mortar and artillery fire that had harassed 4th Marines lines suddenly increased at 2200, when bursts of white phosphorous shells and colored smoke heralded the beginning of the anticipated counterattack. An estimated 700 Japanese struck the positions of Companies K and L of 3/4. As soon as the enemy had showed themselves, they were blasted by the combined destructive force of prepared concentrations fired by six artillery battalions.[13] Gunfire support ships provided constant illumination over the battlefield. Company B was committed to the fight, and "with perfect timing," [14] moved into the line to help blunt the attack.

Star shells and flares gave a surrealistic cast to the wild two-and-a-half hour fracas, fought at close quarters and often hand-to-hand. The fight was over at midnight; the few enemy who had managed to penetrate the Marine lines were either dead or attempting to withdraw. The next morning, unit identification of some of the nearly 500 Japanese dead revealed that fresh units—which included some naval troops—had made the attack. The determination of the attackers to crush the Americans reemphasized the extremely sensitive and immediate Japanese reaction to any American threat against Shuri's western flank.

On 21 May, the main effort of the 6th Division attack was made by the 4th Marines, with the 22d Marines pacing the attack and giving fire support. The objective was the Asato River line. (See Map 15.) Under its new commander, Lieutenant Colonel George B. Bell, 1/4 attacked in the center of the line.[15] Forward progress down the southern slopes of Sugar Loaf towards the easternmost limit of Horseshoe was slowed by both bitter fighting and the

[13] The commanding officer of 4/15, who controlled the artillery fires directed at the counterattack and whose unit was in direct support of the 4th Marines, related an interesting incident of the action. He said that "I was able to talk directly to the front line commander of 3/4, LtCol Bruno Hochmuth, during the entire action. . . . This proved to be very handy indeed, because as usually happens when a large number of artillery units are firing in close proximity to our own front [15 battalions were firing by 2315], 3/4 started receiving an uncomfortably large number of our own rounds. Because of the fact that I was talking directly to LtCol Hochmuth, who was observing the action from his OP, we were able to pick a lull in the fight, straighten out our 'shorts,' and still continue the very heavy artillery support he had requested early in the attack." Col Bruce T. Hemphill ltr to CMC, dtd 28Feb55.

[14] *3/4 SAR, Ph III*, p. 3.

[15] Lieutenant Colonel Beans had been replaced on 1 May, when he resumed his duties as executive officer of the 4th Marines.

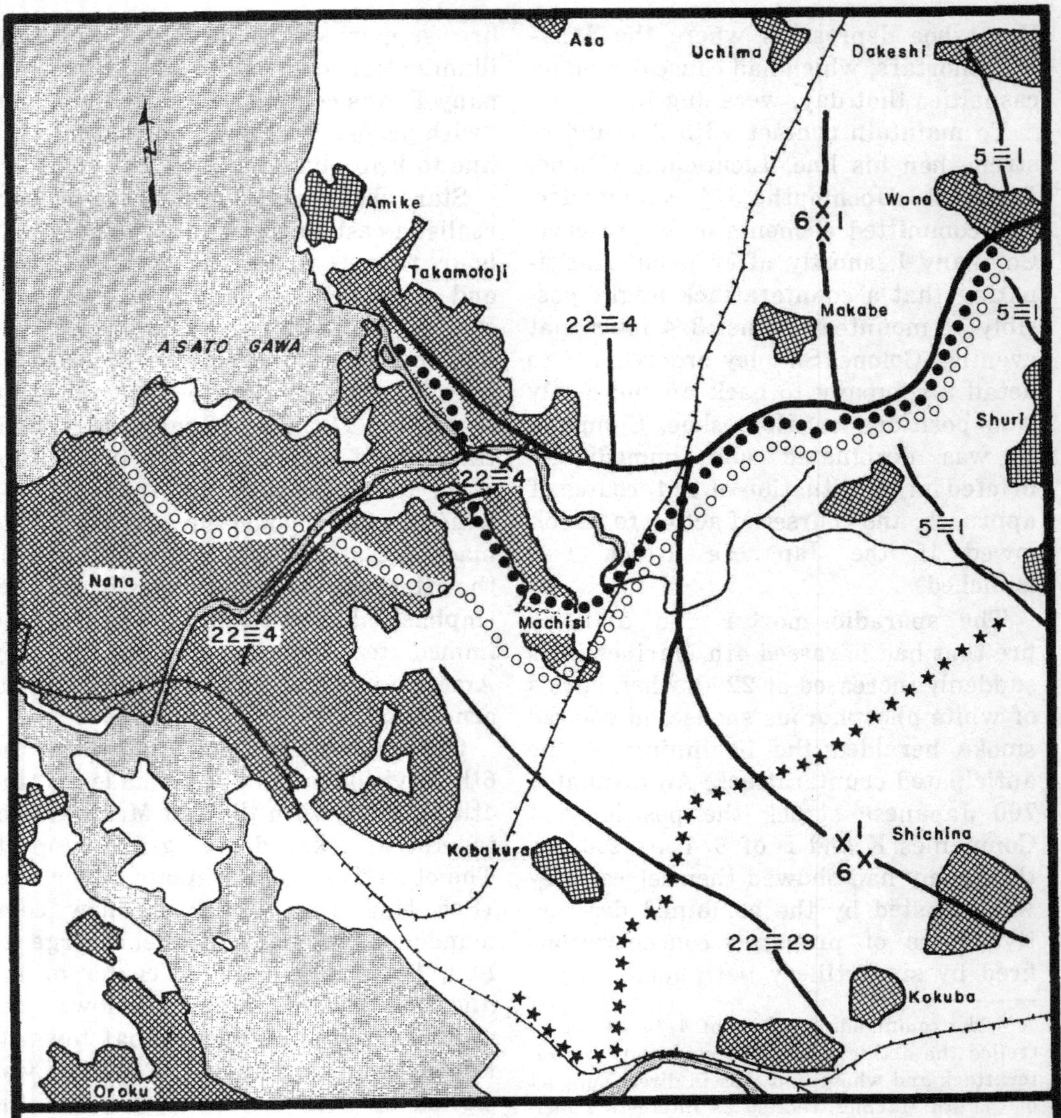

MAP 15 T.L. RUSSELL

rain that fell during the morning and most of the afternoon. This downpour turned the shell-torn slopes into slick mud-chutes, making supply and evacuation over the treacherous footing almost impossible. But the fresh battalion overcame the combination of obstacles placed in its way by the weather, terrain, and numerous remaining enemy pockets all along the river front, to advance 200 yards.

Demolition and flamethrower teams blasted and burned the way in front of 3/4 as it drove into the extensive and well-prepared enemy positions in the interior of Horseshoe. By midafternoon, Companies K and L had destroyed the deadly mortars emplaced there, and were solidly positioned in a defense line that extended approximately halfway between Horseshoe Hill and the Asato Gawa.

Intensive mortar and artillery fire from the heights of Shuri combined with the rugged terrain within the 2/4 zone of action restricted the use of tanks and prevented that battalion from advancing appreciably on 21 May. After five days of furious fighting and limited gains in the Half Moon area, General Shepherd concluded that the bulk of enemy firepower preventing his division from retaining this ground was centered in the Shuri area, outside of the division zone of action.

Thoroughly estimating the situation, he decided to establish a strong reverse slope defense on the division left, to concentrate the efforts of the division on a penetration in the south and southwest, and to make no further attempts at driving to the southeast, where his troops had been meeting withering fire from Shuri. The division commander believed that this new maneuver would both relieve his forces of a threat to their left flank and at the same time give impetus to a drive to envelop Shuri from the west.

The sporadic rain which fell on the 21st, came down even more heavily and steadily that night. Resupply of assault elements and replenishment of forward supply dumps proved almost impossible. The unceasing deluge made southern Okinawa overnight a veritable mudhole and a greater obstacle to all movement than the unrelenting enemy resistance.

THE BATTLE FOR WANA DRAW [16]

When the 1st Marine Division smashed the Japanese outpost line at Dakeshi, the battleground for General del Valle's Marines shifted to the foreboding Wana approaches to the Shuri hill mass. (See Map 14.) All evidence now signified that the main Japanese defenses in southern Okinawa consisted of a nearly regular series of concentric rings whose epicenter was protected by some of the most rugged terrain yet encountered in the drive south. The mission of breaching the Wana defenses fell to the 1st Marine and 77th Infantry Divisions at the same time that the 6th Marine and 96th Infantry Divisions attempted to envelop enemy flanks.

A somewhat crude Japanese propaganda attempt appeared in a leaflet discovered on the body of an infiltrator in the rear of the 1st Division on 14

[16] Unless otherwise noted, the material in this section is derived from *IIIAC AR; IIIAC AR; 1st MarDiv SAR; 1st MarDiv G-3 Jnl; 1st Mar SAR; 5th Mar SAR; 7th Mar SAR.*

May. Purportedly a letter from a wounded 96th Division soldier in enemy hands, it warned in fractured English that:

> . . . the battles here will be 90 times as severe as that of Yusima Island [Iwo Jima]. I am sure that all of you that have landed will lose your lives which will be realized if you come here. The affairs of Okinawa is quite different from the islands that were taken by the Americans.[17]

An analysis made of the Wana positions after the battle showed that the Japanese had "taken advantage of every feature of a terrain so difficult it could not have been better designed if the enemy himself had the power to do so."[18] Utilizing every defense feature provided by nature, General Ushijima had so well organized the area that an assault force attacking to the south would be unable to bypass the main line of resistance guarding Shuri, and would instead have to penetrate directly into the center of the heretofore unassailable defenses of the *Thirty-second Army*.

The terrain within and immediately bordering the division zone was both varied and complex. The southernmost branch of the Asa Kawa meandered along the gradually rising floor of Wana Draw and through the northerly part of Shuri. Low rolling ground on either side of the stream offered neither cover nor concealment against Japanese fire coming from positions along the reverse slope of Wana Ridge and the military crest in the southern portion of the ridge. Approximately 400 yards wide at its mouth, Wana Draw narrowed perceptibly as the stream flowing through it approached the city. Hill 55,[19] a dominating piece of terrain at the southern tip of the ridge, guarded the western entry into the draw. Bristling with nearly every type of Japanese infantry weapon, the positions on the hill had clear fields of fire commanding all approaches to the draw. Manning these guns were troops from the *62d Division's 64th Brigade*, and an ill-assorted lot of stragglers from remnants of the *15th, 23d,* and *273d Independent Infantry Battalions*, the *14th Independent Machine Gun Battalion*, and the *81st Field Antiaircraft Artillery Battalion*, all under command of the *Brigade*.[20]

By 0400 on 15 May, elements of the 5th Marines had relieved 2/1 and most of 3/1. At 0630, the relief was completed and Colonel Griebel assumed command of the former 1st Marines zone west of Wana. The 5th Marines commander placed 2/5 in assault with the 3d Battalion in support and the 1st in reserve. Acting on the recommendations of battalion and regimental commanders of both the 1st and 5th Marines, General del Valle decided to neutralize the high ground on both sides of Wana Draw. Tanks and self-propelled 105mm howitzers were to shell the area thoroughly before 2/5 tried to cross the open ground at the mouth of the draw.

[17] 1st MarDiv G–2 PeriodicRpt No. 44, dtd 15May45.

[18] 1st MarDiv G–2 PeriodicRpt No. 47, dtd 18May45.

[19] Because the first two versions of the official battle map had incorporated incorrect topographical data, this terrain feature was identified first as Hill 57 and then 59. A third and more accurate map, issued in late May designated this point as Hill 55, which shall be used in this text.

[20] 1st MarDiv G–2 PeriodicRpts Nos 44–50, dtd 15-21May45.

WANA RIDGE, *rugged barrier in the path of the 1st Division, is shown looking southeast toward Shuri. (USMC 148651)*

105MM HOWITZER *of the 15th Marines is swamped, but still in firing order after ten days of rain. (USMC 122735)*

Fire teams from Company F protected nine Shermans of the 1st Tank Battalion against possible attacks from suicide-bent enemy soldiers as the tanks worked over the Japanese positions in the mouth of the draw during the morning. Because tanks invariably drew heavy artillery, mortar, and AT fire, the Marines guarding them were forced to take cover. Nevertheless, the open ground of the battle area permitted the infantry teams to cover the tanks with fire from protected positions at long range. The mediums received heavy and intense fire from the sector to their front and from numerous cave positions on both sides of the draw. Some respite was gained when naval gunfire destroyed a 47mm AT gun which had hit three tanks at least five times each.

About midafternoon, the tanks withdrew to clear the way for a carrier-plane strike on the draw. Following this attack, the nine original tanks, now reinforced by six others, continued the process of neutralizing the draw. Another 47mm AT gun opened up late in the afternoon, but it was destroyed before it could damage any of the tanks.

After a day spent probing the mouth of Wana Draw, 2/5 infantry companies set up night defenses east of the railroad, dug in, and established contact all along their front. At the CP that night, the 5th Marines commander observed that "Wana Draw was another gorge like the one at Awacha. . . . It was obvious that the position would have to be thoroughly pounded before it could be taken," [21] and ordered the softening-up operations of the 15th repeated the next day.

Colonel Snedeker's 7th Marines spent the 15th in reorganizing its infantry companies, improving occupied positions, and mopping up in the vicinity of Dakeshi.[22] During the day, air liaison parties, gunfire spotters, and forward observers were kept busy directing concentrated artillery and naval gunfire bombardments and air strikes on known enemy strongpoints on Wana Ridge. At 2100, 1/7 was ordered to prepare a feint attack on 16 May, when all supporting arms were to fire a preparation and troops were to concentrate as though preparing to jump off in an assault.

The battalion was already positioned for the feint when preparatory fires began at 0755. At this time, 4.2-inch and 81mm mortars smoked the area immediately in front of 1/7 to heighten the deception. Fifteen minutes after it had begun, the barrage was lifted for another fifteen-minute period in an attempt to deceive the Japanese. The Marines believed that the enemy, fooled into thinking that an attack was imminent, would rush from covered caves to reoccupy their battle positions, where they would again be blasted. When there was no apparent reaction to the feint, supporting arms resumed firing at 0825 with undetermined results.

[21] *5th Mar SAR*, p. 4.

[22] While returning to his jeep from a visit to the 7th Marines OP near Dakeshi, Brigadier General Smith "passed a battalion coming up to go in the lines. . . . These men were not going into combat for the first time. They were veterans. There was no singing and laughing. They were deadly serious and business like. The men had already been separated from the boys." Smith, *Personal Narrative*, p. 110.

At 0950, regiment notified Lieutenant Colonel John J. Gormley, the 1/7 commander, that an air strike on Wana Ridge was scheduled for 1000, immediately following which he was to send patrols forward to determine what remaining enemy resistance existed on the target. Having learned that the strike was delayed, at 1028 Lieutenant Colonel Gormley requested that the mission be cancelled and sent the patrols out after he had ordered a mortar barrage placed on the ridge.

The Company C patrols moved forward unopposed until they reached the western end of Wana Ridge. Here they received intense grenade and machine gun fire which was answered by their battalion 81mm mortars and supporting fire from the 5th Marines. Rushing forward when this fire had been lifted, the patrols carried and occupied the troublesome objective.

Lieutenant Colonel Gormley then ordered the newly won position held and reinforced by troops he sent forward for this purpose. Once leading elements began to move out again, enemy troops lodged in burial vaults and rugged coral formations showered grenades down upon the advancing Marines. Unsuccessful in halting the advance, the enemy tried but failed to mount a counterattack at 1605. Although supporting arms of 1/7 blunted this attempt, enemy resistance to the Marine attack continued.

Nightfall forced the battalion commander to withdraw the troops spearheading the assault and move them to more secure positions on a plateau almost directly north of the ridge for night defense. Contact was then established on the right with 2/5 and on the left with 3/7. At 2400, 3/7 effected a passage of the lines to relieve the 1st Battalion, which then went into regimental reserve.

During the 16th, 15 tanks, two of them flamethrowers, had supported the attack of 1/7 from positions on Dakeshi Ridge, while a total of 30 tanks—including 4 flamethrowers — supported 2/5 by burning and blasting enemy strongpoints in Wana Draw. At 0900, the 2/5 armored support drew antitank, mortar, and artillery fire that disabled two tanks, and damaged two others, which withdrew after evacuating the crews of the stalled cripples. Two of the AT positions which had been spotted in the morning were destroyed that afternoon when the main battery of the USS *Colorado* was brought to bear on them. Generally, when a Marine tank was damaged and abandoned temporarily, efforts to retrieve it later were usually stymied by enemy fire. Disabled tanks remaining in the field overnight usually were either destroyed by enemy demolition teams or occupied by snipers, who converted them into armored pillboxes.

Before retiring at nightfall on 16 May, the 1st Tank Battalion had expended nearly 5,000 rounds of 75mm and 173,000 rounds of .30 caliber ammunition, and 600 gallons of napalm on targets on Dakeshi Ridge and in Wana Draw that day.[23] Following the two-day process of softening up provided by all supporting arms, the 5th Marines prepared to run the gantlet of Wana Draw on 17 May.

[23] *1st TkBn Summary*, 16May45.

"Under the continued pounding of one of the most concentrated assaults in Pacific Warfare," [24] cracks appeared in the Shuri defenses on 17 May. On that day, 2/5 made the main regimental effort, sending tank-infantry teams to the mouth of Wana Draw, where they worked over the caves and pillboxes lining its sides. The 2/5 attack was made in conjunction with a 7th Marines effort to gain the pinnacle ridge forming the northern side of the draw.

When a terrific mortar and artillery barrage drove the 7th back at 1200, 2/5 assault troops—also under heavy fire—were forced back to their original positions, where they could protect the exposed flank of the 7th Marines battalion.

On the right of 2/5, Company E finally succeeded in penetrating the Japanese defenses. After having been driven back earlier in the day, the company established a platoon-sized strongpoint on its objective, the west nose of Hill 55. Because the low ground lying between this point and battalion frontlines were swept by heavy enemy fire, tanks were pressed into use for supply and evacuation purposes.

Having relieved 1/7 at 0600, 3/7 attacked towards Wana Ridge from Dakeshi Ridge with two companies in assault: Company I on the right, K on the left. A total of 12 gun and 2 flamethrower tanks supported Company K as it attempted to secure the low ridge crest northwest of Wana. Meanwhile, Company I gained and held a plateau that led to the western nose of the Wana Ridge line.

[24] *1st MarDiv SAR*, pt VII, p. 5.

Extremely heavy resistance plagued Company K efforts to move forward, as the enemy concentrated his fire on the leading infantry elements. Attempting to lessen the effectiveness of the Marine tank-infantry tactics, the Japanese employed smoke grenades to blind the tanks and drastically restricted their supporting fires. Before the tanks could be isolated in the smoke and cut off from their infantry protection, and when the flanks of Company K became so threatened as to make them untenable, both tanks and infantry were withdrawn—the latter to Dakeshi to become 3/7 reserve. Late in the afternoon, the 3/7 commander ordered Company L forward to reinforce I for the night and to assist in the attack the next morning.

Following a period of intermittent shelling from enemy mortars and artillery during the night 17–18 May, 3/7 again attacked Wana Ridge. Supporting arms delivered intense fire on the forward slopes and crest of the ridge all morning; the attack itself began at noon. Reinforced by a platoon from L, Company I succeeded in getting troops on the ridge, but furious enemy grenade and mortar fire inflicted such heavy casualties on the assault force that Lieutenant Colonel Hurst was forced to withdraw them to positions held the previous night, where he could consolidate his lines. An abbreviated analysis by the division fairly well summarized that day's fighting: "gains were measured by yards won, lost, then won again." [25]

[25] *1st MarDiv SAR*, pt VII, p. 6.

Pinned down by heavy enemy fire on the reverse slope of its position on Hill 55, the isolated platoon from Company E, 2/5, could neither advance nor withdraw. Tanks again supplied ammunition and rations to the dug-in troops. Six mediums initially supported the early-morning operations of 2/5 by firing into caves and emplacements in the terrain complex comprising the draw. This tank fire was coordinated with that coming from Shermans in the Wana Draw sector. In addition to this day-long tank firing, the artillery battalions expended over 7,000 rounds of 105mm and 75mm artillery ammunition on selected point targets.[26]

Under the cover of tank fire, at 1200 Company F sent one infantry platoon and an attached engineer platoon with flamethrowers and demolitions into the village of Wana to destroy enemy installations there. The party worked effectively until 1700, when it was recalled to Marine lines for the night because Wana Ridge, forming the northern side of the draw behind and overlooking the village, was still strongly infested by the enemy. Before leaving Wana, Marines destroyed numerous grenade dischargers, machine guns, and rifles found in the village and in the tombs on its outskirts.

The 1st Marine Division's bitter contest for possession of Wana Draw continued on 19 May along the same bloody lines it had run on the four previous days. Colonel Snedeker's regiment again made the major effort for the division, and the 5th Marines continued to punish the mouth of Wana Draw. As before, attacking Marines were sorely beset by enemy fire, which answering artillery, tank, mortar, and regimental 105mm howitzer concentrations had failed to neutralize.

The 3d Battalion, 7th Marines, attacked that afternoon in a column of companies, Company I in the lead, followed by Companies L and K in that order. Resistance to the attack was immediate, although the vanguard managed to reach the nose of the coral ridge to its front under a blanket of mortar shells falling all about. Then, because 3/1 was to relieve 3/7 and it was too risky to effect a relief right on the ridge under the conditions then prevailing, the leading elements withdrew about 75 yards to the rear.

Earlier in the day, 1/7 and 2/7 had been relieved in position near Dakeshi by 1/1 and 2/1 respectively. With the relief of the 3d Battalion, the 7th Marines relinquished the responsibility for the capture of Wana Ridge to Colonel Mason's 1st Marines, and Colonel Snedeker's regiment went into division reserve. In the five-day struggle for Wana, the 7th Marines had lost a total of 44 men killed, 387 wounded, 91 non-battle casualties, and 7 missing. Of this number, the 3d Battalion sustained 20 Marines killed and 140 wounded. In a supporting and diversionary role for the five-day period, the 5th Marines suffered 13 men killed and 82 wounded.

Despite the punishment they had received from the 5th Marines and its supporting tanks, the Japanese built new positions in Wana Draw daily, and reconstructed and recamouflaged by night old ones that Marine tank fire had exposed and damaged by day. As the

---

[26] *2/5 SAR*, p. 13.

assault infantry plunged further into the draw, and as the draw itself narrowed, an increasing number of Japanese defensive positions conspired with the rugged terrain to make passage more difficult. Dominating the eastern end of Wana Ridge, on the northwestern outskirts of Shuri, was 110 Meter Hill,[27] commanding a view of the zones of both the 1st Marine and 77th Infantry Divisions. Defensive fire from this position thwarted the final reduction of Japanese positions in Wana Draw and eventual capture of the Shuri redoubt.

Tanks, M–7s (self-propelled 105mm howitzers), 37mm guns, and overhead machine gun fire supported the attacks which jumped off at 0815. The assault troops moved rapidly to the base of the objective, tanks and flamethrowers clearing the way, while enemy mortar and machine gun fire inflicted heavy casualties in the ranks of the onsurging Marines.

Initially, 3/1 moved to the southeast and up the northern slope of Wana Ridge, where it became involved in hand-grenade duels with Japanese defenders. The Marines prevailed and managed to secure approximately 200 yards of this portion of the ridge. By 1538, 2/1 reported to regiment that it was on top of the objective and in contact with 3/1, and had secured all of the rest of the northern slope of the ridge with the exception of the summit of 110 Meter Hill. A considerable gap between the flanks of 2/1 and the 305th Infantry on the Marine left was covered by interlocking bands of machine gun fire and mortar barrages set up by both units. Confronted by intense enemy fire from reverse slope positions, 2/1 riflemen were unable to take the hillcrest and dug in for the night, separated from the enemy by only a few yards of shell-pocked ground.

After it tied in with 2/1 for the night of 19–20 May, Lieutenant Colonel Stephen V. Sabol's 3d Battalion moved out at 0845, and was again within grenade-throwing range of Wana Ridge defenders. Burning and blasting, tanks supported the assault by destroying enemy-held caves and fortified positions blocking the advance. When 3/1 had gained the northern slope of the ridge and could not budge the Japanese troops in reverse-slope defenses, Colonel Mason decided to burn them out by rolling split barrels of napalm down the hill into Japanese emplacements in Wana Draw, and then setting them afire by exploding white phosphorous (WP) grenades on top of the inflammable jellied mixture.

Working parties began manhandling drums of napalm up the hill at 1140, and had managed to position only three of them by 1500. At 1630, these were split open, sent careening down the hill, and set aflame by the WP grenades. An enemy entrenchment about 50 yards down the incline halted the drums; in the end, the Japanese sustained little damage and few injuries from this hastily contrived field expedient. The proximity of the combatants that night led to considerable mortar, hand grenade, and sniper fire, as well as the usually lively and abusive exchange of curses, insults, and threats of violence

[27] Variously named 100 Meter Hill, Knob Hill, and Conical Hill by combat troops who viewed it from different points on the ground, the designation found in the 2/1 SAR will be used hereafter.

that often took place whenever the protagonists were within shouting distance of each other.

On the division right, 2/5 jumped off in attack on 20 May at 0900, supported by artillery and M–7 fire and spearheaded by tanks. The battalion objective was the area running roughly from Hill 55 southwest to the Naha-Shuri road. A continuous artillery barrage was laid on Shuri Ridge, the western extension of the commanding height on which Shuri Castle had been built, as assault units quickly worked their way towards the objective. At 0930, lead elements were engaged in close-in fighting with enemy forces in dug-in positions bordering the road. Under constant and heavy enemy fire, engineer mine-clearing personnel preceded the tanks to make the road safe for the passage of the mediums. Working just in front of the advancing troops, the Shermans flushed a number of enemy soldiers from their hidden positions and then cut them down with machine gun fire. Close engineer-tank-infantry teamwork permitted the Marines to secure the objective by noon.

Heavy small arms and mortar fire poured into the advance 2/5 position, which Company E held all afternoon. The sources of this fire were emplacements located on Shuri Ridge. Continued artillery and pointblank tank fire, and two rocket barrages, finally silenced the enemy weapons. By 2000, Company E had established contact all along the line and dug in for the night. Except for the usual enemy mortar and artillery harassment, there was little activity on the front. Just before dawn, 1/5 relieved the 2d Battalion in place; 2/5 then went into regimental reserve.

Once in position, 1/5 was ordered to patrol aggressively towards Shuri Ridge and on the high ground east of Half Moon Hill. It maintained a sufficient force in the vicinity of Wana Ridge and Hill 55, at the same time, to assist the 1st Marines attack. Tank-infantry teams again reconnoitered the area south of the division line against a hail of machine gun and mortar fire. In addition to providing the tanks protection from Japanese tank-destroyer and suicide units, Marine ground troops directed the tank fire on targets of opportunity. Tank commanders in vehicles that were sometimes forward of foot troops often called down artillery fire on point targets at extremely close ranges. In spite of fierce resistance that became most frenzied as Marines closed in on Shuri, the 5th Marines positions on Hill 55 were advanced slightly in order to give the division more favorable jumping off points for a concerted effort against General Ushijima's headquarters.

At dawn on 21 May, 2/1 moved out against heavy opposition to secure the summit of 110 Meter Hill and the rest of Wana Ridge. Although some small gains were made, the objectives could not be reached. Tank support, which heretofore had been so effective, was limited because of the irregular and steep nature of the ground. Though armor could provide overhead fire, the vehicles were unable to take reverse slope positions under fire because a deep cleft at the head of the draw prevented the Shermans from getting behind the enemy. Reconnaissance reports indicated that as the draw approaced Wana, it walls rose to sheer heights of from 200 to 300 feet. Lining the wall faces were numerous, well-defended caves that were unapproachable to all but the suicidally inclined. It was readily apparent that no assault up the draw would be successful unless preceded by an intense naval gunfire, air, and artillery preparation. Included also in the reports was the fact that the steep terrain forward of Wana did not favor tank operations.

On the left of 2/1, Company G mopped up opposition in the small village on the northern outskirts of Shuri. Resisting the attempts of the company to turn the flank of 110 Meter Hill were elements of the *22d Independent Infantry Battalion*, the sole remaining first-line infantry reserve of the *Thirty-second Army*—thrown into the breach to hold the area around the hill.[28] Advancing down the draw were two companies abreast, E and F, whose attack was initially supported by the massed fires of battalion mortars and then by all other supporting arms.

Darkening skies and intermittent rain squalls obscured the battle scene to friendly and enemy observers alike. Although it was apparent that 2/1 was right in the middle of a preregistered impact area, judging from the accuracy of enemy mortar and artillery fire, the battalion held its forward positions despite mounting casualties. A gap existing between 2/1 and the 77th Division was covered by fire, and Company F linked with Company C of 1/1, which had been temporarily attached to the 3d Battalion for night defense.

[28] 1st MarDiv G–2 PeriodicRpt No. 50, dtd 21May45.

Under its new commander, Lieutenant Colonel Richard F. Ross, Jr.,[29] 3/1 attempted to clear out reverse slope positions on Wana Ridge in a concerted tank-infantry effort. According to the plan, Company L and the tanks—each to be accompanied by one fire team—would attack up Wana Draw. Supporting this assault from the crest of the ridge would be the other two infantry companies in the battalion, prepared to attack straight across the ridge on order. Their objectives were Hill 55 and the ridge line to the east.

Company C, 1/1, was ordered to take over the 3d Battalion positions, when Lieutenant Colonel Ross' men jumped off in the assault. At about 1415, Company L began the slow advance against bitter opposition. Almost immediately, several of the escorting tanks were knocked out by mines and AT guns. Company K moved across the draw to Hill 55 at 1500, followed by I, which was pinned down almost immediately by extremely heavy mortar and machine gun fire and unable to advance beyond the middle of the draw. By 1800, Company K was on Hill 55 and tied in with 1/5, but could not push further east towards Shuri.

Because the rampaging enemy fire prevented Companies I and L from reaching the ridgeline and advancing up Wana Draw, they were withdrawn to that morning's line of departure positions. Company C of 1/1 was placed

[29] Lieutenant Colonel Sabol was relieved on 21 May, when he was transferred to the 7th Marines to become regimental operations officer.

under the operational control of 3/1 for the night and occupied the positions held by Company I on 20 May. Here, it tied in with Company L on the right and on the left with Company F, 2/1.

The miserable weather prevailing all day on 21 May worsened at midnight when the drizzle became a deluge and visibility was severely limited. Taking advantage of these conditions favoring an attacking force, an estimated 200 Japanese scrambled up Wana Ridge to strike all along the Company C line. In the midst of a fierce hand grenade battle, the enemy managed to overrun a few positions. These were recaptured at dawn, when the Marines regrouped, reoccupied the high ground, and restored their lines. In the daylight, approximately 180 enemy dead were counted in front of Marine positions.[30]

Torrential rains beginning the night of 21–22 May continued on for many days thereafter. This downpour almost halted the tortuous 1st Division drive towards Shuri. Seriously limited before by terrain factors and a determined stand by the enemy in the Wana area, tank support became nonexistent when the zone of the 5th Marines, the only ground locally which favored armored tactics, became a sea of mud. Under these conditions aiding the Japanese defense, the 1st Division was faced with the alternatives of moving ahead against all odds or continuing the existing stalemate. To make either choice was difficult, for both presented a bloody prospect.

[30] 1st MarDiv G–2 PeriodicRpt No. 52, dtd 23May45.

THE ARMY'S FIGHT [31]

For IIIAC, the period 15–21 May was marked by the struggles of its divisions to capture two key strongpoints—Wana Draw and Sugar Loaf Hill. During this same seven days, XXIV Corps units fought a series of difficult battles to gain the strongly defended hills and ridges blocking the approaches to Shuri and Yonabaru. (See Map IV, Map Section.) These barriers, incongruously named Chocolate Drop, Flat Top, Hogback, Love, Dick, Oboe, and Sugar, gained fleeting fame when they became the scenes of bitter and prolonged contests. But, when XXIV Corps units had turned the eastern flank of Shuri defenses and anticipated imminent success, the Army attack—like that of the Marines—became bogged down and was brought to a standstill when the rains came.

On 15 May, the 77th Division continued its grinding advance in the middle of the Tenth Army line against the hard core of *Thirty-second Army* defenses at Shuri; 96th Division troops, in coordination with their own assault against Dick Hill, supported the 77th Division attack on Flat Top Hill. Fighting on the left of the 96th, the 383d Infantry found it difficult and dangerous to move from Conical Hill because of overwhelming fire coming from a hill complex southwest of their location. In addition, the *89th Regiment* tenaciously held formidable and well-organized defenses on the reverse slope of Conical,

[31] Unless otherwise noted, the material contained in this section is derived from: *Tenth Army AR; XXIV Corps AR; 77th InfDiv OpRpt*, Okinawa; *96th InfDiv AR*.

and prevented the soldiers from advancing farther south.[32]

On 16 May, 2/383 attacked down the southeast slope of the hill, but murderous enemy crossfire again prevented the soldiers from making any significant gains. A supporting platoon of tanks, however, ran the gantlet of fire sweeping the coastal flat and advanced 1,000 yards to enter the northwestern outskirts of Yonabaru, where the Shermans lashed the ruins of the town with 75mm and machine gun fire. Heavy Japanese fire covering the southern slopes of Conical prevented the infantry from exploiting the rapid armor penetration, however. After having exhausted their ammunition supply, the tanks withdrew to the line of departure.

On the division right flank, the 382d Infantry attempted to expand its hold on Dick Hill. In a violent bayonet and grenade fight, American troops captured some 100 yards of enemy terrain, but heavy machine gun fire from Oboe Hill —500 yards due south of Dick—so completely covered the exposed route of advance, the soldiers were unable to move any farther.

Fire from many of the same enemy positions which had held back 96th Division forces, also effectively prevented the 307th Infantry from successfully pushing the 77th Division attack on Flat Top and Chocolate Drop Hills. Both frontal and flanking movements, spearheaded by tanks, were held up by extremely accurate and vicious Japanese machine gun fire and mortar barrages.

Somewhat more successful on the 16th was the 305th Infantry, which threw the full weight of all of its supporting arms behind the attack of the 3d Battalion. Flamethrower tanks and medium tanks mounting 105mm howitzers slowly edged along the ridges leading to Shuri's high ground. Barring the way in this broken terrain were Okinawan burial vaults which the Japanese had occupied, fortified, and formed into a system of mutually supporting pillboxes. At the end of a ferocious day-long slugging match, this armored vanguard had penetrated 200 yards of enemy territory to bring the 77th Division to within 500 yards of the northernmost outskirts of Shuri.

A very successful predawn attack by the 77th Division on 17 May surprised the Japanese, forcing them to relinquish ground. Substantial gains were made and commanding terrain captured, including Chocolate Drop Hill and other nearby hills. Advancing abreast of each other, 3/305 and 2/307 dug in at the end of the day only a few hundred yards away from Shuri and Ishimmi. Although outflanked by 3/307, Flat Top defenders sent down a heavy volume of machine gun and mortar fire on the soldiers as they attempted to move across exposed country south of the hill. Troops following the assault elements spent daylight hours mopping up, sealing caves and burial vaults, and neutralizing those enemy strongpoints bypassed in the early-morning surprise maneuver.

Practically wiped out that day was the enemy *22d Regiment*, which had defended Chocolate Drop, and whose remnants were still holding the reverse slopes of Flap Top and Dick Hills. Reinforcing these positions was the *1st*

---

[32] *Okinawa Operations Record*, "Record of the 24th Division," p. 177.

Battalion of the *32d Regiment.* On 17 May, this regiment was ordered by the *24th Division* commander, Lieutenant General Tatsumi Amamiya, to take over the ground formerly held by the *22d Regiment,* and to set in a Shuri defense line that would run from Ishimmi to Dick and Oboe Hills. Taking advantage of the natural, fortress-like properties of the region which they were to defend, the depleted *32d Regiment* and survivors of the *22d* were disposed in depth to contain potential American penetrations.[33] Few reserves were available to the defenders should the Americans break through.

On 17 May, the 96th Division ordered the 382d Infantry to attack and capture the hill mass south of Dick Hill and centering about Oboe. The failure of this effort indicated that the ground here needed to be softened up further before the infantry could advance. In the sector of Conical Hill held by the 383d, steady pressure from reverse slope defenders forced the division to commit into the line a third regiment—the 381st Infantry—to maintain the positions already held by the 96th. At this time, 3/381 assumed control of the left portion of the 2/383 sector on the eastern slope of the hill, and brought up its supporting weapons in preparation for a new attack.

While the remainder of 96th Division assault battalions held their lines and tank-infantry, demolition, and flamethrower teams mopped up in their immediate fronts, 3/381 made the division main effort. Operating to the west of the coastal road, medium tanks supported the attack by placing direct fire on machine gun positions on Hogback Ridge, a terrain feature running south from Conical Hill. Hogback's defenders disregarded the tank fire to place heavy machine gun and mortar barrages against the battalion attacking up finger ridges sloping down to the ocean. Although this heavy resistance limited the advance to only 400 yards, the division commander believed he could successfully attack through Yonabaru to outflank Shuri.

Both frontline divisions of XXIV Corps progressed on 18 May. Units of the 77th penetrated deeper into the heart of Shuri defenses by driving 150 yards farther south along the Ginowan-Shuri highway and advancing up to 300 yards towards Ishimmi. On 19 May, the 77th Division began a systematic elimination of Japanese firing positions in 110 Meter Hill, Ishimmi Ridge, and the reverse slopes of Flat Top and Dick Hills. All of these positions provided the enemy with good observation and clear fields of fire, commanding terrain over which the American division was advancing. Every weapon in the 77th arsenal capable of doing so was assigned to place destructive fire on the enemy emplacements. While these missions were being fired, the infantry fought off a series of counterattacks growing in size and fury as darkness fell. The enemy was finally turned back at dawn on 20 May when all available artillery was called down on them.

In the 96th Division zone on the 19th, the left regiment again made the main effort while the center and right regiments destroyed cave positions and gun emplacements in the broken ground be-

[33] *Ibid.*, pp. 179–181, Map 5.

tween Conical and Dick Hills. Hogback Ridge and Sugar Hill, which rose sharply at the southern tip of this ridge to overlook Yonabaru, were bombarded by two platoons of medium tanks, six platoons of LVT(A)s, artillery, and organic infantry supporting weapons. The attack following this preparation failed, however, in the face of overwhelming enemy fire. Destruction of enemy positions spotted the day before did serve, however, to weaken further the faltering *89th Regiment* defense.

Returning to Hogback Ridge on 20 May, the attacking infantry made a grinding, steady advance down the eastern slopes of the ridge and finally reached Sugar Hill. Other 96th Division units also registered some significant gains that day; 383d Infantry assault battalions fought to within 300 yards of Love Hill, destroying those strongpoints that had blocked their progress for a week. The 382d Infantry finally reduced all enemy defenses on the southern and eastern slopes of Dick Hill, while it supported a successful 77th Division attack on Flat Top at the same time.

On gaining Flat Top Hill, the 307th Infantry was then ready to continue the attack south to Ishimmi Ridge and then on to Shuri. Coordinating its attack with the 1st Marines on its right, the 305th advanced down the valley highway 100–150 yards or to within 200 yards of the outskirts of Shuri. As a result of these gains, the 77th Division commander planned another predawn surprise attack, only this time on a coordinated division-wide level across the front.

Assault troops of the 307th Infantry jumped off at 0415 on 21 May in the zone of the 305th, advancing 200 yards without opposition. (See Map VIII, Map Section.) An hour later, leading elements had entered the northern suburbs of Shuri and were fighting their way up the eastern slopes of 110 Meter Hill. The 306th Infantry, which relieved the 305th later that morning, sent its 2d Battalion to the right of the line where visual contact was made with the 1st Marines. By nightfall, having spent most of the day mopping up bypassed positions, the 306th set up a night defense on a line running from the forward slopes of Ishimmi Ridge, through the outskirts of Shuri, to 110 Meter Hill.

The assault battalions of the 307th Infantry, the other 77th Division frontline regiment, jumped off at 0300 to take the regimental objective, a triangularly shaped mass consisting of three hills located in open ground about 350 yards south of Flat Top. The lead elements reached the objective at dawn, but following units were unable to exploit the successful maneuver when they were discovered by the enemy and pinned down by his frontal and flanking fire. Any further move forward was prohibited by this continuous and accurate fire, and the battalion was forced to dig in at nightfall on the ground then held.

Overall, the most important advances on 21 May in the XXIV Corps zone were made by 96th Division units. As 1/383 moved out against moderate opposition to take Oboe Hill, 2/383 paced the advance by attacking over exposed terrain to its southeast to take a hill approximately 400 yards from Shuri. At 1130, when enemy elements were noticed pulling out of their positions in front of the attacking infantry, the Japanese were fired upon as they retreated towards higher ground. Despite this withdrawal of the enemy, American forces were prevented from advancing any further during the day by isolated enemy counterattacks along the regimental lines.

On Love Hill, enemy defenders who had successfully refused to yield ground during the past week again steadfastly maintained their positions on the 21st. They called down heavy and accurate artillery concentrations on American tank-infantry teams reaching the base of the hill and forced them to turn back.

The western slopes of Hogback Ridge were secured by 2/383 as the 3d Battalion, 381st Infantry, fought its way up the eastern slopes to the top of Sugar Hill. Every yard acquired during the day came because of the individual soldier's efforts in the face of fanatic enemy determination to hold. Nevertheless, advance elements of 3/381 were in position about 200 yards from the Naha-Yonabaru highway by nightfall. As a result of this hard-won success, a 700-yard-long corridor down the east coast of Okinawa was secured, giving promise that the final reduction of the Shuri redoubt might be launched from this quarter.

To strengthen the attack on Shuri, which General Hodge believed could be outflanked when he viewed the progress of the 96th Division, he alerted the 7th Infantry Division and ordered it to move to assembly areas immediately north of Conical Hill on 20 May. Two days later, the division was committed in the line and attacked to take the high ground south of Yonabaru.

Intermittent rain beginning on 21 May increased steadily to become soaking torrents before the assault infantry of the 7th Division was in jump-off positions. In no time at all, "the road to Yonabaru from the north—the only supply road from established bases in the 7th Division zone . . . became impassable to wheeled vehicles and within two or three days disappeared entirely and had to be abandoned." [34] Like the Tenth Army divisions on the west coast, those on the east were effectively stymied by the mud and the rain, which now seemed to be allied with General Ushijima and his *Thirty-second Army.*

FIGHTING THE WEATHER [35]

The Naha-Yonabaru valley served as a funnel through which American forces could pass to outflank Shuri. A major obstacle blocking the entrance to this route is the Ozato Hills, a rugged and complex terrain mass paralleling Nakagusuku Wan and lying between Yonabaru and the Chinen Peninsula. Since strong blocking positions were needed in the Ozato Hills to safeguard the left flank and rear of the force assigned to assault Shuri, the 184th Infantry of the 7th Division was ordered to take Yonabaru on 22 May and secure the high ground overlooking the village.

In a surprise attack at 0200, 2/184 spearheaded a silently moving assault force which passed through Yonabaru

[34] *XXIV Corps AR*, p. 31.

[35] Unless otherwise indicated, the material in this section is derived from: *Tenth Army AR; IIIAC AR; XXIV Corps AR; 1st MarDiv SAR; 6th MarDiv SAR, Ph III; 7th InfDiv AR; 77th InfDiv OpRpt*, Okinawa; *96th InfDiv AR; 1st Mar SAR; 4th Mar SAR, Ph III; 5th Mar SAR; 22d Mar SAR, Ph III; 29th Mar SAR, Ph III.*

quickly, and was on the crest of its objective—a hill south of the village—by daylight. When the enemy arose at dawn and emerged from cave shelters to man gun and infantry positions, he met sudden death under American fire. The *Thirty-second Army* was completely taken aback, for an American night attack was totally unexpected in this sector, much less an attack unsupported by armor.[36] When the commander of the 184th saw that his initial effort was successful, he committed a second battalion and drove forward to secure other key points in the zone. By the end of the day, the regiment had advanced 1,400 yards and gained most of its objectives, even though rain and mud drastically hampered all phases of the operation.

While the 7th Division scored for the Tenth Army on the east coast, IIIAC units pushed forward on the west. In the 6th Marine Division zone, the 4th Marines attacked to gain the northern bank of the Asato Gawa. (See Map 15.) The 1st and 3d Battalions advanced as 2/4 maintained positions on Half Moon Hill and kept contact with the 1st Marine Division. Assault troops seized the objective by 1230, when patrols crossed the shallow portion of the Asato and moved 200 yards into the outskirts of Naha before drawing any enemy fire.[37] Frontline Marines dug in reverse slope positions along the northern bank of the river under the sporadic fire of heavy caliber artillery weapons and mortars. At 6th Division headquarters, plans were drawn for a river crossing on 23 May.

Although the flank divisions of the Tenth Army were making encouraging progress, the three divisions in the center of the line found success to be an elusive thing during the week of 22 May. A fanatic Japanese defense compounded the difficulties arising because of the steady rain. Supply, evacuation, and reinforcement were all but forestalled by the sea of mud, which caused the troops to wallow rather than maneuver. Under these conditions, infantry units could only probe and patrol ahead in their immediate zones.

The rain continued for nine days, and ranged from light, scattered showers to driving deluges. In the end, the entire southern front became a morass that bogged down both men and machines. Footing was treacherous in the mud swamps appearing in valley floors, and all slopes—from the gentlest to the most precipitous—were completely untrafficable. Because TAF planes had been grounded and could not fly airdrop missions, all supplies had to be manhandled to the front. Tired, wornout foot troops from both frontline and reserve units were pressed into action and formed into carrying parties.

Despite the unrelenting round-the-clock efforts of engineers to keep the road net between forward supply dumps

---

[36] *Yahara Interrogation.*

[37] By this time, "Naha was no longer tenable by the Japanese as it had been thoroughly flattened by prior bombardment and its flat terrain offered no cover from our fire from the high ground north of Naha. However, the Japanese held the hills south of the narrow harbor of [the city] and to have put any great number of troops in Naha would have invited severe casualties. Therefore, until such time as the hill mass south of the harbor . . . was captured, Naha itself [was] a no-man's land in which only patrols operated." Smith, *Personal Narrative*, p. 105.

and the frontlines in operating condition, continued use by trucks and amtracs finally caused the roads to be closed, but only after the mud itself had bemired and stalled the vehicles. As a result, division commanders found it impossible to build up and maintain reserve stocks of the supplies needed to support a full-scale assault. With the movement of American forces all but stopped, the entire front became stalemated.

During the advance to the south, responsible Tenth Army agencies had reconnoitered both the east and west coasts of Okinawa behind the American lines in an attempt to find suitable landing and unloading sites. When discovered and found secure from enemy fire, they were developed and LSTs and other landing vessels were pressed into use to bring supplies down the coasts from the main beaches and dumps in the north.[38] The two divisions deriving the major benefits from use of the overwater supply routes were the 6th Marine and 7th Infantry Divisions anchored on the open coasts. Behind the 1st Marine and 77th and 96th Infantry Divisions, in the center of the Tenth Army line, was a mired road net which prevented any resupply effort by vehicles coming both from the north and laterally from the coasts.

A sanguine outlook for continued advances by the Tenth Army flanking divisions was dispelled when they ran into resistance of the same type and intensity offered to the center units. A combination of this increased resistance and the appalling weather forestalled the potential envelopment and isolation of the main forces of the enemy, and forced the two coastal divisions into the same sort of deadlock the rest of the Tenth Army was experiencing.

Attacking to the west on 23 May, the 7th Division immediately ran into heavy resistance in the hills just north of the Yonabaru-Naha road. Both division assault regiments met increasingly stiff opposition during the day, because: "The Japanese realized that this advance along the Yonabaru-Naha road threatened to cut off the Shuri defenders. . . ."[39] Even in the midst of the American attacks, the enemy attempted to infiltrate and counterattack.

At the same time that the 184th Infantry moved into the Ozato Hills and towards the mouth of the valley leading to Shuri, the 32d Infantry struck out to the west and southwest through Yonabaru to isolate the forces protecting the *Thirty-second Army* redoubt. Units spearheading the regimental drive were slowed and unable to advance in the face of the considerable machine gun and mortar fire coming from positions in the low hills east of Yonawa. Here, a mile southwest of Yonabaru, the regiment was forced to halt and dig in a line for the night because tanks, urgently needed to sustain the drive, had become immobilized by the mud.

On the west coast, despite the continuing rain during the night of 22–23 May, 6th Division patrols crossed and recrossed the Asato almost at will to feel out the enemy. (See Map 15.) Scouts from the 6th Reconnaissance Company patrolled the south bank of

[38] *Blakelock ltr 1965.*

[39] Smith, *Personal Narrative*, p. 111.

the upper reaches of the river, and reported back at 0718 that the stream was fordable at low tide, resistance was light, and no occupied positions had been found.[40]

Because the patrol reports of this and other units indicated "that it might be feasible to attempt a crossing of the Asato without tank support," [41] early that morning General Shepherd ordered the 4th Marines to increase the number of reconnaissance patrols south of the river, and to be ready to cross it if enemy resistance proved light.

Between dawn and 1000 on the 23d, Marine lines received long-range machine gun and rifle fire from high ground near Machisi, but the patrols met no determined resistance at the river bank. General Shepherd decided to force a crossing here with two assault battalions of the 4th wading through ankle-deep water to the other side. At 1130, a firm bridgehead was established against only light resistance; 1/4 was dug in and prepared to continue the attack on the right, 3/4, on the left.

The regimental objective was a low ridge, running east to west, about 500 yards south of the river in the vicinity of Machisi. The attacking Marines approaching this point began to meet sharply increased opposition. Previous suspicions concerning the nature of the defenses here were confirmed when the infantry neared the height. In addition to reverse-slope mortar emplacements, the face of the height was studded with many Okinawan tombs that had been fortified. Darkness halted the attack 100 yards short of the objective, where the troops were ordered to organize and defend the high ground they held.

Although the Asato could be waded at the time of the assault crossing,[42] strong, steady rain had turned it into a chest-high raging torrent by the next day. Supply and evacuation, difficult enough over the muddy terrain, now became almost impossible. At least 12 men were required to carry a stretcher case safely across the river to the rear.[43] Supplies were sent forward under the same conditions; men stood in the water hour after hour under intense artillery and mortar fire, forming a human chain in a successful attempt to supply the advance companies.[44] By midnight, the 6th Engineer Battalion had constructed two footbridges and was to have begun building a Bailey bridge, but heavy enemy fire during the afternoon prevented the engineers from bringing the components of the bridge forward.

General Geiger shifted the boundary between the two IIIAC divisions slightly to the right (west) to concentrate the corps attack on the right flank. At the same time, 2/4 could both contract its overexpanded lines and better protect the left flank and rear of the regimental bridgehead. As a result of this change, 3/5 relieved 2/4 at 1400, and the latter moved laterally to its right to ease the tension on the strained lines of 1/4.

No forward progress was marked in the center of the Tenth Army line on 23 May by the assault regiments of the 1st

[40] *6th MarDiv G–3 Jnl, Ph III,* 23May45.

[41] *6th MarDiv SAR, Ph III,* p. 12.

[42] *6th MarDiv G–3 Jnl, Ph III,* 23May45.

[43] *3/4 SAR, Ph III,* p. 4.

[44] *Ibid.*

Marine and 77th and 96th Infantry Divisions. Activity in the immediate front of each division was limited to patrol action.

Assessing the threat to Shuri by the American advances down both coasts, General Ushijima, his commanders, and his staff believed that the *Thirty-second Army* was "still able to halt the collapse of all positions by holding positions in depth to the line of Shichina and Kokuba" even though Marines "had broken into the city of Naha." [45] This evaluation was tempered somewhat by the realization that *Thirty-second Army* troops would "be unable to maintain their Shuri front" [46] if the American spearhead in the Naha-Yonabaru valley was not blunted. To stem the tide of the XXIV Corps attack against his positions north and east of Shuri, General Ushijima threw every available man into a defense line that began on the southwest slopes of Conical Hill, ran through Yonawa, and was anchored at the road junction in the village of Chan.

When the 7th Division attempted to expand its hold on the valley and the high ground to the south, the soldiers received ample proof of the presence of enemy reinforcements. Following increased and determined Japanese stands, which reach a climax in a series of counterattacks on the night of 24–25 May, the 7th Division ground to a halt, unable to push any further west. In sharp contrast to the stubborn and immediate reaction aroused by these efforts to gain the road net east of Shuri, a few American patrols progressed slightly towards the Chinen Peninsula against only sporadic interference.

A break in the weather on 24 May was too short-lived to enable the Tenth Army to build up supply reserves, repair roads, or to attack in any great force. The enemy, however, took advantage of the brief respite to counterattack the 7th Infantry Division. In support of this ground action was *Kikusui* No. 7 and the airborne attack against Yontan airfield on 24–25 May.

Although their participation in the air defense of Okinawa was often more glaringly spectacular than destructive, on the night of 24 May, Marine and Army antiaircraft artillery battalions guarding Yontan and Kadena airfields scored heavily against *Kamikaze* raiders over the island. Marines manning 1st Provisional Antiaircraft Artillery Group guns were credited with destroying five planes, damaging six, downing one probable, and assisting in the destruction of another plane during this action.[47]

In the entire month of May, Marine AAA gunners destroyed 8 planes, damaged 15, scored 5 probables, and had 1 assist.[48] For the same period, the 53d AAA Brigade was alerted to 53 air raids in which 88 planes were tracked by its radar, gun directors, and guns. The May score of the brigade overall was 24 planes destroyed, 15 damaged (all by Marine AAA units), and 5 probables.[49]

[45] *Okinawa Operations Record*, p. 111.

[46] *Ibid.*

[47] *53d AA Brig AAR*, pp. 40–43.

[48] *Ibid.*, pp. 39–43.

[49] *Ibid.*, p. 34.

The clear weather permitting the flurry of enemy air activity held for a brief time only. The rains came again. Gravely concerned over the effects the weather was having on his division supply system, General Shepherd believed it necessary to establish firm vehicular and foot crossings over the now-rampaging Asato if 6th Division assault battalions were to be provided with adequate rations, ammunition, and medical supplies. In addition to replenishing Marine forces at the bridgehead, a well-stocked supply reserve would be required to support a continued attack to the south.

At the same time that the 4th Marines sent probing patrols south to the vicinity of Machisi on the 24th, the regimental objective was bombarded by a heavy artillery concentration and an air strike—one of the first to be flown in clearing skies that day in support of ground forces. The 6th Engineer Battalion bridge builders who had labored throughout the night to erect some sort of crossing over the Asato,[50] began putting together a Bailey bridge at dawn. Working all morning and part of the afternoon under enemy artillery and mortar shelling, the Marine engineers finished the bridge at 1400 and opened it for traffic 45 minutes later.[51] A pile bridge, to be utilized as a tank bypass, was completed at 1840. Informing the division commander that the passage was open, the commanding officer of the engineer battalion matter-of-factly added: ". . . tanks should cross as soon as possible as arty [artillery] is falling in area.[52]

To intensify 4th Marines efforts south of the river, 3/22 moved into the line on the division left, relieving 2/4 at 1000 on the 24th. Once in position, Lieutenant Colonel Clair W. Shisler's battalion tied in with 3/5 and 1/4. At this time, the 2d Battalion moved across the river "to relieve the battered 3/4 on the right" of the 4th Marines line.[53] The 3d Battalion became regimental reserve and moved to an assembly area in the rear where it held muster; none of its companies had over 90 men.[54] Major Walker's 6th Reconnaissance Company, which had been attached to the 4th Marines on 23 May as regimental reserve, reverted to division control when the 1st Battalion, 29th Marines, was ordered to an assembly area just south of the Asato. In order to beef up Colonel Shapley's reserve for the continuation of the drive south, 1/29 was attached to the 4th in place of the scouts.

With the rain beating down once again, the 4th Marines attacked on 25 May to capture that part of the regimental objective near Machisi not taken on the 23d. Although severely restricted by the deep mud and limited visibility, assault infantry seized the greater part of the north-south ridgeline lying west of the village. At 1030, the 4th overcame determined Japanese defenders en-

[50] "An attempt . . . made during the night to install a makeshift bridge built on LVTs . . . was unsuccessful due to the large number of enemy mines encountered along the river bank. During the effort two LVTs were seriously damaged." *6th MarDiv SAR, Ph III*, p. 12.

[51] *6th MarDiv G-3 Jnl, Ph III*, 24May45.

[52] *Ibid.*

[53] *4th Mar SAR, Ph III*, p. 5.

[54] *Ibid.*

310-224 O - 69 - 19

trenched on the reverse slope and occupied the objective. Once the captors had reorganized their forces and consolidated the newly won positions, they prepared to carry on further. The attack now progressed slowly as assault troops forged ahead against a storm of frontal and flanking fires.

When the regiment halted for the day at 1630, the battalions established firm contact along the line and dug in night defenses. During the preinvasion rehearsal phase, 6th Division training had emphasized village and street fighting. This training was first put into practical use on the 25th when attacking troops entered the eastern outskirts of Naha and came under fire from Japanese-defended houses. The heavy fire from these as well as from the many burial vaults along the ridges in this area inflicted numerous casualties in the Marine units.

Effectively dividing this section of Naha into two separate zones of action was a canal connecting the Asato River and the estuary of the Kokuba. The waterway was 20 yards wide, had a thick mud bottom, and stone banks 3 to 5 feet high.

As the 4th Marines fought in the eastern portion of Naha, the 6th Reconnaissance Company crossed the Asato to enter the once-urban, now-razed, area of Naha west of the canal. Major Walker's men quickly cleared a sector of snipers and a few disorganized troops, and set up defenses for the night. At 1900, the engineers completed construction of a footbridge across the mouth of the Asato. The bridge was anchored on the southern bank behind the defense perimeter of the scouts, and on the northern bank in front of 2/22, which manned the lines of the 22d Marines. Company G of the battalion was assigned as a clutch unit to relieve or support the reconnaissance company upon order.

Although the reconnaissance company had experienced a relatively quiet night, the two assault battalions of the 4th Marines spent the hours of darkness in fighting off counterattacks. At 2000, 1/4 reported that the enemy was forming up approximately 200–250 yards in front of its positions and had begun smoking Marine lines. A Japanese mortar barrage preceded the counterattack, which was broken up almost immediately under a mixed artillery and mortar shelling that continued on for another two hours without letup. Later, at midnight, when Company E, 2/4, was hit by a counterattack, it was immediately reinforced by a platoon from 1/29. After a two-hour-long hand grenade duel, in which the Marines suffered only light casualties, the enemy was successfully repelled.

Across most of the rest of the Tenth Army front on the 25th, high water and mud limited activity to patrol actions. The 7th Infantry Division, however, which had forced the enemy from commanding terrain features and inserted an opening wedge into the southeastern defenses of Shuri on 24 May, continued its vigorous drive on the 25th. Additional key positions were secured against ever-stiffening resistance and under conditions of terrain and weather that favored the enemy. Limiting the 7th Division attack was the problem of getting supplies forward, after its only supply route to the front sank in a sea of mud under the ravages of continuous use.

In the 96th Division zone, troops holding positions south and west of Conical Hill were all but isolated from rear area facilities of the division. Suffering serious losses under a rash of small counterattacks and continual attempts by the enemy to infiltrate, the depleted infantry companies holding the line were forced to utilize all available manpower from battalion and regimental service and support units. The frontline units put these soldiers into the line as riflemen or assigned them to the tremendously wearing task of hand-carrying supplies forward over the muddy ground. Descriptive of these agonized efforts is a comment made by one of the 96th's officers, who said: "Those on the forward slopes slid down. Those on the reverse slopes slid back. Otherwise, no change." [55]

A happy change in these gloomy reports of a bogged-down campaign—and possibly a favorable portent for the future—occurred on 26 May, when it appeared as though the enemy was pulling out of Shuri. Observers at the 1st and 5th Marines regimental observation posts (OPs) reported that there was a good deal of enemy movement south, and thus prompted the 1st Marine Division G–2 to request air observation of the suspected area at 1200.[56]

Despite hazardous flying conditions in rain and poor visibility, which in themselves would limit the value and amount of information gained, a spotter plane was catapulted from the *New York* almost immediately after receipt of the G–2 request. Upon arrival over the target area, the airborne observer confirmed the presence of a large number of Japanese troops and vehicles clogging the roads leading south from Shuri.

Within 13 minutes after this sighting, the USS *New Orleans* had fired the first salvo in a continued devastating barrage that was brought to bear on the withdrawing enemy by artillery, mortars, the main and secondary batteries of gunfire support ships, and the machine guns and bombs of Marine aircraft that had risen from rain-sodden fields to harry the enemy from above.[57] Commenting on the part played by gunfire support ships at this point, the naval gunfire officer in the IIIAC staff recalled:

> . . . the *New York* was some distance from the beach but the *New Orleans* was close in. The *New Orleans* heard the report of the *New York's* plane and asked the spotter for coordinates. She positioned herself and began adjustment by full salvos of main battery. . . . Other firing ships and support craft with H&I [Harassing and Interdiction] missions or otherwise in the area, noticed the increased activity, sent up planes, and got into the act. Ships without planes asked to be cut in with ships that did have them and often a plane

[55] Quoted in Orlando R. Davidson, *et. al., The Deadeyes: The Story of the 96th Infantry Division* (Washington: Infantry Journal Press, 1947), p. 159, hereafter Davidson, *et. al., 96th InfDiv Hist.*

[56] IIIAC G–2 PeriodicRpt No. 56, dtd 27May45.

[57] *CTF 31 AR*, pt III, p. 27.

spotter was firing two or more ships at the same time.[58]

Enemy hopes for a successful withdrawal under the inclement weather conditions were shattered by the massed fires which caught and blasted some 3,000–4,000 Japanese troops with their tanks, vehicles, and artillery pieces in the open. The pilots of observation planes zoomed through the overcast to treetop height and lower to count and report back an estimated 500 enemy killed.[59]

The continuing stubborn reluctance on the part of some Japanese to give way to the Americans seemed to belie the fact that General Ushijima's forces were indeed withdrawing. The Tenth Army found that only local attacks and patrols could be accomplished in the rain against enemy resistance. Even limited forward movement directed towards the heart of Shuri aroused heavy and immediate response, and indicated that Japanese inner defenses were holding firm. Light resistance was found only along the coasts; in the IIIAC zone on the right, where the 6th Reconnaissance Company held the levelled and deserted Naha, and in the east, where patrols of the 184th Infantry approached Chinen Peninsula.

Following the discovery of the enemy withdrawal and the initial bombardments placed on his movements, artillery batteries and gunfire support ships fired continuous harassing and interdiction missions on all routes, road junctions, and crossroads in the area leading south from Shuri. To keep the enemy disorganized and unable to make a stand, and to exploit the implications inherent in the Japanese withdrawal, General Buckner sent the following message to his two corps commanders on 27 May:

> Indications point to possible enemy retirement to new defensive position with possible counteroffensive against our forces threatening his flank. Initiate without delay strong and unrelenting pressure to ascertain probable intentions and keep him off balance. Enemy must not repeat not be permitted to establish himself securely on new position with only nominal interference.[60]

Continued rains and their subsequent effect on the terrain precluded a full-scale attack all along the front. Therefore, the Tenth Army settled for aggressive patrol action against the remaining Japanese strongpoints facing its lines. Apparently contradicting what influence the previous day's withdrawal

[58] LtCol William M. Gilliam ltr to CMC, dtd 16Mar55. During this and other incidents in the Okinawa campaign, gunfire support from all arms was superb. "It was not uncommon for a battleship, tanks, artillery, and aircraft to be supporting the efforts of a platoon of infantry during the reduction of the Shuri position." *Ibid.*

[59] Commending TAF and the LFASCUs for their day's work, General del Valle sent a message reading "Congratulations and thanks for prompt response this afternoon when Nips were caught on road with kimonos down." 2d MAW WarD, May 45.

[60] *Tenth Army AR*, chap 7, sec III, pp. 24–25. "During the period of May 22d to 30th when the southern front was engulfed in mud, General Buckner chafed at the slow progress being made" and constantly urged his corps commanders to greater speed. "He was under considerable pressure to make faster progress, as the Navy was sustaining heavy casualties by being forced to remain in the vicinity of Okinawa," exposed to the damaging *Kamikaze* raids. Smith, *Personal Narrative*, p. 123.

AERIAL VIEW *of Shuri on 28 April before it was bombed. (USAF 83751)*

SHURI, *one month later. (USAF 83752)*

should have had on the tactical situation, patrol reports reading "Does not appear that resistance has lessened," or "No indication of Japanese withdrawal," implied that Shuri would not yet, if ever, fall easily.[61]

Assault units of the 7th Infantry Division, driving west from Yonabaru, ran into *62d Division* elements hastily committed by General Ushijima to shore up his threatened right flank. On the far left flank of the Tenth Army, advance patrols of the 184th Infantry reached Inasomi, approximately two miles southwest of Yonabaru, without meeting any organized resistance. To contain the overall threat this potentially deep penetration posed to Japanese defenses in the south, the *Thirty-second Army* moved additional troops of the *62d Division* down from Shuri.[62]

On the opposite flank of Tenth Army, early on the 27th, Company G, 2/22, moved across the Asato, passed through reconnaissance company lines, and pressed well into Naha against only slight resistance. At the same time, patrols from the 4th Marines moved 200–300 yards ahead of regimental lines to take advantage of the apparent enemy weakness here. Lieutenant Colonel Woodhouse was ordered at 0915 to move the rest of his battalion across the river, and nearly two hours later 2/22 was directed to attack and seize the high ground approximately 100 yards north of the Kokuba estuary. The 4th Marines received the same order to seize the rest of Naha. Colonel Shapley attacked with 1/4 on the left, 2/4 on the right; 3/4 filled in on the left flank between 3/22 and 1/4.

Following a pre-attack nine-battalion artillery preparation, 6th Division forces advanced abreast at 1230 against light opposition, reached the objective at approximately 1700, and dug in for night defense at 1900, when regimental lines were tied in all around.[63] Actually, occupation of the objective in force amounted to the same thing that had been accomplished by patrols that morning. Badly in need of rest after a 10-day tour in the lines, the 4th Marines was alerted that it would be relieved by the 29th Marines; the changeover was scheduled to begin at 0630 the next morning.

Another important change in the ICEBERG command structure occurring in May took place on the 27th, when the Fifth Fleet once again became the Third Fleet and Admiral Halsey took over the responsibility for supporting the ICEBERG campaign from Admiral Spruance. As in February, when the latter had taken over from Halsey, the ships and men of the fleet remained the same, only the numerical designations of the task groupings changed (*i.e.*, TF 58 became TF 38, etc.). At the same time that this command change was effected, General Buckner became directly responsible for CinCPOA for the defense and development of captured positions in the Ryukyus.[64]

---

[61] *1st MarDiv G–3 Jnl*, 27May45.

[62] *Okinawa Operations Report*, pp. 116–119.

[63] At 1630, Major Walker's company was attached to 2/22 for the night; at 2000, 1/29 reverted to parent control.

[64] ComFifthFlt AR, 1Apr-27May45, Ryukyus Operation, dtd 21Jun45, pt III, p. 7; see also section entitled "The Task Defined" in chap 2, *supra*.

Large-scale movements were still impracticable on 28 May because of the mud, although: "The morning . . . was clear with no rain." [65] Despite the limited trafficability of both roads and terrain, local commanders hewed to the concept of General Buckner's directive of the previous day and maintained constant and continuous pressure on enemy forces. In the XXIV Corps zone, the 184th Infantry deepened the 7th Division salient in the Ozato Hills by moving to within 1,000 yards of Shinzato, a village located where Chinen Peninsula joined the mainland.

Less satisfactory progress was registered for the units driving west, as *62d Division* blocking forces held up other 7th Division troops. Strongly held positions southwest of Conical Hill frustrated 96th Division attempts altogether, and the 77th Division made little or no headway against a determined defense of Shuri heights. At the end of the day, XXIV Corps gains were negligible and inconclusive. Despite reports of increased troop withdrawals, the enemy's obstinate reluctance to yield indicated that General Ushijima either had established a strong rearguard to protect the withdrawal or that he was in fact not abandoning Shuri.

In the IIIAC sector, Marines were no more successful in prosecuting their portion of the war and had no easier time of it than had the Army units. In its attack on 110 Meter Hill, 2/1 was covered by the fires of 3/1 and 3/306. Once 2/1 had gained its objective, 3/1 was to advance down Wana Draw. The 2d Battalion gained the hilltop twice, only to be thrown back by mortar barrages coming from reverse slope emplacements and vicious machine gun fire raking positions on the crest from three directions. Lieutenant Colonel James C. Magee, Jr., believed that, even if his battalion succeeded in carrying the hill, "it could not possibly hold it against a strong counterattack." [66] Heavy casualties had depleted the size of the battalion to a point where its total effective strength was 277 men; no rifle company could muster more than a total of 99 Marines.[67]

At first glance, it would appear as though reinforcement of the battalion at this time might possibly have tipped the scales of victory in favor of the Marines and enabled them to capture 110 Meter Hill. Replacements were available; the 1st Marine Division had received a total of 53 officers and 1,255 enlisted men in the three-day period of 27–29 May, when the 57th, 59th, and 63d Replacement Drafts arrived at Okinawa.[68] The division was precluded by IIIAC orders, however, from inserting these fresh troops into the line during a battle situation, the course of which depended upon close teamwork by experienced veterans. Only after several days of indoctrination and training in reserve areas could the men be sent forward. Infantry replacements were at a premium in any case because of the heavy losses sustained by the rifle regiments. This condition existed even though over 350 emergency replace-

[65] *2/1 SAR*, p. 11.

[66] *Ibid.*

[67] *Ibid.* The authorized T/O strength of a Marine rifle company at this time was 7 officers and 235 enlisted Marines.

[68] *1st MarDiv SAR*, Pers Anx, p. 7.

ments had been assigned from division special and service troops to the infantry regiments in mid-May and the regular "flow of Marine replacements was beyond expectations. . . ." [69] Nonetheless, despite this constant infusion of new Marines, at no time during the drive to the south were 1st Division infantry regiments able to exceed more than 85 percent of effective T/O strength.[70]

To the commanders of undermanned companies and battalions in this period of the campaign, immediate reinforcement was not only desirable—it was of paramount importance. From the point of view of senior commanders, such as the regimental commander of the 1st Marines:

> . . . the existence of a replacement pool which could not, at the moment, be used for combat proved to be extremely valuable. At the end of May, when the rainy period had rendered the roads and the countryside impassable to anything on wheels or tracks, the supply of forward troops became most critical. Something like 500 replacements, if any recollection is correct, were available to the 1st Marines; these men were formed into man-pack trains, under the direction of the executive officer. Their exhausting struggles, heavily laden, through mud which even an unburdened man found difficult to negotiate were the solution to the supply problem at this time, though with no margin to spare.[71]

Even though 2/1 and other assault units could not be reinforced, supply support from non-committed elements permitted them to concentrate on the immediate problem of fighting the enemy. Late on the 28th, after being withdrawn to that morning's jump-off position, organic crew-served weapons of 2/1 worked over the reverse slopes of 110 Meter Hill as the battalion lines were themselves raked by the continual fire of Japanese flat-trajectory cannon. With unabated fury and determined fanaticism, the enemy stood his ground and even sought to infiltrate 2/1 lines after dark when many Japanese soldiers were killed.

Also on 28 May, patrols from 3/1 penetrated some 300 yards into Wana Draw under intense machine gun and rifle fire. At 1600, Colonel Mason ordered the battalion to clear all Japanese troops from the draw, but the battalion commander's request that the attack be delayed until the following morning instead, in order that he might organize and concentrate his forces for the drive, was approved.

The 1st Division concluded that: "The beginning of the end for Shuri came on the 28th." [72] Although rear-guard action continued unrelentingly in the high ground north of the city in the 77th Division zone, the 5th Marines attacked at 0730, captured the village of Asato, and 1/5 patrolled 300 yards beyond that without appreciable opposition. During the same day, the 306th Infantry managed to mop up the area approximately 150 yards in front of its lines. On the corps boundary, 3/306 sent strong patrols forward, coordinating their movements with those of 2/1. By dark, 1/307—to the left of the 306th—attacked the high ground east of Shuri against determined opposition. Despite the fact that it was bitterly opposed and repulsed by heavy mortar fire initially, the bat-

[69] *Ibid.*, p. 8.

[70] *Ibid.*

[71] Col Arthur T. Mason ltr to CMC, dtd 10Mar47.

[72] *1st MarDiv SAR*, p. 7.

talion inched forward under the cover of smoke and managed to dig in for the night.[73]

While the 1st and 77th Divisions were moving slowly forward, the 29th Marines began relieving the 4th in position. Enemy shelling during the relief added to the more-than 1,100 casualties already sustained by Colonel Shapley's regiment. Beginning at daylight, 3/29 relieved 1/4 and 3/4, the elements on the left of the 4th Marines line, 1/29 moved into the western portion of Naha relieving 2/4, and 2/29—in regimental reserve—moved to an assembly area near Colonel Whaling's CP. Upon its relief, the 4th Marines marched and motored to beach areas near Machinato airfield, where it became 6th Division reserve.

Continuing to advance after passage of the 2/4 lines, 1/29 moved abreast of and then paced the attack of the 22d Marines. The direction of the battalion attack changed to the southwest when Company C pivoted on Company A—on the right boundary—and gained 250 yards through the rubble of Naha. Heavy small arms fire and a scattered mortar and artillery shelling followed the Marines, who finally set in a night defense along a line 800 yards from the Kokuba. Here visual contact was established with the 22d Marines across the canal on the right, and with 3/29 on the left flank.[74]

The 22d Marines had moved out on the 28th before dawn. At that time, 1/22 passed through the lines of the 2d Battalion to take up the assault in the direction of the banks of the Kokuba estuary, west of the canal running through Naha. Moving rapidly against only slight resistance, its flanks and rear protected by 2/22, the battalion captured its objective by 0845.

General Shepherd then assigned the task of defending western Naha to his reconnaissance company in order to release the 22d Marines for further offensive action to the east against Japanese positions in the Kokuba hills. Major Walker was first notified of the scouts' new assignment at 1030, when he received the following message from the division G–3, Lieutenant Colonel Victor H. Krulak:

> Reposing great confidence in your integrity and political ability you are hereby named acting mayor of Naha. The appointment effective 281600 carries all pay and emoluments accruing to office. To be collected from Imperial Treasury.[75]

Shortly after midday, the 6th Engineer Battalion was ordered to reconnoiter all bridge crossings over the canal. Also, it was to install a jeep crossing and at least two foot bridges over the canal prior to 0400 on the 29th to facilitate the 22d Marines attack east of the Kokuba. The engineers worked in the dark in front of Marine lines under a constant downpour of rain and shells, as they manhandled the bridge construction material up to the various bridging sites. The task was completed at 0420. At 0430, elements of 1/22 were across the canal and organized on the

[73] *77th InfDiv OpRpt*, Okinawa, p. 61.

[74] *1/29 SAR, Ph III*, p. 3. Lieutenant Colonel Samuel S. Yeaton relieved the temporary battalion commander, Major Robert P. Neuffer, who then became battalion executive officer.

[75] *6th MarDiv G–3 Jnl, Ph III*, 28May45. The new "Naha City Command" was reinforced by 1 officer and 40 Marines from the 6th Motor Transport Battalion. *Ibid.*

eastern shore. This was the division's third opposed river crossing in 20 days.[76]

The attack began a half hour later and the Marines made some immediate progress against scattered machine gun and rifle fire. Supporting the assault was 2/22, while the 3d Battalion followed in reserve. At 0845, 1/22 made contact with 1/29 and the two infantry battalions pushed on abreast of each other. As spearhead elements approached the hills north of Kokuba, increased resistance indicated that the enemy was positioned there in strength. Because of the hard fighting and numerous casualties experienced by 1/22, at 1500 2/22 was alerted to relieve it, when the situation permitted, on the ground then held. Since the 1st Battalion was heavily engaged at the time, its commander advised against the relief being effected then. Instead, Companies E and G were committed into the line at 1800 to reinforce the night defenses of the regiment.

Until darkness fell the 6th Division assault regiments continued the attack in an effort to reduce the strong enemy position encountered earlier that afternoon. Aggravating the situation was the fact that the routes of approach to these Japanese emplacements were across open ground that afforded the attacking Marines little or no cover from the fire of well-dug-in machine guns and mortars. The position itself was centered on a small group of hills on which were located several radio towers that had been demolished in air raids when the rest of Naha was razed earlier. Rain and mud precluded the use of vital armor support. It still remained an infantryman's war.

On the same day, 29 May, the 29th Marines attacked to the south and then cut east in order to come abreast of the 22d Marines; the 29th's objective was the high ground immediately northwest of Shichina. Like the 22d Marines approach route, that of the 29th was over low and open ground, causing the regimental commander to comment that the terrain was "about as suitable to fighting as a billiard table." [77] The 1st Battalion maintained a slow, steady pace against moderate resistance, and dug in at dark slightly to the left rear of 3/22, on the regimental boundary. The 3d Battalion of the 29th paced the advance of 1/29, but 3/5 on the left had advanced so rapidly during the day that 3/29 was forced to bring its lines forward some 600 yards at the end of the day to maintain firm contact with the 5th Marines.

Tuesday, 29 May 1945, is a significant date in the history of the Okinawa campaign, for it was on this day elements of the 1st Marine Division captured Shuri Castle. This ancient redoubt, once the seat of the rulers of Okinawa, had served as General Ushijima's center for controlling the defense of the island.

The 5th Marines attack began at 0730; 1/5 on the left, 3/5 on the right. The 3d Battalion jumped off with Company L in assault. Enemy machine gun, mortar, and small arms fire was placed on the attacking Marines but was unable to prevent them from advancing a total of 600 yards at day's end. Follow-

[76] *6th MarDiv SAR, Ph III*, pt III, p. 14.

[77] *6th MarDiv G–3 Jnl, Ph III*, 29May45.

CORKSCREW: ***Marine assault team attacks a Japanese cave after a satchel charge has exploded. (USMC 120272)***

BLOWTORCH: ***A flame tank burns out Japanese positions in hillside tombs during the drive for Naha. (USMC 122153)***

ing Company L, Companies I and K moved out later in the morning in an attempt to strengthen and protect the left of the battalion line, but enemy mortars positioned west of Shuri fired a furious barrage that seriously limited forward movement. Despite the efforts to destroy them by bazooka fire, the mortars remained active and held the companies back. The battalion night defense set up at dusk showed 3/5 lines cutting back sharply from the left of Company L's exposed position to tie in with 1/5 at Shuri Ridge.

The 1st Battalion moved rapidly over muddy terrain against little opposition and immediately occupied Shuri's ridge crest in close proximity to the castle. From this position, at approximately 0930, Lieutenant Colonel Charles W. Shelburne, the battalion commander, requested permission to send one of his assault companies to storm the apparently lightly manned fortification. Despite the fact that the castle itself was within the zone of the 77th Infantry Division, General del Valle granted the request when it was forwarded to him for approval. He believed that the capture of this enemy strongpoint would favorably effect and shorten the campaign; this opportunity, therefore, had to be seized at once. Shortly after the island was secured, General del Valle offered the opinion that "at that time the position of the 77th Division was such that it would have taken several hard days' fighting through enemy resistance," [78] if he had waited for the tactical situation to unfold normally.

Bowling over the few Japanese that were in their way, Marines from Company A, 1/5, drove east along the ridge and right into the castle itself, securing it at 1015. The 77th Division had programmed an air strike and a heavy artillery bombardment on the bastion for 29 May and had received warning of the attack of the 5th Marines only a few short minutes before it was mounted. Fortunately for the Marines, General Bruce and his staff worked frantically to contact all supporting arms and were just "barely able to avert called strikes in time." [79]

The air and artillery preparation of Shuri by the 77th Division resulted from General Bruce's decision on the 28th to attack the next day, weather permitting. The 1st Marine Division had given him no indication that it planned to enter the zone of the 77th, for: "Had timely notice been given and the move been properly coordinated," the Army commander believed "the 77 Div could have rendered adequate support to the Marines. . . ." [80]

Overshadowing this near tragedy was the fact that Company A success resulted from the close teamwork of Tenth Army support and assault troops who had not permitted the enemy to relax for an instant. Without this unrelenting pressure, the breakthrough would not have been possible.

To profit from the 1/5 gain, General del Valle quickly revised his attack plan and sent 3/1 through the lines of the

[78] MajGen Pedro A. del Valle, "Old Glory on Shuri," *Marine Corps Gazette*, v. 29, no. 8 (Aug45), p. 17.

[79] Myers, *77th InfDiv History*, p. 357; "However, arrangements had been made to see that our own artillery did not come down on these troops." Smith, *Personal Narrative*, p. 114.

[80] *77th InfDiv OpRpt*, Okinawa, p. 62.

5th Marines to relieve 1/5. At 1400, the relief had been effected and 1/5 continued the attack south. The commander of 3/1 immediately set up his battalion in perimeter defense around the battered walls of the castle. Augmenting this defense were two companies from 1/1 which fought their way into assigned positions that faced north, and tied in with the lines of 3/1.

Upon the 5th Marines sweep into Shuri, the 1st Marines was ordered to follow the attack closely; 3/1 was to relieve 1/5, and 1/1, tailing 3/1 around Hill 55, was to attack east into the yet-unoccupied sector north of Shuri. Moving out in a column of companies, the battalion crossed the line of departure in front of Hill 55 where the leading element came under intensive fire from a heavy machine gun hidden in a deep and rugged cut a few hundred yards south of bypassed Wana Draw. Unable to either silence the weapon or attack through its curtain of fire, the battalion axis of advance was angled to the right and the troops eventually made contact with 3/1 south of Shuri. While 1/1 and 3/1 hit Shuri from the west, the 2d Battalion was ordered to hold Wana Ridge, from which it was to provide fire support to regimental assault elements. To augment and increase this effort, "all battalion headquarters personnel, cooks, wiremen, and stretcher bearers were sent forward to help man the lines." [81]

At no time after the capture of Shuri Castle was there any indication that the Japanese defenders of the hills north of the city were either being worn down or concerned with the Americans positioned in their rear. Reports from Tenth Army units all along the line gave proof that Japanese resistance remained undiminished. Only in a goodly portion of Chinen Peninsula, scouted by 7th Division troops during the day, was there little or no opposition.

Offsetting the relatively unhindered advance down the coasts by Tenth Army flanking divisions, a vividly contrasting picture was presented by the massive struggle down the center of the island. Despite the efforts of General Buckner's forces to execute a mass double envelopment successfully and encircle the bulk of General Ushijima's troops at Shuri, all signs pointed up the fact that the Japanese rear guard had accomplished its mission well; the greatest portion of the units defending Shuri had indeed escaped to the south.

---

[81] *2/1 SAR*, p. 12.

Breakout to the South

JAPANESE WITHDRAWAL [1]

Threatened by an American frontal encroachment upon their Shuri defenses and an envelopment of their flanks by the Tenth Army divisions driving down the east and west coasts of Okinawa, the Japanese were forced to reevaluate thoroughly the battle plans adopted in March. On the night of 22 May, General Ushijima convened at his headquarters a conference of his principal commanders and staff officers. The major and only item on the agenda of this momentous meeting was a discussion of how best to prevent—or at least to postpone—the disaster engulfing the *Thirty-second Army*.

Contingency plans calling for a massive defense centering about Shuri had been included in final battle preparations completed before 1 April 1945. All Japanese units located elsewhere on Okinawa and still able to fight would withdraw on order for a last-ditch stand in the vicinity of the *Thirty-second Army* headquarters. Tactical conditions at the time of this conference indicated that, to hold Shuri, approximately 50,000 Japanese soldiers would have to be compressed into a final defense sector less than a mile in diameter. These close quarters would not permit an effective defense, but would, in fact, make the defenders "easy prey" [2] for the massive American fires which would undoubtedly seek them out from all directions. Although many Japanese long-range artillery pieces were still in firing condition, the constricted space factors around Shuri prevented their proper emplacement and subsequently efficient employment.

In a discussion of the alternatives to remaining at Shuri, Ushijima's staff considered two other defensive areas—Chinen Peninsula, and the Kiyamu Peninsula at the southern tip of Okinawa. (See Map 2.) The hills, cliffs, and lack of roads on Chinen presented a group of formidable natural defenses, especially against tank-infantry tactics. Militating against a choice of this area were the lack of sufficient caves and prepared positions to hold the entire *Thirty-second Army* and the inadequacy of ammunition supplies that had been stockpiled there. Another disadvantage ruling out Chinen was the poor road net, which would equally hamper the Japanese and the Americans. Finally, in face of these considerations and the route that would have to be taken to the peninsula, *Thirty-second Army* units would find it difficult to reconcentrate and reorganize speedily at the same time that they would undoubtedly be waging a hard fight during disengagement and withdrawal. The weight

[1] Unless otherwise noted, the material in this section is derived from: *Okinawa Operations Record; Yahara Interrogation: Shimada Interrogation.*

[2] *Okinawa Operations Record,* p. 111.

of the evidence against Chinen ruled it out quickly.

It appeared to the *Thirty-second Army* staff that Kiyamu Peninsula was the best area in which to develop a solid defense for prolonging the battle. This area, dominated by the Yaeju Dake-Yuza Dake [3] Escarpment, contained a sufficient number of natural and man-made caves in which to store supplies and protect troops against American bombardments. The terrain on the peninsula had been defensively organized earlier by the *24th Division*, which also had cached a large store of ammunition and weapons there before it was ordered north into the Shuri defenses. As opposed to the poor road net into Chinen, all roads south led directly to the proposed new positions and would permit the army to make a rapid mass movement. On the other hand, Tenth Army tanks also could move south quickly over these roads, but only to the outpost defenses of the sector. American tanks would be denied passage beyond this point by the sheer cliffs, steep hills, and deep ridges of the region. In this broken terrain, the infantry would be on its own.

Not all of the senior Japanese leaders approved the planned withdrawal. One dissenter was General Fujioka, commander of the *62d Division*. His objection was based on a compassion for the thousands of severely wounded men who could not be taken south. General Fujioka felt most strongly about this point because it was his division that had originally been assigned to defend Shuri, and it was his officers and men who had taken the brunt of American attacks on the city. He contended, therefore, that their desire to fight to the last in their present positions should be fulfilled.

Supporting the anticipated withdrawal was General Suzuki, commander of the *44th Independent Mixed Brigade*, who tempered his approval with the stipulation that the army move into the Chinen Peninsula positions that his forces had previously developed. Fully supporting the move to Kiyamu were Generals Amamiya and Wada, commanders of the *24th Division* and *5th Artillery Command*, respectively. General Amamiya reinforced his argument with information that his *24th Transportation Regiment* had been able both to salvage and to preserve enough trucks to transfer the army ammunition reserve in five nights' time if weather conditions permitted.

Not long after weighing all of these arguments, General Ushijima ordered the move to Kiyamu. The first transportation to head south left Shuri at midnight, 23 May, carrying wounded and a portion of the ammunition supply. The main body of the *Thirty-second Army* was scheduled to begin the trek southward on 29 May.

According to the army plan, the new defensive dispositions would be as follows (See Map 16.):

1. The *44th IMB* was to move from positions on the westernmost flank of the Shuri front to take up defense positions on a line running from Hanagusuku on the east coast to Yaeju Dake.

2. The *24th Division* would occupy the commanding heights of the Yaeju Dake-Yuza Dake escarpment, the ridges of

[3] *Dake* means peak.

Mezado and Kunishi, and Nagusuku, on the west coast.

3. Elements of the above two units would establish and defend an outpost line—and the zone forward of it—which would run from Itoman through Yunagusuku to Gushichan.

4. The heavily depleted forces of the *62d Division* would occupy defenses along the coast in the rear of the main battle line; at these positions the division could reorganize, and, at the same time, be prepared to reinforce threatened sections of the line.

5. The firing batteries of the *5th Artillery Command* were to be emplaced within the confines of a triangularly shaped area formed by Kunishi, Makabe, and Medeera, in direct support of the defense line.

6. Admiral Ota's *Okinawa Naval Base Force* was assigned as reserve and would move on order to an assembly area in the center of the Kiyamu defense sector.

7. Each unit breaking contact with the Americans on the Shuri line was to leave a sufficiently strong force in position to keep the Americans occupied long enough to permit a successful withdrawal.

Posing a threat to this plan was the penetration of 7th Infantry Division elements through Yonabaru. To oppose them, remnants of the *62d Division* were to disengage and pull out of the Shuri front on the night of 25 May, and then move through Tsukasan to counterattack the Tenth Army spearhead. Replacing the *62d* in the line was the comparatively strong and rested *22d Independent Infantry Battalion,* which had been in reserve during the fighting in April and May. The orders to General Fujioka were both explicit and simple: ". . . annihilate the enemy rushing from the Yonabaru area."[4] In addition, a secondary mission laid on, should the primary fail, was to slow or stop the American advance for a period long enough to permit the main body of the *Thirty-second Army* to escape.

To give General Ushijima time to organize his new dispositions on Kiyamu, the force remaining on the Shuri front was to hold until 31 May. Behind it, withdrawing units would leave other rear guard elements to hold a strong defense line running along the Kokumba River to the hills north of Tsukasan and Chan until the night of 2 June. At that time, the line would then cut south through Karadera to the east coast. Approximately 2,000 yards further south, another temporary line—this one centered on Tomusu—would be established and held until the night of 4 June. *Thirty-second Army* staff planners believed that the time gained during holding actions along these lines would permit the organization and manning of the final outpost zone. Before beginning its own retreat south, the composite naval unit on Oroku Peninsula was to guard the western flank of the withdrawal route.

All available replacements were thrown in the disintegrating Shuri front on 23 May, when the onsurging Tenth Army forced the enemy to bolster his defenses. On 24 May, the first Japanese walking wounded began leaving caves that passed as hospitals. Many terminal cases, too seriously wounded to be moved, were either killed with a lethal injection of morphine, or—less mercifully—left behind to suffer a more lingering end without the relief-giving drug. Limited medical care of the wounded because of circumstances rather than willful neglect appears to

[4] *Okinawa Operations Record,* p. 116.

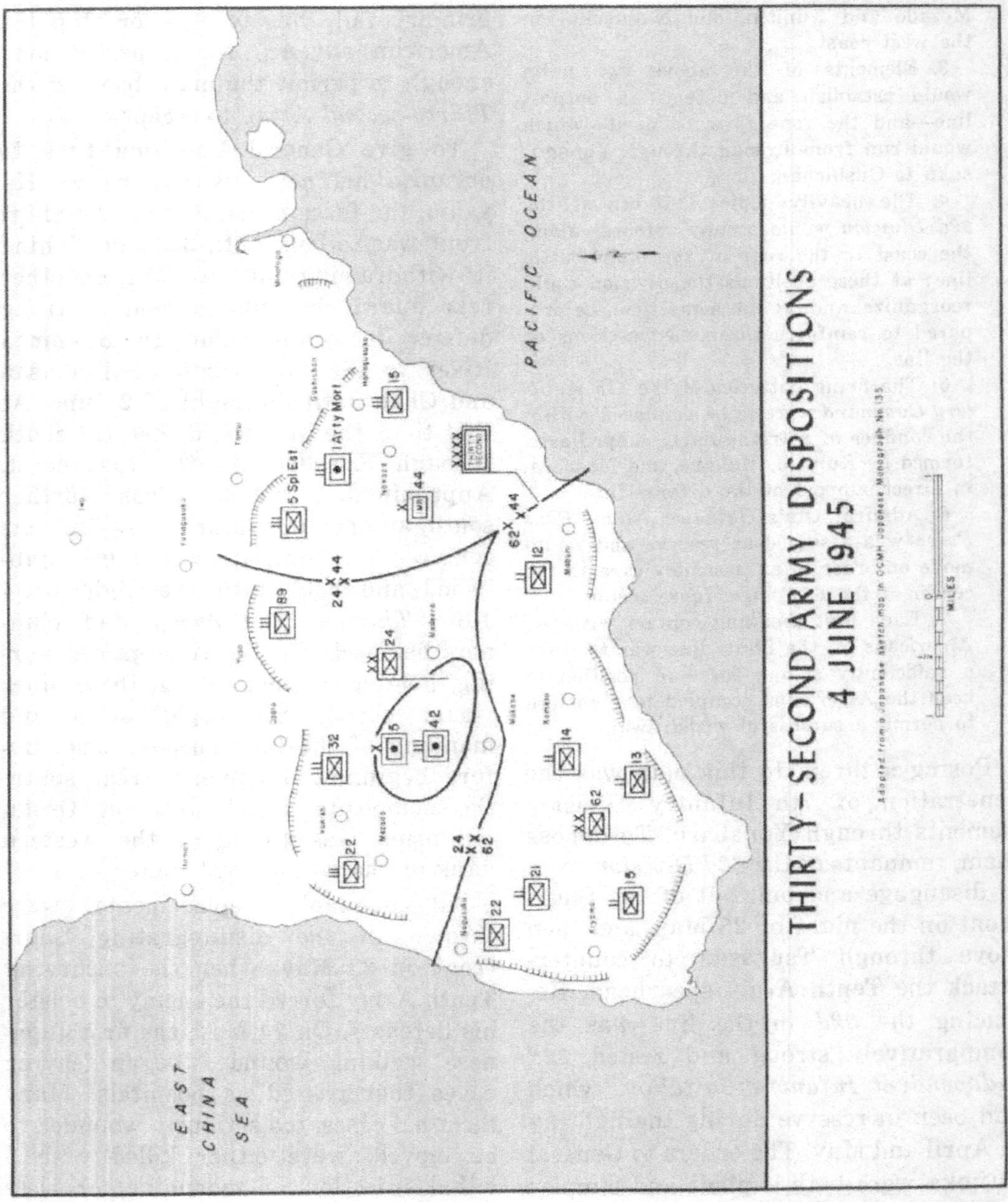
THIRTY-SECOND ARMY DISPOSITIONS
4 JUNE 1945
PACIFIC OCEAN
EAST CHINA SEA
T.L. RUSSELL

MAP 16

have been commonplace, judging from the following description of conditions:

> At one time there were almost 90 men in the cave, lying on the ground in the mud in pitch darkness, except when a doctor or corpsman would come around with a light and ask them how they felt. Medical supplies were very low, so very little could be done to care for the wounded. Men died on all sides. Filth accumulated. In the heavy rains, water poured into the cave and the wounded almost drowned. The smell was so bad that they could hardly breathe.[5]

In accordance with the schedule of the withdrawal plan, on the night of 25 May some 3,000 men remaining in the *62d Division* moved into positions blocking the drive of the 7th Infantry Division. The enemy expected that the continuing bad weather would aid their efforts in this sector considerably.

This expectation was valid, but only to the degree that the rain put greater obstacles in the way of the successful American drive than did the Japanese infantry. Reinforcing the *62d* and coming under its command were the *7th Heavy Artillery* and *23d Shipping Engineer Regiments,* which left their defenses on Chinen Peninsula to occupy holding positions on the right of the division. Owing to the failure of the *62d Division* to accomplish its mission, General Ushijima was convinced that it was necessary to evacuate Shuri while he still had the time. On 28 May, he ordered that the withdrawal was to begin on the following evening. The *Naval Base Force,* misinterpreting the order, jumped the gun by withdrawing to Kiyamu ahead of time. Admiral Ota's force was intercepted and sent back to Oroku Peninsula to take up its assigned positions west of the Japanese escape route.

On the night of 29 May, the *24th Division* moved out of the lines in orderly fashion, leaving one-third of the *32d Regiment* and the *22d IIB* as a covering force. While the division evaded the Tenth Army and headed south, the *44th IMB* remained in blocking positions outside of Naha and the *62d Division* was likewise defensively disposed near Chan and Karadera.

With the arrival of dawn on Wednesday, 30 May 1945, the greater portion of the *Thirty-second Army* had successfully postponed its final reckoning at the hands of the Tenth Army by withdrawing from Shuri and out of the grasp of General Buckner's flanking divisions. Having taken advantage of the heavy rains and the accompanying poor visibility, General Ushijima had executed a "properly deft withdrawal" [6] to establish new army headquarters outside of Mabuni, 11 miles south of Shuri, in a cave deep within Hill 89. By then, his covering forces were in position to slow down a Tenth Army pursuit and thereby give the *Thirty-second Army* a bit more time to organize the defense of Kiyamu Peninsula.

[5] "Interrogation of Superior Pvt, 5th Co, 22d IIB, POW Interrogation Rpt No. 54," in 1st MarDiv G-2 PeriodicRpt No. 63, dtd 3Jun45.

[6] "Matters for the Attention of Unit Commanders During the Change in the Direction of Advance, n. d.," Translation No. 85, in 1st MarDiv G-2 PeriodicRpt No. 64, dtd 6Jun45.

THE PURSUIT [7]

American attempts to exploit the successful breakthrough at Shuri and maintain incessant pressure on the reeling *Thirty-second Army* were frustrated on 30 May by an electrical storm accompanied by torrential rains on an already-saturated landscape. Movement of all Tenth Army units was effectively halted by the mud. Amphibious craft and vehicles were employed on both coasts to give logistic support to the two corps and enabled the ground commanders to maintain at least minimum supply levels.

Out of a total of 916 missions of all sorts flown by VMTB-232 in May, 74 were supply air drops in support of frontline troops and advance patrols. In spite of the submarginal flying conditions and limited visibility on 30 May, the squadron flew 12 air drop missions to the 1st Marine Division. General del Valle expressed his appreciation to Major Feldmeier's pilots in a message which read in part: "Those pilots have guts!" [8] This congratulatory message could have been repeated just as well the next day, when the squadron's Avengers flew 37 missions over southern Okinawa to drop water, medical supplies, rations, and ammunition to the ground forces.[9]

The weather situation changed so abruptly at the end of the month that "For the first time no enemy planes were detected in the area for the 24-hour period" ending at midnight, 30 May.[10] The heavy rain, however, did not completely stop ground activity on the 30th, for attacks were made all along the Tenth Army front. In the XXIV Corps zone, the 96th Division and the right flank regiment of the 7th Division—the 17th Infantry—attacked west and captured the high ground in the rear of the Shuri positions, one and a half miles west of Yonabaru. Other elements of the 7th Division passed through Shinzato and Sashiki against little opposition, and moved deeper into the Chinen Peninsula. The 77th Division ran into determined holding action by the *32d Regiment*, but managed nevertheless to capture all of the high ground and key defensive positions occupied by the enemy immediately to the northeast and east of Shuri.[11]

In the 1st Marine Division zone, a grave situation confronting 3/1 on 30 May was rooted in events that had occurred the day before, when the battalion broke off contact with the enemy in Wana Draw and headed for Shuri Castle. At the time of the disengage-

---

[7] Unless otherwise indicated, the material in this section is derived from: *Tenth Army AR; IIIAC AR; XXIV Corps AR; TAF AR; 1st MarDiv SAR; 6th MarDiv SAR, Ph III; 7th InfDiv AR; 77th InfDiv OpRpt*, Okinawa; *96th InfDiv AR; 1st Mar SAR; 4th Mar SAR, Ph III; 5th Mar SAR; 7th Mar SAR; 22d Mar SAR, Ph III; 29th Mar SAR, Ph III.*

[8] VMTB-232 WarD, May 45.

[9] *Ibid.*

[10] *CTF 31 AR*, pt III, p. 41.

[11] Effective at 1200 on 31 May, the 77th Division was pinched out of the line when the direction of its attack and that of the 96th Division was oriented towards the corps boundary below Shuri. The 77th moved to an assembly area in the rear on the next day, and its 305th Infantry was taken out of division reserve to cover the right rear of the corps advance. The 7th Division was assigned the zone below the Naha-Yonabaru valley and including the Chinen Peninsula, also on the 30th.

AIR DELIVERY SECTION *Marines and ship's crew of USS* Sargent Bay *load a TBM with supplies to be dropped to Tenth Army troops. (USN 80–G–338840)*

VMTB–232 TBM *drops supplies to troops near Shuri. (USMC 126402)*

ment, all battalion assault companies badly needed food, water, and ammunition. An air drop scheduled at 1800 on the 29th at the castle was not made, with the result that the frontline troops were without food and water for 36 hours.[12] Commenting on this later, the battalion commander noted:

> During this period the battalion was operating under the worst imaginable conditions, no food, water, little ammunition, the battalion CP 2500 yards to the rear of the lines, battalion dump 1000 yards to the rear of the CP and all transportation hopelessly mired with the results that no food of any type was available and the men had resorted to drinking water from shell-holes due to their extreme thirst.[13]

Early in the morning of the 30th, Lieutenant Colonel Ross was informed that an air drop was scheduled for 0600, and shortly thereafter he learned that a very low ceiling had grounded the planes. The battalion's Marines had now been without food and water for two days and a night. This situation and an inadequate ammunition supply forced the battalion commander to tell Colonel Mason that unless this logistics problem was solved, it would be most difficult for the battalion to undertake any extensive operations. Finally, at 1335 an air drop was made, but another one had to be scheduled since most of the first had fallen outside of the drop zone and in enemy territory. Enough of the supplies were recovered, however, to issue each man one-third of a K-Ration [14] and one canteen of water.

While attempting to locate the headquarters of the *Thirty-second Army* at Shuri Castle, representatives from the 1st Division G–2 Section had discovered numerous caves containing many enemy documents of intelligence value. Together with these intelligence people, General del Valle had sent the division colors to Lieutenant Colonel Ross with a request that they be raised over the castle. After locating the remnants of a Japanese flagpole, the battalion commander had it erected near the southern wall, raised the American flag, and then ordered all observers to evacuate the area rapidly because he expected the Japanese to use the flag as an aiming point and to fire an artillery concentration on the position almost immediately.

Continuing supply problems prevented the 1st Marines from making any concerted attack on 30 May until the 1st and 2d Battalions had received supply air drops. On the left of the 1st Marines, the 306th Infantry extended to the right, allowing Colonel Mason's left battalion, 2/1, to move its left company to the right, relieving Company C, 1/1. Patrols from 1/1 ventured into the ruins of northern Shuri, but were

[12] *3/1 SAR*, pp. 35–36.

[13] *Ibid.*, p. 36.

[14] This emergency ration, officially described as "Ration, Field K," was packed in a lightweight, compact, wax-wrapped cardboard container measuring 7 inches by 3⅝ inches by approximately 2 inches. This unit could be easily carried in either a haversack or field pack. For variety, there were three different types of units or components—breakfast, dinner, and supper—furnishing about 3,750 calories for the three meals. The ration contained such items as biscuits, enriched chocolate bars, chewing gum, cigarettes, a packet of one of several types of beverage powder, and the main course—a canned protein food consisting of one of the following: chopped ham and egg, pork and veal loaf, cheese, pork luncheon meat, or corned beef hash.

forced back by machine gun and 47mm AT fire from well-entrenched positions in a ravine southwest of Wana Draw. On the right of the division zone, the 5th Marines was also limited to local patrol action by its need to bring supplies forward over muddy routes.

On 30 May, in torrential rains, the 22d and 29th Marines pressed the 6th Division attack east to clear the north bank of the Kokuba. Heavy enemy resistance, built around a framework of mutually supporting machine guns emplaced in the mouths of Okinawan tombs, was made even more effective by the fact that the Marines had no armor support for the greater part of the day. Jump-off time was advanced one hour to 1000 to permit division interpreters and cooperative prisoners to broadcast surrender inducements over loudspeakers to enemy holdouts in front of the 22d Marines. A barrage of small arms and mortar fire signified a negative response to this effort. After a 15-minute artillery, rocket, and naval gunfire preparation, the division attack began at 1010.

On the right of the division line, Lieutenant Colonel Woodhouse was killed by sniper fire while in the van of 2/22 controlling its attack. The battalion executive officer, Major John G. Johnson, assumed command and maintained unit pressure against the caves and the improvised tomb-pillboxes thwarting the Marine advance. By nightfall, after a series of costly local attacks and mopping-up activities, the battalion possessed hill positions overlooking the Kokuba estuary and the trans-island rail line between Naha and Yonabaru.

The 3d Battalion passed through 1/22 and then jumped off abreast of 2/22, meeting the same heavy resistance along the way. Blocking the 3/22 path to the Kokuba was commanding terrain in which Knob Hill, Hill 27, and a number of radio towers were located. At approximately 1530, 3/22 secured this area, but only after the ground troops had fought a number of small arms and grenade-throwing duels while clearing the enemy out of an intricate system of tunnel-connected caves. Following the seizure of this high ground, Lieutenant Colonel Clair W. Shisler reorganized his 3d Battalion and continued the advance to the high ground north of and overlooking the Kokuba. As his troops dug in for the night, they were subjected to intense mortar and artillery fire.

Advancing alongside of 3/22, 1/29 made the main regimental effort. During the attack, a Marine threw a satchel charge or a grenade into one of the tombs along the advance route, setting off an estimated ton of explosives and causing approximately 25 casualties in Company C; B immediately passed through to maintain the attack.[15] Although machine gun and small arms fire from the numerous caves and fortified tombs in the battalion zone slowed the progress of the attack, 1/29 was able to advance under the cover of fire support from 3/29. The latter also advanced slowly, meanwhile maintaining contact with the 5th Marines on its left. At the end of the day, both 6th Division assault regiments had gained 800 yards and were in firm possession of the key

[15] *6th MarDiv G–3 Jnl, Ph III,* 30May45.

high ground overlooking the Kokuba from the south.

During the night of 30–31 May, the volume of enemy artillery and mortar fire on Tenth Army positions was noticeably lighter in comparison to that which had fallen previously in the battle for Shuri. When assault troops surged towards troublesome Japanese pockets remaining about Shuri on the morning of the 31st, they were pleasantly surprised by the almost complete lack of opposition. Only sporadic sniper and machine gun fire broke the weird silence in an area that had just recently been filled with the din and crackle of battle. Adhering to the *Thirty-second Army* withdrawal plan, rearguard forces from the *44th IMB*, *32d Regiment*, and *22d IIB* had pulled out of their positions during the night to occupy the second holding line north of Tsukasan.[16] Another aspect of the completely reversed situation was the break in weather, which changed the seemingly unending period of rain and solid overcast into a day of sunshine and high scattered clouds.

American ground units moved into Shuri, later described as "a perfect final defensive position," [17] and found it to be nearly abandoned. Soldiers from XXIV Corps quickly advanced and occupied assigned objectives, and spent most of the time thereafter mopping up isolated pockets of resistance. Only on the extreme left of the 96th Division line, where attack elements encountered the Tsukasan line defenses, was the corps objective not taken. The encirclement and occupation of Shuri became a reality at 1255 on the 31st, when patrols from 3/383 and 1/5 made contact south of the city.[18]

In a coordinated sweep with the 77th Infantry Division, the 1st Marine Division cleared out the enemy-infested areas immediately surrounding Shuri. Mopping up of bothersome pockets in the northern outskirts of the city and in the stubborn Wana Draw was completed by noon. Later in the day, the 1st Marines was ordered into division reserve and given a primary mission of patrolling Shuri.

Despite supply and evacuation problems, the 5th Marines continued its southeasterly advance towards the hills just north of Shichima, overlooking the Naha-Yonabaru highway. The 3d Battalion made the main effort for the regiment, jumping off at 1445—15 minutes after it had received an air drop of water and ammunition. Upon reaching the hills, rifle and machine gun fire from Japanese blocking units forced the battalion—on the corps boundary—to dig in for the night. A gap existing between the 1st and 3d Battalions was plugged by Company F of 2/5.

The heretofore steady progress of Tenth Army flanking units was slowed on 31 May when enemy resistance to the 6th and 7th Divisions became stronger. General Shepherd's assault regiments jumped off at 0730 and rapidly moved forward for several hundred yards before encountering unyielding Japanese positions in the hill mass west of Shichima and Kokuba. These were occupied by Admiral Ota's ragtag naval troops and units of the *32d Regiment*. The Marine advance was held up until

[16] *Okinawa Operations Record*, p. 120.

[17] *3/1 SAR*, p. 38.

[18] *1/5 SAR*, p. 9.

about 1300, when a coordinated assault was launched under the cover of long-range support fire furnished by a company of tanks situated as close to the line of departure as a minefield and a sea of mud would permit.

Although the division had some evidence that the enemy defense was crumbling here, the Marines had gained only 400 yards by nightfall and were still short of their objective. In night defense lines that were consolidated along a series of low hills immediately west of the objective, the assault battalions made preparations for a coordinated attack on 1 June. All through the night, artillery batteries fired concentrations on suspected Japanese gun positions in an attempt to destroy them.

On the left flank of the Tenth Army, the 7th Division continued its two-pronged attack. One assault force drove up the Naha-Yonabaru valley against a chain of well-defended hills to reach the corps boundary at Chan; the second sent strong combat and reconnaissance patrols into the hills and valleys guarding the neck of Chinen Peninsula. Little opposition was met there.

By the end of May, the Tenth Army had overcome the seemingly impregnable Shuri redoubt, only to run into newly organized defenses positioned along the Kokuba River and north of Tsukasan. Since the initial landings on L-Day, General Buckner's forces had killed an estimated 62,548 Japanese soldiers and captured 465 others in 61 days of bloody endeavor. The Tenth Army had seized all but eight square miles of the island, and that parcel was becoming a pocket of doom into which the remnants of General Ushijima's army were being driven. The battle so far had cost the Americans 5,309 men dead, 23,909 wounded, and 346 missing in action.[19]

On 1 June, as though in anticipation of an imminent end to the fighting but in fact on the date stipulated in the ICEBERG logistics plan, unloading operations off Hagushi changed from the assault to the garrison phase.[20] This same day, the second consecutive clear one, the direction of the attack was reoriented in the XXIV Corps zone. In the 96th Division zone, the 381st and 383d Infantry Regiments relieved the 32d Infantry north of Chan on the line paralleling the corps boundary, where it turned east to end at a point 1,100 yards north of Karadera. A day later, General Griner's two regiments were to attack to the south; their objective, the hill complex approach to the Tomui-Aragusuku-Meka area. Guarding the right rear of the corps advance was 2/305, which had moved out to the boundary when the 77th and 96th Divisions had exchanged zones of responsibility.

With its 32d Infantry in reserve, the 7th Division attacked to the south in a much narrower zone than it had been assigned before. During the previous two days, combat patrols had thoroughly scouted and prepared the way, enabling the division to gain an average of 1,100 yards against steadily rising opposition. Facing the Americans were elements of the *7th Heavy Artillery* and *23d Shipping Engineer Regiments*, which slowly pulled back towards Itokazu during the day.

[19] Tenth Army G-3 Rpt No. 67, dtd 1Jun45.
[20] *CTF 31 AR*, pt III, pp. 44–45.

Maintaining constant pressure, both IIIAC assault divisions made substantial gains on 1 June in a coordinated attack, which resulted in the capture of all the high ground commanding the cross-island highway running through the Kokuba River valley. The 1st Division attack was made by 1/5 and 3/5, which overran enemy positions in the hills east of Shichina to advance 1,500–1,800 yards before halting for the night. (See Map IX, Map Section.)

In the 6th Division Zone, the 22d and 29th Marines broke through the defenses that had held them up the day before and advanced swiftly in a smoothly functioning tank-infantry attack. By late afternoon, the assault regiments possessed the high ground on the northern bank of the Kokuba, and sent patrols across the northern fork of the river to select suitable crossing sites. Having accomplished their mission of slowing the American advance, Japanese holding forces in the second defense line had withdrawn the previous night. Their action paved the way for the Tenth Army to continue the pursuit and to make an unopposed tactical river crossing.

According to oral instructions General Geiger gave him in the early morning of 1 June, General Shepherd was given 36 hours to prepare his division for an amphibious operation. For as complex an operation as this, considerably more preparation time was usually allotted. Nevertheless, division planners were "to study the practicality of a shore-to-shore landing on Oroku." [21]

Major Walker's 6th Reconnaissance Company was to reconnoiter the peninsula after dark on the 1st. The company was to move out at 2100 and cross Naha harbor in rubber boats to the northern part of Oroku. At 1110 that morning, General Shepherd received a IIIAC warning order alerting him that the 6th Division axis of attack would probably be reoriented in the direction of Oroku Peninsula, where the division would land to secure the harbor and seize the airfield.[22] To prevent disclosure of the presence of the reconnaissance company Marines on Oroku, all IIIAC units were directed to restrict the use of illuminating and parachute flares between 2030 and 0300, 1–2 June.[23]

Four scouting teams of four men each spent six hours on enemy-held Oroku, where they heard considerable Japanese activity and were fired upon. On their return, the reconnaissance teams reported that the peninsula was Japanese-occupied, but that the enemy was not there in great strength.

Besides ordering the Oroku attack, General Geiger directed the 1st Division to assume responsibility for and occupy the zone of the 6th, excluding Naha, on 2 June. Colonel Snedeker's 7th Marines relieved the 22d and 29th Marines shortly after noon of this date, and General del Valle assumed control of the overall zone at 1215. On the left, 2/7 took over the 22d Marines line along the north bank of the Kokuba, and 3/7 replaced the 29th Marines in hill positions west of Kokuba village.

---

[21] *6th MarDiv SAR, Ph III*, p. 15.

[22] *6th MarDiv G–3 Jnl, Ph III*, 1Jun45.

[23] *Ibid.*

Immediately after arriving at the Kokuba, 2/7 was ordered across the river and into the hills bordering its south bank. Company E was the first to cross. Utilizing a damaged but still-standing bridge in the battalion zone,[24] the company gained the heights north of Tomigusuku and fought off an estimated 50 to 100 Japanese who tried to turn the Marine right flank while Company G was filing over the bridge. By nightfall, Company F had joined the other two companies in establishing a firm bridgehead south of the river, and thus safeguarded the crossing site.

Early in the morning of 2 June, the 5th Marines crossed the north branch of the Kokuba over a railroad bridge that the retreating Japanese had neglected to blow up. While attempting to advance beyond a seized ridge guarding the approaches to the village of Tomigusuku, the 5th Marines assault units were pinned down by intense frontal and flanking rifle and machine gun fire, which prevented their making even limited gains for the rest of the day. Despite this bitter enemy reaction here, the 5th Marines advance put the final segment of the Naha-Yonabaru highway into Tenth Army hands. Just before midnight, the enemy launched a determined counterattack—the first since his withdrawal from Shuri—against Marine lines. The Japanese were driven back, but left behind 20 dead.

To the left of the IIIAC zone, XXIV Corps units made large gains all along the line. In the process of cleaning out Chan, seizing the high ground north of Tera and Kamizato, and penetrating through Japanese defenses to the west of Kamizato and Karadera, 96th Division assault regiments advanced 800–1,200 yards. Farther west the 7th Division succeeded in pushing forward 2,400 yards against slight opposition from retreating Japanese garrison troops. At the end of 1 June, Army infantry troops were positioned for a final drive to close off Chinen Peninsula entirely. Rain during the night of 1-2 June again resulted in the mud and supply problems experienced by all Tenth Army units earlier, and forced them to accommodate their operations more to the obstacles posed by the rain and mud than the enemy.

By noon of 2 June, the 6th Marine Division had received final instructions regarding the Oroku operation, and General Shepherd's staff had already begun detailed planning. After examining possible courses of action and schemes of maneuver for the landing—and eliminating those that seemed least likely to be successful—the division commander decided to land on the Nishikoku beaches on the northeast coast of the peninsula and drive south, generally following an axis of attack astride the high ground in the center of the peninsula. (See Map 17.)

Governing the acceptance of this landing plan was the fact that Nishikoku had low rolling ground leading from the most suitable beaches on Oroku to the airfield and Naha harbor. In addition, an attack inland from this beachhead would be angled in the best direction for comprehensive, massed artillery support from the mainland.[25] Other landing sites

[24] *7th Mar Hist*, p. 21.

[25] *6th MarDiv SAR, Ph III*, p. 16.

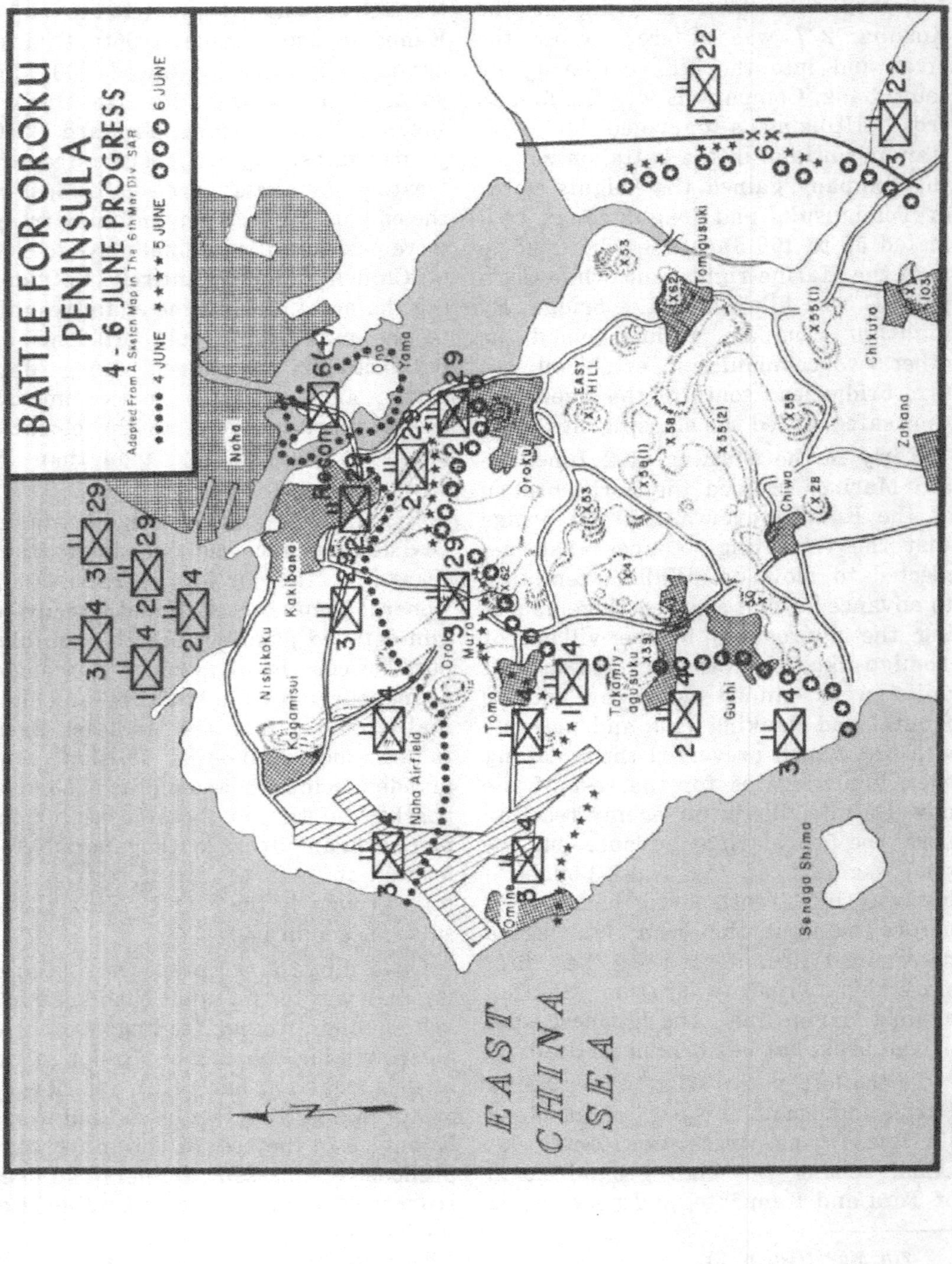
BATTLE FOR OROKU PENINSULA
4 - 6 JUNE PROGRESS
Adapted From A Sketch Map in The 6th Mar Div. SAR
4 JUNE
5 JUNE
6 JUNE
EAST CHINA SEA
Naha
Kakibana
Nishikoku
Kagamisui
Naha Airfield
Oroku
Ono Yama
Reason
Oroku
Muro
Tomo
Omine
Senaga Shima
Takamiyagusuku
Gushi
Chiwa
Zahana
Chikuto
Tomigusuku
EAST HILL
T. L. RUSSELL
MAP 17

on Oroku under consideration were rejected because the high seawall ringing the peninsula would have to be breached, which time limitations prohibited. Also the enemy had direct observation of these other prospective beachhead sites from high ground inland.

Owing to a shortage of amphibious tractors then existing at Okinawa, General Shepherd could count on having only 72 LVTs available for the landing.[26] Most of the other amtracs were in poor condition as a result of continued extensive employment during the heavy rains in ship-to-shore supply operations and in coastal runs supplying the flanking divisions of the Tenth Army, and an almost-complete reliance on the LVTs for overland supply of frontline units. Nevertheless with the LVTs he was given, General Shepherd planned to land his division in a column of regiments, the 4th Marines in assault. Colonel Shapley in turn, chose his 1st and 2d Battalions to spearhead the attack. The regiment was to drive rapidly inland to seize dominating terrain near Kagamisui, just north of the airfield, from which it was to cover the movement ashore of the rest of the division. As soon as the 4th Marines had moved beyond the beachhead area onto its objective, and when LVTs had made the return trip and were available, the 29th Marines would land. After this phase of the assault had been completed, tanks and supplies would be unloaded from landing craft.

The 6th Division assault forces were to mount out from assembly areas near the mouths of the Asato and Asa Rivers, and supplies and tanks would be loaded at a point that had been developed near Machinato and named Loomis Harbor after Colonel Francis B. Loomis, Jr., the G–4 for III Amphibious Corps.[27] Because it would be difficult to maintain a waterborne resupply operation continuously during the peninsular fighting, General Shepherd decided to seize Ono Yama concurrently with the Oroku assault. This small island in the middle of Kokuba Channel, across from the southern end of the Naha Canal, served as an anchor for bridges from the mainland and the peninsula. Once a task force of reconnaissance company Marines reinforced by a company of LVT(A)s had taken the island, it would provide security for an engineer detachment that was to repair the damaged bridges. After its capture, Ono Yama served the 6th Division as a logistic support base that was located fairly close to the fighting.

---

[26] At the beginning of the campaign, the 4th and 9th Amphibian Tractor Battalions with a total of 205 LVTs were attached to the 6th Marine Division. Added to those in the 1st and 8th Battalions attached to the 1st Marine Division, the total number of LVTs available to IIIAC was 421. *IIIAC AR*, chap VII, p. 101. The resupply of spare parts for LVTs was totally inadequate, especially in the case of such vitally needed basic items as tracks, track suspension system parts, front drive assemblies, and transmission parts. The lack of all of these deadlined a good many LVTs and severely limited the amount of support they could have provided during the drive to the south and in the Oroku landing. At the end of the campaign, 75 LVTs had been completely destroyed as a result of enemy action, or, having been badly damaged, they were cannibalized for spare parts. Of the 346 vehicles remaining, 200 were deadlined for lack of spare parts. *Ibid.*, p. 102.

[27] Smith, *Personal Narrative*, p. 126.

Both logistical and personnel preparations for the assault were increasingly complicated by the almost-complete breakdown of the road net as a result of the resumption of heavy rain. Therefore, all division tactical and support movement had to be made over water. "Even the division CP, deploying to a forward location near Amike was required to move entirely by DUKWs." [28] Despite these handicaps, all 6th Division assault and support units made ready for the amphibious landing on K-Day, 4 June; the reinforced reconnaissance Marines were to land at 0500 on Ono Yama, and the main assault force 45 minutes later on Oroku.

While the 6th Division was temporarily out of the fighting and preparing for the Oroku invasion, the attack south increased in impetus and force. By late afternoon of 3 June, the 7th Infantry Division had reached the east coast of Okinawa below Kakibana and cut off the Chinen Peninsula completely. The 32d Infantry then moved into the hill complex of the peninsula to destroy any members of the Japanese garrison still remaining. General Arnold consolidated the lines of the 17th and 184th Infantry in the hills overlooking Itokazu and Toyama, where the soldiers poised for an attack to the southwest against Kiyamu Peninsula positions. (See Map IX, Map Section.)

To the right of the 7th, the 96th Division also scored gains on 3 June. Kamizato, Tera, and then Inasomi fell after only a perfunctory enemy defense. Before halting the attack to set up night defenses, General Bradley's assault regiments had taken 1,400 yards of enemy territory, even though the combination of continuing bad weather and almost insurmountable supply problems seemed to conspire against further American successes. At sunset, 96th Division troops overwhelmed determined enemy defenders to seize commanding terrain in the hill mass north of the road and rail junction at Iwa. Because an already-existing gap between Marine and Army units had been widened by the accelerated pace of the XXIV Corps, the 305th Infantry continued its role of guarding the exposed flank of the corps at the boundary between it and IIIAC.

To the right of the 305th, from midnight, 2 June, to dawn the next day, the 5th Marines frustrated persistent enemy attempts to infiltrate its lines. After sunrise, the Marines spent the morning probing the front with patrols, which soon were pinned down by scattered but well-placed enemy fire from positions south of Tsukasan and west of Gisushi. When the 1st and 3d Battalions could move forward no further, 2/5 was alerted to its possible commitment to ease the situation.

At 1230, Lieutenant Colonel Benedict was ordered to circle around the left battalion—1/5—by moving in a wide arc through the XXIV Corps zone, and to outflank and come up behind enemy strongpoints on the high ground near Tomusu and Gishusi. Taking only equipment that it could carry, the battalion moved out at 1330. At 1800, after having trekked over a difficult, muddy, and circuitous route, it arrived at a jump-off position 400 yards east of its objective. Twenty minutes after arriving here, 2/5 attacked with Companies E and G in

---

[28] *6th MarDiv SAR, Ph III*, p. 16.

the assault and quickly secured the target against only slight enemy resistance.

As 2/5 Marines began to dig in, they found that the entire ridge contained a well-organized cave defense system, and began blowing up cave entrances to seal them. A white phosphorous grenade thrown into one cave set off what apparently was a Japanese ammunition dump; the resulting explosion blew up the entire side of the hill in the Company E sector, killing 3 Marines and wounding 17. The exposed 2/5 position ahead of the 1st Division line presented the battalion with a most difficult task of evacuating casualties, which was accomplished only "by invaluable assistance" provided by 2/383 on its right.[29]

While 2/5 was outflanking the Gisushi position, at 1530 the 1st and 3d Battalions resumed the regimental attack under enemy fire which had lessened considerably since the morning. When these two battalions halted at 1900, they had advanced 1,500 yards and through Tsukasan, placing the 5th Marines south of the former rear command post of the *Thirty-second Army*. The relative ease with which the 1st Division had advanced on 3 June indicated that the Japanese rear guard had once again withdrawn towards fixed positions on Kiyamu.

Spearheaded by 1st Reconnaissance Company patrols, assault units of the 7th Marines rolled up 200 yards in the division right. This advance established Marine control of virtually the entire hill mass south of Kokuba. Scattered enemy holding groups constantly harassed advancing Marines on the right flank during the day, however, with mortar, machine gun, and machine cannon fire from emplacements in the hills guarding the entrance to Oroku Peninsula. This steady enemy fire constantly menaced supply and evacuation parties traveling a well-worn route into the regimental zone from the only bridge over the Kokuba River. A combination of this harassment and the difficulty of negotiating over rain-damaged road nets again forced the Marines to depend upon air drops as a source of supply for rations and ammunition.

To fulfill the logistic requirements of the ground forces, the Avengers of VMTB–232 were kept as busy in June as they had been in May air dropping prepackaged loads to IIIAC and XXIV Corps units that could not be replenished through normal supply channels. The squadron made 24 drops on 1 June; 32 on 2 June; 24 on 3 June.[30] Having received a supply replenishment from planes during the day, both assault battalions of the 7th Marines were across the Kokuba by nightfall of the 3d and solidly set in the hill mass south of the river; 3/7 was in contact with 1/5, and the regiment was tied in across its front.

Well within the period it had been allotted, the 6th Division completed arrangements for its shore-to-shore operation by the end of 3 June. Beacon lights to mark the line of departure were set in place 1,200 yards north of the Nishikoku beaches at 1215 that day, and division assault forces were en route to board the LVTs at embarkation points on the west coast. The 22d Marines were

[29] *2/5 SAR*, p. 16.

[30] VMTB–232 WarD, Jun45.

placed in IIIAC reserve in and around Naha; its regimental weapons company moved to the shore of the Kokuba estuary, where it set up its 37mm and self-propelled assault guns in position to support the 4 June landing. Supplementing the massive fire support provided by artillery, naval gunfire, air, and its own organic weapons, the 4th Marines would have the additional fire support of a company of LVT(A)s, a company of tanks, and a mobile rocket launcher detachment. All preparations were completed at 2300, and the 6th Marine Division stood poised for the Oroku Assault.

THE CAPTURE OF OROKU PENINSULA[31]

As directed in the landing plan, the 4th Marines were to land on beaches designated Red 1 and Red 2 with 1/4 on the right, 2/4 on the left. The total length of the beaches was approximately 600 yards. Offshore, a rough coral shelf about 200 yards long was a not-insurmountable barrier to the landing site. The assault wave was to be followed by the other waves in LVTs. As envisioned, the shore-to-shore movement would be a comparatively simple operation. In addition to the beacon lights marking the line of departure, the only other control measure was to be the normal radio communication between the assault units.

In the early morning darkness of 4 June, troops and equipment loaded aboard the tracked amphibians according to plan. An intense prelanding bombardment was laid down on the target for an hour's duration before H-Hour; over 4,300 rounds of high explosive shells ranging in caliber from 75mm to 14-inch blasted suspected enemy positions on the high ground immediately behind the Nishikoku beaches.[32] (See Map 17.)

Once loaded, the invasion flotilla headed south towards the target in two columns, 400 yards apart; 1/4 in the seaward column since it was to land on Red 2, the westernmost beach. Almost simultaneously with the beginning of the bombardment of Oroku, 3/5 began blasting Ono Yama, and 15 minutes later the 6th Reconnaissance Company landed on schedule, supported by LVT(A)s of the Army 708th Amphibious Tank Battalion.[33]

During their approach to the line of departure, assault Marines were treated to the spectacle of the furious lashing given the beach area by the guns of 1 battleship, 2 heavy cruisers, 1 destroyer, and 15 battalions of artillery which joined in the cannonading. At the line of departure, the lead LVT(A) of each column signalled a column left, whereupon the following LVTs executed the movement, formed up into seven waves, and headed towards the beach. Four LCTs carrying 20 tanks and 10 LCMs carrying elements of the 4th Marines Weapons Company followed in the wake of the assault waves.

Before they had reached the line of departure, mechanical failures forced all

[31] Unless otherwise noted, the material in this section is derived from: *6th MarDiv SAR, Ph III; 6th MarDiv Jnl, Ph III; 4th Mar SAR, Ph III; 22d Mar SAR, Ph III; 29th Mar SAR, Ph III.*

[32] *6th MarDiv SAR, Ph III*, p. 16.

[33] *Ibid.*, p. 17.

of the tractors carrying 1/4 to slow down and some to fall out of the formation. Radio communication as well as overall contact was lost. By the time that the battalion had reached the line of departure, nine tractors had dropped out and only six were able to make the final run into the beach. Some of the cripples got underway again, but troops in the other amtracs had to be picked up by spare LVTs in the wave carrying the regimental headquarters. The battalion commander's request for a delay of H-Hour was refused by Colonel Shapley, who ordered the attack to proceed according to schedule. As a result, only two platoons of the right assault company of 1/4 and one from the left landed on time. The 2d Battalion experienced no difficulty during its approach and landing.

Intelligence estimates had indicated that the peninsula was defended by 1,200–1,500 enemy troops. At 0600, when the first Marines went ashore they saw no Japanese defenders on the beach, however, and were able to rush inland 300 yards to high ground against only scattered machine gun fire. All 2d Battalion units were ashore and reorganized in little more than a half hour after the first elements had landed.

By 0650, all tanks of Company A, 6th Tank Battalion, in support of 2/4, and attached engineer mine removal teams were ashore. Four self-propelled 105mm howitzers of the regimental weapons company also landed at this time. Except for one tank lost in a pothole, Company C tanks landed at 0800 and began supporting those troops of 1/4 already on the peninsula. The rest of the infantry battalion came ashore during the remainder of the morning.

Once ashore, the assault forces found the terrain very open and generally flat, with several 50 to 100-foot tall hillocks breaking up the landscape. As the attack moved inland to the central, southern, and western portions of the peninsula, the Marines encountered many ridges and steep hills—the highest of which was some 183 feet in height. The small hills initially captured were unoccupied by the enemy, but close inspection showed that the terrain was honeycombed with tunnels and numerous firing ports, which, when manned, had given the defenders commanding all-around views of the area.

Following its surge beyond the beachhead, 2/4 met mounting resistance on the left. Extensive minefields on the plateau immediately adjoining the landing site and the rain-soaked ground held up the tank-infantry advance as well. Both obstacles restricted tank operations and forced the Shermans to remain roadbound. Many sections of the roads had been blown up by the enemy and the mediums were unable to bypass these spots over the bemired fields; the ground troops were thus threatened with having their armor support severely curtailed.[34] The job of filling in the cratered roads soon surpassed the capabilities of the tank dozers.

Although the 1/4 reserve, Company B was the only element of the battalion to land in near-full strength. It was then committed at once on the right of 1/4 and ordered to take the high ground overlooking Kagamisui. The company

[34] *6th TkBn SAR, Ph III*, p. 15.

promptly overran the objective maintaining the momentum of the battalion attack and permitting 1/4 to gain its objective, 1,000 yards inland, at 1100.

In regimental reserve, 3/4 was nevertheless committed to the right of 1/4 at approximately 0905, within 20 minutes after it had landed. The 3d Battalion immediately pushed forward to the edge of Oroku airfield. Observers noted that the field was overgrown with rough grass, was swampy, and appeared in very poor condition overall. Large revetments were ranged along the edges of the three runways. Even though they provided excellent concealment and some cover, it would have been dangerous to use them since the enemy seemed to have had them well ranged in with mortars. The wreckage of several planes, apparently strafed and bombed earlier by American aircraft, was strewn over the field. Running along the right (west) edge of the field was a seawall, heavily overgrown with palmetto and brush. On the eastern edge of the airdrome was a series of foothill ridges that were crisscrossed with caves and aircraft revetments.[35]

An hour after 3/4 had landed, General Shepherd believed that the beachhead had been sufficiently enlarged to the point where it could accept the landing of a second regiment. Accordingly, he ordered the 29th Marines to begin moving to Oroku immediately. Two of Colonel Whaling's battalions were quickly transported to the peninsula and moved into the lines on the left of the 4th Marines. The 2d Battalion was ashore and relieving left flank units of 2/4 by 1300; 3/29 took over the rest of the zone at 1430, whereupon 2/4 went into regimental reserve.

At 0900, 6th Division wiremen ferried a four-trunk cable across the mouth of Naha Harbor in rubber boats, strung it over the mast of a sunken ship, and had it tied at the terminals of assault unit switchboards on Oroku at 1100. At the same time that this task was underway, division engineers worked rapidly to repair the bridges between Naha, Ono Yama, and the peninsula. Bridging operations began immediately after the harbor island had been secured at 0600. After 30 minutes of sharp fighting during which it killed an estimated 25 enemy soldiers, the 6th Reconnaissance Company deployed to positions where the scouts could protect engineers who were assembling the Bailey bridge.

Elsewhere on Ono Yama, other engineers inflated rubber pontoons that were to be placed into the water to support a bridge spanning the wide estuary between the island and Oroku. It was not possible to establish this link until Marines on the peninsula had neutralized heavy enemy machine gun fire aimed at puncturing and sinking the pontoons.

By nightfall, 1/29 was landed and in regimental reserve, 2/4 was set up in an assembly area as its regiment's reserve, and the attack had halted for the day. At this time, the invasion force had pushed inland 1,500 yards against steadily increasing resistance. In addition to this Japanese opposition, the attack had been slowed in the afternoon by very heavy rain storms as well as numerous minefields, whose neutralization and destruction taxed overworked mine-disposal teams.

[35] *3/4 SAR, Ph III*, p. 5.

During the first day on Oroku, the assault forces had received a considerable amount of fire from a variety of automatic weapons ranging up to 40mm in caliber. It was later learned that these weapons had been stripped from the damaged aircraft on and around Oroku airfield and distributed to the ground defense force, which then was able to offer a more formidable response to the Marine invasion. From early evening and on through the night of 4–5 June, Marine lines were subjected to sporadic enemy artillery and mortar fire.

A startling new Japanese weapon met by Marines on Iwo Jima[36] was brought into the Okinawa campaign when the 6th Division was introduced to the enemy's 8-inch rocket. Dubbed a "Screaming Mimi" or "Whistling Willie" by the troops,[37] because of the noise it made while tumbling through the air end over end, the projectile was more a source of annoyance than danger and caused few casualties. Its explosion was loud and concussion great, but this rudimentary missile's fragmentation was ineffectual and its accuracy was poor. "It was launched from a pair of horizontal rails about 15 feet long, aiming was strictly hit or miss, a process of sandbags, guesswork, and luck."[38] The rockets continued to fall in rear areas during the night, while enemy snipers and would-be infiltrators were active.

Troops on Ono Yama received machine gun and spigot mortar fire in the darkness. These 320mm mortar shells, nicknamed "flying ashcans" by Americans, had been employed only briefly earlier in the Okinawa campaign and had not appeared again until this time.

At 0700 on 5 June, 1/22 reverted to the 6th Division from corps reserve, and, as division reserve then, was deployed on the division boundary in the right (west) flank of the 7th Marines attacking south. At 0730, the Oroku assault resumed and moved slowly against determined stubborn opposition until noon, when the 4th Marines was halted by an enemy strongpoint near Toma.

Muddy ground on the right of the 4th Marines zone made it impossible to employ tanks, so a platoon of the tracked vehicles skirted the seawall to come up on the airfield behind 3/4—in whose zone Toma lay—to assist the infantry attack. As the armor drew near open terrain on the field, it began receiving enemy artillery fire. The 15th Marines was called upon to provide counterbattery fire against suspected enemy positions revealed by gun flashes; a tank officer adjusted this artillery fire from his forward position.[39] An inspection later disclosed that the 15th Marines had silenced four 120mm dual purpose, one 6-inch, and several field guns of smaller caliber. Blown roads and bridges in the 3/4 zone, not yet repaired by the engineers, forced the battalion to attack Toma without accompanying tanks, which provided direct fire support, however, from positions in the rear of the lines.

The enemy was well dug in in this sector and located in deep, strongly fortified caves that were impervious to all but pointblank fire. Since tanks were bogged down and not available, 37mm

[36] See Bartley, *Iwo Jima*, pp. 13, 13*n*.

[37] *3/29 SAR, Ph III*, p. 4; *3/4 SAR, Ph III*, p. 6.

[38] *Ibid.*

[39] *6th TkBn SAR, Ph III*, p. 16.

guns were brought to the front and employed to good advantage against the enemy positions. After having been stymied through most of the day, 3/4 finally overran the Japanese defenses late in the afternoon with the aid of the fire from M–7s in the 1/4 zone and the support of tanks that had rumbled into position behind the 3d Battalion earlier. By nightfall, the battalion held 75 percent of the airfield and favorable jump-off positions for the resumed attack on the next day.

In the right center of the 4th Marines zone, 1/4 became pinned down by frontal and flanking fire almost immediately after it attacked the morning of the 5th. When 3/4 cracked open the Toma defenses, the 1st Battalion was able to take up the attack again. As it did so, 1/4 moved forward over terrain that was broken by a number of steep hills containing many extensive tunnels in the mouths of which machine guns were emplaced and sited for all-around defense.[40] The 4th Marines' commander noted that the heavy resistance met all along his line was reminiscent of that encountered in the battle for Naha.[41] When the attack for the day ended at 1700, 1/4 held positions on high ground overlooking Ashimine and Toma on the right, and an unnamed village, designated "Oroku Mura,"[42] on the left.

[40] *4th Mar SAR, Ph III*, p. 9.

[41] *Ibid.*

[42] *1/4 SAR, Ph III*, p. 5, Literally translated, "mura" means township, or in this case prefecture; Oroku Mura referred to the political subdivision in which the peninsula resided. The naming of this village was a matter of happenstance since the map designation OROKU MURA was printed directly above the village's location on the map.

Overcoming both bitter enemy resistance and problems of supply and evacuation, the 4th Marines advanced the division line 1,000 yards on the 5th. Frontline units experienced considerable small arms and automatic weapons fire as well as many grenade launcher barrages, "but very little heavy mortar and no artillery fire, which was a relief to all hands."[43] The enemy placed the artillery and mortar concentrations on rear areas instead, however, preventing LVTs from using the tank route leading to 3/4 positions to give that battalion supply and evacuation support. A 50-man working party, organized at regimental headquarters to replace the amphibious tractors, hand-carried urgently needed supplies up to 3/4, and took out evacuees on its return to Colonel Shapley's CP.[44]

The 29th Marines made slow but steady progress on 5 June against enemy opposition that was moderate to heavy. By 1400, the regimental advance was slowed when assault units encountered a strong center of resistance near Hill 57, at the southeast outskirts of Oroku Mura. This strongpoint gave trouble to left flank elements of the 4th Marines also. A Japanese counterattack launched against 3/29 before the battalion had moved forward 1,000 yards was easily blunted, but fire from enemy positions located in the areas of adjacent battalions finally forced the 3rd Battalion to hold up.

The 2nd Battalion continued to push its left flank southeast along the banks of the Kokuba, and finally secured a

[43] *4th Mar SAR, Ph III*, p. 9.

[44] *3/4 SAR, Ph III*, p. 6.

bridgesite area opposite Ono Yama. This permitted the engineers to float a 300-foot pontoon bridge into position. Pausing only to leave security detachments at the bridge as a guard against enemy attempts to destroy it, the battalion continued the attack. At 1810, the first LVT crossed over the bridge to Oroku from Ono Yama, opening a direct ground supply line to the assault troops.[45]

In the course of their operations on 6 June, the two assault regiments of the 6th Division uncovered major enemy defenses that were centered along the axial ridge running northwest-southeast along the length of Oroku Peninsula. The terrain of this hilly region favored the defenders, not only by its complexity but also by a heavy overgrowth of tangled vegetation. Immediately after they had resumed their attacks on the 6th, both the 4th and 29th Marines were held up by determined enemy opposition from concealed and well-camouflaged defenses.

A platoon of tanks supported the attack of the 2/29 with overhead fire at ranges of up to 1,200 yards from a high ridge overlooking the battalion objective —the village of Oroku. Left flank elements of the 2d Battalion pushed forward and captured the high ground in the village itself, but were unable to advance much farther in the face of heavy enemy fire. A second platoon of tanks moved along the river bank and attempted to get into position to subdue this fire, but it was unable to bypass a destroyed bridge in its path.

On the right of the 29th Marines zone, 3/29 moved over terrain that "consisted of a series of small temple-like hills, each of which had been converted into a fortress by construction of innumerable caves, from which mutually supporting automatic weapons could cover adjacent positions and deny the open ground between the hills [to the Americans]." [46] Naval personnel from Admiral Ota's force manned the machine guns and 20mm cannon guarding the sector. After a day of bitter fighting without armored support—the narrow roads in the battalion zone were heavily mined and cratered, and impassable to tanks—the gains of 3/29 were limited to a scant 150 yards.

Immediately fronting the 1/4 line of attack was a hill the Marines called "Little Sugar Loaf," that 3/29 had been unable to take earlier. Lieutenant Colonel George B. Bell planned for his infantry to capture it by means of a double envelopment coordinated with a tank drive up the center of the valley leading to the objective. Assault companies forming the wings of the envelopment were pinned down as soon as they jumped off. The attack did not begin until 1530, when support tanks arrived and were in position. Although the advance began to gain momentum, the battalion commander thought that night would fall before the objective was taken and ordered his assault elements back to the lines occupied that morning, with little to show for the day's efforts. Although 1/4 had demonstrated how the enemy defenses could be breached in this sector, it was not to have the satisfaction of doing it itself; early the next day it was relieved by 2/4.

---

[45] *6th MarDiv Jnl, Ph III*, 5Jun45.

[46] *29th Mar SAR, Ph III*, p. 8.

To the right of 1/4 on 6 June, Lieutenant Colonel Hochmuth's 3d Battalion attacked following an air strike on the many ridges in front of the battalion. As Company I on the right prepared to move out, its right flank was subjected to some 20mm and heavier caliber fire from Senaga Shima, a small rocky island flanking the Marine lines and lying approximately 1,000 yards off the southern coast of the peninsula. Tenth Army artillery and naval gunfire support ships blasted the island, silencing all but the 20mm weapons. An air strike was urgently called for and arrived a half hour later. "As rack after rack of bombs fell," scoring direct hits on the Japanese emplacements, "the troops stood up and cheered."[47] (See Map 17.)

Disregarding the 20mm fire from Senaga as best they could, Marines from 3/4 moved rapidly forward as soon as the last plane in each of a series of air strikes made its final run over a target in front of the battalion. Scattered small arms fire paced the troops attacking over comparatively flat terrain, but 3/4 succeeded in securing the rest of the airfield by the end of the day.

Engineer road-construction crews and mine-disposal teams worked on 6 June in warm and clearing weather. Discovered and disarmed on the main north-south road bisecting the peninsula were 83 mines of all types. At noon, Company B of the tank battalion landed from LCTs with the rest of the battalion's tanks and immediately went into reserve.[48] Also on 6 June, the 22d Marines as a whole reverted from corps reserve; 3/22 joined the 1st Battalion on the division east boundary, adjoining the west flank of the 7th Marines driving south; and 2/22 was alerted to move to a new defense position elsewhere on the division boundary.

Considerable resistance continued to plague the 6th Division as it unrelentingly swept across the peninsula on 7 June. (See Map 18.) The 4th Marines again made the most satisfactory progress of the day, but its right flank, which had advanced against only slight opposition on the previous day, was confronted with a much stronger defense in the vicinity of Gushi. As 3/4 tried to take the last section of high ground on the west coast, its leading company came under a deadly machine gun and rifle crossfire at the same time that extremely accurate and heavy mortar barrages fell on the only route of approach to the battalion goal. Both direct and indirect supporting fires bombarded Japanese positions to no avail. At the end of the day the battalion commander, faced with the prospect of sustaining heavy casualties if he pushed on, decided to pull his forwardmost elements back and hold the ground already taken.

Colonel Shapley's 2d Battalion passed through 1/4 at 0730, and began its attack on Little Sugar Loaf with the supporting fires of 37mm guns, tanks, and self-propelled 105mm howitzers. Left flank elements of 3/4 also supported the attack as Company G maneuvered around to the right of the enemy position and took it at 1100.

Following its capture of Little Sugar Loaf, the 4th Marines pushed ahead slowly against machine gun fire coming from all directions and ever-stiffening

---

[47] *3/4 SAR, Ph III*, p. 6.

[48] *6th TkBn SAR, Ph III*, p. 16.

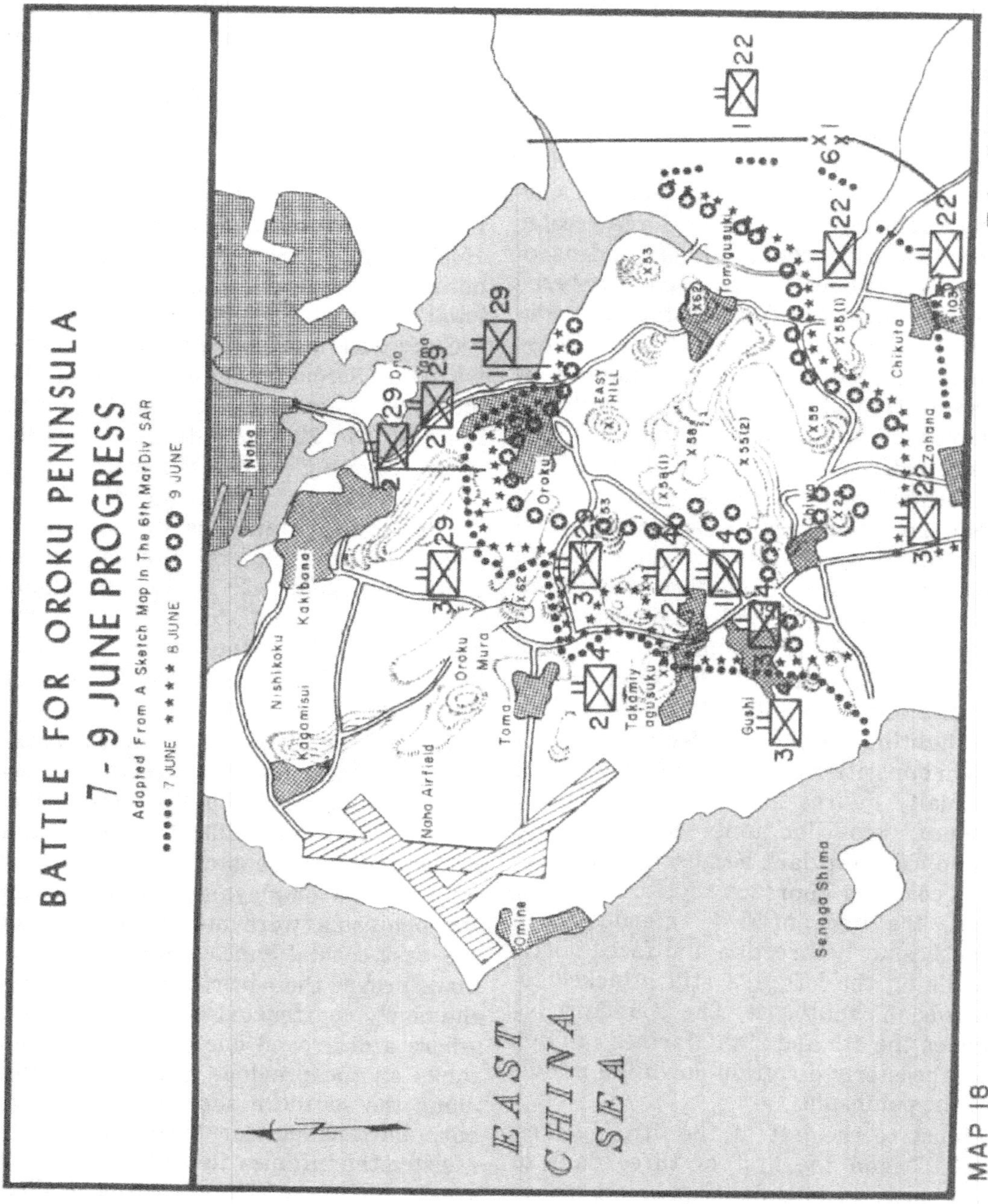
BATTLE FOR OROKU PENINSULA
7 - 9 JUNE PROGRESS
Adapted From A Sketch Map In The 6th MarDiv SAR
7 JUNE
8 JUNE
9 JUNE
Naha
Ono
Kakibana
Nishikoku
Kagamisui
Naha Airfield
Oroku
Mura
Toma
Oroku
EAST HILL
Takamiy agusuku
Gushi
Tomigusuku
Chikuta
Zahana
Omine
Senaga Shima
EAST CHINA SEA
MAP 18
T. L. RUSSELL

enemy opposition. Frontline Marines, already expert in the technique of sealing caves, furthered their expertise while closing the many caves on Oroku with a deadly combination of direct fire, flame, and demolitions. Unit commanders soon surmised that the peninsula was being defended by an enemy force greater than the 1,500–2,000 soldiers and naval troops previously estimated. Captured documents and POWs substantially confirmed the fact that the Oroku defense had been reinforced by a number of naval troops, which had originally moved south to Itoman and then had been ordered back to the peninsula. This information also indicated that the original American strength estimate of naval personnel was faulty and now had to be revised upward because many Okinawan conscripts had been dragooned into the ranks of Admiral Ota's force.

Although many of the Japanese on Oroku had been killed after three days of fighting on the peninsula, the stubborn opposition of those still alive caused casualty figures in the 4th Marines to mount. Frontline units could only be supplied after dark because of the lethal fire covering approach routes. By nightfall, the lines of 3/4 extended in a southeasterly direction and faced north, while on the left, 2/4 still attacked towards the southeast. The boundary between the 4th and 29th Marines ran in a southeasterly direction down the middle of the peninsula.

Just to the left of the 4th Marines, 3/29 began the first of three days of extremely vicious fighting by a grenade and bayonet assault without armored support on the hill to its immediate front. During the period following these three days, the battalion gained little ground, but killed an estimated 500 troops, destroyed a large variety of weapons, and sealed many caves containing enemy soldiers, supplies, and equipment.[49]

Two factors served to restrict the progress of the 29th Marines on 7 June. Hostile enemy concealed in the rocky outcroppings of the coastal ridge paralleling the Kokuba pinned down the attackers with a drumfire from automatic weapons. Secondly, the positions of the enemy in a confined area and the proximity to the 29th of adjacent friendly troops severely limited the employment by the regiment of its supporting fires. To destroy the Japanese weapons positions and the soldiers manning them, gun crews from 2/29 manhandled their 37mm weapons up steep slopes to the ridge overlooking the enemy emplacements and effectively raked them with murderous direct fire.

In the zone of the 2d Battalion, tank-infantry teams made satisfactory progress towards their village objective. The boggy, steep, and difficult terrain and heavy concentrations of minefields that limited tank employment elsewhere on the peninsula were not in evidence in the east coastal zone, where the Shermans proved their worth. After crossing the newly constructed bridge at the site where a destroyed one had held up the tanks on the previous day, and rolling along the southern shore of Naha harbor, the tank platoon attached to 2/29 assisted the infantry in capturing Oroku village. Without pause, the battalion continued its attack and seized the high

---

[49] *3/29 SAR, Ph III*, p. 5.

ground in the immediate vicinity of the village. Accurate and heavy enemy artillery fire and an extensive minefield then held the tanks up.

Along the division boundary, on 7 June the 22d Marines continued sending patrols out into the high ground immediately east of Chikuto. Having fixed the approximate center of enemy strength in this area, 3/22 moved two companies into position to attack the high ground designated Hill 103. By 1400, the Japanese stronghold was overrun, which effectively eliminated fire from that area on the 1st Division west flank, and gave it an additional measure of security.

Hill 103 proved to be an important enemy observation post occupied by a large number of Japanese troops. By choosing to remain in their caves, these soldiers sealed their own doom since this ineffectual defensive tactic confined their fields of fire and permitted the Marine attackers to outflank the position over covered routes of approach.[50]

According to the original scheme of maneuver established for the Oroku invasion, the 4th and 29th Marines would drive towards the base of the peninsula in a southeasterly direction. But, the rapid pace of the division attack during its first four days on the peninsula had forced the enemy to withdraw to the south of Oroku village and, with his back to the Kokuba Gawa, into the hills which were honeycombed with strong defensive positions. General Shepherd's order on 8 June reorienting the axis of attack to the northeast was a formal recognition of the course that the battle was taking. By this time, the 4th Marines on the right had advanced much further than the stalled 29th had in its zone on the left. Colonel Shapley's regiment was in the process of pivoting on the right flank elements of the 29th Marines in a counterclockwise movement that, when ended, would head the 4th in the direction of the hard core of Japanese resistance. In effect, the elements on the right wing of the 4th would sweep in front of the 22d Marines and continue on to the northeast. Neither the 22d nor the 29th Marine lines would remain static, however, for at this point all three infantry regiments were moving and inexorably tightening the circle around Admiral Ota's hapless force.

During the evening of 7 June, 1/4 was alerted and prepared to enter the line the next morning on the right of 3/4. For the 8 June attack, battalion boundaries were changed to reflect the new direction in which these two units were to head. Early on the 8th, Marine mortars laid a smoke screen over the route 1/4 was to take as it skirted along the eastern edge of the airfield while getting into jump-off positions. The 1st Battalion's objective was the high ground located approximately midway between Uibaru and Gushi. At 1030, the assault elements attacked and immediately were pinned down by a hail of fire from rifles, machine guns, and mortars.

Bitter enemy reaction to the Marine assault was unallayed despite the massive fires of American tanks, M–7s and organic infantry crew-served weapons. The attacks of the 1st and 3d Battalions were so coordinated that one could aid the other at any given time. Because the

---

[50] *3/22 SAR, Ph III*, pp. 8–9.

tanks were unable to deliver direct support fire from their masked positions, they lumbered forward into the open shortly after midday and blasted the 1/4 objective for 20 minutes. After this preparation, Company A again attacked the high ground, this time overrunning enemy machine gun and mortar emplacements. At 1430, Company C jumped off to the south with armor support and proceeded to clean the enemy out of the high ground in its sector and down to the seawall. Meanwhile, Company B entered the battalion line to the left of Company A and swung north, tying in for the night with 3/4. After clearing the ground in the battalion rear, Company C moved into position on the right rear of A to cover the exposed battalion flank overlooking the north-south Itoman road. Thus, the 1st Battalion commander had the unique experience of having his three infantry companies make successful attacks in as many different directions.

While the 1st Battalion headed for the seawall, 3/4 began a cross-peninsular attack over extremely rugged terrain that was marked by a maze of interlacing ridges. "Every slope had its allotment of caves, each covering the other from flank and rear."[51] Many of these caves were filled with enormous stores of explosives, which created a hazardous condition for the demolition teams attempting to seal them. Nonetheless, the indomitable teams set off hundreds of pounds of demolitions to destroy the honeycomb of cave entrances.

At 1300 on 8 June, an hour and a half after it had resumed its attack, 2/4 was just 200 yards short of that day's objective. Taking time out only to regroup, the battalion continued its advance, but was slowed by ever-increasing Japanese fire from well-constructed positions in the mouths of caves. Nevertheless, by 1530, 2/4 assault elements had captured the objective and began organizing for night defense. Before dark, patrols were sent back to mop up bypassed positions in the battalion rear.

To effect a junction with the 4th Marines, the 22d Marines pivoted on its right flank unit, while the 3d Battalion on the left moved in a clockwise direction to tie in with 3/4. A 3/22 patrol moved to the seawall and made contact with the right flank element of the 4th at 1550. Shortly thereafter, another battalion patrol scouted potential LVT landing beaches on the East China Sea coast north of Itoman. Reinforced by an infantry company from the 2d Battalion, 1/22 sent out strong combat patrols to take two hills. One, Hill 55, was approximately 500 yards east of Chiwa, and the other, designated Hill 55–1, was almost the same distance east of the first. Throughout the day, the patrols received light small arms fire which increased in intensity as the hill objectives were neared. At 1800, the easternmost height, Hill 55–1, was in possession of the Marines, who were forced to withdraw under cover of darkness because of an ammunition shortage coupled with heavy incoming enemy mortar fire.

Originally scheduled to jump off at 0830, on 9 June, the 22d Marines attack was delayed until 0900. The 1st Battalion was to retake Hill 55–1, 2/22 was to seize Hill 55, while 3/22 was ordered to capture Hill 28 on the outskirts of

[51] *3/4 SAR, Ph III*, p. 7.

Chiwa. The plan of attack called for the 1st Battalion to seize its objective, and for the 2d Battalion to pass through and capture its target. Not until late afternoon was 1/22 able to complete its mission, and the few daylight hours remaining did not give 2/22 enough time to capture its objective. As a result, Colonel Harold C. Roberts concurred in the battalion commander's recommendation to postpone the attack.

Intense fire coming from Hill 55 prevented Lieutenant Colonel Shisler's 3/22 from outposting Hill 28 until after dark. But Hill 26, just south of the primary battalion objective, was secured and occupied at 1000 by Company I, which soon made firm contact with 4th Marines patrols after the latter had cleaned out Chiwa.

In the course of its fighting on 9 June, the 4th Marines found little that was different from previous days' experiences on the peninsula, for:

> The advance was still slow and tedious against bitter resistance. Every Jap seemed to be armed with a machine gun, and there was still some light and heavy mortar fire. Casualties continued to mount and the number of Japs killed soared over the maximum of 1500 which were supposed to have been defending, and there were still plenty left.[52]

The 1st Battalion was ordered to seize high ground near Hill 55–2—the third hill so designated in the 6th Division zone—in the vicinity of Uibaru. The Marine attack was delayed until supporting armor could get into firing positions on the road paralleling the right flank of 1/4. Once ready to fire, the tanks were driven off by a bombardment from an enemy rail-mounted 75mm gun, firing from cave ports on the side of a cliff near Chiwa.[53]

Despite the temporary loss of its supporting armor, 1/4 attacked in the face of intense machine gun and mortar fire. Progress was slow and casualties increased steadily as the battalion advanced over ground that was honeycombed with caves, all of which had to be blown before they could be passed. At dusk, the right flank of 1/4 was anchored on a ridge northwest of Chiwa, while the battalion left flank extended to the outskirts of Uibaru, which had been taken earlier that day by 3/4.

A rocket barrage preceded the morning attack of 3/4 on 9 June. After the 3d Battalion moved out, difficult terrain prevented the battalion commander from maintaining unit control as his men worked closely with 2/4 to take the latter's objective, Uibaru. Upon occupying the village, the 3d Battalion received 20 casualties when a heavy enemy mortar concentration blasted its positions.

As the three infantry regiments of the 6th Division converged on the Oroku garrison from different directions, and completely isolated it from the main body of the *Thirty-second Army* at Kiyamu, Admiral Ota's mixed defense force was slowly compressed into a small pocket in the southeast region of the peninsula. On all levels, Marine com-

[52] *4th Mar SAR, Ph III*, p. 12.

[53] *1/4 SAR, Ph III*, p. 6; "The gun seemed as surprised by the arrival of the tanks as the tanks by fire from the gun, because it fired only HE [High Explosive projectiles] and no AP [Armor Piercing] and the tanks were able to get to cover without loss." Maj John R. Kerman ltr to CMC, dtd 7Jan48, hereafter *Kerman ltr.*

6TH DIVISION MARINES *land on Oroku Peninsula. Note the medium tank with flotation gear. (USMC 122601)*

FINAL SWEEP *of Oroku Peninsula in the last stages of the battle. (USMC 126176)*

manders found it increasingly difficult to maintain unit control and to coordinate the employment of their supporting fires with those of adjacent friendly units because of the limitations imposed by restricted zones of action. These conditions conspired with the stubborn terrain and the no-less yielding defense to slow to some degree all of the attacking Marine battalions.

One of these units, 2/4, was ordered to capture the last remaining Japanese-held high ground in its zone. To complete this mission, the 2d Battalion was required to mount a frontal attack up a 400-yard wide valley over terrain that offered little cover or concealment. After the battalion jumped off at 1145 on 9 June, supported by tanks, M–7s, and 37mm guns, its initial progress was slow. Further inhibiting the advance was the fact that a lack of tank approaches to the objective lessened the amount of close armor support given to the infantry. Also, 2/4 had to move ahead cautiously, for it was attacking in the direction of its own artillery and across the front of the 29th Marines.

Supplementing the natural tank obstacles in this sector, the Japanese had constructed a tank trap in front of their well-prepared ridge position and further safeguarded the area by a liberal sprinkling of mines. Since the Marines had no armored bulldozers or tank dozers immediately available, they were unable to construct a bypass in time to permit tanks to move ahead to support 2/4. At 1530, therefore, the battalion commander decided to halt the attack for the day.

Late in the afternoon, after 1/4 had pushed through Gushi, a tank managed to move through the now-demolished village and on to the road leading south to Itoman. Once in position on the flank of the cliff-emplaced enemy 75mm gun, it knocked the Japanese field piece out of commission. Only two shots were fired—one from the enemy gun, which missed and one from the tank, which didn't.[54]

On 10 June, the momentum of the 6th Division attack was accelerated. (See Map 19.) Early that morning, heavy construction equipment began clearing all tank approaches to the 2/4 frontlines, and by 0815 tanks and self-propelled howitzers were moving into position to support the infantry attack. In coordination with the 29th Marines on the left, 2/4 jumped off at 0945 with three companies abreast in assault. Less than an hour later, all attack elements were on the objective and organizing defensive positions from which they were to support the attack of the 29th Marines for the next two days.

From all appearances, the end of the battle for Oroku was near. At the same time that 2/4 had broken through the Japanese lines, the 1st and 3d Battalions advanced against lessening resistance. By 1400 on 10 June, the battalion boundaries of the 4th Marines had converged to squeeze 3/4 out of the line, and it went into regimental reserve.

While the 4th Marines pressed eastward, the 22d Marines drove northeast towards Tomigusuki, with 2/22 making the main regimental effort. This 1st and 3d Battalions provided fire support from positions they then held. When 2/22 had seized its objective, 3/22 was ordered forward and coordinated its attack with

[54] *Ibid.*

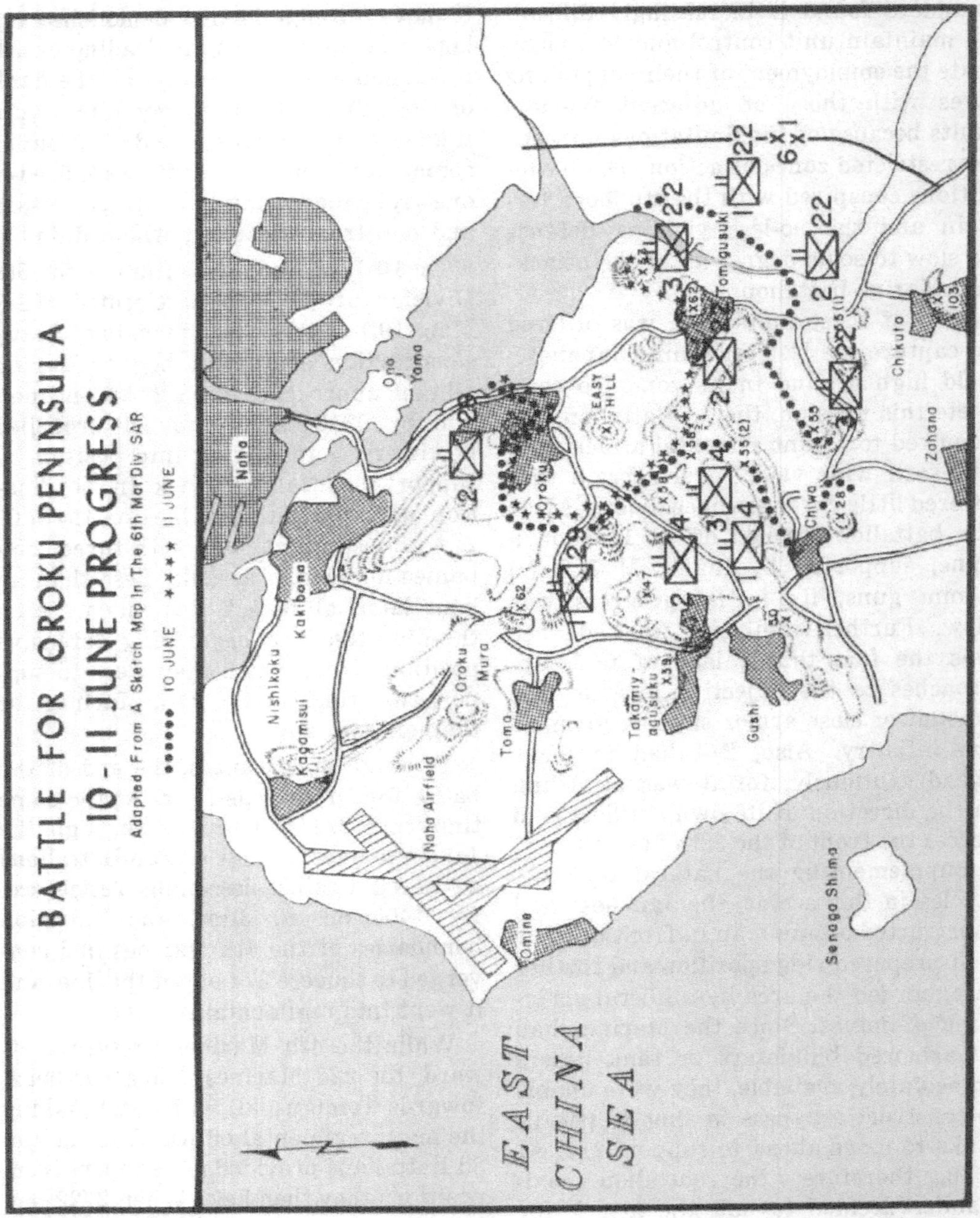

MAP 19 T.L. RUSSELL

that of 1/4 on the left. The 4th and 22d Marines made slow but steady progress on 10 June, but 29th Marines battalions continued to meet stubborn resistance and could report only limited gains. Moving slowly through Oroku village behind flame tanks, 2/29 was held up and its way blocked when the lead tank was destroyed by a direct hit from a Japanese 8-inch shell. The regiment, therefore, was unable to reach the last major enemy defense pocket in the sector, which was located on the high ground west of Oroku village. The Japanese troops trapped here began a number of frantic attempts to break out.

During the night 10–11 June, a series of local counterattacks hit all along the front. The heaviest of these took place in the sector of 1/4, which counted over 200 enemy dead in front of its lines after dawn. In reaction to the unfaltering and determined opposition of the Japanese defenders, General Shepherd launched an all-out armor-supported attack, committing the greater portion of eight infantry battalions to destroy the last vestiges of enemy resistance on Oroku.

In the 4th Marines zone, 3/4 resumed its attack at 0730, passing through the right elements of 1/4; the latter along with 2/4 remained in position to support the attack by fire. As the leading company began moving forward over a route that ran between Hill 58 (east of Uibaru) and Tomigusuki, it was held up by a hail of fire coming from Hill 62, on the right front. Covered by sniper fire from 1/4 on the left and tank supporting fire from the rear, Company I spent the better part of the day attempting to overcome the fortified hill blocking its path, and captured it before dark. By this time, Company K on the battalion right still was 300 yards short of establishing contact with 3/22.

The 22d Marines, led by 2/22, attacked Hill 62—north of Tomigusuki—following an intense 30-minute artillery preparation fired by six battalions of 105mm and one battalion of 155mm howitzers.[55] Once 2/22 had seized Hill 62, 3/22 was to support the 4th Marines until the latter masked its fires, after which it would pass through 2/22 and capture Hill 53, overlooking Kokuba Estuary. The 2d Battalion was unable to carry the hill in its first attempt and did not, in fact, seize the hilltop until 1220, after a heavy fire fight.

At 1300, 3/22 effected a passage of the 2d Battalion lines and was in position to attack 45 minutes later. Despite the lack of an artillery preparation on the objective,[56] the assault elements attacked, following a heavy mortar concentration. One factor preventing the tanks from gaining more favorable firing positions or even advancing with the infantry was the presence of well-concealed minefields along the route they were to travel. A mine-removal team worked under direct enemy fire and finally cleared a lane through which the tanks could pass to provide limited support. At 1450, Company L occupied Hill 53, giving the 6th Division high ground that overlooked not only the Kokuba Estuary, but also the entire Oroku area

---

[55] *22d Mar SAR, Ph III*, p. 12.

[56] "As the three regiments came closer together, it became dangerous and finally impossible to use even 60mm mortars, and some casualties were incurred from friendly fire on both flanks." *Bergren ltr.*

to the north where the 29th Marines had been unable to make any headway. That regiment attacked to seize the commanding terrain west of Oroku village repeatedly throughout the day, but was unable to find a way to overcome the series of small, mutually supporting hill positions that comprised the defense system here.

Undoubtedly aware that his end and that of the Oroku garrison force was not far distant, Admiral Ota had sent the following communique to his superiors in Tokyo on 6 June:

> More than two months have passed since we engaged the invaders. In complete unity and harmony with the Army, we have made every effort to crush the enemy.
>
> Despite our efforts the battle is going against us. My own troops are at a disadvantage since all available heavy guns and four crack battalions of naval landing forces were allocated to Army command. Also, enemy equipment is superior to our own.
>
> I tender herewith my deepest apology to the Emperor for my failure to better defend the Empire, the grave task with which I was entrusted.
>
> The troops under my command have fought gallantly, in the finest tradition of the Japanese Navy. Fierce bombing and bombardments may deform the mountains of Okinawa but cannot alter the loyal spirit of our men. We hope and pray for the perpetuation of the Empire and gladly give our lives for that goal.
>
> To the Navy Minister and all my superior officers I tender sincerest appreciation and gratitude for their kindness of many years. At the same time, I earnestly beg you to give thoughtful consideration to the families of my men who fall at this outpost as soldiers of the Emperor.
>
> With my officers and men I give three cheers for the Emperor and pray for the everlasting peace of the Empire.
>
> Though my body decay in remote
> Okinawa,
> My spirit will persist in defense of the
> homeland.
>
> Minoru Ota
> Naval Commander[57]

Four days after the transmission of the above, Admiral Ota released his last dispatch to his immediate commander, General Ushijima:

> Enemy tank groups are now attacking our cave headquarters. The Naval Base Force is dying gloriously at this moment. . . . We are grateful for your past kindnesses and pray for the success of the Army.[58]

Marine artillerymen killed or dispersed a group of Japanese soldiers attempting to break out of their entrapment during the night 11–12 June,[59] and the 22d Marines dispatched 51 of the enemy attempting to infiltrate the regimental line. Obvious signs of a break in the enemy's stubborn and well-coordinated defense appeared on 12 June, when the 4th and 29th Marines compressed an already compact enemy pocket west of Tomigusuki, while the 22d pressed to the north in the direction of Oroku village. (See Map 20.)

The 4th Marines advanced slowly under heavy machine gun fire from positions in the hills and draws surrounding Hill 62 and from well-concealed caves on the hill itself. At 1225, the hill was captured and the attack continued with the Marines systematically cleaning out all pockets of resistance as they advanced. Three hours later, 3/4 tied in

---

[57] Inoguchi, Nakajima, and Pineau, *Divine Wind*, p. 147.

[58] *Okinawa Operations Record*, p. 126.

[59] *IIIAC Arty AR*, p. 30.

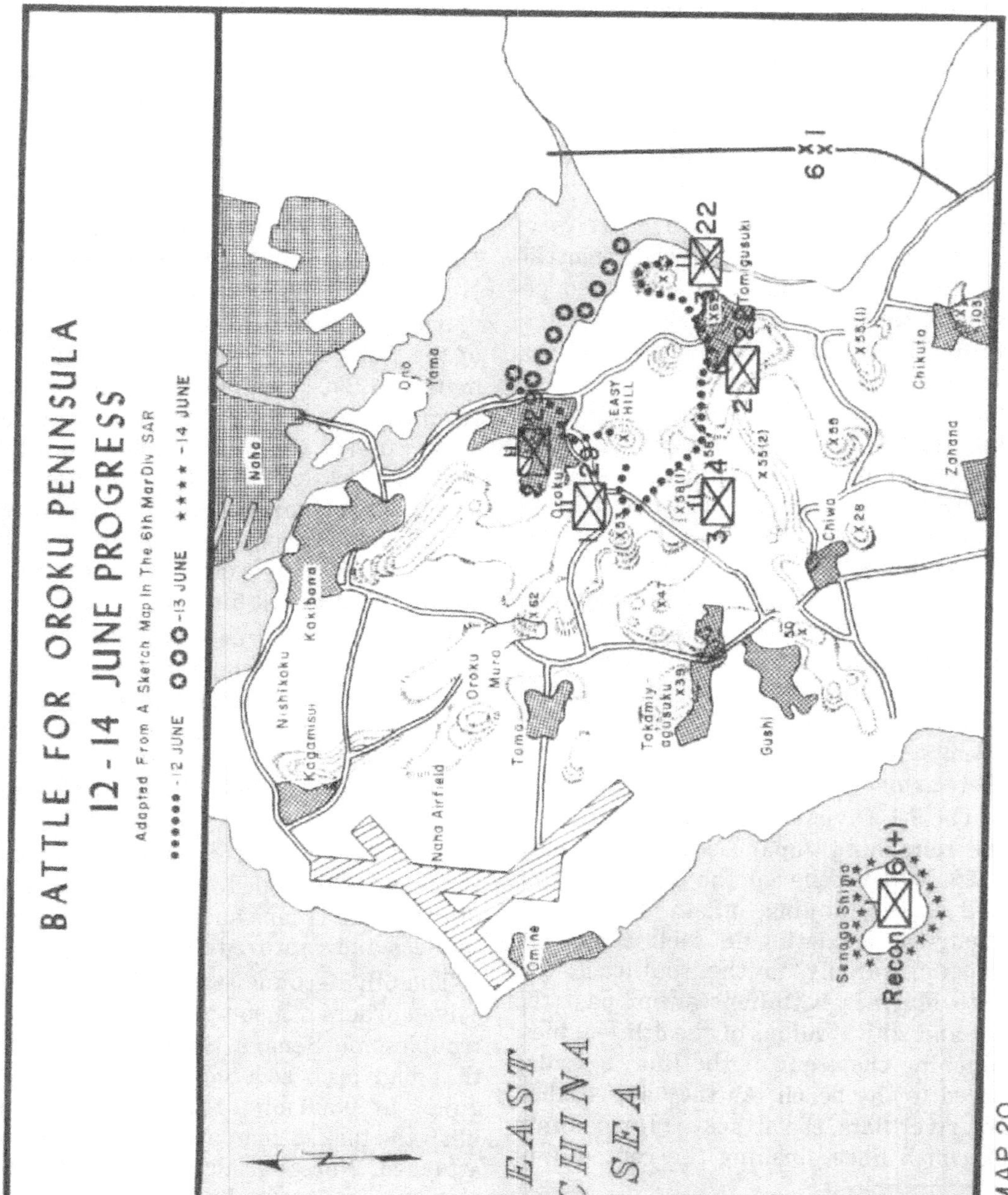

T. L. RUSSELL

MAP 20

with the 22d Marines for the night, 500 yards from Naha Bay and with only one more large hill to be seized.

The same day, the 29th Marines cracked open the firm enemy defenses that had held it up for almost a week. Oroku village was cleaned out by 2/29, as 1/29 began the first in a series of coordinated attacks at dawn to neutralize the enemy's mutually supported positions west of Oroku.[60] By late afternoon, Easy Hill—the last Japanese strongpoint in the zone of the 29th Marines—was taken. Having lost this key terrain feature, enemy troops were forced to flee to the alluvial flatlands along the river coast between Hill 53 and Oroku. At this time, they "began displaying flags of surrender. Language officers equipped with loudspeaker systems were dispatched to the front line areas to assist in the surrender of those Japanese who desired to [do so]. The attempt was partially successful, 86 enemy soldiers voluntarily laid down their arms."[61]

The 6th Division made a final sweep of the remaining Japanese-held area with 3/29, which relieved the 1st Battalion, and 2/29 jumping off to destroy all enemy still existing in their zone. Advancing rapidly to the southeast, the 29th Marines battalions swept past the 1st and 2d Battalions of the 4th Marines, pinching them out of the line; 3/4 also raced to the beach. As they approached the river flats, the attackers formed into skirmish lines, flushing the routed Japanese from the marshy grasslands along the river bank. A number of enemy soldiers gave themselves up, some committed suicide, others fought to the bitter end, and a few stoically awaited their deaths.

During the liquidation of this pocket, Colonel John C. McQueen, 6th Marine Division Chief of Staff, and Colonel Roy N. Hillyer, Tenth Army Chaplain, viewed the fighting from the north shore of the Naha Estuary at a point approximately 1,000 yards across the water from Oroku. "They saw the Marines come up over the high ground from the south and close in on the Japanese . . . The last survivor was a Japanese officer who calmly walked over to the seawall, sat down, lit a cigarette, and waited for the Marines to kill him."[62]

Marine assault troops reached the seawall at noon and spent the rest of the day ferreting out small enemy groups attempting to evade death or capture by hiding in the cane fields and rice paddies near the river. At 1750, General Shepherd reported to General Geiger that all organized resistance on Oroku Peninsula had ended.[63] During the day, 6th Division troops had killed 861 enemy soldiers and captured 78 prisoners.[64]

The 6th Reconnaissance Company received orders at noon on 13 June to seize troublesome Senaga Shima—the island that had been scouted the night of 10 June—at 0500 on 14 June. To accomplish the task, a company from 1/29 was attached. For four days preceding the assault, the island had been subjected

[60] After the 29th Marines first attack on 12 June, the last company of the 3d Battlaion still on the line was squeezed out and passed to 1/29 as reserve. *3/29 SAR, Ph III*, p. 5.

[61] *6th MarDiv SAR, Ph III*, p. 21.

[62] Smith, *Personal Narrative*, p. 127.

[63] *6th MarDiv Jnl, Ph III*, 13Jun45.

[64] *Ibid.; 6th MarDiv SAR, Ph III*, p. 21.

310-224 O - 69 - 22

to a heavy and continuous bombardment. At the scheduled time, the LVT-(A)-borne attack was launched and proceeded according to plan. There was no resistance to the landing. As the reconnaissance Marines combed the island, they found only dead bodies and silenced guns—all victims of the intense prelanding preparation.

The battle of Oroku ended on 14 June. General Shepherd noted that:

> The ten-day battle was a bitter one, from its inception to the destruction of the last organized resistance. The enemy had taken full advantage of the terrain which adapted itself extraordinarily well to a deliberate defense in depth. The rugged coral outcroppings and the many small precipitous hills had obviously been organized for defense over a long period of time. Cave and tunnel systems of a most elaborate nature had been cut into each terrain feature of importance, and heavy weapons were sited for defense against attack from any direction.
>
> Despite the powerful converging attack of three regiments, the advance was slow, laborious, and bitterly opposed. The capture of each defensive locality was a problem in itself, involving carefully thought out planning and painstaking execution. During ten days' fighting, almost 5,000 Japanese were killed and nearly 200 taken prisoner. Thirty of our tanks were disabled, many by mines. One tank was destroyed by two direct hits from an 8" naval gun fired at point blank range. Finally, 1,608 Marines were killed or wounded.[65]

A most noteworthy aspect of the Oroku operation was the ability of the Tenth Army to exploit the amphibious capability of one of its Marine divisions during a critical phase of the Okinawa campaign despite the extremely limited time available for assault preparations. Overcoming most obstacles and discounting others, the 6th Marine Division planned and launched an amphibious assault within the 36-hour period allotted to it.[66] In an after-action analysis of the operation, General Shepherd stated that "with trained troops and competent staffs in all echelons, the amphibious landing of a division is not of excessive complexity."[67]

---

[65] *Ibid.*, p. 22.

[66] *Shepherd memo II.*

[67] *6th MarDiv SAR, Ph III*, p. 60.

Battle's End

ON TO KUNISHI RIDGE[1]

At the same time the 6th Marine Division was landing on Oroku Peninsula, the 1st Marine Division was rolling up gains totalling 1,800 yards in its drive south from the Naha-Yonabaru valley. (See Map IX, Map Section.) General del Valle's regiments made this advance while a faltering division supply system behind them threatened to break down completely because of the mud and the rain. The roads had become such quagmires that even tractors and bulldozers became stalled when they attempted to drag division vehicles out of or over the mud. Tanks and trucks were unable to cross the Kokuba; the approaches to the bridge at the mouth of the river were untrafficable for a distance of over 500 yards.[2] In an effort to facilitate resupply and evacuation operations, tanks were ordered off the roads. In general, forward units were logistically supported by Marines who hand carried supplies up to dumps behind the lines; the "trails were only negotiable for foot troops—vehicles could not have been used if we [2/7] could have gotten them across the inlet." [3]

On the division right on 4 June, the 7th Marines pushed forward to close off the neck of Oroku Peninsula and further entrap Japanese forces there. The hill mass at the base of Oroku in the division zone held the commanding terrain feature of the area, Hill 108. This height overlooked the East China Sea and the next major division objective, Itoman. Although the exposed right flank of the 7th Marines came under constant harassing fire from high ground to the right of the division boundary, the division left flank was generally secure since the adjacent 96th Infantry Division had moved forward steadily since its advance from the Kokuba River line.

After the Japanese defenses at Shuri had collapsed, the 1st Marines remained behind in the vicinity of the city to patrol and mop up, and the 5th Marines pursued the fleeing enemy. Before the dawn of 4 June, the 1st joined in the pursuit; 3/1 made a wide swing through the zone of the 96th Division in order to take the high ground north of Iwa and Shindawaku while 1/1 passed through the lines of the 5th Marines and took up positions in front of Hills 57 and 107.

By 0930, 3/1 had reached the small village of Tera, just north of Chan.[4] At

---

[1] Unless otherwise noted, the material contained in this section is derived from: *Tenth Army AR; IIIAC AR; 1st MarDiv SAR; 1st MarDiv G–3 Jnl; 1st Mar SAR; 5th Mar SAR; 7th Mar SAR; 22d Mar SAR, Ph III; 7th Mar Hist;* MajGen Pedro A. del Valle, "Southward From Shuri," *Marine Corps Gazette*, v. 29, no. 10 (Oct45), hereafter del Valle, "Southward From Shuri."

[2] *1st MarDiv G–3 Jnl*, 4Jun45.

[3] *2/7 SAR*, p. 7.

[4] This Tera is not to be confused with a second village of the same name located near Itoman on the east coast.

1300, the battalion point was pinned down by fire coming from high ground just west of the Tomusu-Iwa road, and the advance guard attempted without success to clean out the enemy position. Just before 1400, the time scheduled for 3/1 to make its coordinated attack with 1/1, a cloudburst occurred. The supply problems here were further aggravated by the rain, and because of a communications blackout between the battalion and its artillery and naval gunfire support, 3/1 broke contact with the enemy and withdrew to a bivouac area in a draw behind the 383d Infantry.

At 1730, Lieutenant Colonel Ross' entire battalion was in defilade, protected from enemy artillery fire. Contact with the artillery battalions and naval gunfire support ships was still lacking at this time and a mortar ammunition shortage existed. The afternoon downpours had turned the roads into morasses and the fields into calf-deep mud wallows in which the suction of the ooze pulled the soles off of the shoes of men walking in it.

Since food as well as mortar ammunition was in short supply, the 383d Infantry generously supplied the battalion with enough K-rations to enable 3/1 to issue two meals to each Marine.[5] It was the general consensus of the members of 3/1 that "taking all things into consideration, this day probably was the most miserable spent on Okinawa by the men of this battalion."[6] In addition, 3/1 found itself all but isolated from its regiment, since there was neither communication with nor a supply route to the 1st Marines CP, some 11,000 yards to the rear.[7]

The 1st Battalion, 1st Marines, passed through the lines of 2/5 at approximately 1000 on the 4th. This was nearly three hours after Company F of 2/5 had attacked and seized Hill 107 without opposition, and completed its occupation of the high ground across the entire front of the regiment. When 1/1 took over from 2/5 at noon, the latter passed into corps reserve with the rest of the 5th Marines, but maintained its positions as a secondary line.[8]

Although the downpour on 4 June had forced General del Valle to cancel the attack of 1/1 scheduled for 1400 that afternoon, the 7th Marines on the right had already jumped off. An hour later, the cancellation order was rescinded, and Lieutenant Colonel Shofner's battalion was again ordered to attack, to contact the 7th Marines, and to seize its original objective—the high ground north of Iwa and Shindawaku.

At 1630, the assault companies of 1/1 moved out to secure their target, some 1,500 yards away. The route of attack was up a valley floor, at the end of which a number of lesser hills rose in front of the objective. A creek that ran east to west across the valley was not visible from the LD; but a map reconnaissance indicated that the assault

[5] The commanding general of the 96th Division believed that these Marines "were not equipped or organized for a protracted campaign. I was glad to assist in supply, air drops, and the care of their wounded. They were fine comrades and cooperated to the fullest extent." MajGen James L. Bradley ltr to CMC, dtd 22Oct54.

[6] *3/1 SAR*, p. 41.

[7] *Ibid.*

[8] *2/5 SAR*, p. 16; *5th Mar SAR*, p. 9. At this time, the 5th Marines CP was moved to the vicinity of Giushi.

forces would be able to cross it with little difficulty. The Marines met no opposition after jumping off until reaching the "creek," now swollen into a raging torrent by the day's rains. It presented a formidable barrier to further progress. A reconnaissance of the stream banks uncovered a rudimentary bridge for carts to the left of the battalion position.[9] The assault troops were ordered to move upstream, cross the bridge, and redeploy on the other side.

As soon as the first Marine elements had crossed and were wallowing in mud towards firm ground, the heretofore-silent Japanese opened up with mortars and point-blank machine gun fire, sweeping the ranks of the onsurging troops. The Marines pushed on, nonetheless, and two platoons made it across to the south bank of the creek, only to become pinned down.

The 7th Marines on the right was unable to negotiate the swiftly flowing waters and was held up on the north bank,[10] and the bridge-crossing site was fully covered by enemy defensive fires coming from a 200-foot-high ridge in front of 1/1. Therefore, the battalion commander ordered his troops to withdraw to the sector of 2/5 for the night.[11] Because 1/1 had sustained a number of casualties, a covering force remained behind to evacuate the wounded after dark. The next morning, the 1st Battalion was ordered to bypass the enemy strongpoint by swinging into the zone of the 96th Division and follow closely in the trace of the 3/1 attack on Iwa.[12]

Colonel Mason anticipated the problem of maintaining radio and wire contact with his battalions as they raced south. His movement order provided that, in case of a complete communications breakdown between regiment and the assault battalions, the most senior battalion commander of the committed units would assume tactical command overall until contact was established with regiment once more.[13] Following a mud-slogging and wearying march south on 5 June, 50 men from 1/1 dropped out of ranks from exhaustion. During the trek, the battalion lost contact with regimental headquarters for a brief time and temporarily came under control of Lieutenant Colonel Ross.[14]

Out of contact with regimental headquarters from the time he had led his battalion south from Shuri, and with the battalion objective yet uncaptured, Lieutenant Colonel Ross decided to complete his mission nevertheless. He took his command group forward early in the morning of 5 June to make a visual reconnaissance of the target. While this inspection was taking place, the Marines of 3/1 built fires in an attempt to warm themselves and dry as much of their clothing as possible before mounting the attack. Prior to the jumpoff, 3/1

[9] LtCol Austin C. Shofner interview by HistDiv, HQMC, dtd 19Mar47 hereafter *Shofner interview*.

[10] "There was no bridge in the 3/7 ZofA [Zone of Action]. Several men were drowned attempting to carry lines across the stream in an attempt to get troops across." LtCol Walter Holomon ltr to CMC, dtd 22Mar55, hereafter *Holomon ltr*.

[11] *Shofner interview*.

[12] *1/1 SAR*, p. 21.

[13] Col Arthur T. Mason ltr to CMC, dtd 10Mar47.

[14] *1/1 SAR*, p. 21.

received 19 supply air drops from VMTB–232 aircraft. In between the day's intermittent showers, the squadron flew a total of 41 resupply sorties; its all-time high to that date.[15] The battalion departed its bivouac area at 1030 and arrived at the assembly area shortly thereafter; Lieutenant Colonel Ross then issued his attack order for the capture of the Iwa-Shindawaku ridge.

Before jumping off at 1230, 3/1 learned that patrols from 2/383 had passed through Iwa without opposition. As soon as the Marine attack began, lead elements were held up for a short time by sporadic machine gun and sniper fire, but took the ridge before dark. In the two days spent to envelop the objective, the battalion had travelled more than 3,000 yards. By this time, the advance CP of the regiment had moved far enough forward to enable Colonel Mason to issue attack orders personally to his battalion commanders.

The plan for the next day's attack called for 3/1 to continue the advance and seize Shindawaku. The 1st Battalion would destroy all bypassed enemy pockets in the regimental zone and to the rear of 3/1, and would backtrack to the stream where the 4 June attack had been stymied.

To the relief of all, the rain stopped during the night of 5–6 June. At dawn of the 6th, 1/1 moved out of its bivouac east of Iwa, swung down to the village, and then turned north. At this point, all three of its infantry companies formed a battalion skirmish line over an extremely wide front. The Marines then swept northward and past the zone where 3/7 was preparing to attack in a southwesterly direction. Lieutenant Colonel Shofner's troops accomplished their sweep at 1400 and then attacked and seized the ridge overlooking the stream. The few enemy soldiers still manning positions on this objective, not expecting an attack from the rear, were surprised while changing into civilian clothes. After taking the position with little effort, 1/1 went into reserve near Tomusu.[16]

Because 3/1 had not been resupplied before its attack at 0900 on the 6th, the 383d Infantry again issued the Marines K-rations; this time, enough to provide each man in the battalion with one and a half meals.[17] After jumping off, 3/1 advanced west and reached the outskirts of Shindawaku at 1030, when enemy troops were discovered occupying commanding ground on the ridge running northwest from the village. By 1800, however, the battalion had secured the ridge after a brief fight and 2/1 had moved to an area northwest of Iwa. Although the left flank of 3/1 was tied in with 2/383 for the night, the Marine battalion had not been able to contact the 7th Marines on the right. Early the next morning, 2/1 was moved into position to plug this gap.

During its drive south, the 1st Marine Division was sporadically halted for brief periods before a number of blocking positions organized and manned by small enemy groups. Each of these

[15] VMTB–232 WarD, Jun45. The squadron parachuted more than 20 tons of supplies to 1st Marine Division troops on 5 June. *1st MarDiv SAR*, chap VII, OpAnx, p. 10.

[16] *1/1 SAR*, p. 21.

[17] *3/1 SAR*, p. 43.

groups was generally the size of a company, and all of them together comprised a force equaling no more than two battalions. The Japanese holding units had been ordered and were determined to delay the Tenth Army as long as possible. The tactical situation and the nature of their mission, however, prohibited their setting up anything more permanent and stronger than hastily contrived defensive positions, which were unable to hold back the aggressive Marine offensive for long.

When reconnaissance patrols uncovered these strong points, infantry commanders deployed their forces to take the objective by a combination of fire and maneuver. In most cases, the major attack force maneuvered into position to assault the objective from its flank or rear. At this time, Marine elements in front of the target supported the attack by firing on the objective to keep the enemy fixed in position. At times, the enveloping force provided fire support for a frontal attack. Regardless of the methods employed, the weather situation, and the condition of the terrain, General del Valle felt that "it was refreshing to be able to maneuver again, even on a modest scale." [18]

On the critical right flank, the 7th Marines paced the division advance on 5 June with the 2d and 3d Battalions attacking against increasing opposition; 1/7 followed behind, mopping up the rear area. Acting as a screen to the right of 2/7 along the division right boundary, the 1st Reconnaissance Company dispatched patrols far ahead of the battalion advance, which sent back invaluable information. The company, however, found that its operations were severely restricted by its limited communications system and supply organization.[19]

Like the 1st Marines, the 7th found the enemy less difficult than such other problems as those caused by the weather and the terrain. Marine wounded were evacuated in the rain over a five-mile sea of mud; sniper fire generally harassed the 8 to 10 litter bearers required for each casualty during the entire trip to the rear. Each day's attack was usually delayed until the weather was clear enough for land- and carrier-based [20] planes to make a supply drop; so many sorties were flown for the 7th Marines as it trekked southwards that the trail of the regiment was blazed with brightly colored cargo parachutes.[21]

The initial attack of the 7th Marines southwards from the Kokuba River bridgehead on 4 June gained the regiment approximately 1,100 yards. That same day, 2/7 captured Takanyuta. On the next day, the formerly raging torrent in front of 3/7 had receded to uncover a causeway over which part of the battalion crossed; the remainder moved to the zone of the 2d Battalion and crossed the stream from there.[22] Once beyond the south bank, the assault

[18] del Valle, "Southward From Shuri," p. 39.

[19] *Snedeker ltr 1947.*

[20] During the course of the entire campaign, carrier-based planes often joined the Avengers of the two TAF VMTBs in making supply drops. Air Delivery Section, H&S Bn, Corps Troops, IIIAC, ICEBERG Op AR, dtd 24Jun45, encl A, hereafter *AirDelSec AR.*

[21] *7th Mar Hist*, pp. 22–23.

[22] *Holomon ltr.*

battalions of the 7th drove forward 1,000 yards to a point just north of Hanja village.

When furious machine gun and mortar fire from a hill mass in the zone of the 6th Division held up the 7th Marines, General del Valle received permission from IIIAC to lay the artillery fire of the 11th Marines on the suspected enemy positions. General Shepherd was authorized to cancel the fire when it threatened his troops. As the 1st Division continued its drive past the neck of the Oroku Peninsula, expanding the already-lengthy right flank of the division, 1/22 was ordered into defensive positions along this flank.

On the next day, 6 June, the 22d Marines battalion had not yet occupied its assigned flank security positions. It became necessary, therefore, to order 1st Division troops into the 6th Division zone to capture Hill 103 and destroy the enemy automatic weapons and mortars harassing the right flank of the 7th Marines. Lieutenant Colonel Berger's battalion had already attacked and was, in fact, within a few yards of the crest of the hill when elements of the 22d Marines arrived. Reorienting the direction of its attack to the south towards Hill 108, 2/7 advanced 1,000 yards before encountering stiff opposition near Dakiton, where it dug in for the night. On the left, the 3d Battalion pushed to the high ground southeast of the same town and likewise dug in.

Clearing skies on 7 June heralded a 1st Division success in breaking through to the coast that day and isolating Admiral Ota and his ill-fated troops on Oroku from the rest of the doomed *Thirty-second Army* in the south. Following up a thorough combined arms preparation, 2/7 overran Hill 108 to command a view of the island south to Kunishi. The former defenders of 108 were seen fleeing south in small groups ranging in size from 10–20 men each. The fire of Marine support weapons and machine guns relentlessly pursued the Japanese troops, killing many. After receiving an air drop of supplies, 3/7 attacked at 1430, overran Hanja, made contact with 2/1 on its left, and dug in for the night on a ridge just north of Zawa.

Following receipt of still another supply air drop early on 8 June, 3/7 resumed its attack with a sweep through Zawa as advance elements of 2/7 probed the Japanese positions guarding Itoman. Besides positioning the division for a final drive south, the breakthrough to the seacoast uncovered beaches on which LVTs could land when a waterborne supply system was established. When the first LVTs touched down on the coast approximately 500 yards north of Itoman shortly after noon on 8 June, General Hodge congratulated General del Valle "for cutting the island in two." [23] Use of this new water route brought in enough rations to permit distribution of the first full issue to 7th Marines troops in more than a week. As the weather improved, some vehicular traffic appeared over slowly drying roads in the south. A few new bridges were constructed across the onceswollen streams in the north to help speed supplies of all sorts to assault troops driving to the southern tip of the island.

---

[23] *1st MarDiv G–3 Jnl,* 8Jun45.

Advancing abreast of and pacing the march of the 7th Marines to the sea on 7 June, the 1st Marines also reported substantial gains. Early in the morning, 2/1 filled the gap existing on the right between 3/1 and 3/7, while 3/1 maintained contact with 3/383. By 1800, 2/1 was in possession of the height overlooking Zawa, and the 3d Battalion had moved 1,200 yards along the corps boundary to occupy the high ground 1,000 yards north of Yuza; 1/1 had moved to Iwa preparatory to relieving 3/1.[24]

All infantry battalions had been plagued by the supply situation, but it seemed to members of 2/1 that they had been especially dogged since leaving Shuri. Their only source of supply had been the air drops, and by the time that one could be made, the assault companies were several thousand yards forward of the drop zone.[25] The supplies then were recovered by headquarters personnel, who carried them to forward dumps. At this point, Marines from the reserve company would pick up the supplies and carry them to assault units.[26]

Sustained by supplies brought ashore by the LVTs, in the continuing good weather of 8 June, 1st Division troops pushed ahead against perceptibly stiffening resistance. The 1st Marines rolled forward; 3/1 secured its objective near Yuza at 1600, when it was relieved by 1/1 and went into regimental reserve near Shindawaku. Slightly later that day, 2/1 secured the high ground overlooking the Mukue River. On 9 June, division assault units spent the day probing enemy positions to their front in preparation for a major attack on the 10th.

Improved weather conditions and correspondingly better road nets over which supply convoys could travel served to release the VMTBs for other assignments. Following 6 June, when VMTB–232 made 49 drops, ground units requested paradrop missions on only eight other days in the rest of the month.[27] By this stage of the campaign, the Marine pilots had become quite proficient and accurate in paradrop operations. In reference to a drop Major Allen L. Feldmeier's VMTB–232 had made on 8 June to its soldiers, the 383d Infantry sent him the following message: "Your drops have excellent results. We received 95 of the 97 packs which you dropped." [28] Later in the month, VMTB–131 flew 3 missions totalling 20 sorties in which 70 packs—each averaging 1,000 pounds of food and ammunition—were dropped. Ground units receiving the supplies re-

[24] *1st Mar SAR*, p. 21.

[25] *2/1 SAR*, p. 13.

[26] Since a means of delivering water rations by air drop had not been perfected, the troops were given permission to drink stream water only after each Marine had made the contents of his filled canteen potable by treatment with halazone. Halazone is a white crystalline chemical compound used to disinfect water suspected of being or known to be impure. The compound was compressed into tablets and became a major medical supply item in the war.

[27] VMTB–232 WarD, Jun45.

[28] "A History of MAG–22 in the Okinawa Campaign," Anx A in MAG–22 WarD, Aug45. VMTB–131 arrived on Okinawa on 29 May when it began antisubmarine warfare patrolling, its primary mission. During the later stages of the campaign the squadron made some supply air drops.

ported that they had recovered 90 percent or more of the packs.

Increased enemy opposition arose on 9 June as 1st Marine Division units approached the Tera-Ozato area, which had been outposted by the *Thirty-second Army*. Patrols from both the 1st and 7th Marine received heavy rifle and machine gun fire while attempting to cross to the south bank of the Mukue Gawa. Small infiltration groups finally forded the stream, but were unable to advance beyond the bank. The 2d Battalion, 1st Marines, sustained moderate casualties during this day's fighting, but was unable to evacuate them until after dark because accurate enemy fires covered evacuation routes.

In the 7th Marines zone, 1/7 relieved 3/7, which then went into regimental reserve. No appreciable gains were made as 1/7 companies mounted two unsuccessful attempts to seize high ground overlooking Tera. The second effort was repulsed by extremely heavy small arms fire, which forced the assault elements to withdraw under the cover of a smoke screen. On the extreme right of the division, 2/7 patrols crossed the Mukue Gawa and attempted to seize the ridge north of Itoman,[29] but were thwarted by enemy fire coming from emplacements fronting the 1st Battalion zone. One platoon of Company E was able to get to the far side of the river where it was pinned down immediately by accurate frontal and flanking fire.

Operating a combination CP-OP while aboard an LVT(A) floating 100–200 yards offshore of the battalion flank, Lieutenant Colonel Berger had a grandstand view of the fighting. When he saw that the advance platoon was pinned down, he went ashore to order the rest of the company to cross at the river mouth and reinforce the stricken unit. Steady Japanese machine gun fire prevented the Marines from wading across and shortly thereafter denied passage to troop-laden LVTs attempting the same route. At nightfall, the battalion commander ordered the exposed units to withdraw to the northern bank of the Mukue River under the cover of LVT-(A) fire.

The 2d Battalion jumped off on the 10th with Companies F and G passing through the night defenses of Company E, dropping onto the beach from the top of a 10-foot-high seawall, and wading 400 yards across the stream mouth to a point on the south bank opposite the ridge. Shells from LVT(A)s pounded this high ground and Itoman beyond it. Following this preparation, the assault troops scaled the seawall to attack these two objectives. Although the battalion lost five officers in the first seven minutes of fighting,[30] the onsurging Marines swept over the ridge, through the ruins of Itoman, and on to the high ground beyond the southern edge of the town.

At the same time that the 2d Battalion had crashed through Itoman, 1/7, spear-

[29] Actually, the town of Itoman was south of the Mukue Gawa, although the 1:25,000 battle map used during the campaign indicated that the large inhabited area north of the river was Itoman. *Snedeker ltr 1947*, encl D; *2/7 SAR*, p. 8. The latter spot was an undefended suburb of Itoman and did not hold up the advance of 2/7. Capt Verle E. Ludwig interview by HistBr, G–3 Div, HQMC, dtd 27Jan55, hereafter *Ludwig interview*.

[30] *2/7 SAR*, p. 8.

headed by Company A, made a rapid and unopposed rush to the crest of the hill north of Tera, where the Japanese strongpoint that had opposed the 7th Marines on the previous day was located. From this newly gained height, the battalion called for an artillery concentration on the high ground immediately east of Tera. Battalion 81mm mortars blistered the village with a barrage of white phosphorous shells, burning to the ground all buildings still standing. Although few Japanese troops were found in the area, numerous dazed civilians, who had miraculously escaped death in the bombardment, were discovered wandering aimlessly among the ruins. After sending the Okinawans to stockades in the rear, 1/7 prepared night positions and organized for the scheduled 11 June attack on Kunishi Ridge. (See Map 21.)

By 10 June, the rains had ended and the transportation problem, although not so critical as it had been previously, was still not completely alleviated. The ground was drying and once-overflowing streams had lowered to a point where the road-construction and bridge-building efforts of the engineers could open the way to tracked and wheeled vehicles. Division engineers had converted oil drums into culverts and built tank fords over the fire-swept Mukue Gawa. In addition, the engineers provided round-the-clock maintenance for these fords so that the heavy traffic south could continue unabated. One especially important ford was built at the point where the Zawa-Tera highway crossed the Mukue.[31]

The task assigned to 1/1 was the capture of Yuza Hill, the high ground approximately 700 yards west of Yuza and about 450 yards directly north of Ozato; 2/1 was to support the attack by fire, while units of the 96th Division were to provide security to the left flank of the 1st Marines. When the tank fords over the Mukue were opened on 10 June, Shermans also moved forward to support the 1st Marines attack.

Another support element, the 11th Marines, assisted the infantry assault. Following a rolling barrage, tank-infantry teams from 1/1 swept onto the western nose of the hill and Company C swarmed up to its crest in the face of blazing enemy machine gun and artillery fire. The company lost 70 of its 175 men in this charge.[32] Lieutenant Colonel Shofner's attack plan called for Company B to follow in the left rear of the lead elements and then to attack straight up the hill after first having worked its way through Yuza into jump-off positions. Upon reaching the crest, B was to tie in with Company C on its right and with the Army units on the left. Although the latter had begun the attack abreast of 1/1, Japanese troops entrenched in the extremely well-fortified Yuza Dake escarpment prevented the soldiers from advancing beyond their line of departure. Company B was unable to move forward because of the intense artillery and mortar fire coming from the front of the Army zone, and could not help Company C, which was isolated in an exposed and extremely tenuous position.

Late in the afternoon, Shofner sent Company B around to the right to join

[31] *1st MarDiv SAR*, chap VII, pt III, p. 13.

[32] *1/1 SAR*, p. 21.

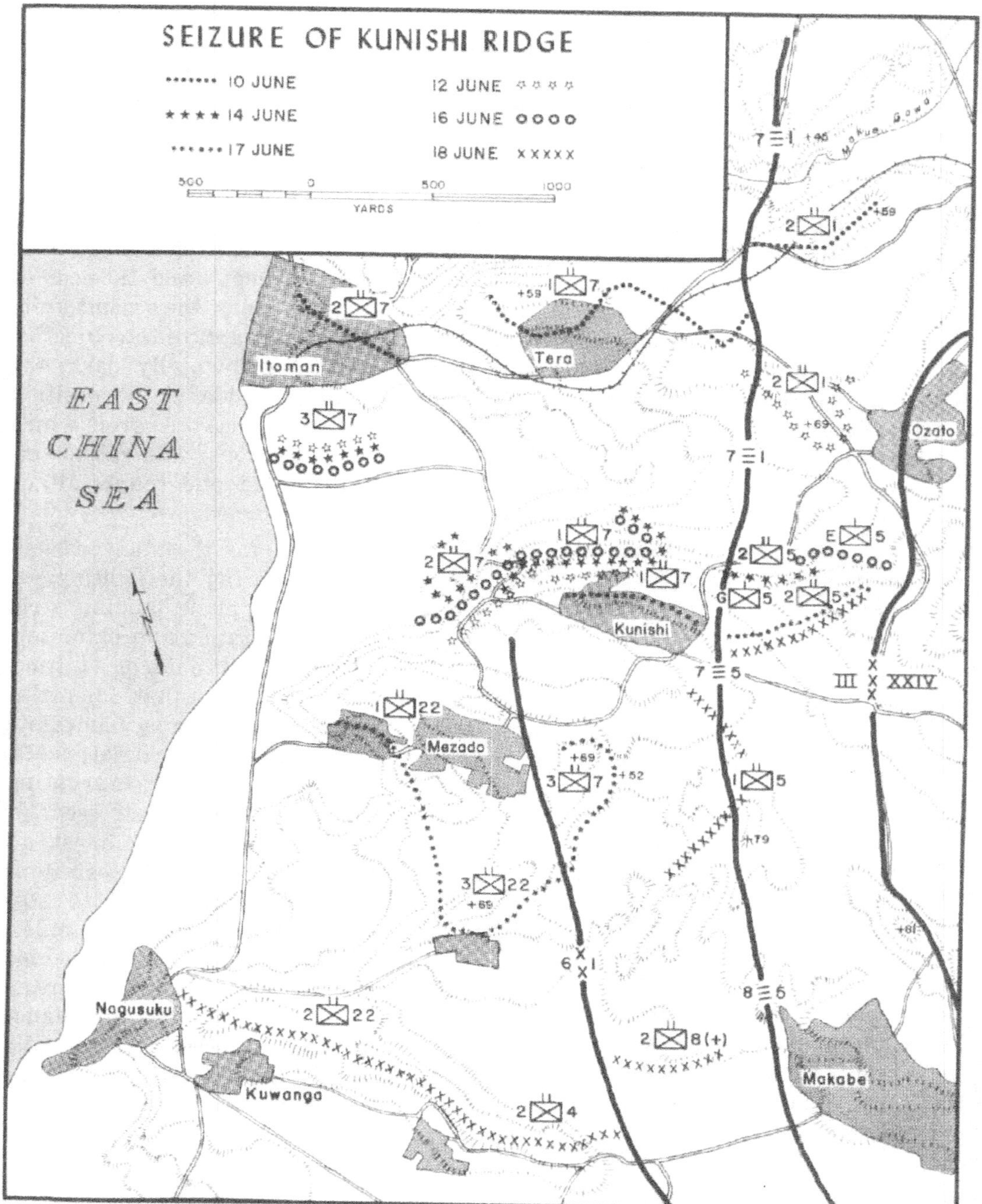

MAP 21

T.L. RUSSELL

the company on the hill. Both assault companies had sustained heavy casualties in the fighting, but C was hit hardest; all of its officers were either dead or wounded. More Marines were lost during the night, 10–11 June, as a result of the constant deluge of enemy mortar and artillery fire placed on the hill. Added to this heavy toll were the casualties caused by grazing machine gun fire coming from weapons emplaced on Yuza Dake. Twenty more men were wounded in the dawn of 11 June, when at 0400, the Japanese mounted an unsuccessful counterattack. For the next two days, 1/1 waited on Yuza Hill for the 96th Division infantry to reduce the escarpment to the east. The Marine battalion had little trouble in maintaining its hill position despite the persistent Japanese artillery fire harassing it the entire time.

While the 1st Battalion fought to gain its objective on the 10th, 2/1 with the help of armor support successfully cleaned the enemy out of the commanding ridge between Tera and Yuza. The next day, because 1/1 had been stymied on Yuza Hill, Lieutenant Colonel Magee's battalion was ordered to capture Hill 69, the commanding terrain feature directly west of Ozato. At 1030, the closely coordinated tank-infantry-artillery attack began when the battalion moved out in a column of companies. Initial progress was rapid, but when the infantry vanguard entered the valley leading to Ozato, well-placed Japanese mortar and artillery concentrations caused many Marine casualties. As the left flank of the spearhead approached Ozato, enemy machine gun and rifle fire began mowing down the attackers.

These mounting losses gradually slowed the momentum of the Marine assault, which the battalion commander attempted to revive by placing a second company in the line of attack to the right of the first. Despite the increasing volume of the massed fires provided by 2/1 supporting arms, enemy fire continued unabated. Disregarding their slowly ebbing strength and the loss of three supporting tanks, the assault units surged forward to capture the objective shortly before sundown. By dark, the battalion had consolidated the position and set in night defense lines from which it repulsed numerous infiltration attempts and blunted one counterattack before 11 June dawned.

In preparation for the many casualties anticipated during the fighting yet remaining, a light plane landing strip was placed into operation approximately 2,000 yards north of Itoman on 11 June. It was noted at the time that, for maintaining morale and obtaining immediate medical treatment for critical cases, "the value of this means of evacuating casualties cannot be overstressed."[33] Now casualties were flown almost directly from medical clearing stations immediately behind the front to hospitals in the rear, a distance of 12 miles, in an average time of 8 minutes. This brief flight obviated a long and often body-racking haul in an ambulance jeep over roads that were practically nonexistent. From 11 to 22 June, VMO-3 and -7 flew out 641 casualties from this strip.[34]

---

[33] *1st MarDiv SAR*, chap VII, Logistics Anx, p. 9.

[34] *Ibid.*

MARINE CASUALTIES *are evacuated by a spotter plane from a temporary airstrip north of Itoman. (USN 80-G-498161)*

8TH MARINES *landing on Iheya Shima. (USMC 126988)*

On the same day the landing strip became operational, Colonel Snedeker's 7th Marines advanced 400 to 1,000 yards against ever-stiffening enemy opposition. The 1st Battalion, having cleared Tera, attacked to gain the high ground immediately south of the village. After mopping up in Itoman, 2/7 pushed 500 yards southward. Confronting the regiment now, approximately 800 yards equidistant from the outskirts of Tera and Itoman, was Kunishi Ridge, to be "the scene of the most frantic, bewildering, and costly close-in battle on the southern tip of Okinawa." [35]

BATTLE FOR KUNISHI RIDGE [36]

Running from the northeast to the southwest for a distance of perhaps 1,500 yards, the sheer coral escarpment of Kunishi Ridge held Japanese positions which comprised the western anchor of the last heavily defended line in front of Kiyamu. Both the forward and reverse slopes of the ridge were replete with caves, weapons emplacements, and fortified tombs, all of which reinforced natural defenses provided by the complex and difficult terrain features of the ridge itself. In front of the 7th Marines line, a broad valley containing grassland and rice paddies led to this crag and afforded the defenders unobstructed lanes of fire and the attackers little cover and concealment. Approaching tanks would fare no better than the infantry since they were restricted to two routes leading into the objective area—both covered extremely well by Japanese antitank guns. One road followed the coast line; the second cut across the center of the ridge at a right angle, dividing it. (See Map 21.)

Having pushed through Itoman and Tera during the morning of 11 June, 2/7 and 1/7 prepared to continue on to Kunishi Ridge. Immediately after midday, tank-infantry teams from both assault battalions moved out towards the objective. Two hours later, withering frontal fire from the ridge, enfilade fire from the yet-uncaptured Hill 69, and accurate artillery concentrations on the tanks forced the attack to a halt. Because of these fires, and more coming from Japanese-held Yuza Dake, the commander of the 7th determined that it would be too costly to continue the attack in the daylight, so at 1447, he ordered the assault forces to withdraw. After making an aerial reconnaissance of the ridge in a low-flying observation plane,[37] Colonel Snedeker concluded that a night attack would be the course of action most likely to succeed.[38]

That afternoon, as the battalions dug in a night defense and prepared to continue the attack on the following day, the two assault battalion commanders were thoroughly oriented on the general

[35] Appleman, *et. al.*, *Okinawa Battle*, p. 451.

[36] Unless otherwise noted, the material contained in this section is derived from: *1st MarDiv SAR; 1st MarDiv G–3 Jnl; 1st Mar SAR; 5th Mar SAR; 7th Mar SAR; 7th Mar Hist; 22d Mar SAR, Ph III.*

[37] Maj John S. Hudson ltr to CMC, dtd 27Mar47.

[38] General del Valle had visited Snedeker's CP that day "in order to see how we could break the deadlock, stop our heavy losses and get on with the war. When he suggested a battalion night attack for a limited objective, I agreed." LtGen Pedro A. del Valle ltr to Asst G–3, HQMC, dtd 1Oct65, hereafter *del Valle ltr 1965*.

scheme of maneuver at the regimental CP. Colonel Snedeker decided to attack straight across the valley, using the road leading into the ridge as the boundary separating the battalion zones and the telephone poles bordering the road as a guide. The assaulting battalions were to penetrate the enemy defenses at the point where the road entered the ridge. There the battalions were to peel off to their zones of attack and roll up the enemy's line. Until the hour of attack, 0330, on 12 June, normal artillery fires would be placed alternately on Kunishi Ridge and then Mezado Ridge (500–600 yards southwest of Kunishi), and thereafter only on the latter. In order to maintain deception and guarantee that the enemy would be surprised, the division issued an order prohibiting the use of flares and illumination of any kind—except in emergencies—after 0245.[39]

Before the night attack began, however, the Tenth Army decided to employ another type of weapon. Prior to and following the 1 April assault landings, the Japanese on the island had been subjected to a massive psychological warfare effort in which propaganda leaflets were delivered by aircraft and artillery shells. Also, Japanese-language broadcasts were directed at the enemy over loudspeakers placed near the front lines.[40] For a period of several days preceding 11 June, this war of paper and words had been accelerated and an emphasis placed on the hopelessness of the Japanese position and the futility of continued fighting. Both the leaflets and the broadcasts called upon General Ushijima to surrender.

[39] *1st MarDiv G–3 Jnl,* 11Jun45.

[40] *Tenth Army AR,* chap 11, pt II, p. 21.

On the afternoon of 11 June, General Buckner sent a Tenth Army reception party, fully empowered to negotiate with any Japanese parley group, to the 2/7 observation post overlooking Itoman. At 1700, all American firing ceased in the 7th Marines zone in dubious but hopeful anticipation of an enemy party bearing white flags. No such group appeared, although six Japanese soldiers did surrender about an hour later to Marines in the lines. The battlefield's unnatural silence was shattered at 1804 when hostile mortar fire fell on the surrender point and American artillery resumed fire on Kunishi in answer.[41]

Both battalions were poised to attack Kunishi Ridge with one company leading the assault. At H-Hour minus 1 (0230), Companies C and F proceeded to assembly areas and contacted each other on the line of departure. At 0500, when reinforcing Companies B and G moved out in their trace, the two assault units had already reached the crest of the ridge, achieving complete surprise. Company C, for example, destroyed several small enemy groups in the act of preparing their breakfasts.

[41] *1/7 SAR,* p. 20. General Ushijima did not immediately receive the message affording him an opportunity to surrender to General Buckner. It arrived at *Thirty-second Army* headquarters on 17 June, a week after it first had been dropped behind Japanese lines. Colonel Yahara stated that this was the normal amount of time required for a message to be passed from the front to the rear at this stage of the campaign. *Yahara Interrogation.* Upon delivery of the Buckner communique, "Cho and Ushijima both laughed and declared that, as *Samurai,* it would not be consonant with their honor to entertain such a proposal." *Shimada Interrogation.*

At daybreak, while en route through the valley to reinforce the Marines digging in on the ridge, intense enemy fire caught the two follow-up companies and pinned them down. It became painfully apparent that the Japanese defenders had quickly recovered from their initial setback. In no mood to relinquish their hold on Kunishi Ridge without a last-ditch fight, they began lobbing hand grenades on the Marines situated in the forward positions. Here, Company F was consolidating at a point approximately 400 yards due north of Mezado village and was tied in on its left with Company C, whose line was extended some 450 yards to the northeast.

Under the cover of smoke and with the assistance of tanks, the companies stalled in the valley made three attempts to reach the ridge during daylight on the 12th. Meanwhile, the troops already there needed rations, medical supplies, ammunition, and reinforcements; there were wounded to be evacuated also. Tanks attempting to get into firing positions south of Tera to silence the enemy weapons and relieve the companies pinned down were themselves fired upon, and in fact were unable even to leave the cover of the village.

In midafternoon, the Shermans were pressed into service to carry rations and personnel up to the ridge. At 1555, concealed in Tera from enemy observation, the tanks were loaded with supplies and Marines for the trip forward. Before dark, a total of nine tankloads in three runs had carried a reinforced platoon of 54 Marines and critical replenishment items up to the line. By displacing the assistant driver of each tank, it was possible to cram six riflemen inside instead. On arrival at the ridge, men and supplies were unloaded through tank escape hatches and casualty evacuees embarked in their place.

No further trips to the ridge were possible because of approaching darkness. In addition, the road leading to the front lines had caved in under the last tank in the column returning from the third trip. This tank bellied up, and neither the crew nor the casualties inside could leave through the escape hatch. After Marine mortars had fired a smoke cover for the tank, another came alongside to evacuate the troops it held. The disabled tank was then disarmed and abandoned for the night. In all, 1st Tank Battalion vehicles evacuated 22 wounded from the ridge.[42]

The darkness precluding further tank operations enabled the remainder of the 1st and 2d Battalion of the 7th Marines to move to the ridge without incident. In regimental reserve, the 3d Battalion patrolled to the rear and guarded the flanks of the other two. With three companies now up front, each of the two forward battalions extended its lines further. By midnight, the battalion commanders were convinced that their positions were reasonably secure,[43] and reassured that "the large amount of artillery support available could destroy any enemy counterattack which might be made against the initial ridgehead. . . ." [44] As General del Valle described it, "The situation was one of the tactical oddities of this peculiar warfare. We were *on* the ridge.

---

[42] *1st TkBn Summary*, 12Jun45.

[43] *7th Mar Hist*, p. 30.

[44] *Snedeker ltr 1955.*

The Japs were *in* it, both on the forward and reverse slopes." [45]

Patrols from the 1st Marines ranged south along the corps boundary and into the outskirts of Ozato on 12 June. Although furious fighting was then going on in the 7th Marines zone, the 1st encountered relatively little opposition except for sporadic fire from Kunishi Ridge that was placed on cave-sealing and mopping-up teams working in the vicinity of Hill 69. On the following day, combat patrols began reconnoitering towards Kunishi Ridge in preparation for a predawn attack scheduled for the 14th.

Throughout the division zone, all efforts on 13 June were concentrated on preparing for this large-scale operation. The incessant cannonading of artillery pieces and naval rifles gave the forces of General Ushijima in the southern part of the island no respite. Two rocket launching craft took positions off the southern tip of Okinawa to rake reverse slope defenses of the *Thirty-second Army*. More than 800 5-inch rockets ripped into the towns of Makabe and Komesu in an hour's time alone.

In the four days following its seizure of Kunishi Ridge, the 7th Marines was somewhat isolated from other friendly ground units by "No Man's Valley," [46] the 800-yard approach to its positions. This broad expanse was thoroughly covered by the fire of Japanese soldiers infesting the lower slopes and crests flanking the ridge. Supplies were either paradropped or brought in by tanks. Some air drops fell in the valley, "but they were in the minority." [47] The rest were right on target and fell into a drop zone under Marine control. Sometimes it was even dangerous for the Marines to recover supply containers in these supposedly safe areas because of the many enemy snipers awaiting such targets of opportunity. One Japanese sharpshooter alone killed and wounded 22 Marines before he was finally located and eliminated.[48]

Despite the inviting target their sheer bulk offered, tanks had to be used and did yeoman work in hauling supplies forward to the ridge. On their return trip, they evacuated casualties, some of whom were strapped to the side of the Shermans and then sandbagged as protection against enemy fire. In the morning of the 13th, a tank dozer constructed a bypass around the place where the road had caved in on the previous day. Upon completion of the detour, the lumbering mediums began shuttling back and forth to the ridge. Some of the tanks placed point-blank fire on enemy ridge positions covering the supply route in the 1/7 zone, and other tanks operated in the 2/7 zone, working over the western end of Kunishi Ridge. On the supply/evacuation runs, tanks lifted some 50 Marines from Company A to reinforce the rest of 1/7 on the ridge, and took out 35 casualties on the return trip.[49]

During the course of the day, the assault battalions continued consolidating their holdings on Kunishi, and 1/7 sent patrols east along the ridge to contact

[45] del Valle, "Southward from Shuri," p. 40.

[46] *7th Mar Hist, loc. cit.*

[47] *Snedeker ltr 1947.*

[48] *7th Mar Hist, loc. cit.*

[49] *1st TkBn Summary,* 13Jun45.

the enemy and uncover his positions. The battalion advanced only slightly. When a Japanese smokescreen obscured Kunishi village to friendly observation, 81mm mortars hammered the area to disrupt enemy activity suspected there. Shortly after twilight, a group of enemy troops was sighted on Mezado Ridge and was quickly dispersed by a heavy concentration of mortar and small arms fire. Marines from 2/7 patrolled along the west coast but were soon pinned down by long-range enemy fire coming from the eastern part of the ridge; they had to be withdrawn under the cover of smoke.

An increasing number of tanks became disabled by the accurate fire of AT guns well hidden in the ridge. A salvo from the main battery of a supporting battleship scored four observed direct hits on enemy emplacements, but did little to subdue other Japanese positions in the area. This particular barrage exploded on targets within 250 yards of friendly troops.[50]

During the night of 13 June, the 1st Marines was ordered to attack the front of Kunishi Ridge in its zone before dawn the next day; H-Hour was set for 0330. Following a 30-minute artillery preparation, 2/1 jumped off with two companies in assault. Despite an earlier division order prohibiting the use of flares by units adjacent to the assaulting force before and during an attack at night, the attack area and the attackers were nonetheless illuminated. Many urgent calls to higher echelons for an immediate ban placed on the firing of flares proved fruitless.[51] Fortunately, the Marines advanced undiscovered by the enemy and initial progress was unopposed. By 0500, two platoons from Company E had reached the topographical crest of the ridge; the support platoon and company headquarters were stopped well below this point by extremely severe enemy fire.

Half an hour later, Company G had worked one of its platoons up to a point on the ridge where it tied in with the left flank of Company E. At daybreak, increasingly active enemy sniping and intense fire on the flanks and rear of the assault companies served to isolate these platoons from the rest of their battalion. Tanks then rumbled forward to support the beleaguered Marines, whose casualties were mounting rapidly. One company lost six of its seven officers.[52] Because routes to the advance positions were under accurate and direct enemy fire, the mediums were again pressed into action to haul supplies up to the line and carry casualties to the rear, much in the same manner as they had for the 7th Marines elsewhere on Kunishi Ridge. At the end of the day, an estimated 110 wounded Marines had been evacuated by the tanks.[53]

Even though the enemy attempted to dislodge them, the Marines held onto their tenuous position. After dark, the reserve company was moved up and a perimeter defense was established for the night. Japanese small arms, mortar, and artillery fire, and recurring showers of hand grenades poured on the Marine

[50] Marginal comment by the battalion commander in *2/7 SAR*, p. 9.

[51] *2/1 SAR*, p. 15.

[52] *1st Mar SAR*, p. 22.

[53] *1st TkBn Summary*, 14Jun45.

positions throughout the hours of darkness. In addition, the enemy made many attempts to infiltrate all along the line.

As the sun rose on 15 June, 2/1 found enemy pressure to be as constant as it had been the day before, and battalion casualties reaching alarming proportions. Although the tanks continued to carry supplies and evacuate the wounded, 2/1 critically needed ammunition and rations. A requested air drop scheduled for 0900 was delayed until the middle of the afternoon, and then more than two-thirds of the packs dropped into enemy territory and could not be recovered. The efforts of 2/1 assault companies notwithstanding, Kunishi Ridge was far from secured when 2/5 relieved 2/1 after dark on the 15th.[54]

Earlier that day, 3/5 [55] had relieved 1/1 on Yuza Hill, following which, the latter moved to an assembly area near Dakiton; 3/1 had already set up in the vicinity of Shindawaku. With 2/5 on the line, Colonel Griebel had complete responsibility for the zone formerly held by the 1st Marines.

During the afternoon before 2/5 was to go into the lines, its company commanders were taken by tank to the front lines to make a personal reconnaissance of the area they were to take over. When they arrived at the positions of 2/1, they discovered that the tactical situation precluded a daylight relief. They also found that 2/1 held only a 75-yard section on the crest of Kunishi Ridge, with a portion of the battalion occupying a small pocket on the forward slopes of the ridge. As a result of this situation, the commanders of 2/1 and 2/5 agreed that the relief should begin only after it had become dark. In order to maintain tight control over the move and prevent matters from becoming confused in the restricted area where the relief was to take place, Lieutenant Colonel Benedict decided to commit only one of his companies initially. The relief of 2/1 was completed at 2030, a half hour after it had begun. The 1st Marines as a whole went into division reserve at 2300, after having been in the division line for 12 straight days during which it suffered nearly 500 casualties.

On the day before fresh troops had joined in the fight for Kunishi, the 7th Marines resumed its grinding advance by "the slow, methodical destruction of enemy emplacements on the ridge, to which the descriptive word 'processing' had come to be applied." [56] The 2d Battalion was ordered to seize the rest of the ridge in its zone and to be prepared to continue the attack to Mezado Ridge. Company A attacked east to seize the remainder of the reverse slope of Kunishi Ridge within the 7th Marines zone, while B and C provided fire support and mopped up behind the assault company. Despite difficult terrain and an unrelenting enemy opposition, Company A succeeded in closing to the outskirts of Kunishi village.

At 1247, Company B was ordered to continue the advance through Kunishi, and then to attack north to secure the forward slope of the easternmost sector of Kunishi Ridge. Although the company passed through the village and

[54] *2/1 SAR*, p. 15.

[55] Lieutenant Colonel Robert E. Hill assumed command of the battalion on 8 June, when 3/5 had been in reserve.

[56] *7th Mar Hist*, p. 31.

began heading for the high ground with only slight interference, withering machine gun fire soon pinned down all but two rifle squads, which were able to climb the height. Once the Marines had gained the crest of the ridge, the Japanese launched a strong counterattack, forcing the squads from their temporary holding. The company as a whole then withdrew to lines held the previous night.

To the right of 1/7, the 2d Battalion was subjected to increasingly intense enemy fire despite suppressive American counter-barrages; 2/7 reported only limited gains that day. At 1530, the logistic and tactical support of the 7th Marines by tanks ended when the armor began assisting the 1st Marines. On the 14th, the tracked vehicles had carried 48 men of the 7th Marines forward and evacuated 160.[57]

During the following two days, 2/7 was supported by naval gunfire, artillery, air, rockets, and 81mm mortars, which mercilessly pounded the enemy. Both gun and flame tanks furnished direct close-in support, but could make no appreciable dent in Japanese defenses. A stubborn enemy notwithstanding, 2/7 moved its lines some 500–700 yards to the right and in front of the first high ground leading to the Mezado Hill mass, the division objective after Kunishi.

The 1st Battalion fared no better in its attempt to seize the rest of Kunishi Ridge in its zone on 15 June than it had on the 14th. Notified that 15 artillery battalions were on call for supporting fires, 1/7 moved out at 0945 following an artillery preparation and preliminary patrolling. Company C attacked directly east along the ridge while B moved through Kunishi village and then turned north towards the high ground again.

Heavy Japanese fire from prepared emplacements prevented the Marines from advancing across the open ground between the village and the ridge line, and Company C was unable to relieve the pressure on B. At 1600, the two units were withdrawn once again to positions held 13–14 June.

During the night 15–16 June, small enemy groups were active in front of 1/7 lines harassing the Marines with small arms fire and lobbing hand grenades into their foxholes. Before dawn on 16 June, the troops on the left (east) flank were pulled back to the west approximately 200 yards to permit a massive artillery preparation on the objective which had stymied 1/7 the preceding two days. An extremely heavy concentration of artillery, mortar, and rocket fire drummed that day's target for nearly three hours before the assault forces jumped off. By 1345, 1/7 had completely seized the rest of the ridge in its zone and immediately began mopping up and consolidating its newly won ground. Shortly thereafter, battalion troops "repeatedly encountered and destroyed numerous groups of the enemy wandering through the town of Kunishi in a confused, disorganized, and bewildered state. It was evident that the end was not far off." [58]

One other major accomplishment that afternoon was the capture by Company A of "The Pinnacle," a particularly difficult enemy strongpoint situated so that

[57] *1st TkBn Summary*, 14Jun45.

[58] *7th Mar Hist*, p. 32.

it could be neither destroyed nor neutralized by any type of support weapon immediately available. It was in this area that the enemy sniper who had shot 22 Marines earlier was hunted down and killed. Approaches to The Pinnacle were swept by Japanese fire, and its seizure by the infantry was slow, tortured, and costly.[59]

To the right of 1/7, the 2d Battalion lines were extended some 400 yards further west to where the battalion held the first high terrain approaching the Mezado hills. This progress was accomplished even while the battalion had sustained heavy casualties and lost its valuable armored support, which fell victim to Japanese land mines and 47mm AT guns. Expert employment of its supporting arms enabled 2/7 to make slight gains on the 16th. For example, salvos from the main battery of USS *Idaho* were called down on targets located within 400 yards of frontline troops. In addition, air liaison parties controlled air strikes, often consisting of 25–30 planes each, which successfully destroyed stubborn pockets holding up the advance.[60]

No longer was Kunishi Ridge a major obstacle in the way of the 1st Marine Division, for the terrain that the Japanese had so doggedly defended here, including the approaches to Mezado, had been virtually cleaned out by the end of 16 June. Only that portion of the ridge on the far left of the division, in the 5th Marines zone, still presented some problems. With the reduction of enemy opposition on The Pinnacle, the 7th Marines was able to make physical contact with the 5th.

As 1st Division troops prepared for the final drive south, mopping up operations on Oroku Peninsula neared an end. Concurrently, General Shepherd's staff drew up plans for the eventual commitment of the 6th Division in the southern front. Initially, the 22d Marines was to pass through right flank elements of the 7th Marines on 17 June to relieve 2/7; 3/7 would come out of reserve to relieve the 1st Battalion. (See Map 21.)

In the 5th Marines zone on 16 June, 2/5 attacked at 0730 and spent the day working over that portion of the regimental area that lay between Kunishi Ridge and Hill 69. At approximately 1800, a reinforced company reached the crest of the ridge and tied in with the left flank unit of the 7th Marines. Bitter, close-quarter fighting had been the order of the day for 2/5, whose assault companies had received continuously heavy small arms fire. Rising casualty figures again required tanks to be employed as evacuation vehicles; this task was in addition to their shuttling ammunition and rations forward. In face of Japanese holding action to its front, 2/5 made slow but steady progress.

Although enemy infiltrators attempted to breach 5th Marines lines during the night 16–17 June, they were thoroughly discouraged. As 2/5 resumed the attack on the 17th, oppressive enemy small arms fire coming from the vicinity of Aragachi in the XXIV Corps zone punished the front and flank of the battalion. Its task was to seize that portion of Kunishi Ridge still held by the enemy. Attacking with a two-company front, the 2d Battalion faced the problem of coping

---

[59] *Ibid.*, p. 31.

[60] *2/7 SAR*, p. 10.

with Japanese reverse-slope positions and destroying them. To smooth the way somewhat, a rocket barrage was laid on the objective. A short time later, at 1030, tanks moved out and clambered over the ridge route, which had been opened earlier by an armored bulldozer.

Murderous enemy fire criss-crossed the crest of the ridge as 2/5 grimly pushed on. All tanks were pressed into action as armored ambulances once again, but only the walking wounded could be taken inside of the vehicles and evacuated. Stretcher cases presented a serious problem because they could not be taken up through the tank escape hatches, but had to be lashed to their rear decks. Often, wounded were hit a second and third time on their trip to the aid station.

Throughout the afternoon, the volume and intensity of enemy fire as well as the ferocity of the enemy opposition remained undiminished. Tank, artillery, and mortar fire, and the ripple fire of several rocket barrages were directed at suspected Japanese strongpoints and weapons emplacements in an attempt to open the way for Marine tank-infantry teams. At 1700, Lieutenant Colonel Benedict decided to commit his reserve company and further strengthen the assault companies by sending forward 133 replacements, which had been assigned to the battalion three days earlier.[61]

With this infusion of fresh troops in its line, 2/5 surged eastward along the ridge; by nightfall, the battalion held approximately three-fourths of the 1,200 yards of Kunishi Ridge in the regimental zone.[62] Because the position of the battalion was somewhat precarious, at dusk Colonel Griebel attached Company K, 3/5, to the 2d Battalion with a mission of protecting the battalion rear. At 2315, an estimated company-sized counterattack hit 2/5 positions, but was thoroughly blunted; Company K troops killed the few Japanese that succeeded in penetrating the lines.

On the coastal flank of the IIIAC zone, 6th Division Marines had become fully involved in the drive to the south by the end of 17 June. Moving forward during the afternoon of the previous day to relieve 2/7, the 1st and 3d Battalions of the 22d Marines attempted to cross the valley between Itoman and Kunishi, but were forestalled by the severe enemy fire covering this route. Forced to turn back to their previous positions, the two battalions waited until dark to begin the relief.[63] The uneventful passage of the lines began at 0300, 17 June, and by dawn assault elements were in jump-off positions at the base of the northern slope of Mezado Ridge, prepared to attack at 0730 in coordination with 3/7 on the left.

An artillery, naval gunfire, and air bombardment of Mezado Ridge, and of Hill 69 [64] and Kuwanga Ridge beyond it, preceded the attack. Once the fires had lifted, the 22d Marines moved out with

[61] *2/5 SAR*, p. 21. On 11 June, the division received 369 officers and enlisted from the 55th Replacement Draft, which arrived that day. With the arrival of the 62d Replacement Draft five days later, the division absorbed 295 more Marines. *1st MarDiv SAR*, PersAnx, p. 7.

[62] 2/5 SAR, *loc. cit.*

[63] *7th Mar Hist*, p. 32.

[64] This is not to be confused with the Hill 69 west of Ozato, or the Hill 69 east of Mezado.

two assault battalions abreast—3/22 on the left. Machine gun and intermittent mortar fire paced the advance up the slope of the ridge, but as the morning wore on, the Marine progress became increasingly difficult in the face of stiffening resistance.

To support the attack of 1/22, 6th Division tanks moved around the right flank of the regiment and through the water towards an off-shore reef to gain firing positions commanding direct observation of the caves on the western tip of Mezado Ridge. As one armored platoon began to negotiate the route, the unexpected depth of the water prevented it from working its way forward far enough to enfilade the ridge, and its tanks were forced to deliver supporting fire from the most advanced points that they had been able to reach.[65] Tank weapons could not suppress the heavy machine gun fire coming from the reverse slope of the hill mass holding up 1/22. As a result, the battalion was unable to gain more than a foothold on the forward slope of Mezado Ridge until 1700, when it positioned two companies on the crest of the ridge for night defense.

Inadequate maneuvering room to the front also limited the employment of supporting armor. Besides the flank route through the water, the only other suitable tank road ran through a rice paddy which had been cratered in four places and heavily mined as well. Once the mines were removed or neutralized, tanks lumbered up to these craters and dumped into them bundles of large logs that had been attached to their front slope plates. Tanks and logs instead of dump trucks and fill dirt were used to plug the craters because only armored-plated vehicles could weather the severe enemy fires.

After two craters in the road had been filled, it was discovered that the approaches to a small bridge further up the road had been mined. Sniper and machine gun fire prevented engineer clearing teams from neutralizing the mined area, and the road project was abandoned temporarily. Nonetheless, the tanks advanced as far forward as possible to deliver overhead supporting fires.

By noon, 3/22 had secured the highest point on the ridge and maintained the momentum of its attack to clean out the town of Mezado as well. Before dusk, the battalion had captured the key terrain around Hill 69 and was in command of the ground overlooking the next objective, Kuwanga Ridge. With the exception of an attempted enemy counterattack in the sector of 1/22 at 2210, a generally quiet night was passed by the 22d Marines.

When 1st Division troops jumped off on the 17th, 3/7 attacked in a column of companies, Company K leading, to take the Hill 69 east of Mezado. Company I maintained contact with the 22d Marines, and Company L took up positions to protect the left flank of K. Following an unopposed 1,400-yard drive across the plateau just east of Mezado to seize Hills 69 and 52, 3/7 halted for a short time to reorganize, and then attempted to continue the drive to the crest of Hill 79—the last remaining barrier before Makabe. Heavy Japanese fire from positions on the high ground commanding the Kuwanga-Makabe road forced the

[65] *6th TkBn SAR*, Ph III, p. 20.

battalion to dig in for the night before it could gain the hill. Once dug in, 3/7 Marines quickly organized to blunt all enemy attempts to infiltrate and counterattack in the darkness.

When the Japanese *22d Regiment* did launch its counterattack, it was directed against 1/22. This determined effort born of despair was doomed from its inception because that portion of the enemy regiment scheduled to exploit the counterattack had been almost completely destroyed that afternoon. In effect, the near annihilation of the *22d Regiment* meant that the left flank of the Japanese outpost line had all but collapsed, and that the *32d Regiment*, holding positions near Makabe, was faced with the threat of having its left flank rolled up.[66]

The Marines were prepared to turn this threat into reality by exploiting the successes of 17 June with the commitment of fresh troops into the battle on the next day. While the 7th Marines finished "processing" Kunishi Ridge, the 8th Marines (Reinforced), commanded by Colonel Clarence R. Wallace, prepared to relieve 3/7 to continue the attack southward. Before the dawn of 18 June, this 2d Marine Division infantry regiment, now attached to the 1st Division, entered the lines.

IHEYA-AGUNI OPERATIONS[67]

After its feint landings on the southeastern coast of Okinawa on L-Day and L plus 1, the 2d Marine Division remained on board its transports which steamed in the vicinity of the target area until 11 April. On that date, the Demonstration Group set out for Saipan, arriving there four days later. On 14 May, CinCPac ordered the division (less one RCT) detached from the operational control of the Tenth Army and designated it as the area reserve under control of CinCPOA. In addition, General Watson received an alert for his division to conduct Phase III(d) of the ICEBERG Plan, the landing on Kikai Jima. Once captured, this small island north of Okinawa was to be utilized as a northern outpost for the Ryukyus area, and was to base four fighter groups, two night-fighter squadrons, and one torpedo-bomber squadron.[68] The 8th Marines (Reinforced) remained under the control of General Buckner for the impending landings on Iheya and Aguni Shimas, and was ordered to reembark immediately.

For the entire month after its arrival at Saipan, the division remained on board the transports. A warning order for the Kikai invasion had been issued on 6 May,[69] but this alert was reduced in urgency four days later by a message that indicated that Phase III(d) might not be conducted. With the arrival of the 14 May message releasing the division from Tenth Army control, the Marines began unloading and rehabilitating their equipment ashore in preparation for the time when it was to be

[66] *Okinawa Operations Record*, p. 134.

[67] Unless otherwise noted, the material in this section is derived from: *Tenth Army AR;* 2d MarDiv WarDs, Apr-Jun45; Combat Team 8 AR, Iheya-Aguni Operations, n.d., hereafter *8th Mar AR, Iheya-Aguni;* Combat Team 8 AR, Okinawa Operation (11-22Jun45), n.d., hereafter *8th Mar AR.*

[68] *USAFMidPac G-5 Hist*, p. 245.

[69] *Ibid.*, p. 258.

ordered to mount out for Kikai. On 3 June, the landing was deferred for an indefinite period, and on the 19th, the 2d Marine Division (less RCT 8) was released from its role as Ryukyus area reserve and reverted to the control of FMFPac.

Once again, on 24 May, the 8th Marines departed Saipan headed for Okinawa; its eventual target, the islands of Iheya and Aguni. Because of the heavy damage that had been sustained by the fleet and especially the radar pickets during *Kamikaze* raids, early in May Admiral Turner asked General Buckner to begin a study of outlying islands to determine where long-range radar and fighter director facilities could be installed. Resulting from this study was the decision that Tori, Aguni, Iheya, and Kume Shimas could be captured in that order. A special landing force, a reinforced company from the 165th Infantry, made an unopposed landing on Tori on 12 May and a detachment from Air Warning Squadron 1 began operations almost immediately. (See Map 22.)

Since the Okinawa campaign was now reaching a crucial stage, General Buckner believed that the forces already committed in the fight southward should not be diverted to such secondary actions as the proposed landings on the other outlying islands noted above. He requested, therefore, that the reinforced 8th Marines be returned to Okinawa to effect the Iheya-Aguni landings. Brigadier General LeRoy P. Hunt, ADC of the 2d Marine Division, was designated the landing force commander for these operations. Flying to Okinawa with key members of his staff on 15 May, General Hunt spent the 16th and part of the 17th conferring with Tenth Army staff officers about the proposed operation plan. By 30 May, when the 8th Marines arrived at Okinawa, a complete naval gunfire and air support schedule had already been established, and detailed contingency plans drawn up to meet any situation that might arise from enemy sea or air action.

The attack force, commanded by Admiral Reifsnider, steamed from the Hagushi transport area early on 2 June and set a course for the target, located 15 miles northwest of Hedo Misaki. The bombardment prior to the H-Hour of 1015 proceeded as scheduled;[70] 2/8 and 3/8 landed on Iheya 27 minutes later.[71] Neither enemy opposition nor enemy troops were encountered. The Marines only found some 3,000 confused but docile natives who were taken under tow by military government teams supplied by the Tenth Army Island Command. Late in the afternoon of the 3d, the troops began general unloading and the island was officially declared secure the next day.

The landing on Aguni Shima, 30 miles west of Okinawa, was delayed until 9 June by inclement weather. On that day, 1/8 went ashore under circumstances similar to those found at Iheya. The only Marine casualties of the two amphibious assaults were sustained at Iheya; 2 Marines were killed and 16 wounded by aerial rockets and short rounds of naval gunfire. In accordance with the instructions it had received from Tenth Army before the operation, the 8th Marines stood ready for im-

[70] ComPhibGru 4 AR, Iheya-Aguni, Ser 0327, dtd 3Aug45, pt VI, p. 8.

[71] *Ibid.*, pt I, p. 1.

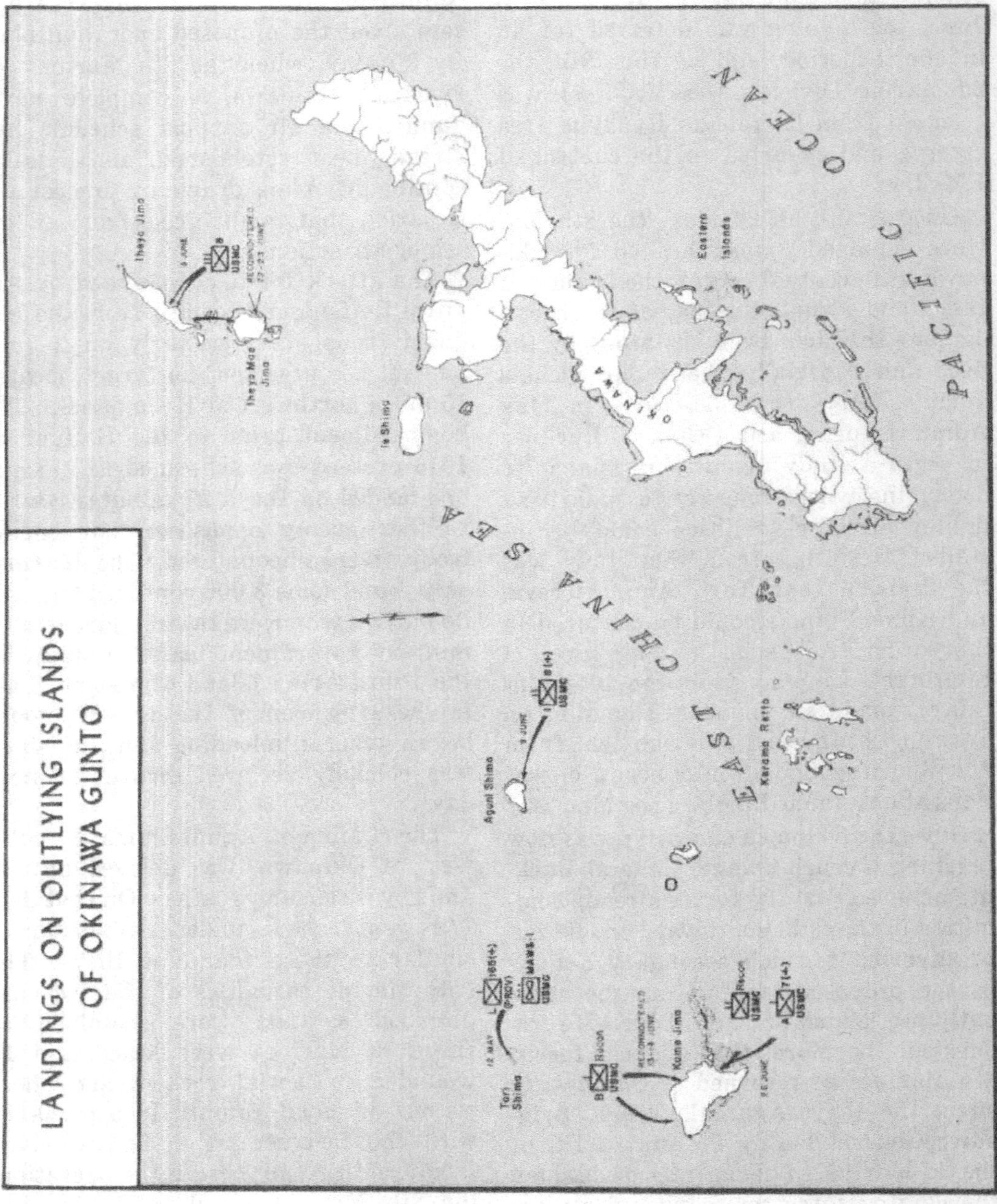
LANDINGS ON OUTLYING ISLANDS
OF OKINAWA GUNTO
PACIFIC OCEAN
EAST CHINA SEA
Iheya Jima
Iheya Mae Jima
Ie Shima
Eastern Islands
OKINAWA
Aguni Shima
Kerama Retto
Tori Shima
Kume Jima
T. L. RUSSELL

MAP 22

mediate commitment on Okinawa upon completion of the two landings. When fresh units were needed for the final thrust against the Japanese dug in on Kiyamu Peninsula. Colonel Wallace and his troops were available.

THE FINAL PUSH [72]

By 4 June, the remnants of the *Thirty-second Army* had fully manned the outpost line of Kiyamu Peninsula. Concentrated in this area were approximately 30,000 Japanese troops, distributed as follows: *24th Division* and attachments, 12,000; *62d Division* and attached units, 7,000; *44th IMB* and attached units, 3,000; *5th Artillery Command* and attached units, 3,000; and troops attached directly to *Thirty-second Army Headquarters*, and the command itself, 5,000. "Attrition during retirement operations," [73] was the official Japanese explanation for the 20,000-man differential between their estimated strength figure of 50,000 in late May and the total number of effectives available at the beginning of June.

Of General Ushijima's remaining forces, approximately 20 percent were survivors of the original, first-rate infantry and artillery defense garrison; the rest were either untrained rear-echelon personnel or *Boeitai*. Leading this motley force at battalion level and above were many of the original senior commanders who had remained alive and were still capable of arousing a fighting spirt in their men.

Their unflagging belief in a final Japanese victory was unrealistic in view of the alarming losses of weapons and equipment that the *Thirty-second Army* had sustained since the American landing on 1 April. Hand grenades and explosives either were in short supply, or in the case of some units, non-existent. Only 20 percent of the original number of heavy machine guns owned by the army remained, and few of its heavy infantry cannon and mortars were still firing. Although the army ammunition supply along with 2 150mm guns, 16 150mm howitzers, and 10 antiaircraft artillery pieces had been transported south to Kiyamu when Shuri was abandoned, the stock levels of artillery ammunition precluded more than 10 days of sustained firing.

Despite these outward signs of its imminent defeat and impoverished condition, the belief held by General Ushijima's army in ultimate victory was derived from deep-seated tradition, strongly enforced discipline, and the historically pervasive influence of Japanese military doctrine throughout the Empire. These intangibles, almost completely alien and incomprehensible to Americans, promised that Kiyamu Peninsula was not to fall and the battle for Okinawa was not to end before a final, violent climax.

Influenced by the location and relative strength of enemy strongpoints facing the Tenth Army, and the availability and status of his assault forces, General Buckner had shifted the corps boundary west on 4 June. In the now-narrower IIIAC zone, General Shepherd's division sought to capture the Oroku Pen-

---

[72] Unless otherwise noted, the material contained in this section is derived from: *CTF 31 AR; Tenth Army AR; XXIV Corps AR; 7th InfDiv AR; 77th InfDiv OpRpt*, Okinawa; *96th InfDiv AR; Okinawa Operations Record.*

[73] *Okinawa Operations Record*, p. 122.

insula while the 1st Marine Division was to cut off the peninsula from the rest of the island, capture Itoman, seize both Kunishi and Mezado Ridges, and drive to Ara Saki, the southernmost point of the island. The assignment given XXIV Corps included the capture of the Yuza Dake-Yaeju Dake Escarpment as a primary objective. On line facing this foreboding terrain were the 96th and 7th Divisions.

Nearly two weeks of punishing and brutal fighting were to ensue before the two army divisions could eliminate all enemy resistance in this *Thirty-second Army* defense sector. (See Map IX, Map Section.) XXIV Corps units spent the period 4–8 June in regrouping and attempting to gain favorable jump-off positions for the attack on the escarpment on the 9th. All supporting arms were employed to soften the well-organized enemy defense system. Armored flamethrower, tank, assault gun, and artillery fires were added to the point-blank blasts of experimental 57mm and 75mm recoilless rifles [74] in an effort to reduce the natural bastion.

The defense of the Yuza Dake-Yaeju Dake outpost line had been assigned to two units. Guarding the escarpment from Hill 95 on the east coast to Yaeju Dake was the *44th IMB;* the remainder of the high ground, including Yuza Dake, was the responsibility of the *24th Division.* Added to the tenacious determination of the foe was the natural, fortress-like quality of the terrain he guarded. This combination enabled the Japanese to defend the Yuza Dake area with only one regiment, the *89th.*

Facing the 7th Division were enemy troops who compared unfavorably with the veterans defending Yuza Dake. Coming from miscellaneous shipping engineer, sea raiding, mortar, and line of communication units, the soldiers were loosely organized into provisional infantry regiments and put into the *44th IMB* line. The vital Hill 95-Nakaza Valley area was held by survivors of the *15th Independent Mixed Regiment,* which first began to give way under the repeated pounding of the 7th Division attack. General Arnold's soldiers relentlessly pushed forward on 11 June, the second day of the all-out corps assault on the escarpment, and threatened the rest of the *Thirty-second Army* line by breaking into the *44th IMB* defenses. An attempt by General Ushijima to shore up this section of his rapidly crumbling outpost by committing jerry-built infantry units comprised of service and support troops proved to be "as ineffective as throwing water on parched soil." [75]

The *89th Regiment* continued to withstand the inroads of 96th Division in-

[74] These newly developed weapons had been sent to Okinawa in late May by the War Department for test firing under combat conditions. After the limited supply of ammunition accompanying the weapons had been expended, air shipments of the special ammunition were rushed to the island in time for employment by XXIV Corps units at the escarpment. Field commanders who had the opportunity to use these weapons were unanimous in their praise and recommended that the recoilless rifle be adopted as a standard infantry weapon. According to the former commander of the 7th Marines, these weapons ". . . could have been used to great advantage by the 7th in its seizure and processing of Kunishi ridge. They were asked for but the reply came back [that] all their ammunition had been used up." *Snedeker ltr 1965.*

[75] *Okinawa Operations Record,* p. 129.

fantry on 12 June, but this day marked the beginning of the end for the *44th Independent Mixed Brigade.* Although it had been reinforced with two battalions from the *62d Division* as a result of the brigade commander's urgent pleas, the time for decision was already past, as was the chance for these newly committed units to affect the ultimate course of the battle.

Clear weather on 13 June, following a night of abortive enemy counterattacks, permitted General Hodge to employ fully all of his supporting arms. Units of the *62d Division* attempting to reach and revive the hapless *44th IMB* were themselves blasted by American air, artillery, and naval gunfire. Although the *89th Regiment*—reinforced by the *24th Reconnaissance Regiment*—still maintained its hold on Yuza Dake, its rear and flank were threatened this day by the impending penetration sout of Yaeju Dake. (See Map X, Map Section.) Further advances on 14 June forced General Ushijima to commit the *13th Independent Infantry,* which was almost immediately smashed by 7th Division troops. Also committed and destroyed on the 14th were the remaining reserve battalions of the *62d Division.*

Elsewhere, as Japanese positions began to give way under the pressure of the American onslaught, *Thirty-second Army* headquarters lost all contact with the *15th IMR*—the last infantry element of the *44th IMB* able to maintain unit integrity. To stave off the last stages of a crushing defeat, General Ushijima ordered the *62d Division* into the deteriorating Japanese line from reserve positions southwest of Makabe, but a savage lashing from American artillery, naval guns, and air-delivered napalm and bombs thoroughly disrupted the deployment. Few, if any, of the enemy troops arrived at their destination.

The 96th Division took advantage of this confused situation to rush its infantry through the Yuza Dake perimeter. On the left, the 7th Division surged down the coast. By the end of 17 June, XXIV Corps regiments held firm control of all commanding ground on the Yuza Dake-Yaeju Dake Escarpment. Compressed between the front lines of the corps and the southern tip of Okinawa were the remnants of the *Thirty-second Army*—a hodge-podge of units and individuals from the *62d Division, 44th IMB,* and *24th Division.* Before the island had been secured by the Tenth Army, most of these Japanese troops would die violently in a forlorn attempt to protect the headquarters of General Ushijima.

DEATH OF AN ARMY [76]

The death throes of the *Thirty-second Army* became even more obvious as the Tenth Army advanced against steadily lessening resistance on 18 June. Although most sections of the Japanese line proved softer than before, two isolated centers of opposition developed during the day—one around Medeera and the other in the area of Mabuni. The first was held by the remnants of the

[76] Unless otherwise noted, the material in this section is derived from: *CTF 31 AR; Tenth Army AR; IIIAC AR; XXIV Corps AR; 1st MarDiv SAR; 6th MarDiv SAR, Ph III; 7th InfDiv AR; 77th InfDiv OpRpt,* Okinawa; *96th InfDiv AR; 4th Mar SAR, Ph III; 5th Mar SAR; 8th Mar AR; 29th Mar SAR, Ph III; Okinawa Operations Record.*

24th Division, and the second, around Hill 89, was defended by elements of the headquarters and troops of the remaining *Thirty-second Army* units. (See Map 16.)

Leading the 1st Marine Division attack was the 8th Marines, which had relieved the 7th Marines the previous night. At 0730, 2/8 (Lieutenant Colonel Harry A. Waldorf) jumped off from Mezado Ridge to head south and occupy a line west of Makabe from which it could launch a "quick decisive thrust" to the sea.[77] Light machine gun and rifle fire, later mixed with sporadic mortar and artillery rounds, hit the left front and flank of the battalion as it made a rapid 1,400-yard advance to cap its first day in the lines. By dark, the battalion had secured its objective and began digging in for the night. Since its left flank was well forward of 1/5, Company B, 1/8, was attached to fill the gap.

Early on 18 June, General Buckner had gone forward to witness the fighting, and "probably chose the 1st Division front on this date because he wanted to see the 8th Marines in action," as he thought well of the regiment.[78] As General Oliver P. Smith recalled:

> On his way to the front [to the 3/8 OP], General Buckner met Bob Roberts (Colonel Harold C. Roberts, commanding Officer of the 22d Marines). Roberts urged General Buckner not to go to the front at this particular point as the rapid advance had bypassed a good many Japanese, and, further, there was considerable flanking fire coming from the high ground in front of the 96th Division. General Buckner did not heed this advice. (Roberts was killed an hour or so later on another part of the front.) The General got up on a ridge where Lieutenant Colonel Paul E. Wallace [commanding 3/8] had an OP. Tanks and infantry were operating ahead. A rifle company was on the ridge preparing to move forward. General Buckner took position behind two coral boulders separated by a slit through which he could look. His position was slightly forward of the crest. He had not been in this position long when a Japanese 47mm shell hit the base of the boulders. The first shell was followed by five more in rapid succession. Either a fragment of the first shell or a piece of coral rock thrown out by the detonation hit General Buckner in the chest. This wound was mortal. Hubbard [General Buckner's aide], with the assistance of others in the vicinity, dragged General Buckner over the crest to a defiladed position. A Navy hospital corpsman was there and a doctor arrived within three minutes. Plasma was available, but the General had lost blood so rapidly that plasma could not save his life.[79]

Upon being informed of General Buckner's death, Brigadier General Elwyn D. Post, Tenth Army Chief of Staff, sent a message to CinCPOA reporting the death. In addition, General Post, knowing General Buckner's expressed desires concerning the succession of command, recommended in the message that General Geiger be designated the new Tenth

[77] LtGen Pedro A. del Valle ltr to CMC, dtd 9Mar55, hereafter *del Valle ltr 1955*. "When General Geiger gave me the 8th Marines, a fresh regiment at full strength, I conferred with Colonel Wallace and we agreed upon his leapfrog attack, 3 battalions in column, with limited objectives, while I would cover his flanks with my exhausted troops. This was the 'coup de morte' [death blow] which broke the remnant of the 32d Japanese Army." *del Valle ltr 1965.*

[78] Smith, *Personal Narrative*, p. 135.

[79] *Ibid.*

Army commander.[80] On 19 June, General Geiger was appointed a lieutenant general and was officially designated Commanding General, Tenth Army (CTF 56), the same day, making him the senior officer present on Okinawa. This was the first time that a Marine officer had commanded a unit of this size. General Joseph W. Stilwell, U. S. Army, former deputy commander of the Southeast Asia Command, arrived on the island at 0700, 23 June, succeeding General Geiger the same day,[81] after the Marine general had successfully directed the final combat operations on Okinawa.

Early in the morning on which General Buckner died, the 5th Marines was to take Hill 79, northwest of Makabe. At dawn, 1/5 moved out around the western nose of Kunishi Ridge and then south through the 8th Marines zone in order to get into positions to jump off at 0730.[82] As soon as the attack began, the assault units were pinned down by fire coming from the objective and unable to move until 1100, when tanks arrived and rumbled into support positions. A coordinated tank-infantry assault was launched soon after, and the Marines were on the hill by noon. Following in the wake of the attack was 3/5, up from reserve, which moved into support positions behind the 1st Battalion for night defense.

Enemy AT fire forced the Marine tanks to operate cautiously, but did not slow them down. Artillery-delivered smoke on a suspected antitank gun position on Hill 81 just north of Makabe blinded the enemy gunners and permitted the Shermans to operate without being fired upon. Other AT guns were destroyed during the day after having been spotted from the air by an experienced tank officer, who was flown over the battlefield for that purpose.[83] By dark, 1/5 had gained the lower slopes of Hill 79 with armored assistance, but could not advance beyond that point because of heavy enemy fire from high ground in the 96th Division zone and Hills 79 and 81. At the end of the day, tank-infantry teams from 2/5 eliminated the last large pockets of enemy

[80] *Ibid.*, pp. 135–136. During the planning phase of ICEBERG, General Buckner stated that a second-in-command of the army should be designated prior to the operation, and that he felt General Geiger capable of handling a field army. A letter recommending this appointment was sent to CinCPOA—the strategic commander of the area in which the operation was to be held—via Lieutenant General Robert C. Richardson, Jr., the Commanding General, U. S. Army Forces, Pacific Ocean Areas. The latter returned the letter with a note stating that this assignment was one for the War Department to decide. No further action was taken until the time of General Buckner's death. *Ibid.*, pp. 34–35. The right of Admiral Nimitz to make this assignment evolved from the fact that the Tenth Army, Task Force 56 for the Okinawa campaign, was a task unit within the Fifth Fleet, and not under the control of General Richardson's command.

[81] LtGen Merwin H. Silverthorn comments to HistBr, HQMC, dtd 25Oct65.

[82] The right (east) flank of the 5th Marines had been exposed by the rapid advance of the 8th Marines. Therefore, "the battle-weary and decimated infantry of the 1st MarDiv had to be employed in attacking the various hill positions along the [eastern] flank simply to cover it. . . . We knew that the XXIV Corps could not keep up with the penetration executed by fresh troops, so we planned our maneuver to provide for this expected contingency." *del Valle ltr 1955.*

[83] *Ibid.*

MARINE TANKS *shelling positions in the south, where the enemy is holed up in a last-ditch stand. (USMC 126411)*

COOPERATIVE PRISONERS *call upon other Japanese to surrender. Note man at water's edge preparing to swim out to the LCI. (USA SC209533)*

resistance on Kunishi Ridge, and dug in on commanding ground for the night.

In the 6th Marine Division zone, 2/22 passed through the lines of 3/22 on 18 June to attack Kuwanga Ridge. Moving rapidly ahead despite steady automatic weapons fire, the battalion gained a foothold on the high ground and began simultaneous drives to the east and west to clear the ridge of enemy. Although fired upon by rifles, machine guns, and mortars, the battalion possessed the greater part of the ridge before midafternoon. At this time, General Shepherd saw that the understrength 2/22 was spread too thinly over the 1,800-yard-wide ridge to withstand a concerted enemy counterattack, so he ordered the 4th Marines to attach one battalion to the 22d for night defense; Colonel Shapley ordered 3/4 forward.

The other two battalions of the 22d Marines spent the day hunting down and destroying numerous enemy groups infesting the reverse slopes of Mezado Ridge. Forward observation posts became especially plagued by all sorts of fire coming from these bypassed Japanese soldiers. Colonel Roberts, the regimental commander of the 22d Marines, was killed at his OP by sniper fire at 1430. The regimental executive officer, Lieutenant Colonel August Larson, assumed command.

Assault forces of XXIV Corps also made important gains on the 18th. The 96th Division push on Medeera positions from the east was coordinated with the 1st Division attack on the same objective from the west. The 7th Division continued its drive with a two-pronged attack. One assault element dashed down the reverse slope of Hill 153 to sweep past Medeera and ended its attack at the corps boundary near Komesu. Three battalions abreast spearheaded the second prong of the attack, advancing slowly down the coast towards Mabuni. During the night of 18–19 June, at least 340 Japanese soldiers were killed in attempted infiltrations and scattered attacks all along the Tenth Army front.[84]

With the realization that "his Army's fate had been sealed," [85] General Ushijima began spiritual and physical preparations for a *Samurai's* death. On 16 June, he sent the first of his farewell messages, this a report to *IGHQ* in Tokyo, which read:

> With a burning desire to destroy the arrogant enemy, the men in my command have fought the invaders for almost three months. We have failed to crush the enemy, despite our death-defying resistance, and now we are doomed.
>
> Since taking over this island our forces have, with the devoted support of the local population, exerted every effort to build up defenses. Since the enemy landing, our air and land forces, working in concert, have done everything possible to defend the island.
>
> To my great regret we are no longer able to continue the fight. For this failure I tender deepest apologies to the emperor and the people of the homeland. . . . I pray for the souls of men killed in battle and for the prosperity of the Imperial Family.
>
> Death will not quell the desire of my spirit to defend the homeland.
>
> With deepest appreciation of the kindness and cooperation of my superiors and my colleagues in arms, I bid farewell to all of you forever.
>
> Mitsuru Ushijima [86]

---

[84] Tenth Army G–2 Rpt No. 86, dtd 20Jun45.

[85] *Okinawa Operations Record*, p. 133.

[86] Inoguchi, Nakajima, and Pineau, *Divine Wind*, pp. 148–149.

Three days later, he sent a last message to all *Thirty-second Army* units with which he still had contact, congratulating the survivors on having performed their "assigned mission in a manner which leaves nothing to regret" and calling upon them "to fight to the last and die for the eternal cause of loyalty to the Emperor." [87] General Ushijima then directed most of his staff officers to leave the Mabuni command post, to disguise themselves as island natives, and to infiltrate the American lines in order to escape to northern Okinawa. Some of his key advisors, like Colonel Yahara, were assigned the mission of reaching Japan in order to report to *Imperial General Headquarters;* others were ordered to organize guerrilla operations in the rear of Tenth Army tactical units and the Island Command.[88]

Despite their having been thoroughly indoctrinated with the tenets of Japanese military tradition, there were some enemy soldiers who did not particularly wish to die for Emperor and Homeland. Psychological warfare teams had interpreters and cooperative prisoners [89] broadcast surrender inducements in Japanese over loudspeakers mounted on tanks operating at the 7th Division front and on LCIs cruising up and down the southern coast. These broadcasts successfully convinced 3,000 civilians to surrender.

A more significant result of these messages occurred on 19 June, for instance, when 106 Japanese soldiers and 283 *Boeitai* voluntarily laid down their arms and gave up in the face of the 7th Division advance.[90] At this stage of the campaign, the broadcasts influenced increasing numbers of the enemy to surrender as the conviction that all was lost and their cause was hopeless sank into their war-weary minds.

Their forward progress now slowed by fleeing civilians as well as the entrenched enemy, 7th Division troops, nonetheless, advanced to within 200 yards of the outskirts of Mabuni by nightfall of 19 June. Tanks accompanying the assault infantry placed direct fire on caves fronting Hill 89, not knowing that at that very time, General Ushijima was giving a farewell dinner for his departing staff officers.

Farther inland, on the right of the division zone, 184th and 381st Infantry units drove towards Medeera from the south and east against considerably lessened fire and resistance. Nevertheless, small fanatic groups, defending the complex terrain protecting the *24th Division* headquarters, had to be overcome before the major objective could be seized. To the northwest of Medeera, 96th Division soldiers pushing towards Aragachi from the north found the same enemy reluctance to withdraw, encountered elsewhere along the Tenth Army front, before they could reach the high ground overlooking the village. While observing the 384th Infantry fighting

[87] *Okinawa Operations Record,* p. 134.

[88] *Yahara Interrogation; Shimada Interrogation.*

[89] Commenting on this after the campaign, the 6th Division noted that "By far, the most potent propaganda messages are those composed by local civilians and prisoners of war, and the most satisfactory broadcasts are those in which such persons are utilized." *6th MarDiv SAR, Ph III,* chap VI, p. 13.

[90] Tenth Army G-2 Rpt No. 86, dtd 20Jun45.

to gain these heights, the ADC of the 96th Division, Brigadier General Claudius M. Easely, was killed by enemy machine gun fire.[91]

The advance of IIIAC assault forces on 19 June was highlighted when the 8th Marines completely penetrated Japanese defensive positions to reach the sea. Less successful, however, were the efforts of the 5th Marines in a day-long attack on Hills 79 and 81. With a company of tanks in support, 1/5 jumped off at 0730 to take Hill 79 first and then 81. Despite the direct fire placed on the initial objective by the Shermans and M–7s, the battalion was unable to take Hill 79 and was forced to return to positions held the previous night.

As he observed the course of the fighting and judged that neither Hill 79 nor 81 were going to be taken, Colonel Griebel ordered 2/5 to take the latter from the south in order to lift some of the enemy pressure on 1/5. Lieutenant Colonel Benedict's 2d Battalion, which had been relieved on Kunishi Ridge at 1315 that day by 3/7, moved out in a march column at 1515, made a wide swing to the southwest through the 8th Marines zone, and halted at a point some 300 yards southwest of Hill 79 at 1700. Moving out some 15 minutes later, the battalion headed towards Makabe preparatory to attacking Hill 81. As the battalion cleared the southern slope of Hill 79 and began to maneuver across the 1,000 yards of exposed flat terrain lying between that hill and Makabe, the entire column was taken under sniper fire from the hill. Company G, in the lead, was forced to double time over the entire route in order to reach some cover in Makabe. During this race for life, the company sustained some casualties from the fire as well as 20 exhaustion cases.

To maintain the momentum of the attack, the battalion commander passed Company F through G at 1950 and he himself accompanied the assault platoon, which was pinned down as soon as it attempted to move up the slope of Hill 81. The condition of his men, the lateness of the hour, and the intensity of the enemy fire compelled him to call off the attack and organize his battalion into a defense perimeter near Makabe.

More satisfactory progress in the 1st Division advance was made by the 8th Marines. (See Map 23.) After moving through 2/8 at 0800, 3/8 continued south to attack Ibaru Ridge following an hour-long artillery preparation and a 15-minute smoking of the target. At 1024, the battalion was on the ridge. Quickly it reorganized and resumed its drive by passing Company K, 3/8 reserve, through the initial assault elements more "for the experience rather than for any tactical necessity."[92] By 1634, the entire battalion line was in place on the seacoast in its zone. The 3d Battalion, 5th Marines, which had taken Makabe that morning, kept pace with the rapid 2,500-yard advance of 3/8 and reached the coast in its zone at approximately the same time. With 3/5 attached, Colonel Wallace's regiment took charge of the night defense of the coastal zone, and tied in with the 5th Marines and the 4th Marines along a

[91] Davidson, *et. al.*, *96th InfDiv Hist*, pp. 182–183.

[92] 3/8 AR, p. 6, encl to *8th Mar AR*.

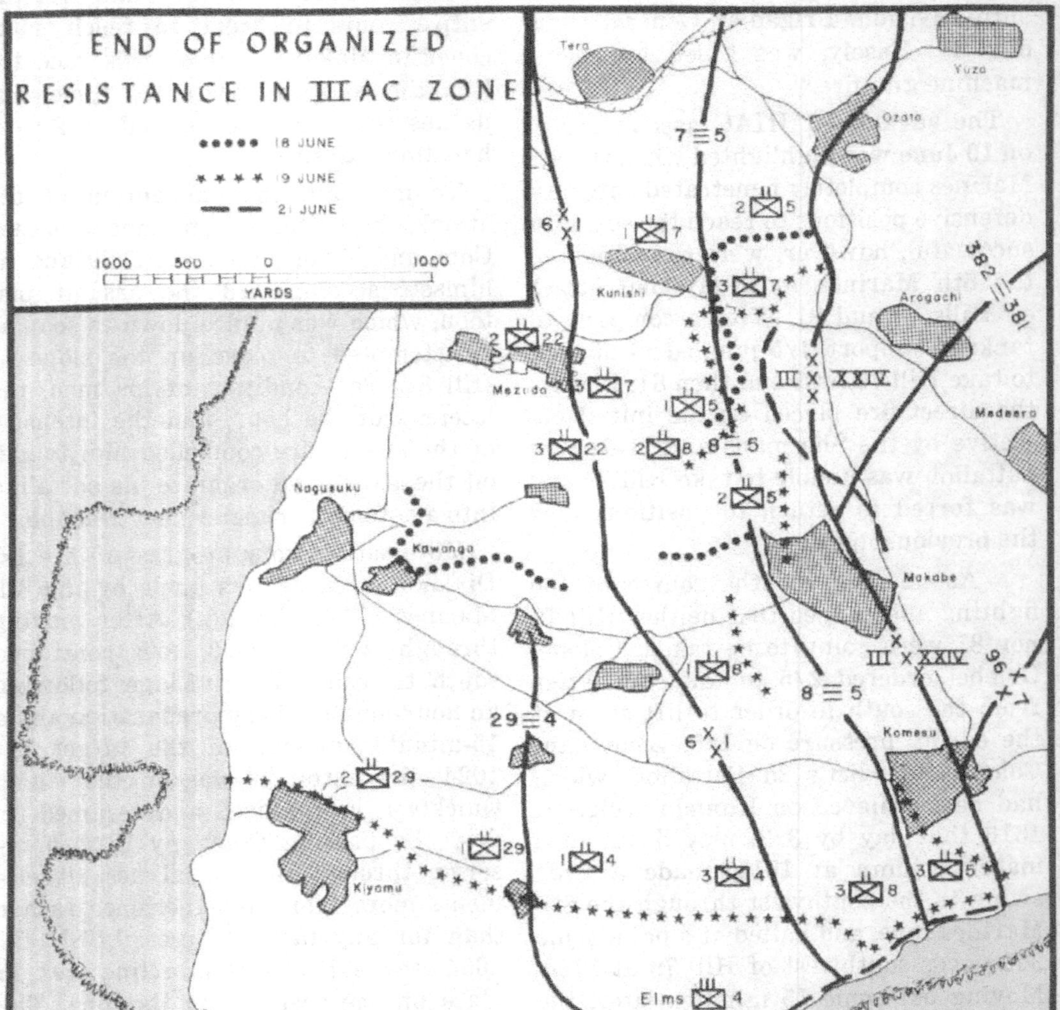

MAP 23 T. L. RUSSELL

curving line reaching from Komesu to the boundary between the Marine divisions.

The 4th Marines made the major effort of the 6th Division on 19 June, with the 22d Marines mopping up behind. Colonel Shapley's battalions kept pace with the 8th Marines most of the day, but strongly defended enemy positions in the Kiyamu-Gusuku hill mass prevented the 4th from reaching the coast on the 19th. Mortar fire from defiladed emplacements behind the hill, and machine gun fire as well, increased in volume as the 1st and 3d Battalions moved into the low ground leading from Ibaru Ridge to the Kiyamu-Gusuku plateau. With the approach of night, the two battalions dug in at the foot of the steep rise leading to the hilltop. At 1845, 2/4, which had covered the open right flank of the regiment and had taken part in the attack on the ridge, was relieved by 1/29.[93]

Before moving into the 6th Division line on 20 June, the 29th Marines began marching south from Oroku Peninsula at 0800 on the 19th. Its former sector was then occupied by the 6th Reconnaissance Company. At 1415, Colonel Whaling received orders to attack immediately in coordination with the 4th Marines. The regiment jumped off from Kuwanga Ridge at 1705 with 1/29 on the left, 2/29 on the right. Moving rapidly against light enemy resistance, the troops reached the Kiyamu-Gusuku Hill mass before dark and immediately tied in with the 4th Marines for the night.

Unperturbed by night-long disorganized enemy infiltration attempts all along its front, the 6th Division jumped off with four infantry battalions abreast —3/4, 1/4, 1/29, and 2/29 from left to right—to take the hill complex on 20 June. Again making the division main effort was the 4th Marines, in whose zone lay Hills 72 and 80, the key terrain features on the objective. Directly in front of 1/4 line of departure was Hill 72, but the battalion could not place enough men on the crest of that height at one time to maintain a solid foothold. Japanese defenders hidden among the brush and boulders lining the narrow approach to the top frustrated all efforts to gain the hill. Tanks attempted to cut a road to the crest from the flank of the position, but this scheme was foiled when an armored dozer was completely destroyed by a satchel charge thrown from a distance of 15 feet. After a day of bitter fighting at hand-grenade range, the battalion dug in for the night at the same place it had been the night before, less than 20 yards away from the enemy on the ridge above.

[93] At 1600, a POW revealed that there was an enemy force of about 20 entrenched on the small island approximately 300 yards off Nagusuku, and that these men were armed with mortars and light machine guns, which had fired on the 4th Marines flank that day. Following a short but intense Marine artillery concentration on the islet, five soldiers waded to the mainland and surrendered. One was sent back to induce the others to do the same, but without success. "The answer that came back was a definite no, and also included a remark that was not exactly complimentary." *4th Mar SAR, Ph III*, p. 15. A task force, hastily organized from the regimental weapons company, LVT(A)s, and the 1st War Dog Platoon, stormed the island. Five prisoners were taken, 20 enemy killed, and several machine guns and mortars destroyed without a single Marine casualty.

Steep rock cliffs, ranging from 50 to 200 feet in height and covered with heavy undergrowth, faced 3/4. Since a frontal attack was clearly infeasible, the battalion commander sent a company to the left through the 8th Marines zone to take the ridge by attacking up its nose on the east; this sector appeared to be the one most susceptible to attack. Clearing out several bunkers and numerous caves to make way for Company L following in its wake, Company I mopped up the eastern slope while Company L tied in with the 8th Marines in order to extend the battalion's hold on the ridge. By late afternoon, 3/4 held strong positions on the left flank of Hill 72 and was ready to close in on that strong point.

Although it was in regimental reserve when the attack had begun, 2/4 was alerted to support either of the other two assault battalions. At 1040, it was committed on the right of 1/4 with orders to take Hill 80. Attacking with two companies abreast, the battalion reached its objective at approximately 1230, when Company G fought its way to the top against only moderate resistance. Company E, the other assault element, was held up at the base of the hill by an enemy pocket, which the battalion commander decided to bypass, leaving a Company E platoon behind to guard it. At 1520, the two-platoon company passed through the right element of Company G to seize the remainder of Hill 80 from the west. By 1645, all units of the battalion were on the hill and digging in. Possession of Hill 80 gave the battalion terrain commanding the right flank of the stubbornly held Hill 72.

On the extreme right of the division, the 29th Marines advanced to the coast on 20 June against little opposition except for heavy fire received on the left flank of 1/29 from enfiladed positions on the reverse slope of Hill 72. Later in the afternoon, when General Shepherd decided to envelope the Kiyamu-Gusuku sector from the left (east), he shifted the boundary of the 29th Marines to the east to include all of Ara Saki. The regimental line was then tied in with the 4th Marines for the night. The 29th Marines positions barred escape to the sea from the tip of the island.

On 20 June, psychological warfare detachments on board a LCI equipped with a loudspeaker broadcast surrender inducements to the many civilian and military personnel hiding in inaccessible cave refuges lining the coastal cliffs. A feeling that further resistance was futile as well as a sense of impending doom impelled over 4,000 island natives and some 800 soldiers to heed the message and to surrender. These POWs were then herded through the front lines before dark to stockades in the rear.[94]

By 20 June, 1st Marine Division action centered about Hills 79 and 81. While 1/5 and 2/5 concentrated their efforts

---

[94] IIIAC G–2 PeriodicRpt No. 81, dtd 21Jun45. A large number of the enemy surrendered in the zone of 3/4, which detailed some of its Marines to strip and search the military prisoners. Before they were escorted to the rear, the POWs were employed to distribute the last supply air-drop made during the campaign. *3/4 SAR, Ph III*, p. 10; VMTB–232 WarD, Jun45; *AirDelSec AR*. Some of the supplies in the drop were for 6th Division tanks, also. Pilots flying the mission reported that the tanks were so close to enemy troops that the Marine armor was "obliged to use their flame throwers to keep the Japanese away from the provisions." MAG–33 WarD, Jun45.

in this area, extensive mopping up operations were conducted by the 7th Marines at Kunishi Ridge, the 8th Marines along the coast north of Ara Saki, and 3/5 around Komesu. These exercises added approximately 50 military and 2,000 civilian POWs to those already captured by IIIAC forces.[95]

After 3/5 gave fire support to the 7th Division from positions on Komesu Ridge, its patrols linked up with 1/184 at 1520. Physical contact was not maintained for the night, but both battalions occupied high ground near Komesu and Udo and were able to cover the gap between battalions by fire.

A brief but soaking downpour before dawn turned the roads around Makabe into knee-deep quagmires, and the tanks and M–7s supporting 2/5 were prevented from moving into position until shortly before noon. A more favorable situation existed in the 1/5 zone, where tanks lumbered forward at 0730 to join the infantry in the attack on Hill 79. The battalion commander swung the axis of attack from the northwest to the southeast and assaulted the objective with three companies abreast. By 1300, Company C on the right flank was 75 yards from the hillcrest, while the other two companies, A in the center and B on the left, were destroying snipers and machine gun nests on the hillside with the aid of flame and gun tanks. At 1635, Company A announced that some of its troops were on the hill, but less than two hours later it reported that heavy small arms fire had prevented it from consolidating its slight hold with the few men available; it was forced, therefore, to withdraw. In possession of most of Hill 79, 1/5 dug in for the night, fully expecting to secure the entire objective the next day.

At 1230, when the 2/5 tank-infantry assault on Hill 81 began, the tracked vehicles reported that road blocks in Makabe denied them passage to the hill. An armored dozer cleared the way by 1400, and tanks moved along the road on the corps boundary to positions where they could fire into the right of Hill 81. The infantry battalion moved to and jumped off from the northern edge of Makabe at 1520 with Companies E on the right, F on the left, and G in reserve. Twenty-five minutes later, Company F was pinned down in the low ground south of the hill; a smoke screen was required to cover the evacuation of casualties. Company E, attacking from the southeast, pushed forward for about 100 yards along the eastern slope of the hill before it too was pinned down. First Company F, and then G was ordered to pass through E and continue the attack. Enemy machine gun and mortar fire pinned down these two companies also. When tanks supporting the attack ran out of ammunition at 1910 and withdrew, the assault companies attempted without success to garner more ground on their own. His troops stymied, the battalion commander pulled them back to more favorable positions for night defense.

To the left, in the XXIV Corps zone, only two strong enemy pockets remained at the end of 20 June. One was centered about the caves containing the *Thirty-second Army* headquarters in Hill 89, and the other was in Medeera and west of the village on Hills 79 and 85, which

---

[95] IIIAC G–2 PeriodicRpt, *supra.*

together with Hill 81 in the 1st Marine Division zone formed the Makabe Ridge defenses. The last courier contact between the two strongpoints was made on the night of 20 June, after the commander of the *24th Division*, Lieutenant General Amamiya, urged his soldiers "to fight to the last man in their present positions." [96]

This exhortation fell on deaf ears for the general had few live men remaining to defend the Medeera sector at the time of proclamation. The 1st Marine Division had just about annihilated the *22d* and *32d Regiments* during its march to the coast, and the 96th Division had destroyed the *89th Regiment* and its reinforcements when taking Yuza Dake and Aragachi. The only troops left to General Amamiya were a motley conglomeration of artillerists, drivers, medical attendants, engineers, *Boeitai*, and personnel from almost every headquarters unit of the forces that had made up the island garrison on L-Day. Despite the growing numbers of the enemy which surrendered and others who committed suicide, the Tenth Army still had to contend with some Japanese who fought to the last with fanatic determination. An attack to destroy these soldiers holding the Makabe Ridge defenses was scheduled for noon of 21 June.

At 1027 that day, General Shepherd notified the Tenth Army commander, General Geiger,[97] that organized resistance had ended in the 6th Marine Division zone of action. Beginning this last official day of the Okinawa campaign, the 4th Marines enveloped troublesome Hill 72. While 2/4 and 3/4 worked around to the south of the ridge, 1/4 held its position to support the attack by fire. Linking up at 0930, the two assault battalions and supporting armor worked north to the objective, and then drove over its top and down the reverse slope. By 1020, the Marines and both flame and gun tanks were mopping up the last vestiges of enemy resistance on the hill. At the tip of the island, the 29th Marines met only light opposition during its sweep of Ara Saki; Company G, 2/22, attached to 1/29, raised the division colors on the southernmost point of the island later in the day.[98]

Both the 7th and 8th Marines were assigned the task of flushing out enemy holdouts in the IIIAC zone and of accepting the surrender of an ever-increasing number of soldiers and civilians. Hill 79 was finally taken by 1/5 at 1735; more difficult, however, was the capture of Hill 81.

Although scheduled to jump off at 0900, the attack of 2/5 was delayed until 1104 in order that tank routes could be prepared and so that the battalion could take immediate advantage of a blistering rocket barrage on the hill ob-

[96] *Okinawa Operations Record*, "Record of the 24th Division," p. 194.

[97] General Geiger retained command of IIIAC during the period he temporarily commanded the Tenth Army. LtGen Merwin H. Silverthorn comments to HistBr, HQMC, dtd 25Oct65.

[98] This company thus duplicated what it had done earlier in the campaign when it had been the first unit to raise the same flag over the northernmost point on Okinawa, at Hedo Misaki. Cass, *6th MarDiv Hist*, pp. 174–175. It appears as though the 22d Marines had a predilection for securing the extremities of island objectives, for it had done the same thing during the retaking of Guam. Lodge, *The Recapture of Guam*, p. 154.

jective.[99] The attack plan called for Company E to lead the assault on the hill, and to be followed successively by Companies F and G, which were to be fed in from the left until Hill 81 was taken. Company E encountered only light and scattered small arms fire as it jumped off, and finally fought to and occupied its assigned objective after having destroyed two machine gun positions that had halted it on the way up. Almost immediately, Company F began fighting its way up the slope to the hilltop, burning out and sealing caves along the route. Shortly thereafter, Company G made its tortuous trek up the incline to join the other two at the top, all companies received heavy fire from caves, which honeycombed the enemy position.

The effort to secure the objective was spurred on by information received at the battalion CP of that Hill 81 was the last organized enemy position on Okinawa; this story later proved untrue. After having made several unsuccessful requests for reinforcements, and been ordered in turn to continue the attack with the forces at hand, at 1430 Lieutenant Colonel Benedict was relieved and ordered to report to the regimental commander. He then turned over command of the battalion to his executive officer, Major Richard T. Washburn. At 1500, the commander of 3/5 reported in at the 2/5 OP and assumed joint command of the two battalions; his Company L began moving to Makabe soon after to support the attack on Hill 81.

All companies advanced slowly during the afternoon, and as 2/5 reached the crest of the hill, enemy fire slackened noticeably. At 1700, all companies reported their portion of the objective secured; all organized enemy resistance in the IIIAC zone had ended.[100]

In the XXIV Corps zone, a heavy 4.2-inch mortar concentration on Hill 79 preceded the attack of 305th Infantry elements at 1200. The crew-served weapons organic to the infantry battalions supported the tank-led attack. At 1630, following an afternoon of withering rifle and machine gun fire coming from caves and pillboxes on Makabe Ridge, the infantry launched a final, successful surge to the top of the hill. Before XXIV Corps units could report the end of organized resistance in the army zone, they had to come to grips with a bitter, last-ditch Japanese defense; objectives were captured only after enemy defenders had been killed to the last man. The soldiers first secured Mabuni and then Hill 89. General Buckner's doctrine of "corkscrew and blowtorch" was employed effectively by flame tanks and demolition teams burning and blasting the "palace guard" defending the cave entrances leading to General Ushijima's headquarters. By the end of the day, Hill 89 had been secured, and its inhabitants were frantically attempting to escape a death by entombment.[101]

[99] LtCol William E. Benedict ltr to CMC, dtd 27Mar47.

[100] *2/5 SAR*, p. 25.

[101] It was surprising that there were any Japanese alive in the vicinity of Hill 89 at this point. Discovered earlier in the month, the *Thirty-second Army* CP had been a prime target for TAF aircraft ever since. On 13 June, for instance, a total of 64 planes—23 from VMF-323, 24 from VMF-441, and 17 from VMF-314—burned and blasted the hill and its environs with 124 napalm bombs and 335 5-inch rockets in less than an hour. *ADC IntelSums*, Jun45.

After 82 days of bloody and bitter fighting, the rapid advance of the Tenth Army in the final stages of the campaign brought about irrevocable collapse of all major Japanese opposition. General Geiger could thus announce at 1305 on 21 June that the island of Okinawa had been secured by American forces. The official end of the Okinawa campaign was marked by a formal flag-raising ceremony at Tenth Army headquarters at 1000, 22 June, attended by representatives of all units in that command. As described by General Smith: "A large metal flagpole had already been erected at Army Headquarters. . . . The only band available[102] was that of the 2d Marine Aircraft Wing. Prior to playing the National Anthem,"[103] the band played "Anchors Aweigh," "The Marines Hymn," and an appropriate Army tune. Brigadier General Lawrence E. Schick, USA, Tenth Army Deputy Chief of Staff, read the official dispatch declaring the end of organized resistance, and General Geiger then gave the signal for the flag to be raised.

[102] Each Tenth Army division had its own band, which participated in the campaign according to the combat functions assigned it by the parent division. None of the bands functioned as musical units until after Okinawa had been secured. During the fighting, the bandsmen performed as stretcher bearers.

[103] Smith, *Personal Narrative*, p. 138.

Following the official announcement on 21 June of the ending of organized resistance on Okinawa, Tenth Army headquarters began receiving congratulatory messages from statesmen and military commanders throughout the world. Though heartfelt and sincere, none of these commendations to the men who had fought the Battle of Okinawa could match the simple accolade bestowed on Marines of the IIIAC by the commander who had led them, for as General Geiger wrote:

> This has been a hard campaign. The officers and men have simply been marvelous. They have carried on day and night, mud and battle, without a murmur and could have continued had it been necessary. They have carried out every mission assigned by the Tenth Army and have broken through every position of the Japanese defenses which stood in their way in a minimum of time. The Marine Corps can ever be proud of the two divisions which fought on this island. The cost has been high, but the time element was essential and I am sure you will be happy to know that the Marines required no urging to attack, attack, and again attack, until the Japanese were completely annihilated. You will never know how I regret leaving the III Corps.[104]

[104] LtGen Roy S. Geiger ltr to CMC, dtd 26Jun45 (Vandegrift Personal Correspondence File, HistBr, HQMC). Shortly thereafter, General Geiger became the Commanding General, FMFPac.

MEN *of the Tenth Army pay homage to their fallen leader, Lieutenant General Simon B. Buckner, Jr. (USN 80–G–498161)*

SURRENDER *of all Japanese forces in the Ryukyus takes place at Tenth Army headquarters on 7 September 1945. (USA SC211950)*

ICEBERG Dissolves

MOPPING UP[1]

With his defenses overrun and forces shattered, there was little hope of diverting or lengthening the path leading to the inevitable fate of his *Thirty-second Army*. Lieutenant General Mitsuru Ushijima decided, therefore, to end his life according to the dictates that governed his living of it, the traditional way of the *Samurai*. Joining him in fulfilling his obligation to the Emperor and dying in the symbolic way of *bushido* was the army chief of staff, Lieutenant General Isamu Cho.

Following a meal late on the night of 21 June, Cho and Ushijima composed their last farewell messages and the following valedictory poems written in the classic Japanese style:

> The green grass of this isle
> Withers untimely before fall,
> Yet it will grow again
> In the warm spring of the Empire.
>
> Smearing heaven and earth with our blood,
> We leave this world with our ammunition gone,
> Yet our souls shall come back again and again
> To guard the Empire forever.[2]

At noon on 22 June, Ushijima dressed himself in his full field uniform and Cho donned a white kimono on which he had written "The offering of one's life is to fulfill the duties towards the Emperor and the Country. Cho, Isamu."[3] As the two led a party of aides and staff officers out to a ledge at the mouth of the cave headquarters, Cho was quoted as saying, "Well, Commanding General Ushijima, as the way may be dark, I, Cho, will lead the way." Ushijima replied, "Please do so, and I'll take along my fan since it is getting warm."[4]

Ten minutes after leaving the cave, first Ushijima and then Cho died in the Japanese time-honored ritual of *hara-kiri*. Each in turn bared his abdomen to the knife used in the ceremonial disembowelment and thrust inward; as each did so, there was a simultaneous shout and flash of a sword as the headquarters adjutant decapitated first one general and then the other. The bodies were then secretly buried in graves prepared earlier. Three days later, 32d Infantry patrols discovered them at the foot of the cliff of Hill 89 where it faced the sea. On the white bedding

[1] Unless otherwise noted, the material in this section is derived from: *Tenth Army AR; IIIAC AR; XXIV Corps AR; 1st MarDiv SAR; 6th MarDiv SAR, Ph III; 7th InfDiv AR; 77th InfDiv OpRpt*, Okinawa; *96th InfDiv AR;* Capt Russell A. Gugeler, USA, The Operations of the 7th InfDiv on Okinawa, 1Apr-22Jun45, 3 vols, n.d., (OCMH), hereafter Gugeler, *7th InfDiv Hist.*

[2] Quoted in Hattori, *War History*, p. 131.

[3] Gugeler, *7th InfDiv Hist*, p. 497. The account of the deaths of the two generals in this work was derived from a POW interrogation. The Japanese officer who told this story had learned the details from other prisoners and prepared this account.

[4] *Ibid.*, p. 498.

cover which served as his winding sheet after death, General Cho had written:

> 22d Day of June, 20th Year of Showa
> I depart without regret, fear, shame, or obligations.
> Army Chief of Staff, Army Lt. Gen. Cho, Isamu
> Age at departure, 51 years.
>
> At this time and place, I hereby certify the foregoing.[5]

On 25 June also, the *Imperial General Headquarters* in Tokyo announced the end of Japanese operations on Okinawa, and, in effect, of the *Thirty-second Army*.[6] *IGHQ* then put all of its efforts into preparations for the defense of the Home Islands against an anticipated American invasion.

Although the commander and chief of staff of the *Thirty-second Army* were dead, and many Japanese officers and enlisted men were surrendering, other enemy soldiers both in groups and individually continued a fanatic, last-ditch stand until they were destroyed. General Stilwell believed it necessary to eliminate these isolated Japanese pockets to safeguard the Island Command forces that were developing the additional supply, training, airfield, and port facilities required to convert Okinawa into a massive base for further operations against Japan. He ordered, therefore, the Tenth Army to begin an intensive, coordinated mop-up of southern Okinawa on 23 June; 10 days were allotted to this task.

The 1st Marines and 307th Infantry were deployed in a line of blocking positions paralleling the Naha-Yonabaru highway to bar the way to enemy soldiers who were attempting an escape to northern Okinawa. The American sweep northwards was mounted by the five assault divisions that had made the final drive in the south and had been on line when the war ended; they began the sweep by merely making an about-face in position. As the soldiers and Marines drove towards the Tenth Army blocking positions, they smashed all remaining enemy opposition, blew and sealed Japanese caves, buried all Japanese dead, and retrieved all salvageable enemy and friendly equipment along the way. To coordinate and pace the 10-day sweep, three phase lines were established. Flanking divisions were to guide on the 96th Division as it progressed up the center of the island. General Stilwell retained control of the entire operation.

On 30 June, in less than the time allotted, the mop up was successfully completed. Elements of the 77th Division reduced the final defensive positions of the *24th Division* near the ruins of Medeera; the 96th Division thoroughly cleaned out enemy pockets in the Medeera-Aragachi sector; the 1st and 6th Marine Divisions worked over Japanese survivors in the Kiyamu-Gusuku and Komesu Ridges; and the 7th Division did the same to the Hill 89-Mabuni area. Several brief but bloody fire fights flared during the methodical, workmanlike sweep of the objective area when strongly armed enemy bands tried futilely to break through the American line and were smashed.

Results of the sweep indicated that an estimated 8,975 Japanese had been killed and 2,902 military prisoners and 906 labor troops had been added to those

[5] *Ibid.*

[6] Hattori, *loc. cit.*

already in Tenth Army stockades. Enemy losses for the entire Okinawa campaign, were placed at 107,539 counted dead and an estimated 23,764 more which were assumed to have been sealed in caves or buried by the Japanese themselves. In addition, a total of 10,755 of the enemy had been captured; some of this number had surrendered. As the overall Japanese casualty total of 142,058 was "far above a reasonable estimate of military strength on the island," Tenth Army intelligence agencies presumed that approximately 42,000 of these casualties were civilians that had been unfortunately killed or wounded in American artillery, naval gunfire, and air attacks on enemy troops and installations while the natives had been in the proximity.[7]

American losses were heavy also. The total reported Tenth Army casualty figures were 7,374 killed or died of wounds, 31,807 wounded or injured in action, and 239 missing. There were 26,221 nonbattle casualties in addition. The combat divisions alone reported a total of 38,006 casualties of all types.[8] Between 1 April and 30 June, Army units received 12,277 replacements; Marine units joined 11,147 Marines and naval corpsmen in the same period.

Both British and American naval forces took heavy casualties while supporting and maintaining the Tenth Army. During the 82 days of ground operations, 34 ships and craft were sunk and 368 damaged; 763 carrier-based aircraft were lost to all causes. In addition, 4,907 sailors were killed or missing in action and 4,824 were wounded. At the time that these losses were sustained, ships and ground antiaircraft artillery and planes controlled or coordinated by the Navy claimed the destruction of 7,830 Japanese aircraft and 16 combatant ships.[9]

In accordance with the planned succession of operational control established for ICEBERG,[10] Headquarters, Ryukyus Area superseded the Tenth Army on 1 July 1945. At that time, General Stilwell became a joint task force commander directly responsible to Admiral Nimitz for the defense and development of all captured islands and the defense of the waters within 25 miles of Okinawa. Concurrently, after

[7] *IntelMono*, pt I, sec B, chaps 2, 3. One Japanese source indicated that approximately 75,000 soldiers and 50,000 Okinawan noncombatant civilians were killed during the battle for the island and that half of the survivors were wounded. *Okinawa Operations Record*, p. 152. A second and more recent Japanese account, contradicting the casualty figures noted above as well as those cited by American sources, stated that about 90,000 Japanese troops and Okinawan volunteers were killed and that there were as many as 150,000 island natives killed. This source said also that slightly more than 7,800 troops had survived the battle, but half of these were wounded. The other half continued to resist from underground positions and by operating as guerrillas. Hattori, *War History*, p. 132.

[8] Sources for these figures are the appropriate annexes in the *Tenth Army AR* and the ARs of the combat divisions. See Appendix M for the final compilation by unit of all Marine casualties.

[9] *USSBS Campaigns*, p. 331. A Japanese source states that 7,852 aircraft (2,393 *Kamikazes*) in both *Kikusui* and smallscale attacks were sent against American forces between 6 April and 22 June. Of this number, 2,258 never returned. Hattori, *War History*, table facing p. 132.

[10] *Tenth Army Tntv OPlan 1-45*, Anx 1, p. 4.

CinCPac had dissolved Task Force 31, Admiral Hill and his staff departed for Pearl Harbor and Rear Admiral Calvin H. Cobb took over as Commander Naval Forces, Ryukyus, under General Stilwell. TAF at this time came under the Ryukyus command. All of these forces, and others that were to be sent to Okinawa, were to be commanded by General Stilwell. He was to coordinate and control the massive effort supporting the impending operations against the center of the Japanese Empire. Slated to become a major force in carrying the air war to Japan was the Tactical Air Force.

TAF FIGHTS ON[11]

Only five days had intervened between the eighth mass *Kamikaze* raid of 27–28 May—*Kikusui* No. 8—and the ninth, which began on the evening of 3 June and lasted until 7 June. As before, TAF fighter aircraft rose from fields on Okinawa and Ie Shima to meet approximately 245 Japanese planes coming from the Home Islands. American pilots and antiaircraft artillery units claimed a total of 118 enemy planes downed during *Kikusui* No. 9; the Marine pilots of TAF claimed 35 of this number.[12]

At the same time that Generals Cho and Ushijima began their suicide preparations, Japanese pilots flying the final mass *Kamikaze* raid of the Okinawa campaign arrived over the island, prepared to die according to the philosophy of the *Samurai*, but in a more modern fashion. Approximately 68 of the 257 aircraft launched in *Kikusui* No. 10 were suiciders. The first group of raiders appeared over Kerama Retto on 21 June at 1830, and correctly replied to friendly recognition signals.[13] One *Kamikaze* dived headlong into the seaplane tender *Curtis* to start night-long fires that severely damaged the ship. Shortly after, planes from this flight attacked LSM–*59* as it was towing the hulk of decommissioned *Barry*[14] away from the Kerama anchorage to act as a *Kamikaze* decoy, and both vessels were sunk.

On 22 June, Marine pilots from MAG–22 were flying a barrier combat air patrol over Amami O Shima, when they were jumped by approximately 60 enemy planes heading for Okinawa along the well-travelled *Kamikaze* air route

[11] Unless otherwise noted, the material in this section is derived from: *CNO Record; AirIntelBuls*, May-Sep45; *CTF 51 AR; Tenth Army Tntv OPlan 1–45; Tenth Army AR;* TAF WarD, Jun45; TAF PeriodicRpt, Jun45; ADC WarDs, Jun-Jul45; *ADC AcftOpAnalysis*, Jun45; *ADC IntelSums*, Jun45; *IIIAC AR; XXIV Corps AR; 1st MarDiv SAR;* 2d MAW WarDs, Jun-Jul45; *6th MarDiv SAR, Ph III; 1st Prov AAA Gru AR; 53d AAA Brig AR;* MAG–14 WarD, Jun45; MAG–22 WarD, Jun45; MAG–33 WarDs, Jun–Jul45; VMTB–131 WarD, Jun45; VMTB–232 WarD, Jun45; *AirDelSec AR;* Hattori, *War History;* Sherrod, *Marine Air Hist.*

[12] *CTF 31 AR*, pt III, pp. 51–61; TAF WarD, Jun45. Japanese sources claim that only 54 planes did not return. Hattori, *War History*, table facing p. 132.

[13] CinCPac WarD, Jun45, p. 75.

[14] The APD *Barry* had been crashed by a *Kamikaze* and gravely damaged late in May. A Navy Board of Inspection and Survey recommended that it be decommissioned and cannibalized. *Ibid.*

from Kyushu. The skies immediately buzzed with a frenzy of darting and diving aircraft. One pilot was later heard to say over the radio, "Come on up and help me, I've got a Frank and two Zekes cornered." [15] No further word was heard from him, and he was later listed as missing.

During the debriefing after this engagement, the MAG–22 fliers reported that the enemy had tried to decoy them into unfavorable positions. Four of the Japanese planes were first sighted at 20,000 feet, and as a division of Corsairs went after them, the decoy planes made a run for safety, but pulled up "and dropped their belly tanks in front of and above the Marine planes. Our pilots had to [maneuver violently] in order to evade the falling tanks. The F4Us turned to press home their attack when the larger force of enemy planes jumped in and a general melee resulted." [16] In evaluating the enemy, the Marines reported that the Japanese pilots flew a good, tight division formation of four planes abreast, and "they seemed to be good pilots but maneuvered poorly." [17] Of the 51 planes Americans claimed to have shot down in this encounter, TAF pilots listed 44.

Although MAG–14 (VMF–212, –222, –223), commanded by Colonel Edward A. Montgomery, did not arrive on Okinawa until 8 June, too late to participate in the "turkey shoots" against the *Kamikaze* attacks, once the group began operations on the 11th, its pilots and planes took part in the stepped-up tempo of TAF strikes on such scattered targets as Sakashima Gunto to the south of Okinawa, Kyushu to the north, and the coast of China to the west. On 22 June, Captain Kenneth A. Walsh, an ace at Guadalcanal and winner of the Medal of Honor for achievements during the same campaign, shot down his 21st enemy plane. In its brief combat tour in the Ryukyus, the group as a whole claimed nine kills.

Whenever the weather permitted in June, TAF greatly expanded its offensive operations and strikes on outlying targets. The primary mission of the far-ranging American planes was to seek out and destroy enemy planes and support installations. These operations involved flights of large numbers of single engine aircraft over water for distances nearly equalling their maximum ranges. Because of their long-range capability, the P–47 Thunderbolts of the AAF fighter squadrons attached to TAF [18] performed a dual role as both fighters and bombers. On some missions, the P–47s bombed and strafed targets of opportunity as well as assigned targets; they escorted light, medium, and heavy

[15] MAG–22 WarD, Jun45. The Zeke, known earlier in the war as the Zero, was a single engine fighter manufactured by Mitsubishi. The Frank, likewise a single engine fighter but manufactured by Nakajima, was a newer and faster plane that appeared during the late stages of the war.

[16] *Ibid.*

[17] *Ibid.*

[18] The 318th Fighter Group joined TAF about 30 April and began operations from the field on Ie Shima almost immediately thereafter. The other two groups of the 301st Fighter Wing arrived in Okinawa in succeeding months; the 413th on 19 May and the 507th on 24 June.

310-224 O - 69 - 25

bomber missions after Bomber Command joined TAF in June and July.[19]

At this time, existing airfields on Okinawa were expanded, and new ones built at Awase on the east coast and Chimu in the north in accordance with base development planning. The influx to these and the other fields of newly joining squadrons increased ADC aircraft strength from 432 planes at the beginning of June to 711 at the end.[20] With these additional aircraft, TAF mounted increasingly stronger air attacks against the Japanese Home Islands. Marine fighter planes from ADC hit Kyushu installations for the first time on 10 June, the day before Major General Louis E. Woods relieved Major General Mulcahy as TAF commander.[21]

There was little change in the missions of TAF, Ryukyus Command, from those it had fulfilled as an agency of the Tenth Army. On 1 July, when the command change occurred, ADC assumed complete responsibility from TF 31 for the air defense of the Ryukyus. At this time, TAF aircraft strength was substantially increased, especially by the bomber squadrons, and General Woods could send his planes to better objectives further away from Okinawa than those attacked previously. In its first raid under TAF, on 1 July the 41st Bombardment Group sent its Mitchell bombers to blast Kyushu. On that same day, TAF inaugurated a combat air patrol over Kyushu in hope that Japanese pilots would take off from island airdromes to engage the American planes. Few enemy pilots rose to the occasion.

In another phase of TAF operations, Thunderbolts began hitting Japanese installations on the China coast near the Yangtze Estuary on 1 July. A landmark

[19] Four AAF bomber groups of the Bomber Command joined TAF, beginning with the 41st which arrived in Okinawa on 7 June. The 494th Bombardment Group arrived on 24 June, the 319th about 2 July, and the 11th actually on 2 July.

[20] ADC WarD, Jun45.

[21] "My assignment as General Mulcahy's relief was a most unexpected one for me. I was in Pearl waiting to see [Major] General [Ralph J.] Mitchell, whom I was relieving as CG of the First Air Wing, then at Bougainville, when AirFMFPac gave me immediate orders to go there. I took off with double crews and flew continuously until arrival. Upon arriving, I reported to General MacArthur's Headquarters by dispatch. Several hours after my arrival I received immediate orders to proceed to Guam and to report to Admiral Nimitz. I did as ordered and Admiral Nimitz personally briefed me, and told me to take the necessary steps to have the ADC of TAF take over the air defense of the area in the very near future as the losses from the Kamikaze attacks were heavy and he was going to withdraw all ships as soon as possible. Upon my arrival [at Okinawa], I found that Admiral Hill and his flagship were responsible for the air defense of the area, and that the ADC of TAF was really only a fighter command and not responsible for keeping track of friendly aircraft in the area. When I took command . . . I ordered the ADC to get set up to take over the complete responsibility for the air defense . . . without delay. . . . Why the ADC hadn't been called upon to [keep track] of the friendly planes in the area long before this has always been a mystery to me, for how can an outfit provide air defense properly unless it has full information about all friendly and enemy aircraft in the area?

To deepen the mystery was the fact that it has been reliably reported to me that neither AirFMFPac nor Headquarters Marine Corps knew of my being sent to Okinawa until after my arrival." *Wood ltr I.*

in TAF operations occurred on the 9th, when B–24s attacked Japan from Okinawa. All together, the 47 heavy bombers—and the 25 Mitchells and 32 Thunderbolts acting as bombers accompanying them—spread 1,880 clusters of fragmentary bombs and 280 clusters of incendiary bombs over dispersal areas and field installations of Omura airfield on Kyushu. Another 92 Thunderbolts escorting the mission acted merely as spectators; no enemy interceptors appeared.

In accordance with orders from CinCPOA, TAF, Ryukyus was dissolved on 14 July.[22] On that date, all Marine air units reported to the 2d MAW, which was then designated Task Group 99.2, and assigned to the Ryukyus Command. AAF squadrons and groups that had been temporarily assigned to TAF were transferred to the Far East Air Forces (FEAF), which assumed control of the mounting number of air attacks against Japan.

Under the Ryukyus Command, Marine squadrons continued flying the types of missions they had flown previously, but they now ranged much further away from the island than when they had been committed to the air defense of ICEBERG forces. On 19 July, ADC flyers made their first visit to the China coast, when 59 F4Us flew cover for TF 95, then operating off the enemy-held littoral.[23] At 0001, 1 August, the 2d MAW and all of its squadrons with the exception of VMTB–131 and –232, and VMB–612,[24] passed to the operational control of FEAF; the three other squadrons were assigned to the control of Fleet Air Wing 1.[25]

In the period 7 April through 13 July, TAF amassed a creditable record. A Marine aviator himself, General Geiger wrote General Woods that the air support provided by TAF pilots was "outstanding and contributed materially to a speedy and successful completion of the campaign." [26] By the end of 13 July, TAF claimed a total of 625 Japanese planes destroyed in the air and 29 probables; MAG–33 pilots were the high scor-

[22] TAF WarD, 1-13Jul45.

[23] 2d MAW PeriodicRpt No. 1, 14-20Jul45.

[24] VMB–612, commanded by Lieutenant Colonel Jack R. Cram, was the first Marine bomber squadron trained to operate at night employing radar-operated bomb sights and search gear, and to launch rockets from its PBJs (Mitchells, twin engine bombers), the Navy-Marine Corps designation for the B-25. The squadron arrived on Okinawa in July, and operated from Chimu airfield until the end of the war.

[25] 2d MAW WarD, Jul45. Before this change took place, General Stilwell had received a different set of orders from CinCPac headquarters concerning the future disposition of the 2d MAW. According to Admiral Nimitz' first directive, which assigned ". . . all the Fighter Squadrons to the [FEAF] Fighter Command, and the VMTB squadrons to the Fleet Air Wing. That left only my [General Woods] command headquarters without any units . . . under General Stilwell. On the last day I was under his command, I went . . . to see him, and told him I was a General 'without portfolio' and asked permission to fly to Guam. I told him that I would be back in command of the units of the Wing or I wouldn't be back. He wished me luck and I left. When I got to Guam, I saw Vice Admiral [Charles H.] McMorris [CinCPac Chief of Staff] and after some discussion all units were returned to my control and new orders were issued." *Woods ltr I.*

[26] 2d MAW WarD, Jun45.

ers with claims of having shot down 214 enemy aircraft.[27]

Of particular interest is the fact that Marine night fighters came into their own in the air above Okinawa; VMF(N)–533 registered claims of 35 enemy planes downed,[28] while VMF(N)–542 claimed 17, and –543, 11.[29] Some overwhelming statistics appeared in the course of the Okinawa air operations. For example, while flying 118,982 hours and 38,192 sorties, TAF pilots expended 4,102,559 rounds of .50 caliber ammunition and 445,748 rounds of 20mm. In addition, the flyers released 499 tons of napalm, 4,161 tons of bombs, and 15,691 rockets.[30]

The pilots and planes of VMTB–131 and –232 recorded some amazing statistics during their supply drop operations to ground troops. In addition to the 70 supply sorties carrier-based aircraft flew in support of IIIAC ground units, the two TAF squadrons flew 760 sorties for the Tenth Army—80 of these went to XXIV Corps, the rest to IIIAC. The total weight that the TBMs carried on these missions was 668,984 pounds; the supplies weighed 495,257 pounds, cargo paracutes and air delivery containers took up the rest of the weight.[31] Handling these supplies on the carriers first and at the airfields on Okinawa later was the IIIAC Air Delivery Section. Consisting of 1 officer and 82 enlisted Marines, the section was attached to Tenth Army and worked very closely with the TAF squadrons.

Although seemingly prosaic when compared to combat air patrols, supply drop missions were very often just as hazardous. For optimum results, the Avenger pilot had to maintain an air speed of 95 knots, very close to a stall, at an altitude of about 250 feet while trying to spot a drop zone that was supposedly marked by colored smoke, WP grenades, or panels, either separately or together. At the same time, he was being fired upon by Japanese weapons of all sorts—antiaircraft guns as well as small arms. Some pilots had to fly under an arc of friendly artillery and naval gunfire.[32]

In attempting to drop supplies on targets, Marine aviators often found that the drop zone had not been properly marked or correctly identified, or that the Japanese were using the same color of smoke that Tenth Army ground units were supposed to have employed. As a result, the drop mission either was aborted until the zone could be properly identified, or the pilots made an educated guess—in which case, the supplies sometimes were dropped into enemy territory.

When the drop zone was particularly difficult to spot, Air Liaison Parties from

[27] The pilots of the "Death Rattlers," VMF–323 commanded by Major George C. Axtell, Jr., contributed over half, 124½ planes, to the MAG–33 score. One unusual feat in the air campaign of Okinawa took place during the *Kamikaze* attack on 22 April, when, in a 20-minutes period, Axtell shot down five raiders and his two wingmen—Major Jefferson D. Dorroh, squadron executive officer, and First Lieutenant Jerry J. O'Keefe—destroyed six and five enemy aircraft, respectively.

[28] VMF(N)–533 WarDs, May-Jul45.

[29] *Ibid.*, Jul45.

[30] *Ibid.*

[31] *AirDelSec AR*, p. 2.

[32] See discussion of LFASCU operations below for further comments on this problem.

the Joint Assault Signal Companies[33] attached to frontline infantry units coached the TBM pilots to their target by radio. The primary mission of the ALPs was to direct TAF and carrier-based aircraft to the target. Coordinating the requests from lower echelons were the three Marine Landing Force Air Support Control Units (LFASCUs) commanded by Colonel Vernon E. Megee. Colonel Megee wore two other hats: he was representative ashore of the Navy Close Air Support Control Unit (CASCU) that was on board *Eldorado*, and he commanded LFASCU-3 which was the control unit at Tenth Army headquarters. LFASCU-3 coordinated the air requests forwarded from the IIIAC infantry regiments by LFASCU-1 and from XXIV Corps units by LFASCU—2. Each of these control units operated at the headquarters of the corps to which it was attached.

Although close air support techniques and the methods for their control were rudimentary at the beginning of World War II, during the latter stages of the war and especially on Okinawa, improved aircraft, proven control procedures, and pilots skilled in providing close air support served together to make this supporting arm one of the most powerful that was available to the infantry. On Okinawa, ground troops developed great trust and confidence in the ability of close air support to strengthen attacks on particularly stubborn enemy strongpoints and to clear the way for assaults in general. Surprisingly enough in view of the many support sorties flown, there were but few instances when friendly troops were bombed, strafed, or rocketed by accident, even though strikes were often conducted less than 100 yards away from friendly lines. After getting their first taste of what close air support could do for them, Army units were soon "insatiable in their demands."[34]

Throughout the course of the war in the Pacific, senior Marine commanders became and remained staunch adherents to and supporters of the close air support doctrine. As it developed, they became convinced that more extensive use of the ALPs at the division, regimental, and battalion levels would increase the quality and quantity as well as the effectiveness of air support.

After the Okinawa campaign, the consensus of the Marine commanders present there was that, with proper communications equipment and more intensive and complete training, ALPs could easily take over control of strike missions from LFASCUs and "talk" the pilots directly to their targets.[35] This

[33] The JASCO was a joint Navy-Marine Corps organization; the naval component contained the shore fire control parties, which operated with the frontline infantry battalions, spotting targets for and controlling the naval gunfire support of these ground units. The Marine ALPs functioned similarly, but controlled the close air support provided the infantry.

[34] Col Vernon E. Megee quoted in Sherrod, *Marine Air Hist*, p. 411.

[35] *IIIAC AR*, pp. 197–198; *1st MarDiv SAR*, Air Support Anx; *6th MarDiv SAR, Ph I&II*, pt X, pp. 24–26; CG, AirFMFPac ltr to CMC, ser 002233, A16–3/G–3/eas, Subj: Air Support, dtd 28Aug45, with encls A–H, Rpts of 1st MarDiv (Rein), 6th MarDiv, XXIV Corps, 7th InfDiv, 27th InfDiv, 77th InfDiv, 96th InfDiv, and IIIAC, hereafter *CG, AirFMFPac ltr*. These reports duplicate the comments concerning air support each command made in its Okinawa action reports.

procedure of direct air-ground control between ALPs and the planes above them had been developed by the Marine Corps prior to the Okinawa invasion and was used in the 1st and 6th Marine Division training cycles. Colonel Megee later explained that this system was not used at Okinawa because:

> . . . to have permitted each battalion air liaison party to control striking aircraft on a corps front of only ten miles, when many simultaneous air strikes were being run, would obviously have led only to pandemonium and grave hazard for all those concerned. On the other hand, where conditions approximated those in the Philippines, i.e., battalion or regimental actions in an uncrowded area, actual control of aircraft was frequently delegated to the air liaison party.[36]

After having read the comments and recommendations of both Army and Marine commanders concerning the air support they received in the Okinawa campaign, Major General James T. Moore, commander of Aircraft, Fleet Marine Force, Pacific, forwarded them to the Commandant of the Marine Corps. In a covering letter, General Moore recommended "that Marine Air and Ground be organized and combined under one command with the primary mission of Marine Air being the support of Marine Ground Forces." [37] This might very well be interpreted as the first definitive recommendation made by a senior Marine general for the establishment of the balanced air-ground amphibious force in readiness which has become the hallmark of the present-day Marine Corps.

That the ready acceptance of Marine aviation by Marine ground forces as an equal or supporting partner in amphibious operations was not an immediate thing is indicated by General Woods, who said:

> All senior ground generals in World War II believed in the air-ground team but when in the combat area, they were never able to keep aviation under their command. Maybe it is because they gave only lip service to the doctrine. Even as early as Guadalcanal, the First Wing was not under the command of General Vandegrift, and when the First Division left the combat area, all [Marine] aviation units were left behind! [38]

A review of Marine air activities is not complete without mention of the Marine observation squadrons, the VMOs. Although their exploits were not so spectacular as those of the fighter and torpedo bomber squadrons, nor their planes so swift and deadly, the VMOs attached to IIIAC performed as vitally important a role in the successful prosecution of the Okinawa campaign. Assigned to the Marine components of the ICEBERG forces as artillery spotters, VMO–2,[39] –3 and –6 and their OY "Grasshoppers" were ashore and operating from Okinawa fields by 3 April. The squadrons were soon flying other types of missions, however, and not neces-

[36] Quoted in Isely and Crowl, *Marines and Amphibious War*, p. 567. For the story of Marine close air support operations in the recapture of the Philippines, see Maj Charles W. Boggs, Jr., *Marine Aviation in the Philippines* (Washington: HistDiv, HQMC, 1951); see also Garand and Strobridge, "Western Pacific Operations," *passim*.

[37] *CG, AirFMFPac ltr.*

[38] *Wood ltr I.*

[39] Until 15 April, when they returned to Saipan, VMO–2 pilots flew missions for the 11th Marines with VMO–3, but used VMO–6 planes. *Tenth Army AR*, chap 11, sec VI, p. 59.

sarily for the Marine artillery regiments. Within two days, for example, both VMO–2 and –3 were serving a total of 11 Army and Marine artillery battalions—the equivalent of nearly three full regiments.[40] As soon as VMO–7 arrived in early May, it, too, was kept busy.

In addition to spotting missions, Grasshopper pilots and their aerial observers flew photographic and reconnaissance missions. Sometimes, line routes for ground communications were selected after the observers had reported the number and location of telephone poles still standing.[41] In early June and until the end of the campaign, the VMOs made many evacuation flights. During the 12-day period from 11–22 June inclusive, VMO–7 made a total of 369 evacuation flights from the strip behind 1st Division lines; these were in addition to the 243 spotting and 17 photo-reconnaissance missions flown in the same period.[42]

By the end of the Okinawa battle, the four VMOs had flown 3,486 missions.[43] The most valuable of these, in the view of artillery commanders, were the spotting missions. As the G–3 of IIIAC Artillery noted later:

> If there was any group of indispensable officers in IIIAC Artillery on Okinawa, it was our air spotters. The nature of the terrain in southern Okinawa seriously limited ground observation—especially while we were fighting our way uphill on the Shuri massif. Without our AOs [Aerial Observers], IIIAC Artillery would have been blind.[44]

Colonel Henderson continued:

> The courage and daring of our AOs and the VMO pilots was an outstanding feature of the campaign. I think that VMO pilots are the unsung heroes of Marine Aviation. . . .
>
> When they wanted to really investigate something . . . they would go right down on the deck. Often they would fly past cave openings at the same level so they could look in and see if there was a gun there.[45]

This tactic was most important because of the difficulty that often arose in locating Japanese artillery positions, especially those sited in cave mouths.

Considered more a hindrance than a safeguard by both artillery and air support units, restrictive fire Plans Negat and Victor[46] greatly diminished the effect of artillery and naval gunfire bombardments during the early part of the campaign. Colonel Henderson noted that "They were supposed to protect our own close air support planes from friendly artillery fire, but more often served to protect the Japanese from our fire."[47] In addition, the plans were invoked too often, and then remained in effect far

[40] *Air IntelBul,* Jun45, p. 11.

[41] *Ibid.*

[42] *Ibid.,* Aug-Sep45, p. 22; VMO–7 WarD, Jun45.

[43] *Tenth Army AR,* chap 11, sec VI, p. 59. Included in this number are the ones flown by VMO–2 from 3–15 April.

[44] *Henderson ltr.*

[45] *Ibid.*

[46] Containing safety factors that were contingent on specific danger conditions, Plans Victor and Negat were invoked either separately or jointly whenever an air strike was to be delivered on the same target being fired upon by artillery and naval guns. The effect of the plans was to protect the pilots and planes flying the close support mission. *Tenth Army Tntv OPlan 1–45,* Anx 5, p. 5.

[47] *Henderson ltr.*

too long. Colonel Kenneth H. Weir, commander of LFASCU–1, agreed in principle with this complaint. He said that "if air support units could have been given the maximum ordinates and azimuths of the artillery and naval gunfire falling into an area in which air strikes were to be made," [48] in many instances the aircraft could have attacked or continued an attack without invoking the restrictive fire plans. This controversial point was settled on 16 May when Tenth Army cancelled the use of Plans Negat and Victor, except in unusual circumstances.[49]

ISLAND COMMAND ACTIVITIES [50]

The tasks to be carried out by Island Command during both the combat and the garrison phases of the Okinawa campaign were more complex and staggering in many ways than those assigned to other Tenth Army combat organizations. Major General Fred C. Wallace was responsible for providing administrative and logistic support to combat units, executing the CinCPOA base development plan, and assuming—when directed by Tenth Army—the responsibility for the garrison and defense of Okinawa and its outlying islands. To achieve the objectives required in these various assignments, Island Command had been organized so that it would direct, control, and coordinate a joint task force comprised of a large portion of the service and support troops in the Tenth Army. As Tenth Army noted later: "In effect, Island Command [served] as a combined Army Service Area and advance section of a Communication Zone." [51]

The degree and scope of the functions delegated by Tenth Army to General Wallace increased in an almost direct proportion to the decrease in fighting and subsequent narrowing of the combat zone. Before the beginning of July, Island Command controlled some 153,000 men and had become responsible for the defense and development of every major island in the entire Okinawan chain of islands.[52] Subordinate and reporting to General Wallace were the commanders of Naval Operating Base, Ryukyus; Joint Communication Activities; Hydrographic Survey; Army and Navy Air Bases; Construction Troops; Military Government; and Ground Defense Forces. Additionally, General Wallace exercised control over a large number of service troops which had been assigned directly to his headquarters.

When ICEBERG Plan Phase III operations against Miyako and Kikai were cancelled in late April, all base development efforts, and troops scheduled for employment on these and other islands of the Ryukyus, were reassigned to Okinawa. In the planned revisions,

[48] Quoted in *AirIntelBul,* Aug-Sep45, p. 24.

[49] *Henderson ltr.*

[50] Unless otherwise indicated, the material in this section is derived from: *IsCom AR;* MilGovtSec, IsCom, Histories of MilGovt Ops on Okinawa, Apr-Aug45 rpts; *27th InfDiv OpRpt.*

[51] *Tenth Army AR,* chap 11, sec XXVI, p. 1.

[52] Patrols of the 2d Marine Division reconnaissance company scouted Izena Shima—to the north of Okinawa—from their base on Iheya Shima, on 23 June. No enemy soldiers were found; the island held some 4,000 natives, however, who were friendly to the Americans. On 29 June, Island Command took over Kume Shima, which had been captured by the FMF Reconnaissance Battalion earlier in the month.

the number of airfields originally scheduled for development on the island was doubled, and a corresponding increase in supply installations and troop staging, rehabilitation, and training areas was envisioned. All of these impending developments, however, were held in abeyance until remnants of the *Thirty-second Army* had been destroyed.

As an example of his single-minded determination to pursue the basic objective, General Buckner had ordered all airfield construction units to concentrate on maintaining and reconstructing supply roads to frontline organizations when the heavy rains and resultant mud of late May and early June threatened to bog down but failed to halt the Tenth Army attack. In spite of the weather and incident delays, the first American-built airstrip on Okinawa—a 7,000-foot runway at Yontan—was completed by 17 June. Before the end of the month, 5 airfields were operational on the island, and 8 of the 18 proposed fields were sited and were in the midst of being rehabilitated or constructed to meet the needs of the increased numbers of newly arriving B-29s.

Besides air base development and road maintenance, the Island Command engineering troops fulfilled other important tasks. They widened over 160 miles of existing native roads into two-, three-, and four-lane highways to accommodate the burgeoning load of supply and troop traffic. Island Command also opened new beaches, constructed piers, and cleared dump areas to handle the influx of supplies to be used in the impending operations against Japan proper. Engineers developed a massive water system capable of answering the needs of hundreds of thousands of soldiers and civilians. Other pipelines were laid and tank farms built to handle the tankerloads of aviation fuel necessary to maintain current and act as a reserve for projected air operations. Construction of the hundreds of storage, administration, and hospital buildings to be used by invasion-bound troops paced the buildup elsewhere on the island.

As the end of organized resistance on Okinawa neared, Island Command shifted the weight of its logistical support from Tenth Army to preparations for approaching operations. One base development phase influenced by planned future operations resulted in the preemption of vast areas of arable land in southern Okinawa and on the Motobu Peninsula. Not only was the topography of the island altered, but the way of life, means of subsistence, and sources of sustenance of island natives were irrevocably changed. Ejection of the natives from generations-old family holdings and removal of other islanders from more populated areas meant that they became, in effect, wards of the Island Command.

The agency responsible to Island Command and taking over its role as guardian for the displaced Okinawans was Military Government. Like so many of the other agencies directed by General Wallace, this one was a joint service effort. Even during the initial stages of the battle, military government teams functioned as though they were conducting a "disaster relief operation," [53] in which they had to clear the islanders out of the way of the course of the fighting

[53] *Tenth Army AR*, chap 11, sec XXVII, pp. 3–4.

for reasons of mercy as well as for the purpose of keeping them from hampering Tenth Army operations. In this period, as the native population became concentrated in stockades and resettlement areas in northern Okinawa, the Americans gave assistance to the Okinawans as the natives reconstituted the normal functions of civil government and developed a self-sustaining local economy. Primary emphasis was on increased Okinawan participation in both areas. An idea of the magnitude of the job that was performed by a relatively small group of military government personnel is reflected by the fact that it was in charge of 261,115 civilians on 30 June, and 100,000 more by the end of the war.

Complementing the sweep that Tenth Army forces made in the south after the end of organized resistance, Island Command garrison forces in occupied areas of northern Okinawa conducted mopping-up operations, which lasted well into August and assumed the proportion of pitched battles at times. The majority of the flare-ups occurred north of the Ishikawa Isthmus, garrisoned by the 27th Infantry Division on 2 May. Army forces on Kerama Retto also felt the backlash from survivors of Japanese units that had been defeated but did not know it.[54]

[54] A provisional infantry battalion, formed from the 870th Antiaircraft Artillery (Automatic Weapons) Battalion, relieved 2/305 as the Kerama Retto garrison on 23 May. The former AAA gunners were given some rudimentary infantry training by experienced 27th Infantry Division officers and noncommissioned officers, and then began operations which lasted until the end of the war to destroy the numerous survivors of the sea raiding battalions hidden out in the rugged hills of the island.

Upon passing to Island Command control and moving to the areas in northern Okinawa formerly occupied by the 1st and 6th Marine Divisions, the 27th Division began patrolling extensively, assisting the military government collection teams, and blowing caves as well as fortified and prepared positions found in its assigned zone of responsibility. When the toll of enemy dead rose from an average of 3 or 4 to 15 a day and Army troops found evidence of increasing numbers of recently occupied and prepared bivouac positions, General Wallace decided to make a thorough sweep of northern Okinawa to kill or capture the Japanese remaining there.

On 19 May, the division began a sweep northwards from the base of Ishikawa Isthmus with three regiments abreast. Within five days, the soldiers met heavy resistance at Onna Take, the heavily forested hill mass rising to 1,000 feet from the center of the isthmus. Here, 1st Division Marines had fought guerrillas in April while the 6th Division was fighting the battle on Motobu Peninsula. Since that time, the enemy had added to the natural defenses of the area and extensively fortified the region. The soldiers fought a 10-day pitched battle here without benefit of air or artillery support. After it was over, there was evidence that a sizable number of Japanese had escaped the trap and headed further north. The 27th Division continued its sweep and followed the Japanese. The mop up was finally completed on 4 August, when Army troops reached Hedo Misaki. The division reported at the end of the nearly three-month drive that it had killed over 1,000 Japanese and captured 500.[55]

As the fighting on Okinawa drew to a climax, preparations for another off-island operation began. Like Tori, Ibeya, and Aguni Shimas, Kume Shima had been one of the targets originally selected for capture during Phase III (d) of the ICEBERG operations.[56] The priority of these targets was downgraded later as the ground campaign unfolded, and this phase of the ICEBERG operation was finally cancelled. A Tenth Army study in late May resulted in the choice of these islands as radar and fighter director sites. The first three were captured in early June, and Kume was targeted for seizure during the mop up phase on Okinawa. Largest of the outlying islands selected for early-warning facilities—some 40 square miles in size—it is approximately 55 miles west of Naha. Assigned to capture the island was the FMF Amphibious Reconnaissance Battalion, which had been attached to Island Command for garrison duty in the Eastern Islands after the Marines had seized them. On 21 June, the battalion was released to Tenth Army control for the Kume Shima assault landing. (See Map 22.)

Kume was scouted in the night of 13–14 June by Company B patrols. Information received from captured civilians indicated that only a 50-man enemy garrison held the island. This intelligence proved correct after the landing on 26 June, but Company A and the 81mm Mortar Platoon from 1/7 were attached to Major Jones' 252-man battalion in case the Japanese force encountered was larger than expected.

Leaving the company from 1/7 behind to guard the beachhead, Major Jones and his battalion set out to contact the enemy. After five days of intensive patrolling, no Japanese were found and no opposition was developed. On 30 June, Jones declared the island secure.[57]

Although the Kume assault force had encountered no enemy in the late June operation, the garrison troops only several days later became involved in two fire fights with Japanese soldiers. Six of the enemy were killed and three of their four machine guns were captured. Constant aggressive patrolling forced the survivors to scatter into the hills in the interior of the island, where they offered no threat to the successful operation of air warning facilities. Air Warning Squadron 11 arrived at Okinawa on 4 July, and its units were set up on Kume Shima two days later. By 12 July, the radar and fighter director sections of the squadron had begun operations and been integrated into the system controlled overall by the Air Defense Control Center on Okinawa.[58]

EVALUATION OF OPERATIONS[59]

As some scholars in the field of military history and tactics have noted, the

[55] Love, *27th InfDiv Hist*, p. 649.

[56] *ICEBERG Study*, App H.

[57] *PhibReconBn AR*, Ph III, encl A, The Assault and Capture of Kume Shima, dtd 15Aug45.

[58] *ADC Hist*, p. 6.

[59] Unless otherwise noted, the material in this section is derived from: *Tenth Army AR; IIIAC AR; XXIV Corps AR; 1st MarDiv SAR; 6th MarDiv SAR, Ph III;* Isely and Crowl, *Marines and Amphibious War.*

Okinawa operation represents "the culmination of amphibious development in the Pacific war." [60] Shortly after the initial landings, British observers accompanying the ICEBERG force reported that "This operation was the most audacious and complex enterprise which has yet been undertaken by the American Amphibious Forces, . . ." [61] And they were undoubtedly right, for "more ships were used, more troops put ashore, more supplies transported, more bombs dropped, more naval guns fired against short targets" [62] than in any previous campaign in the Pacific. Despite the immensity of all of the factors involved in the ICEBERG operation, the Okinawa landing realistically demonstrated the soundness of the fundamental amphibious doctrine that the Navy and the Marine Corps had developed over the years and had tempered in the Pacific fighting. This thesis was amplified by General Geiger, who pointed out that the battle for Okinawa "reemphasized most clearly that our basic principles of tactics and technique are sound, 'in the book,' and need only to be followed in combat." [63]

The touchstone to success at Okinawa was interservice cooperation, where "Army artillery supported Marine infantry and vice versa," and "Marine and Army planes were used interchangeably and operated under the same tactical command," and "each contiguous infantry unit was mutually supporting and interdependent," and finally, when "the Navy's participation was vital to both throughout." [64] The target information center (TIC) was the primary Tenth Army agency that coordinated the request for and assignment of supporting arms. In the TICs existing at division, regimental, and battalion levels throughout the Tenth Army, a centralized target information and weapons assignment system gave unit commanders the ability to mass the maximum amount of firepower on both assigned targets and targets of opportunity.

At each infantry echelon down to battalion level, the artillery liaison officer was also in charge of the TIC and worked very closely with the operations officer. Utilizing previously collated intelligence pinpointing enemy positions and screening support requests, the TIC section head—an artillery liaison officer—and the naval gunfire and air liaison officers allocated fire missions to each of the three support elements which they represented. A primary consideration in making each assignment was the capability of the weapon or weapons to be employed.

The target information center at IIIAC headquarters was controlled by the Corps Artillery commander—who made it one of his special staff agencies—and its mission was to provide supporting arms with target information. Colonel Henderson, the operations officer of IIIAC Corps Artillery, described the TIC as General Nimmer's S–2 Section:

> . . . expanded to meet the needs of artillery, NGF and CAS [close air support] on a 24 hour basis. The Corps Arty S–2 was the IIIAC TIO [target information

[60] Isely and Crowl, *Marines and Amphibious War*, p. 551.

[61] *British Observers rpt.*

[62] Isely and Crowl, *loc. cit.*

[63] *IIIAC AR*, p. 194.

[64] Isely and Crowl, *op. cit.*, p. 578.

> officer]. The working responsibility for coordinating arty, NGF and air lay with the Corps Arty S–3 for both planned fires and targets of opportunity.
>
> The Corps Arty S–3, S–2 (TIO) and Corps AirO and NGFO were all located in a big hospital tent adjacent to IIIAC headquarters most of the time. The S–3 and S–2 (TIO) had 'hot line' phones to Corps G–3 and G–2. The Corps Arty FDC [fire direction center] and the Corps fire support operations center were one and the same facility—with NGF and air added.[65]

The TIC was given radio jeeps and operators from the Corps Signal Battalion and Corps Artillery to man the Support Air Request, Support Air Observation, and Support Air Direction (SAD) radio nets. As all division and corps commanders commented favorably on the TIC system, Tenth Army recommended that it be adopted for all future operations.

In writing about the fire support functions of the TIC, the commander of the 11th Marines noted:

> For the first time in the Pacific, coordination of naval gunfire and air support with artillery was prescribed in army orders, a forerunner of the present FSCC [Fire Support Coordination System]. Examination of the record will show that each division and corps, Army and Marine, used a different modification of it. It is worthy of note that the system used by the First Marine Division was most like what we have today.[66]

Until the *Kamikaze* threat waned in late May and early June, most of the close air support missions were flown by carrier-based planes rather than the TAF aircraft on Okinawa. The latter were too fully committed flying combat air patrols and intercepting Japanese planes to fly strike missions until the later stages of the campaign; Marine-piloted Avengers on supply drops were an exception. The majority of the close support missions in the Okinawa campaign were pre-planned; strike requests were submitted to the LFASCUs, which assigned them well enough in advance so that the strike pilots could be thoroughly briefed before the mission was flown. When a ground element urgently needed close air support, its air liaison party submitted a request through the chain of command to the LFASCU at corps headquarters, which approved the request or turned it down, if, in fact, this action had not taken place earlier at regimental or division level.

Tenth Army unit commanders were favorably impressed also by the aerial supply drop system that was of such vital assistance to the attackers when supply routes had become bogged down. They recommended that a unit similar to the IIIAC Air Delivery Section be formed to work with each field army or independent corps. Tenth Army also recommended that the JASCOs assigned to each combat division be disbanded.[67] Motivating this proposal was the feeling that when the marked dissimilarity in the training and functions of the various components of the JASCOs were taken into account, separate air liaison, shore fire control, and shore party communications parties would operate more effi-

[65] *Henderson ltr 1965.*

[66] *Brown ltr.*

[67] It will be noted in the discussion of the Marine division in pt VI, chap 2, *infra*, that this is what in fact took place with the publication of the G-Series Tables of Organization in late 1945. An Assault Signal Company (ASCO) was made organic to the Marine division and placed in the division headquarters battalion.

ciently. The naval gunfire spotting and liaison teams were specially commended for competently handling the staggering volume of naval shells fired in support of the land forces.

The shore bombardment of Okinawa on L-Day was "the heaviest concentration of naval gunfire ever delivered in the support of the landing of troops." [68] Some 3,800 tons of shells poured in from battleships, cruisers, and destroyers, and from the rocket racks and mortars of the support vessels to explode on enemy shore targets. During most of the campaign, each frontline regiment was assigned one call fire ship and one illumination ship. In certain instances, such as during the 6th Marine Division drive to the Motobu Peninsula, each assault battalion had a destroyer on call. Most fire support ships remained on station for the entire campaign and were not rotated to other duties. As the operation progressed, the quality and results of their shooting improved immeasurably.

On certain occasions, however, the ground units encountered intricately sited and deeply dug-in enemy positions which were impregnable to even the weight of naval gunfire salvos. At these times, the Japanese positions would withstand the fires of individual supporting arms or all of them together. Then, assault forces began a wearing-down process involving the employment of flame and gun tanks, demolitions, and infantry all together in what General Buckner referred to as "the corkscrew and blowtorch" method. Although artillery utilized every expedient conceivable, including the use of antiaircraft artillery guns and LVT(A) howitzers to supplement their regular fires, the Shuri and Kiyamu defenses remained invulnerable for long periods at a time.

One artillery weapon that was organic to the infantry regiments and immediately available for employment under optimum frontline conditions was the 105mm self-propelled howitzer, the M–7. This field piece was found in the 105mm howitzer (self-propelled) platoon containing four gun sections, which replaced the 75mm howitzer (self-propelled) platoon, in the regimental weapons company when it was reorganized on 1 May 1945 according to the G-series Table of Organization (T/O). The 1st and 6th Marine Divisions had received the T/O change, revamped their weapons companies, and were supplied with the M–7s before embarking for Okinawa.[69]

No other Tenth Army units remained continuously on line so long a period as the artillery battalions of both Marine and Army divisions during the battle in southern Okinawa. In this period, the artillery of all six infantry divisions supported the attack. Marine and Army corps artillery units supplemented the fires of the 24 divisional battalions with 12 of their own in general support.

Augmenting the Marine artillery were the guns of two LVT(A) battalions, which had been organized and trained as field artillery before the landing. Be-

[68] *Tenth Army AR*, chap 11, sec V, p. 6.

[69] Prior to the T/O change, the regimental weapons company consisted of a 75mm howitzer (self-propelled) platoon and three 37mm gun platoons. After the new T/O went into effect, the company was comprised of the 105mm self-propelled howitzer platoon and two 37mm gun platoons.

cause of its organization, each LVT(A) battalion had the fire support capability of a four-battalion regiment of 75mm howitzers.

Prior to Okinawa, General Geiger had become convinced that the armored amtracs could be trained as field artillery and used as such immediately after landing on L-Day at H-Hour and until direct support battalions arrived ashore. Thereafter, the LVT(A)s would reinforce corps and divisional artillery. After landing on L-Day, the LVT(A)s had their "batteries laid and ready to shoot for forward observers as early as H plus 30 minutes—but the Japanese wouldn't accommodate us with targets." [70]

A total of 2,246,452 rounds were fired in support of the infantry by tanks, LVT(A)s, M-7s, and field artillery pieces; this was more than triple the 707,500 rockets, mortars, and rounds of 5-inch shells or larger fired by the gunfire support ships.[71] In either case, the figure is staggering. Because Tenth Army had established a centralized system of target assignment and fire direction, unit artillery commanders were able to mass the fires of all their guns that were within the range of a specific target with little effort in a minimum of time.

In an analysis of Marine artillery operations on Okinawa, General Geiger discovered that there had been instances when 155mm guns and howitzers were unable to destroy certain well-built Japanese defenses when called upon to do so. Further, both corps and division artillery often found it difficult to reduce natural cave positions, which fell only under the direct fire of self-propelled guns or when artillery of a larger caliber than that found in Marine artillery battalions were employed.

The expectation that the invasion of Japan would require a vastly increased fire potential in the existing Marine artillery organizations led General Geiger to recommend changes in its makeup. Accordingly, he proposed a new setup consisting of a field artillery observation battalion and four group headquarters and headquarters batteries, and the following firing batteries: one 105mm howitzer (self-propelled); three 155mm howitzer; two 155mm gun; one 155mm gun (self-propelled); two 8-inch howitzer; one 8-inch howitzer (self-propelled); and one 240mm howitzer.

---

[70] *Henderson ltr.*

[71] A breakdown of the first figure by type of shell fired reveals: 75mm guns (tanks)—199,522; 75mm howitzers (including LVT(A)s)—387,112; 105mm howitzers (including M-7s)—1,119,210; 155m howitzers—375,241; 155mm guns—146,359; 8-inch howitzers—19,008. Because of discrepancies appearing between the figures given in the appropriate ammunition expenditure sections of the section reports of the Tenth Army, the two corps, and the divisions, these totals represent those given by the unit in which a specific weapon was organic; the reports of the LVT(A) battalions, attached to the 1st and 6th Marine Divisions, were incorporated within the division reports. The 1st Battalion, 11th Marines, alone fired a total of 100,339 rounds of 75mm ammunition, "which was the largest number of rounds fired by Marine 75mm pack howitzer battalion in any of the Pacific campaigns during World War II." LtCol Robert C. Hilliard comment to HistBr, G-3 Div, HQMC, 6Apr65. Colonel Hilliard is an artillery officer who served with the 11th Marines at Okinawa.

General Geiger was particularly impressed by the penetrating and destructive power of the 200-pound shell of the 8-inch howitzer when compared with the results achieved by the 95-pound projectile fired by 155mm guns and howitzers, the largest caliber pieces organic to Marine artillery units. The Marine commander asked that some of these 8-inch battalions be included when task organizations were formed for future scheduled Marine operations against Japan scheduled for the future.

Teamwork was a most important ingredient in the formula for reduction of heavily fortified Japanese positions. During the course of the Okinawa campaign, the work of supporting arms, infantry-engineer, air-ground, and tank-infantry teams played a vital role in the defeat of the enemy. Ground assault operations, however, were the especial province of the tanks and the infantry. Concerning the armored support of 6th Division Marines on Okinawa, General Shepherd wrote that "if any one supporting arm can be singled out as having contributed more than any others during the progress of the campaign, the tank would certainly be selected." [72] In a battle lesson issued to the *Thirty-second Army*, General Ushijima supported this opinion, stating that "the enemy's power lies in his tanks. It has become obvious that our general battle against the American forces is a battle against their M–1 and M–4 tanks." [73]

In comparison with the factors limiting armored support during some of the other Pacific island battles, tanks were more widely employed on Okinawa because its terrain, for the most part, favored armored operations. Tenth Army units lost a total of 153 tanks to accurate enemy AT fire, vast and thickly sown minefields, and demolitions-laden Japanese soldiers who attempted to destroy both the tanks and themselves, but who failed in their efforts for the most part, however, because of the accurate fire of the infantrymen protecting the tanks. Individual Japanese damaged seven tanks from the five Army battalions, disabled one from the 6th Marine Division, and none in the 1st Marine Division where "the alertness of the covering infantry and the tank crews prevented the successful completion of these attacks." [74]

Tanks from the Army 713th Armored Flamethrower Battalion, the first unit of its type to be formed and take part in sustained action, supported Army and Marine units alike. After the campaign, the battalion was highly praised for "a consistently outstanding record of performance." [75]

While covered by infantrymen and standard tanks, flame tanks were par-

[72] *6th MarDiv SAR, Ph III*, pt III, p. 28.

[73] Quoted in the preface to the Tank Support Anx, *1st MarDiv SAR*.

[74] *Ibid.*, pp. 23–24.

[75] *Ibid.*, p. 41. A few tanks in each of the standard tank battalions on Okinawa were equipped with small, limited-range flamethrowers, which were mounted either on the periscope mounts or where the bow machine gun had been. Both were inferior to the gun tube flamethrower of the 713th Armored Flamethrower Battalion, which had "a greater range in addition to being an all-around better weapon and the most practicable of the three." *Tenth Army AR*, chap 11, sec IX, p. 12. See also pt VI, chap 2, *infra*, for a discussion of armored flamethrowers.

ticularly successful in burning the enemy out of rocky outcroppings, reverse slope positions, and ruins. The commanders of both the XXIV Corps and the IIIAC favored the increased employment of flame tanks. General Hodge suggested the addition of two battalions to each corps in future operations; General Geiger recommended that one company of these tanks be made organic to each Marine tank battalion.

Both Marine combat divisions had Army 4.2-inch chemical mortar companies attached for the campaign. The division commanders reported that they were very satisfied with the performance of the large-caliber mortars, which could furnish high angle fire on targets not otherwise suitable for 81mm mortars and artillery howitzers. After noting the successful results that had followed employment of the 4.2-inch mortars attached to his division, General del Valle was convinced that their accuracy, long range, and tremendous destructive power were such that he recommended the inclusion of this type of company in the T/O of a Marine division.

Two other new Army support weapons impressed Marine leaders for the same reasons as had the heavy mortars; they were the 57mm and 75mm recoilless rifles. Although neither had been issued for testing by Marine units, nor were the rifles employed extensively by the Army, after viewing a combat demonstration of the effectiveness of the new weapons, IIIAC observers reported that the recoilless rifles held considerable promise for tactical employment. General Geiger acted on this information and recommended that the Marine Corps thoroughly field test both weapons with a view of adopting them in place of the 37mm guns and 2.36-inch bazookas in the infantry regiments at that time.[76]

Few startling innovations to accepted infantry tactical methods appeared out of the Okinawa fighting. Concerning this, General Geiger commented: "No new or unusual features of infantry combat were disclosed or developed during the campaign on Okinawa which would tend to modify or annul current standard principles or doctrines."[77] Those facets of the battle sometimes cited as having reflected the emergence of new concepts in the Pacific war—such as the employment of night attacks and refinement of tank-infantry tactics—were actually just the logical outgrowth of existing tactical doctrine that evolved after the Americans had become familiar with the enemy and his way of fighting.

For the most part, in the early years of the war, there was little inclination toward night offensive action; Marines were too intent on tying in their lines before darkness in order to blunt inevitable Japanese counterattacks and infiltration of the lines. During the Okinawa campaign, however, Marine units took part in night operations more extensively than ever before, and with a great degree of success. Approximately 21 patrols and attacks were mounted at night by Marines; 13 of this number were conducted by the Amphib-

[76] *IIIAC AR*, p. 214.

[77] *Ibid.*, p. 103.

ious Reconnaissance Battalion.[78] In commenting on this aspect of Marine tactics on Okinawa, General Geiger said:

> All night operations were characterized by the fact that they were performed in an orthodox manner. Previous training in such maneuvers and existing doctrines on the subject were employed and proved sound. Daylight reconnaissance, a limited objective of a prominent terrain feature, explicit orders for all echelons, noise discipline, and contact were as prescribed in the training manuals. In every case surprise was achieved and the night attack or movement was successful.[79]

Regarding the American night attacks, Colonel Yahara commented that they were:

> . . . particularly effective, taking the Japanese completely by surprise. The Japanese had so accustomed themselves to ceasing organized hostilities at nightfall, and . . . reorganizing and relaxing during the night that attacks in these hours caught them both physically and psychologically off-guard.[80]

In general, a study of the Marine conduct of night operations on Okinawa revealed no new, startling doctrine, for it indicated the following:

> 1. Orthodox methods are good methods.
> 2. A correct estimate of the situation is a major contributing factor toward success.
> 3. Night operations need not be confined to highly specialized units.
> 4. Such operations afford echelon commanders with an excellent tactical device.
> 5. Present doctrine is quite satisfactory for the training and indoctrination of troops.[81]

In reviewing the success of those night attacks launched during the Okinawa campaign, it seems surprising that American commanders did not employ this offensive tactic more often.

Immediately after the fighting for Shuri had intensified, severe gaps appeared in the ranks of the assault elements. Although replacements were fed to Tenth Army continually during the course of battle, they were often too poorly or incompletely trained to go into the frontlines immediately. Yet, they were needed to beef up the strength of the hard-hit units. Nevertheless, Tenth Army issued an order to the corps commanders directing that newly arriving personnel were to be indoctrinated and oriented before assignment to frontline units. It was very often difficult to adhere to this directive, especially when the situation demanded

[78] Major Jones' Marines conducted the following night patrols and operations as follows: Keise Shima, Aware Saki, Mae Shima, and Kuro Shima, all pre-L-Day; Tsugen Shima, 5–6 April; Ike Shima, Taka Banare, Heanza Shima, and Hamahiki Shima, all 6–7 April; Kutaka Shima, 7 April; Minna Shima, 7–8 April; Yagachi, 20 April; and Sesoko, 21 April. The following IIIAC units conducted night operations on Okinawa proper as noted: Asa Kawa crossing by 22d Marines, 10 May; Relief of 1st Marines at Dakeshi-Wana by 5th Marines, 13–14 May; Naha Canal crossing by 22d Marines, 29 May; Reconnaissance of Oroku by 6th Reconnaissance Company, 2 June; Kunishi Ridge attack by 7th Marines, 12 June; Kunishi Ridge attack by 1st Marines, 14 June; Kunishi Ridge relief by 22d Marines, 17 June; and Mezado Ridge relief by 8th Marines, 18 June.

[79] *IIIAC AR*, p. 103.

[80] *Yahara Interrogation.*

[81] Capt James R. Stockman, "Night Operations on Okinawa," *Marine Corps Gazette*, v. 30, no. 9 (Sep46), p. 28.

that the replacements be committed into the lines before they were completely "shaken down."

General Geiger "had only two divisions to fight" on Okinawa and found it impossible to guarantee the "relief of front line divisions for rest and assimilation of replacements." To remedy this, he suggested that a corps on extended operations should have a triangular organization much like that of the Marine divisions to provide for an "automatic reserve." Without this, his two Marine divisions had to remain constantly on line until the end of the operation. Based on the knowledge gained at Okinawa, a corps of at least three divisions was considered a must for future joint operations of a similar nature.[82]

Some serious personnel problems arose before and during the campaign for Army and Marine divisions alike. Most deeply concerned was XXIV Corps, which had been deeply involved in the Philippines operation during the time that preparations for ICEBERG were underway. General Hodge favored the Marine replacement system in which Marine replacements were attached to and trained with infantry units during the preinvasion phases, and then travelled with these units to the target area, where they worked as shore party labor units until needed in the lines to replace infantry casualties.

Including the replacements they had received before departing Pavuvu and Guadalcanal respectively, the 1st Marine Division landed at Okinawa approximately 10 percent over T/O strength, and the 6th Marine Division arrived at the target with a 5 percent overage. Because they had participated in the training and rehearsal phases of ICEBERG, the replacements could be assigned to line regiments when required. Most of the replacements who arrived at Okinawa during the later stages of the battle had come directly from Stateside. Since they were not so well trained as the earlier replacements, the infantry units to which they were assigned had to divert some of their efforts to indoctrinate and train the new arrivals for battle rapidly.

Possibly influenced by the Marine replacement system, Tenth Army recommended that, in future operations, a large-sized replacement company should be assigned to and train with an infantry division before an invasion, and then accompany that division to the target area. General Hodge suggested that infantry battalions be permitted to carry a 25-percent-strength overage to the target, and that balanced infantry replacement battalions, each consisting of 1,000 men, be attached to and loaded out with every invasion-bound infantry division.

Both the 1st and 6th Marine Divisions contained a large number of combat veterans who had participated in two or more campaigns in the Pacific.[83] As of 30 June 1945, the 1st Marine Division had 205 officers that had served overseas 24 months or more; over half of these had been in the Pacific area for more than 30 months. Nearly 3,200 enlisted Marines had been in the field

[82] *IIIAC AR*, p. 195.

[83] See pt II, chap 3 above, section entitled "Training and Rehearsals."

for two years or more; almost 800 of these had been in a combat zone for 30 or more months. General del Valle considered that these facts reflected the approach of a serious personnel and morale problem in the division. By the fall of 1945, 1st Marine Division personnel already in or entering the two-year category "will have spent their entire time in a coconut grove or jungle with not a single opportunity for leave or liberty." [84] Steps were taken later, however, which alleviated the situation before it reached a crucial point.[85]

The immediate replacement of infantry losses was a problem common to commanders of all assault echelons. They believed that the solution was to be found in the establishment of a smoothly working replacement systems, wherein replacements would be attached to and train with an infantry unit before an invasion. Experienced troop leaders knew that long hours of closely coordinated training were needed before assault and replacement organizations could be considered combat ready. Arduous hours of team training served as the basis of American successes at Okinawa. The final action report of the Tenth Army noted:

> The support rendered the infantry by naval gunfire, artillery, air and tanks was adequate in every respect. Without such magnificent support, little progress could have been made by the infantry in their advance against the heavily organized enemy positions in southern Okinawa. Supporting fires enabled the infantry to carry out the tremendous task of repeated assaults against strongly fortified positions.[86]

Logistical planning too required teamwork, and the problems facing the logistics planners reflected the magnitude of the Okinawa operation. Consider for example, that in Phase I of ICEBERG alone, a total of almost 183,000 troops and 746,850 measurement tons of cargo had to be loaded into 433 assault transports and landing ships by 8 different subordinate embarkation commands at 11 widely separated ports from Seattle to Leyte over a distance of some 6,000 miles.

The Joint Expeditionary Force alone contained 1,200 ships of all kinds.[87] By the time that the island was secured, "About 548,000 men of the Army, Navy, and Marine Corps took part, with 318 combatant vessels and 1,139 auxiliary vessels exclusive of personnel landing craft of all types." [88] These figures coupled with the long distances over which supplies had to travel, created logistics problems of an immense nature beyond all that which had transpired in earlier Pacific operations.

Some concept of the size of the unloading job at Okinawa may be seen in the table in chapter 7, p. 240, which depicts the amounts of assault and first echelon cargo unloaded in all the Central Pacific campaigns from the Gilberts through Okinawa. This chart dramatically indicates that in the overall tonnage of supplies and equipment unloaded, the total for Okinawa was almost double that for the entire

[84] *1st MarDiv SAR*, PersAnx, p. 4.

[85] See pt III below, section dealing with the rotation home of combat veterans.

[86] *Tenth Army AR*, chap 11, sec IX, p. 1.

[87] *War Reports*, p. 665.

[88] *Ibid.*, p. 664.

Marianas operation and three times that for the Iwo Jima campaign. Errors of omission and commission in the logistics program seemed critical at the time that they appeared, but none was grave enough to effect the fighting for long. Some problems arising from the nature of operations began before L-Day and continued thereafter; they were important enough, however, to cause unit commanders to comment on them and make recommendations for improvement in their action reports.

In the logistics planning phase, embarkation officers too often found that ships' characteristics data for assigned ships was incorrect or out of date; at times, it was either not furnished or unavailable. When division staffs began completing loading plans, they found that, for the uninitiated and non-specialist, there were too many forms. These were too complicated and often repetitive. During the loading phase, ships' captains often received confusing and contradictory orders, which on several occasions resulted in their ships arriving in loading areas or appearing at places other than those to which they were to have gone. In most cases, the confusion arose from poor coordination between Marine and Navy staffs.

A sequel to this liaison gap at times appeared in the improper loading of assault transports. The commander of the transport group that lifted the 1st Marine Division to the target from the Russells reported that plans for loading some of his ships were not even begun until the vessels were alongside waiting to take a load. In reference to the loading of his entire group, he also said:

> It can be fairly stated that these ships were not combat loaded. It is true that cargo was landed according to priority. However, the 60 per cent combat load as expressed in Transport Doctrine was greatly exceeded. All ships were, in the opinion of the squadron [TQM] 'commercial loaded, according to a definite priority.' This was due to the fact that an inadequate number of vessels were assigned by higher command to lift the First Marine Division.[89]

During the preinvasion preparatory period, Marine divisions, especially the 6th, found the Marine Corps supply system on the Pacific overly cumbersome. Two basic factors aggravated the situation. One was the fact that the relations of the Marine Supply Service, FMFPac, to the several combat and service commands in the Guadalcanal area—where the greater portion of IIIAC strength was based—caused many delays because of the many agencies through which supply requisitions had to pass before the requestor received the items requisitioned. In addition the 6th Division was located too far away from the stocking agency, which in this case was the 4th Base Depot on Banika in the Russells.

General Shepherd believed:

> Supply problems, many requiring written correspondence and decisions by high authority, were not simplified by the addition of another senior echelon, the South Pacific Echelon, Fleet Marine Force, Pacific. The recent change in the concept of operations of the Corps, by which administration of divisions is theoretically divorced from the Corps, has not benefited the Division. Supply and administration

---

[89] Commander, Task Group 53.2 (Transport Group Baker) rpt in *CNO Record*, chap 7, p. 32.

> cannot, in practice, be separated from command.[90]

A built-in problem, inherent in the nature of the organization and equipment of a Marine division, appeared on L-Day. The initial lack of resistance beyond the beachhead permitted the landing of many Marines who would otherwise not have gone ashore until scheduled. This caused a shortage of landing craft slated to move cargo to the beaches and in turn brought about a delay in the landing of such selected items of division cargo as motor transport and prime movers, which were ticketed for unloading on L-Day.

> The truth is that neither Marine division ever had enough motor transportation either to supply itself adequately or to move its artillery. An allotment of motor vehicles and prime movers which might have been sufficient to the normal small island type of fighting to which Marines were accustomed was insufficient for a long operation such as Okinawa.[91]

At the end of the campaign, General del Valle recommended that each infantry regiment be furnished five prime movers with trailers to supplement motor transport already organic to the division. He also recommended that the infantry regiments be given in addition two bulldozers for "initial road, trail, dump clearance. . . ." [92] The 1st Division commander noted that motor transport, tractors, and engineering equipment, urgently needed for combat operations were often deadlined for lack of spare parts. To alleviate this situation, he recommended that, in future logistical planning, provisions should be made for the inclusion of an ample supply of spare parts in resupply shipments.[93]

According to an officer who was deeply involved in shore party and supply operations at Okinawa:

> Logistically, the touchstone of success was . . . interservice cooperation. In many instances, shortages of . . . supplies suffered by one service was made up by another service. It was a unique example of the unification that was developed throughout the campaign through the Central Pacific.[94]

In the end, hasty field expedients and the overwhelming superiority of American materiel strength, as well as the interservice collaboration, overcame any obstacle to the capture of Okinawa that logistical problems may have caused.

The story of the Okinawa campaign is incomplete without a brief investigation of enemy tactics. Contrary to the Japanese beachhead defense doctrine encountered in earlier Pacific landings, when the enemy strongly defended his beaches or ferociously attacked the invader before he could organize the beachhead, at Okinawa, the Tenth Army met a resistance in depth similar to that experienced by Americans in the Philippines invasion. *IGHQ* had ordered General Ushijima to fight a long holding action to buy the time necessary for Japan to complete Homeland defenses. If the Americans sustained a high attrition rate while attempting to batter down the *Thirty-second Army* defenses, so much the better; there would be that fewer Americans in the anticipated in-

---

[90] *6th MarDiv SAR, Ph I&II*, chap III, p. 6.

[91] Isely and Crowl, *Marines and Amphibious War*, p. 574.

[92] *1st MarDiv SAR*, chap X, p. 19.

[93] *Ibid.*

[94] *Blakelock ltr 1965.*

vasion of Japan. From the time that the Tenth Army landed over the Hagushi beaches until it encountered the northern outposts of the Shuri line, it was harassed, harried, and delayed by small provisional units and somewhat stronger blocking forces, the latter comprised of veteran regulars.

The fall of Saipan in 1944, if nothing else, brought home to *IGHQ* the military potential of the United States. This loss caused the Japanese command to accelerate the construction of defense positions in Japan as well as on Iwo Jima and Okinawa. The fast carrier task force air raids on Okinawa beginning in October 1944 spurred General Ushijima's *Thirty-second Army* on to strengthen Okinawa defenses further. Beginning in mid-April, when the Tenth Army encountered the maze of concentric defense rings encircling Shuri, Americans became painfully aware of the results of these efforts.

The rugged and complex ridgelines in the Shuri area were defended from vast entrenchments, from a wide variety of fortified caves employed as pillboxes, and from elaborate, multi-storied weapons positions and gun emplacements that had been gouged out of the ridges and hills and connected by tunnels, which usually opened on the reverse slopes. "The continued development and improvement of cave warfare was the most outstanding feature of the enemy's tactics on Okinawa." [95]

Among other outstanding features of *Thirty-second Army* defense tactics was the use of a considerable amount of reinforcing artillery, mortar, and machine gun fire. Also, the Japanese made mass *Banzai* charges only infrequently, but with a hopeful view either of exploiting a successful attack or of just keeping the Americans off-balance. The enemy did, however, fritter away his strength and dwindling forces in small-sized counterattacks, which had little chance of success and which were, in most cases, blunted easily by the Americans.

Despite the obvious fact that his *Thirty-second Army* was decisively beaten, General Ushijima must be credited with having successfully accomplished his assigned mission. He did provide Japan with valuable time to complete the homeland defense.

The final act of the Okinawa story unfolded on 26 August 1945, when General of the Army Douglas MacArthur—appointed earlier as Supreme Commander for the Allied Powers (SCAP)—authorized General Stilwell to negotiate the surrender of enemy garrisons in the Ryukyus. Responding to orders issued by Stilwell, top enemy commanders reported at the headquarters of the Ryukyus command on 7 September to sign "unconditional surrender documents representing the complete capitulation of the Ryukyus Islands and over 105,000 Army and Navy forces." [96] Witnessing the ten-minute ceremony in addition to those officiating were Army and Marine infantry units and tank platoons, while above it all hundreds of planes flashed by.

[95] *Tenth Army AR*, chap 8, p. 4.

[96] 1st Information and Historical Service, Documents Relating to the Surrender of Japanese Garrison in the Ryukyus and the Occupation of that Area by Elements of the Tenth Army, September and October 1945, dtd Dec45, attached news release.

In a report to the Secretary of the Navy, Fleet Admiral Ernest J. King, CominCh, stated that "the outstanding development of this war, in the field of joint undertakings, was the perfection of amphibious operations, the most difficult of all operations in modern warfare."[97] As the next to last giant step leading to the defeat of Japan, the Okinawa invasion was a prime example of a successful amphibious operation, and the culmination of all that Americans had learned in the Pacific War in the art of mounting a seaborne assault against an enemy-held land mass. This knowledge was to serve well in preparing for the invasion of Japan.

[97] *War Reports*, p. 658.

PART III

The End of the War

Future Operations

Following its participation in the sweep of the southern portion of Okinawa in June, on 2 July IIIAC was released from taking part in further mop-up activities. Thereafter, corps units not already in the process of doing so, moved to rehabilitation areas that were either on the island or at bases elsewhere in the Pacific.

IIIAC was detached from Tenth Army and came under the operational control of FMFPac on 15 July, the same day that the corps CP closed on Okinawa and opened on Guam.[1] During the period 13–18 July, Corps Troops, less the 1st Armored Amphibian Battalion, a portion of IIIAC Signal Battalion, and a small detachment from the IIIAC staff redeployed to Guam. General Shepherd's 6th Marine Division had begun the move to Guam on 4 July and completed it on the 11th. The 8th Marines (Reinforced) had left Okinawa on the first of the month and by 12 July the whole of the regiment and its reinforcements had rejoined the 2d Marine Division on Saipan. IIIAC Corps Artillery and the 1st Marine Division remained on Okinawa and set up rehabilitation camps on Motobu Peninsula.

About a thousand of General del Valle's 1st Division Marines had been sent to Motobu during the last week of June to begin building campsites for the rest of the division, which had remained in southern Okinawa until the early part of July, when the new camp areas were ready for occupancy. Between 1 and 15 July, some troops and light equipment were moved overland to Motobu; heavy equipment was moved over a water route by way of Oroku in the same period. On the 20th General del Valle opened his CP at the Motobu camp area.[2]

Originally, the division was to have been rehabilitated in the Hawaiian Islands and a number of units of the rear echelon, which had remained on Pavuvu, had embarked for the new rest area before the Okinawa campaign was over; a few had actually arrived in Hawaii before the remainder was diverted to Okinawa. When the rumor that the division would remain on Okinawa and build its own bivouac area was confirmed, "there was outright dismay and discouragement in high and low ranks." [3] Of the six Marine divi-

[1] IIIAC WarD, Jul45. On 30 June, Major General Keller E. Rockey took over as commander of IIIAC, relieving General Geiger, who shortly thereafter relieved Lieutenant General Holland M. Smith, Commanding General, FMFPac. Brigadier General William A. Worton relieved Brigadier General Merwin H. Silverthorn as IIIAC Chief of Staff. *Ibid.* On the 25th, Major General Thomas E. Bourke had relieved General Rockey as commanding general of the 5th Marine Division. VAC WarD, Jun45.

[2] 1st MarDiv WarD, Jul45. Major General DeWitt Peck assumed command of the division on 9 August, relieving General del Valle. *Ibid.*, Aug45.

[3] McMillan, *The Old Breed*, p. 424.

sions, the 1st had been away from civilization and in the Pacific for the longest period. As the troops began the displacement to the north:

> . . . the feeling of persecution had begun to go through its classic transformation. 'Well, dammit,' said one man above the rumble of the truck, 'if they can dish it out, I can take it!'
>
> And as he straightened the straps of his pack and turned to look out toward the sea, there were grunts of belligerent agreement behind him.[4]

Even while IIIAC Marines took a breather before preparations for the final operation of the war, the size of the Corps continued to grow. Marine Corps strength on 30 June 1945 was 476,709 men and women, nearly a seventeenfold increase over the size of the Corps in July 1940.[5] Over half of the Marines represented by the 1945 total were serving overseas, most of them in the Pacific; 184,800 alone were in FMF ground units. By June, final plans had been initiated to commit them and other forces in the massive assault against Japan.

OPERATIONS OLYMPIC AND CORONET[6]

While en route to Hawaii from Iwo Jima, General Harry Schmidt's VAC was ordered by CinCPac to begin planning for Phase III(c) of Operation ICEBERG, the invasion of Miyako Jima in the Sakishima Gunto, southwest of Okinawa. The VAC command post opened on Maui on 29 March, and on the next day its staff officers flew to Pearl Harbor for a conference at FMFPac headquarters concerning future VAC operations, primarily the Miyako Jima landing. Following their return from Iwo Jima to rest camps on Guam and in the Hawaiian Islands, the 3d, 4th, and 5th Marine Divisions began a period of rehabilitation and filled in their depleted ranks with replacements. Some replacement drafts that had been slated to restore VAC infantry regiments to full strength were diverted to Okinawa, where there was an even greater need for fresh troops. Within a short period of time, other replacements arrived, however, and the three VAC divisions were steadily built up.

When the Iwo Jima veterans were rested and most infantry regiments were again near-T/O strength, General Schmidt's three divisions embarked upon an extensive training program based upon a VAC schedule for the period 23 April to 1 July. By 1 July, all VAC units were to be ready for "amphibious operations involving opposed landings on hostile shores. . . ."[7] Despite the cancellation on 26 April of the Sakishima Gunto operation,[8] VAC adhered to the 1 July readiness date, and its divisions keyed their preparations to the requirements of Operation LONGTOM, which directed assault landings

[4] *Ibid.*

[5] *CMC Rpt*, 1945.

[6] Unless otherwise noted, the material in this section is derived from: CinCPac WarDs, Jan–Aug45 (OAB, NHD); *USAFMidPac G–5 Hist;* IIIAC WarDs, Jun–Aug45; VAC WarDs, Apr–Aug45; *War Reports;* McMillan, *The Old Breed;* Johnston, *2d MarDiv Hist;* Aurthur and Cohlmia, *3d MarDiv Hist;* Proehl, *4th MarDiv Hist;* Conner, *5th MarDiv Hist;* Cass, *6th MarDiv Hist.*

[7] VAC TrngO 1–45, dtd 15Apr45.

[8] *USAFMidPac G–5 Hist*, p. 257.

on the China coast south of Shanghai. The FMFPac warning order issued VAC gave a target date of 20 August, and also said that should this operation be cancelled, another would take place at some date after October.

As reflected in the CinCPOA Joint Staff Study LONGTOM, issued on 27 February 1945,[9] Admiral Nimitz' planners assumed that ICEBERG had been completed, Luzon had been captured, and that necessary service forces would be made available for LONGTOM from the United States or sources outside of the Pacific Ocean Areas. The purposes of LONGTOM were "to intensify air attacks against Japan" and "to seize approaches to increase the effectiveness of the blockade against Japan. . . ."[10] The assault forces assigned to LONGTOM consisted of three Marine divisions in VAC, three Army infantry divisions in IX Corps, and two Army parachute regiments. Once their objective was captured, a garrison air force consisting of three Marine aircraft groups, two Army Air Forces medium bombardment groups, and other naval and Marine aviation organizations would be established. On 16 May, CinCPac assigned VAC the duty of developing plans for a future operation based on a Joint Staff study; on 27 May CinCPac informed all of the commands concerned that the JCS had ordered Operation OLYMPIC—the invasion of southern Kyushu — executed on 1 November 1945, and consequently LONGTOM was deferred for an indefinite period. It was finally cancelled. On 31 May, CinCPac ordered General Schmidt to report by dispatch to the Commanding General, Sixth Army, for purposes of planning for a future operation—OLYMPIC.

The many changes in strategic planning during the final year of the war with Japan, and particularly in the last six months, reflected the constantly changing aspects of the conflict itself during those 12 months. None of the adjustments that were made, however, deflected from the aims of the Cairo Declaration of 1943, in which the Allies stated their determination to end the war by forcing Japan to surrender unconditionally. At the various major conferences and in their innumerable meetings, the wartime heads of government and the Combined Chiefs of Staff were faced with the problem of deciding just how Japan was to be defeated.

Events during the winter of 1944 and spring of 1945 provided a variety of indications of the course that the war might take in the Pacific in the summer of 1945. The atomic bomb project was near completion but its future was uncertain. It seemed possible that Russia would enter the Pacific battle because Stalin had committed his country to this action at the Yalta Conference in February 1945, but this matter was equally uncertain. Although the successful invasion of Luzon had undoubtedly made the collapse of Japan that much more imminent, American planners were faced with the urgent requirement of deciding the strategy by which the enemy was to be brought to his knees. Two of many alternatives particularly favored by the JCS were: first, to rush the defeat of the Japanese by continuing and intensifying the

[9] *Ibid.*, p. 263.

[10] *Ibid.*

existing blockade and bombardment of the Home Islands, and, second, to invade Japan and force the enemy to capitulate when he was left with no other resource. Actually, these two concepts were not so much alternatives as they were parallel steps by which the Allies planned the defeat of Japan.

In July 1944, the JCS had submitted a proposed revision of the SEXTANT timetable of operations against Japan to the Combined Chiefs of Staff for consideration at the OCTAGON Conference to be held at Quebec in September.[11] The American leaders had based their recommendations on one concept, among others, that envisioned an invasion into the industrial heart of Japan following the capture of Formosa. The JCS stated:

> While it may be possible to defeat Japan by sustained aerial bombardment and the destruction of her sea and air forces, this would probably involve an unacceptable delay.[12]

The JCS recommendations were accepted at OCTAGON, where the Combined Chiefs approved for planning purposes a new schedule of operations for 1945; Kyushu was to be invaded in October and the Tokyo Plain in December.[13]

The Navy view of an invasion of Japan has been stated succinctly by Fleet Admiral William D. Leahy, Chief of Staff to Presidents Roosevelt and Truman, who wrote of the Quebec Conference:

> . . . here the coming Battle of Japan was at the top of the agenda. Nothing had happened to alter my conviction that the United States could bring about the surrender of Japan without a costly invasion of the home islands . . . although the Army believed such an offensive necessary to insure victory.[14]

Leahy never was in agreement with the proposition that an invasion of Japan was a prerequisite to a final Allied victory, reasoning that:

> A large part of the Japanese Navy was already on the bottom of the sea. The same was true of Japanese shipping. There was every indication that our Navy would soon have the rest of Tokyo's warships sunk or out of action. The combined Navy surface and air force action by this time had forced Japan into a position that made her early surrender inevitable. None of us then knew the potentialities of the atomic bomb, but it was my opinion, and I urged it strongly in the Joint Chiefs, that no major land invasion of the Japanese mainland was necessary to win the war.[15]

Leahy credits the early pressure for an invasion of Japan to the Army, which:

> . . . did not appear to be able to understand that the Navy, with some Army air assistance, already had defeated Japan. The Army not only was planning a huge land invasion of Japan, but was convinced that we needed Russian assistance as well to bring the war against Japan to a successful conclusion.

---

[11] See pt I, chap 1, *supra*, for a discussion of Allied strategic planning in 1944.

[12] CCS 417/3, 11Jul44, title: Over-All Objective in War Against Japan, cited in Cline, *Washington Command Post*, p. 339.

[13] CCS 417/8, 9Sep44, title: Op for Defeat of Japan; CCS 417/9, OCTAGON, 11Sep44, title: Over-all Objective in War Against Japan; Min 173d Meeting CCS, 13Sep44, all cited in *Ibid.*

[14] FAdm William D. Leahy, *I Was There* (New York: Whittlesey House, 1950), p. 2, hereafter Leahy, *I Was There*

[15] *Ibid.*, p. 245.

> It did not appear to me that under the then existing conditions there was any necessity for the great expenditure of life involved in a ground force attack on the numerically superior Japanese Army in its home territory. My conclusion, with which the naval representatives [on JCS and JCS planning staffs] agreed, was that America's least expensive course of action was to continue and intensify the air and sea blockade and at the same time to occupy the Philippines.
>
> I believed that a completely blockaded Japan would then fall by its own weight. Consensus of opinion of the Chiefs of Staff supported this proposed strategy, and President Roosevelt approved.[16]

Leaders of the Army Air Forces took the Navy view that the Japanese could be forced to surrender—without an invasion of the Home Islands—under the "persuasive powers of the aerial attack and the blockade."[17] It appeared that other military planners, however, ". . . while not discounting the possibility of a sudden collapse, believed that such a cheap victory was not probable, at least within the eighteen months allotted in the planning tables" established in the revised strategy agreed upon at OCTAGON.[18] In the end, the concept that an invasion was necessary prevailed and vigorous efforts were applied in planning and preparing for it.

As the basic directive ordering the invasion of Japan took shape, it became obvious that the command relationship established between MacArthur and Nimitz in the Pacific in 1942 needed revision. Recognizing this need, the JCS designated MacArthur Commander in Chief, Army Forces in the Pacific (CinCAFPac) on 3 April 1945 and gave him control of all Army units in the Pacific except those in the North and Southeast Pacific Areas. This new appointment was in addition to the position he held as commander of SWPA. Admiral Nimitz was to retain his position and title as CinCPac–CinCPOA, and would have under his control all naval resources in the Pacific with the exception of those in the Southeast Pacific. The Twentieth Air Force, with its B–29s based in the China-Burma-India Theater as well as in the areas controlled by Nimitz and MacArthur, was to be subject to the requirements of both commanders under the new setup, but would remain under the direction of the JCS.[19] In addition, the directive stipulated that the JCS would issue directions for future operations, assign missions, and fix the command responsibility for major operations and campaigns. The JCS also stated that hereafter MacArthur would be responsible for the conduct of land campaigns and Nimitz for naval operations.

On the same day that this directive went out to the Pacific commanders, the JCS sent them another operational directive which among other things, in-

[16] *Ibid.*, p. 259.

[17] Craven and Cate, *Matterhorn to Nagaski*, p. 703.

[18] *Ibid.*

[19] JCS 1259/4, dtd 3Apr45, cited in *Ibid*, p. 682; JCS msg to CinCSWPA, CinCPac, and CG, Twentieth AF, dtd 3Apr45 in *Ibid*, and in LtCol Henry G. Morgan, Jr., "Planning the Defeat of Japan: A Study of Total War Strategy" (unpublished MS, OCMH), p. 152. The Twentieth was a strategic air force established in 1944 to operate under the direct control of the JCS. General H. H. Arnold, commanding the AAF, was the executive agent of the JCS to implement its directives concerning the deployment of the very-long-range bombers of the Twentieth.

structed Nimitz to continue and complete the Ryukyus operations in accordance with his earlier orders. Included in this followup message was the provision that the two theater commanders would continue to command forces of the other services then allotted to them and would not transfer these forces except by mutual consent or by order of the JCS. MacArthur was directed to complete the liberation of the Philippines, to plan to occupy North Borneo, and to provide Army forces needed by Nimitz. Both MacArthur and Nimitz were to make plans for the invasion of Japan, cooperating with each other in the task.[20]

On 30 April, Admiral King proposed to the JCS that they issue the order for the land-sea-air assault of Kyushu.[21] Less than a month later, on 25 May, the JCS issued the order setting forth the provisions for Operation OLYMPIC and assigned a target date of 1 November 1945 for the invasion.[22]

This document presupposed that Japan would be forced to surrender unconditionally as the Allies lowered both its will and its ability to resist, and as the Allies later seized objectives in the industrial heartland of Honshu. The defeat of Japan would be accomplished in two steps. The first, OLYMPIC, was the invasion of Kyushu on 1 November, which was designed to isolate this southernmost island of the Japanese chain, destroy the enemy forces there, and capture the airfields and bases required to support the second step, Operation CORONET—the invasion of Honshu, tentatively set for March 1946.

At a series of White House conferences following the issuance of the 25 May directive, its contents were discussed but not altered appreciably. The JCS determined at their 14 June meeting that, pending the approval of President Truman, the invasion and seizure of objectives in the Home Islands would constitute the major effort in OLYMPIC and that no other operations would be considered if they did not contribute substantially to the success of the Kyushu landings. On the other hand, the JCS agreed that while preparations for the invasion were taking place, aerial and naval blockades and bombardments of Japan were "to be maintained with all possible vigor." [23]

In their meeting with the President on 18 June, both Marshall and King strongly recommended an invasion of Kyushu at the earliest possible date. Admiral King had evidently modified his preference for an invasion of the China coast in the vicinity of Amoy, and decided to go along with Marshall in recommending the landings on Kyushu.[24] In accepting Marshall's views, King noted that the more he had studied the matter, the more he was impressed with the strategic location of Kyushu, whose capture he deemed a necessary prerequisite to any siege operations against the rest of Japan.[25]

[20] JCS msg to CinCSWPA, CinCPac, and CG, Twentieth AF, dtd 3Apr45, cited in *supra*.

[21] CominCh–CNO Memo to JCS, "Proposed Issue of 'OLYMPIC' Directive," dtd 30Apr45. (OAB, NHD)

[22] JCS 1331/3, dtd 25May45, cited in Craven and Cate, *Matterhorn to Nagasaki*, p. 686.

[23] King and Whitehill, *King's Naval Record*, p. 605.

[24] Leahy, *I Was There*, p. 384.

[25] King and Whitehill, *op. cit.*, p. 606.

Despite his concurrence in the plans for the Kyushu landings, King retained his earlier belief that Japan could be defeated by the sea-air power combination and without the necessity of invasion. He was fully aware of the fact that planning for an amphibious operation was a slow and painstaking process, and posed no objection to the preparation of contingency plans for the invasions of Kyushu and the Tokyo Plain. It was apparently for this reason that in June 1945 he joined in the majority decision of the JCS "to make plans for the invasion and seizure of objectives in the Japanese home islands without sharing the Army conviction that such operations were necessary." [26]

Marshall advanced the opinion that OLYMPIC "would not cost us more than 63,000 casualties of the 193,000 combatant troops estimated as necessary for the operation." [27] After hearing all arguments and absorbing the estimates, President Truman approved the Kyushu operation, but withheld his approval of a general invasion of Japan for consideration at a later date. He also said that he was in complete favor with any plan that would defeat the enemy with the smallest loss possible of American lives. "It wasn't a matter of dollars. It might require more time—and more dollars—if we did not invade Japan. But it would cost *fewer lives*." [28] On 29 June, the JCS met again to prepare the military agenda for the impending Potsdam Conference, and firmly set 1 November as the date for OLYMPIC.[29]

The JCS charged General of the Army MacArthur,[30] in his capacity as CinCAFPac/CinCSWPA, with the primary responsibility for conducting Operation OLYMPIC including control, in case of exigencies, of the actual amphibious assault through the appropriate naval commander. In addition, MacArthur was to make plans and preparations for continuing the campaign in Japan and to cooperate with Fleet Admiral Nimitz in planning and preparing for the naval and amphibious phases of this aspect of OLYMPIC.

On his part, CinCPac was responsible for the conduct of the naval and amphibious phases of OLYMPIC, subject to the JCS-imposed provision concerning exigencies. Nimitz was required to correlate with and assist MacArthur in the preparation and planning for the campaign in Japan and its conduct. The JCS directive of 25 May enjoined both senior commanders to remember that "The land campaign and requirements . . . are primary in the OLYMPIC Operation. Account of this will be taken in the preparation, coordination and execution of plans." [31]

Prior to publication of this JCS order, representatives of MacArthur and Nimitz had met in Manila on 16 May 1945 to

[26] *Ibid.*, p. 605*n*.

[27] Leahy, *op. cit.*, p. 384.

[28] *Ibid.*, p. 385.

[29] King and Whitehill, *op. cit.*, p. 606.

[30] On 11 December 1944, Congress had authorized President Roosevelt to appoint four Fleet Admirals and four Generals of the Army. The President immediately named King, Leahy, and Nimitz to the naval five-star rank, and Marshall, MacArthur, Eisenhower, and Arnold to the corresponding Army grade. Halsey was named later as the fourth fleet admiral. *Ibid.*, p. 582.

[31] JCS 1331/3, dtd 25May45, cited in Craven and Cate, *Matterhorn to Nagasaki*, p. 686.

310-224 O - 69 - 27

discuss OLYMPIC plans and preparations. Especially, they were to establish for the record a set of principles or division of responsibilities that would govern whatever action would be taken by either commander or his deputies in organizing for the invasion of Japan. Primarily, these principles concerned impending and future deployments, attachments and detachments from both commands, and logistical plans and troop buildup tasks charged to each commander.[32] Upon mutual agreement of these coordinating decisions and after the publication of the JCS implementing order for OLYMPIC, the stage was fairly well set for the moves that would lead to the invasion itself.

Assigned to conduct the Kyushu landings was General Walter Krueger, USA, and his Sixth Army. The following units comprised the OLYMPIC assault force: I Corps, V Amphibious Corps, IX Corps, XI Corps, 40th Infantry Division, 11th Airborne Division, 158th Regimental Combat Team, Sixth Army Troops, and Army Service Command OLYMPIC. Including the personnel in aviation and follow-up echelons, a total of 815,548 troops was to participate in the operation.[33]

The fast carrier task groups of Admiral Halsey's Third Fleet were to provide strategic support for the landings while Admiral Spruance's Fifth Fleet conducted the operations immediately concerned with the seizure and occupation of beaches in southern Kyushu.[34] Quite simply, the overall OLYMPIC scheme of maneuver called for three of the four corps assigned to the Sixth Army to make separate landings on the east and west coasts of the southern tip of Kyushu on 1 November, X-Day. (See Map 24.) The fourth corps would not land until at least X plus 3; it was to prepare to make a contingent landing or to reinforce other landing forces on order. After they had captured the beachheads, the landing forces were to fan out, link up, and drive northward to form a line from Sendai to Tsuno. After this deployment had been accomplished, further operations were to be based on expediency influenced by the course of events.[35] Because the primary objectives of OLYMPIC forces were airfields and sites for the establishment of bases to be used for staging and mounting CORONET, the conquest of the entire island of Kyushu did not appear necessary. On the other hand, because plans for the conduct of the campaign after the establishment of the Sendai-Tsuno line were fluid, the possibility that all of Kyushu could or should be captured was not excluded.

The most critical part of the amphibious phase of OLYMPIC, aside from the

---

[32] CinCPac WarD, May45.

[33] Sixth Army FldO 74, dtd 28May45, Anx 3; CinCAFPac Staff Study—OLYMPIC Operations—Southern Kyushu, dtd 28May45.

[34] CinCPac-CinCPOA Joint Staff Study Kyushu Island for OLYMPIC, Ser 0005081, dtd 18Jun45, App C. For OLYMPIC, the Third and Fifth Fleets, which previously had been alternative organizational titles for much the same groupment of ships, became separate entities.

[35] A chief source, in addition to the OLYMPIC plan and the operation orders based on this plan, has been Drs. K. Jack Bauer and Alvin D. Coox, "Olympic vs Ketsu-Go," *Marine Corps Gazette*, v. 49, no. 8 (Aug65), p. 32 *ff*.

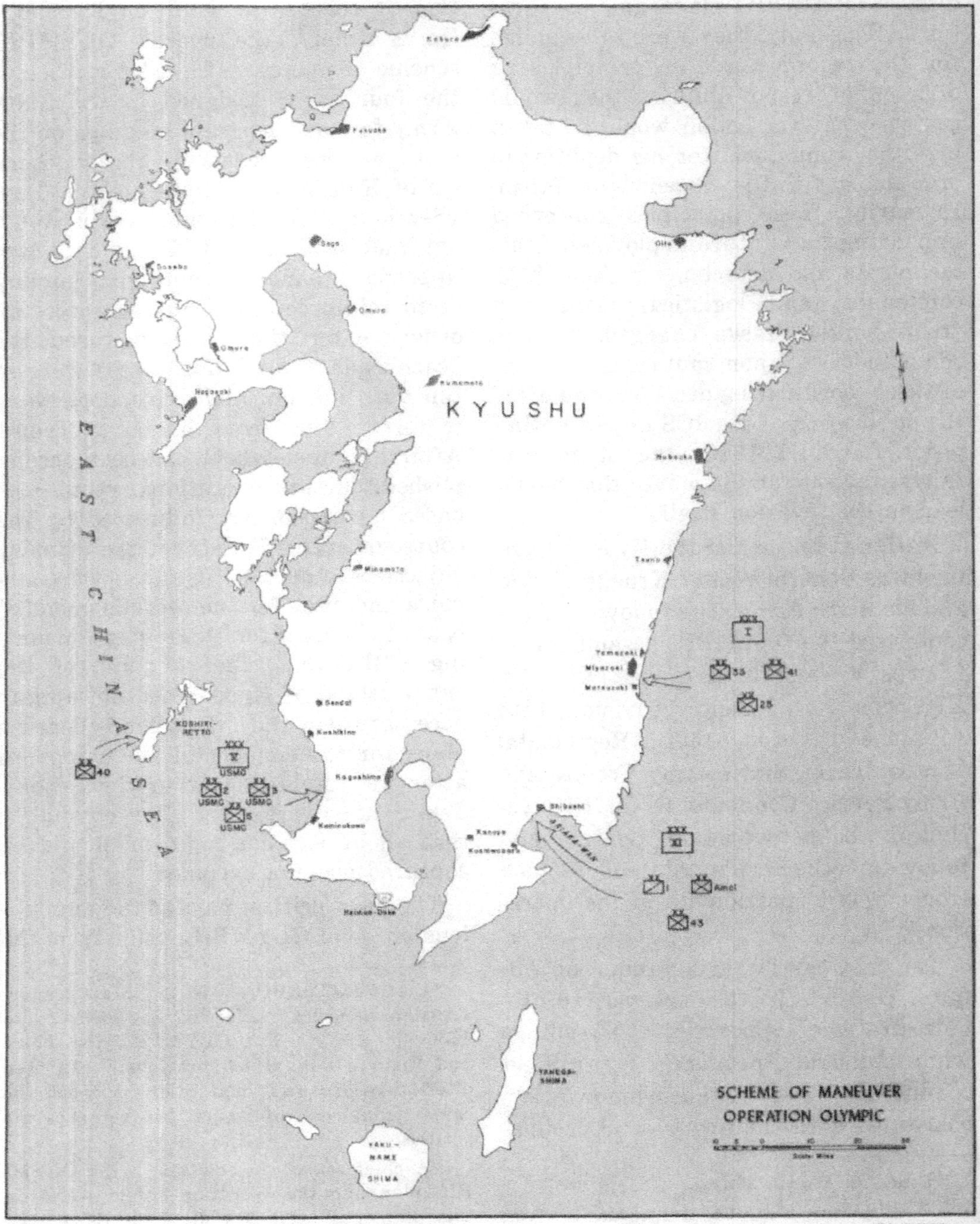

MAP 24

T.L. RUSSELL

assault itself, was thought to be the capture of Kagoshima, near the southwestern tip of Kyushu. The importance of this objective lay in the fact that the American planners had selected enormous Kagoshima Bay to become the primary port through which troops and supplies intended for the buildup of CORONET would pass. In addition, Kagoshima and its landlocked bay were to serve as an advanced naval base.

The Kyushu landings were not expected to be easy, for all intelligence estimates had predicted that the island would be heavily defended. Japanese strength on Kyushu was placed at 450,000 troops, of whom nearly half were deployed south of a line between Minamata and Nobeoka. Intelligence agencies believed that another three or four divisions were in the northernmost portion of the island and available as reinforcements, and that other troops could be brought over from Honshu. OLYMPIC planners did not expect that this reinforcement would be a factor to be concerned with, since the enemy would undoubtedly be reluctant to release any of the forces needed on Honshu to defend against future landings there. The Americans also expected that the Japanese would exploit the complex mountainous terrain inland of Kyushu and build formidable defenses to be held by the existing garrison force.

Besides the resistance anticipated from Japanese ground forces, OLYMPIC intelligence estimates indicated that the enemy had approximately 5,000 *Kamikaze* planes and pilots available with which to attack the landing forces. Although most of the *Special Attack* squadrons, like a portion of the ground units, would be withheld for the defense of Honshu, 60 airfields—and 5 more under construction—had been spotted on Kyushu; 22 of these were located south of the Sendai-Tsuno line. The Japanese response to American air attacks on the Home Islands in the latter period of the Okinawa campaign and following its conclusion indicated that the enemy either did not have the available air strength or the will to fight, or that he lacked both. Another possibility was that Japan was husbanding its resources for a massive air attack on the anticipated American invasion forces. The reported Japanese air strength and the number of fields on which it was based led OLYMPIC planners to believe that enemy air posed a real threat to the landings. Reports of the presence of numerous suicide submarine and boat bases on the coasts of Kyushu led the Americans to expect trouble from these craft during the assault phase of the operation.

Admiral Spruance's Fifth Fleet, as composed for OLYMPIC, was the largest and most formidable array of its kind yet to appear in the Pacific war. It contained two groups of fast carriers, a gunfire support and covering force, an escort carrier force, and a composite force containing the Third, Fifth, and Seventh Amphibious Forces. This was the first time in the war that three amphibious forces had been assigned to a single operation. The Fifth Fleet also contained a minecraft group and a large assortment of service units.

Admiral Halsey's Third Fleet was comprised of the Second Carrier Task Force and the British Carrier Task Force as well as numerous supporting elements. First of the OLYMPIC forces to go into action before the invasion, the Third Fleet was to make widespread attacks on all of the Home Islands in the period between 28 July and 23 October to destroy the Japanese air potential, interdict communications between Kyushu and Honshu, and to sink anything that was afloat. For 10 days in this preinvasion period, the British contingent would strike at the Hong Kong-Canton area. From X minus 14 to X minus 8, the Third Fleet would concentrate on targets in and around Honshu, Kyushu, and Shikoku to create a diversion and to isolate the scene of the impending invasion. On 23 October, aircraft from two of Halsey's carrier task groups were to join Fifth Fleet planes in a series of last-minute strikes against targets in the landing zone while the rest of the Third Fleet would continue to pound installations and targets of opportunity along the Japanese coastlines. When directed by CinCPac after X-Day, Halsey's two groups would be returned to the Third Fleet.[36]

Third Fleet aircraft were to operate generally east of a boundary drawn down Honshu to the eastern tip of Shikoku. Attacks west of that line and diversionary strikes along the China coast would be flown by the Far East Air Forces (which included the Fifth, Seventh, and Thirteenth Air Forces and the 2d Marine Aircraft Wing) commanded by General George C. Kenney from a forward CP established on Okinawa. In the period following X minus 10, FEAF was to cut communications between the target area and northern Kyushu. When the ground forces had seized the beachhead and had seized or built airfields inland, and when a sufficiently large garrison force had landed and was ready to maintain airfields and aircraft, Kenney would take over air support of the ground forces from the Navy.

Differing only in the size of the forces and the area involved, preinvasion operations would be conducted along the same successful patterns that had evolved from other Marine and Army landings in the Pacific. Because the invasion of Kerama Retto prior to that of Okinawa had demonstrated the value of obtaining a base on islands that were offshore of the major target, OLYMPIC plans provided for the seizure of Koshiki Retto and other small islands west of Kyushu on X minus 5 by the 40th Infantry Division. The OLYMPIC directive provided also for the capture of the northern portion of Tanegashima—south of Kyushu—by the 158th RCT on or after X minus 5 if Japanese guns on the island threatened minesweeping operations. If they did not, the RCT would land as a reinforcing element on X plus 3.

Most of the Army troops assigned to land on X-Day would mount out of, rehearse, and stage in the Philippines; the Marines would hold rehearsals in the Marianas after their units mounted out of that area and the Hawaiian Islands. Because the assault forces were to land on three different and widely separated beaches, there were to be three different H-Hours.

At 0600 on X-Day, Admiral Turner's advance force, after conducting preinvasion operations was to be absorbed by Task Force 40, OLYMPIC Amphib-

[36] CinCPac OPlan 1E-45, dtd 8Aug45, p. 8 (OAB, NHD).

ious Force, also commanded by Turner. At some time shortly thereafter,[37] the three attack forces would land their landing forces.

The Third Attack Force (Third Amphibious Force, Vice Admiral Theodore S. Wilkinson) was to land XI Corps (1st Cavalry, 43d Infantry, and Americal Divisions) on the east coast of Kyushu on the beaches at the head of Ariake Wan in the Shibushi-Koshiwabaru area. After the consolidation of the beachhead and the capture of Shibushi and its airfield, XI Corps was to drive inland and to the north to make contact with I Corps. Upon establishing contact, the two corps, and VAC also, would form a line and advance northwards to establish the Sendai-Tsuno line. (See Map 24.)

North of the XI Corps beachhead, I Corps (25th, 33d, and 41st Infantry Divisions) was to be landed by the Seventh Attack Force (Seventh Amphibious Force, Vice Admiral Daniel E. Barbey) on beaches in the vicinity of Miyazaki to secure a beachhead in the Yamazaki-Matsuzaki area. I Corps was then to push inland and to the north together with XI Corps.

The third OLYMPIC landing, and the only one to be conducted on the west coast, was that of General Schmidt's VAC (2d, 3d, and 5th Marine Divisions), which was to be lifted to the target by the Fifth Attack Force (Fifth Amphibious Force, Vice Admiral Harry W. Hill). Following its seizure of the Kushikino-Kaminokawa beaches, the assault Marines would fan out to extend the beachhead to Sendai. In addition, VAC forces were to set up a line between Sendai and Kagoshima to block any Japanese drive down the west coast and the southwestern leg of Kyushu. After consolidating the beachhead, VAC would join the other two corps in the general drive to the north.

Besides the X-Day landings, OLYMPIC plans called for the reserve force, IX Corps (77th, 81st, and 98th Infantry Divisions), to land in the vicinity of Kaimon-Dake on X plus 3 or later, depending on the situation ashore. Once the whole of IX Corps had landed, it would clear the southwestern shore of Kagoshima Wan and prepare FEAF facilities and installations for other OLYMPIC garrison units.

The 13 assault divisions of the OLYMPIC force would carry the full burden of the fighting until on or after X plus 22 (23 November), when the 11th Airborne Division, Sixth Army reserve afloat, was scheduled to be off Kyushu and ready to land. General Krueger's planners believed that, based on intelligence estimates of the size of the Japanese defense forces, Sixth Army combat strength would be superior to that of the enemy and would be able to advance to the Sendai-Tsuno line.

The war ended while CORONET was still in the planning stages and there was little material distributed on that operation. A broad outline had been drawn, however, which established that two armies, the Eighth and Tenth, numbering nine infantry and two armored divisions and three Marine Divisions of IIIAC, would land on the Pacific beaches of Honshu leading to the Kanto Plain—between Choshi and Ichinomaya—in

[37] The exact time for the three major landings had not yet been determined when OLYMPIC planning ended.

March 1946. Immediately following the assault armies ashore would be the First Army, redeployed from Europe, with one airborne and ten infantry divisions. The primary objectives of CORONET were to crush Japanese resistance on the plain and to occupy the Tokyo-Yokohama area. If the accomplishment of these objectives did not force the enemy to surrender, the three armies were to fan out and secure the rest of the Home Islands. Ultimately, an air garrison equalling 50 groups was to support these final operations.

On 2 June 1945, shortly after receipt of the OLYMPIC plan, VAC reported to the commander of the Sixth Army for further orders in the impending Kyushu operation.[38] Even before the actual operation order had been published, enough of the proposed plan had been known and made available to the assault forces to permit them to begin preparations for the landings. Planning and training for OLYMPIC on a division level continued through June and July. Because the tentative mission of the 5th Marine Division, for instance, originally called for it to be in either the assault or the reserve, General Bourke formed nine BLTs in his division and prepared each one to land at any stage of OLYMPIC. Division artillery and tanks also trained to land on short notice on any designated beach in the VAC target area and to operate under division control. During the summer, 5th Division BLTs rehearsed intensively and made many practice landings. Their training for operations inland stressed the assault of fortified positions, village and street fighting, and the removal of mines and demolitions.[39]

The 2d and 3d Marine Divisions also prepared vigorously, and like the 5th, rotated their combat veterans home when due, brought their regiments up to strength as they received replacement drafts, including many second-timers, and refurbished their equipment for OLYMPIC. While VAC prepared for this operation and IIIAC awaited its further orders, Marine carrier planes and pilots, along with the Third Fleet and FEAF squadrons, carried the war straight to the heart of the Empire.

[38] CG, FMFPac ltr to CMC, dtd 13May46, encl A, Administrative History of FMFPac, p. 7.

[39] Conner, *5th MarDiv Hist*, p. 130.

The Closing Days

MARINE AIR ON CARRIERS [1]

On 18 January 1939, the Secretary of the Navy approved the following mission and organization of Marine Corps aviation:

> Marine Corps aviation is to be equipped, organized and trained primarily for the support of the Fleet Marine Force in landing operations and in support of troop activities in the field; and secondarily as replacement squadrons for carrier-based naval aircraft;
>
> The organization, personnel complements, and other details of Marine Corps aviation are to conform as closely as practicable to similar naval aviation organizations;
>
> The Bureau of Aeronautics is to exercise supervision over their respective activities connected with Marine Corps aviation in the manner provided for similar naval aviation units.[2]

Until carrier-based Marine squadrons supported Tenth Army landings on Okinawa in 1945, Leatherneck pilots had been in a position to support an amphibious assault from its beginning only twice in World War II: at New Georgia and Bougainville. And not until the latter part of 1944, when a few squadrons were assigned to carriers did Marine aviation fulfill its secondary mission.

In all other operations, landings were made so far away from the nearest air base that Marine squadrons had to wait for an airstrip to be completed or a captured one to be put into operation again before they could fly in to begin supporting the ground troops. Assignment to carriers was the only solution by which Marine aviation could carry out the principal missions assigned to it. As soon as it was feasible, Marine squadrons landed on Henderson Field on Guadalcanal, but their basic role was in the air defense of the island, with a secondary emphasis placed on air support. When the American offensive began climbing up the Solomons ladder, Marine pilots flew missions under the control of the Strike Command, Commander Air Solomons.

With the beginning of the Central Pacific campaigns in late 1943, Marine ground commanders became increasingly dissatisfied with the type and amount of air support they received. At Tarawa, defense against air attack and the close support of ground troops were both entrusted to carrier planes flown by Navy pilots. In the opinion of both Navy and Marine officers, the air support at Tarawa left much to be desired in the way of accomplishment. Many apparent shortcomings in this operation indicated that, among other things, truly effective air support was impossible unless the pilots and ground troops had trained as a team.

[1] Unless otherwise noted, the material in this section is derived from: *AirIntelBul*, Aug–Sep45; Isely and Crowl, *Marines and Amphibious War;* Sherrod, *Marine Air Hist.*

[2] *DivAvn OpD*, dtd 1Jun45.

After the operation General Holland M. Smith recommended that Marine aviators, thoroughly schooled in the principles of direct air support, should be assigned to escort carriers and included in any future amphibious operation undertaken by a Marine division. If this request could not be granted, he continued, the Navy airmen selected for the task should be carefully indoctrinated in the tactics they would employ.[3]

Granting the validity of General Smith's recommendations concerning the assignment of Marine squadrons to carriers, no one else in the Marine Corps seemed disposed to push for such a program at that time. Earlier in the war, many factors, such as the shortage of manpower and the need to send increasing numbers of air units to inland bases in the Solomons, militated against the employment of Marine air in support of ground operations. The pressure for the employment of Marine Corps planes and pilots in ground support operations increased as the war progressed and the need for such support became apparent.

The criticism of the conduct of air support at Tarawa was later echoed following the end of the Marianas campaign, where Marines believed that the Navy system of controlling close air support missions was too rigid and time consuming. This matter was made an agenda item to be discussed during one of the King-Nimitz Pacific conferences. The item noted that:

> During the Saipan operation T.F. 58 was necessarily withdrawn from the immediate area, leaving 8 CVE's to perform a multiplicity of missions, including direct support of ground troops. What are CinCPac's views as to the following plan to avoid this situation in the future:
>
> (a) Embark in CVE's Marine aircraft squadrons whose sole duty will be direct support of ground troops. (Training in carrier operations will obviously be a preliminary requisite.)
>
> (b) As soon as airfields are available ashore, transfer those same squadrons ashore to continue direct support of ground troops.[4]

During the 13–22 July 1944 conference, Admiral Nimitz addressed the question of assigning Marines to carriers. In essence, he did not consider the proposal desirable because he believed that "it would require a great deal of extra training and equipping of Marine squadrons for carrier operations, anti-submarine warfare, navigation, etc." [5] In addition, Nimitz believed that the personnel and equipment of the squadrons would have to be revised considerably to make the Marine units suited for both ship-based and advance base operations. He also believed that, if Marines were to be assigned to carriers, there would be a surplus of Navy CVE squadrons. As a final thought, CinCPac stated that the Navy CVE pilots were rapidly gaining experience in ground support operations, and therefore, there was no real need for Marine CVE squadrons.

Admiral King then stated that, in his opinion, the Marine ground forces could be supported adequately without employing Marine aviation squadrons—and "thus prevent two air forces in the Navy." [6] For some time, CominCh had

[3] VAC AR GALVANIC, dtd 11Jan44, p. 16.

[4] Item 68, CominCh Agenda for conference with Adm Nimitz, dtd 10Jul44 (OAB, NHD).

[5] *Minutes, CominCh–CinCPac Conference*, 13–22Jul44.

[6] *Ibid.*

been concerned that the expansion of Marine Corps aviation strength had exceeded the point where it could be gainfully employed, because there were not that many missions available for Leatherneck pilots. In view of the location of Marine squadrons and the nature and location of U. S. operations in the Pacific, he was right. As of 30 June 1944, Marine Corps aviation strength consisted of 5 wings, 28 groups, 128 squadrons, and 108,578 personnel, of which slightly more than 10,000 were pilots.[7]

Lieutenant General Alexander A. Vandegrift, who became Commandant of the Marine Corps on 1 January 1944, also was concerned with the status of his air units and had consulted with Admiral King regarding the future employment of those squadrons and pilots sitting in the backwash of the war on South Pacific islands. He proposed that one of the five wings be eliminated, but also argued that to employ the fliers and planes based in the rear areas of the South and Central Pacific gainfully Marine pilots should be assigned to carriers. King agreed in principle to this compromise, but stated that Nimitz' approval had to be gained before any final action could be taken.[8]

A desire to visit his Marines in the Pacific as well as to determine at first hand the facts surrounding the Saipan command controversy[9] impelled General Vandegrift in late July 1944 to make an inspection trip, in which he covered:

> . . . 22,000 miles in eighteen days, saw all the force, corps, and division commanders and practically all the regimental and battalion commanders in the field. I went to Saipan, Tinian, and Guam, getting to Guam just before the show was over. Our people did a superb job on all three of those islands, the fighting on all three of them being entirely different. . . .
>
> I went from Guam to Kwajalein to Guadalcanal then up to the Russells to see my old division and to Bougainville to see Ralph Mitchell and his crowd. Then back to Pearl for a three-day session with Nimitz.[10]

Accompanying the Commandant were Brigadier Generals Field Harris—the newly appointed Director of Aviation—and Gerald C. Thomas, the Director, Division of Plans and Policies. Upon their return to Pearl, they went into conference with Nimitz, Vice Admiral John H. Tower, Deputy CinCPac–CinCPOA, Rear Admiral Forrest P. Sherman, Nimitz' deputy chief of staff and head of his War Plans Division, and Major General Ross E. Rowell, since 1941 head of MAWPac (the forerunner of AirFMFPac). The decisions they made in the course of these talks determined the course that Marine aviation was to take in the Pacific for the remainder of the war.

Vandegrift broached the subject of the future employment of Marine squadrons, and informed Nimitz of what had been said in the course of conversations

[7] *FMF Air Status Rpt*, 30Jun44.

[8] Gen Alexander A. Vandegrift comments to Robert Sherrod, dtd Sep49, cited in Sherrod, *Marine Air Hist*, p. 327.

[9] For the story of the relief of the 27th Infantry Division commander by Lieutenant General Holland M. Smith, see Shaw, Nalty, and Turnbladh, *The Central Pacific Drive*, pt IV, chap 5.

[10] LtGen Alexander A. Vandegrift ltr to Gen Thomas Holcomb dtd 5Sep44 (Vandegrift Personal Correspondence File, HistBr, HQMC).

about the matter with CominCh, especially the recommendation that Marines be assigned to carriers. Although Towers argued that there had been no indication in the past that Marines wanted to operate from carriers, Vandegrift and Harris persuaded him that times and attitudes had changed.

It was agreed at this conference that the primary mission of Marine Corps aviation was to support the Marine ground forces and to participate in amphibious assaults. Therefore, in order to focus the activities of Marine aviation on its mission more effectively, the following package of proposals submitted to Admiral King were concurred in by Nimitz with an endorsement stating that it would "more firmly integrate Marine Corps aviation within the Marine Corps and is therefore in the interest of the naval service." [11]

Essentially, it was recommended that a complement of Marine squadrons to be employed in the close support of amphibious operations be assigned to one CVE division of six *Commencement Bay*-class carriers. This complement was to consist of six 18-plane fighter (F4U or F6F aircraft) and six 12-plane torpedo bomber squadrons whose pilots were to be specially trained in the use of rockets with which their planes were to be armed. It was further recommended that a Marine aviator of suitable rank be directed to organize and prepare these squadrons for carrier operations. He later would be assigned to duty on the carrier division staff.

Concerning another aspect of the Marine aviation problem, the conferees agreed that Marine aviation should gradually take over the responsibility for controlling aircraft in direct support of ground troops in amphibious operations. Gradually, and as practicably as possible without impairing the conduct of combat operations then in process, Marine Corps personnel would replace their Navy counterparts in the existing Air Support Control Unit organizations.

One other recommended change was to effect the reorganization of Marine aviation in the Pacific, wherein Marine Aircraft Wings, Pacific, would become Aircraft, Fleet Marine Force, Pacific, in order to identify the Marine air components more closely with the ground elements. This proposal also established the relationship of AirFMFPac with ComAirPac and FMFPac under the overall command of CinCPac.

In addition, the strength and composition of Marine aviation forces in the Pacific was to undergo change. Hereafter, MAGs would be comprised of three 24-plane squadrons instead of four 18-plane squadrons, and the number of Air Warning Squadrons would be cut from 32 to 24 or less in view of the number of Army units of the same type that were scheduled to arrive in the Pacific for future operations.

General Vandegrift signed the basic memorandum listing the proposed recommendations and stated in the last paragraph of this report that "Every effort will be made to increase the mobility and effectiveness of marine aviation by accomplishing such readjustments of personnel and equipment among Headquarters, Service and Tac-

---

[11] CinCPac enforsement, ser 002567, dtd 22Aug44, on CMC secret memo to CominCh, Subj: Marine Aviation in the Pacific, n.d. (OAB, NHD).

tical Squadrons as may be indicated." [12] Admiral King approved the proposals on 10 September 1944 with the proviso that when Army Air Forces units were available in the Pacific to replace certain Marine Corps aviation squadrons, Marine Corps aviation strength would be reduced by or up to the equivalent of one wing.[13] In a bucktag comment on the conference proposals, Admiral King wrote: "Good, but does not go far enough towards reducing MarCorps aviation. K." [14] In notifying General Holcomb of what had transpired regarding the future of Marine Corps aviation, General Vandegrift wrote:

> Another thing we have done, which I pinch myself now and then to see if I am still awake, we have gotten both Nimitz and King to approve a division of the larger CVEs for use of Marines. That will give us four carriers with a carrier group of Marines aboard, and I can assure you that took some days of hard talking.[15]

Following up the approval of CominCh for placing Marine planes on carriers, on 28 October 1944, the Chief of Naval Operations directed the formation of the Marine Air Support Division.[16] To comprise this organization, the Commandant of the Marine Corps selected MAG–51, MBDAG–48, and the following squadrons: VMO–351,[17] VMF–112, –511, –512, –513, and –514, and VMTB–132, –143, –144, –233, –234, and –454. All of these units were attached to Marine Fleet Air, West Coast, (MarFAirWest) at San Diego and were redesignated as follows: MAG–51 became MASG–48 (Marine Air Support Group), and MBDAG–48 similarly became MASG–51. All of the squadrons attached to these groups were further identified with the following letters "CVS," meaning Carrier Support, as VMF(CVS)–112.

The overall designation given to the all-Marine carrier force was Marine Carrier Groups, Aircraft, Fleet Marine Force, Pacific. The next subordinate echelon to this was the MASG, which was comprised of the fighter and torpedo bomber squadrons for a CVE division of six ships. Each of the escort carriers, in turn, was to have as its air complement a Marine Carrier Group (MCVG) consisting of a Marine Carrier Aircraft

[12] CMC memo to CinCUS [CominCh], Subj: Marine Aviation in the Pacific, n.d. (OAB, NHD).

[13] CominCh ltr FF1/A4–3 serial 002624 to CinCPac and CMC, Subj: Marine Aviation in the Pacific, dtd 10Sep44 (OAB, NHD).

[14] CominCh bucktag comment dtd 10Sep44 on CinCPac End to *Ibid* (OAB, NHD).

[15] LtGen Alexander A. Vandegrift ltr to Gen Thomas Holcomb dtd 5Sep44 (Vandegrift Personal Correspondence File; HistBr, HQMC). General Vandegrift was perhaps, being unduly modest, for as General Gerald C. Thomas, former Director of the Division of Plans and Policies and later Assistant CMC, recalled: "General Vandegrift's relations [with King] were pleasant, not familiar, but always on a really good sound basis, and never in my three and a half years with him there [Headquarters Marine Corps], did I see him lose a battle. He won every one of them." Gen Gerald C. Thomas interview with HistBr, HQMC, dtd 26Sep66. (Oral History Collection, HistBr, HQMC).

[16] CNO ltr Op–37–C–fgd serial 08837 to ComAirPac, Chief BuAer, ComFairWest, CGAirFMFPac, and MarFAirWest, Subj: Marine Air Support Division, formation of, dtd 28Oct44 (OAB, NHD).

[17] Although VMO–351 was an observation squadron earlier in the war, its mission was changed later and it became a fighter squadron without a redesignation in its unit identification.

Service Detachment (MCASD), a VMF-(CVS), and a VMTB(CVS).

MASG–51 was given four VMFs and four VMTBs to form four active groups, and MASG–48 was given the VMO, a VMF, and two VMTBs, which were to comprise the two replacement groups. Marine Carrier Groups, AirFMFPac, was officially activated on 21 October 1944 at Marine Corps Air Station (MCAS), Santa Barbara, California, with Colonel Albert D. Cooley as commanding officer.

For more than a month before the first MASG squadron went on board its CVE, however, other Marine squadrons had been flying combat missions from the decks of fast carriers on a temporary basis. The appearance of the *Kamikaze* menace during the Leyte operation in the fall of 1944 created the need for additional fighter-type aircraft aboard the carriers of the Third Fleet. Brigadier General Frank G. Dailey, then a colonel assigned to TF 58 as Vice Admiral Marc A. Mitscher's Marine Air Officer,[18] commented that in addition to the requirement for more fighters on the CVs:

> . . . another primary consideration in putting Marine squadrons aboard with F4Us was due to the fact that the Navy squadrons with their F6Fs did not have the speed or altitude to intercept a Japanese light bomber designated 'Betty,' which appeared about this time. Consequently the Marine squadrons were also used as fleet combat air patrols; in fact, for a time, this was their primary duty. I think it should be emphasized that, prior to this time, the Navy did not consider the F4U a suitable aircraft for carrier operations because of the known difficulty in take-offs and landings on CVs. When you consider that these [Marine] squadrons were literally picked off the beach with very little CV training to operate under war time conditions, our operational losses were expected and accepted. It is necessary to have experience in carrier operations to appreciate the magnitude of making this transition in such a short time and especially in this type aircraft. The plane crews should also be mentioned, as they were operating under conditions foreign to many of them and kept a high aircraft availability, even by Navy standards.
>
> When the decision was made to put Marine squadrons with F4Us on board the CVEs, it was thought that the operational losses, in view of our experience on fast carriers, would be prohibitive. Here again, we were using aircraft which were not initially considered suitable for the large CVs. And now, we were expected to operate from CVEs. Colonel Albert Cooley was the officer responsible for the successful operations of this venture and proved it could be done.[19]

It was not until the end of 1944 that the first of the VMFs boarded a big carrier in the Pacific. Between January and June 1945, 10 Marine fighter squadrons flew from the decks of 5 CVs in major fast carrier task force operations. On 28 December 1944, VMF–124 (Lieutenant Colonel William A. Millington) and –213 (Major Donald P. Frame) boarded the *Essex* at Ulithi, "equipped with F4U–1D Corsair fighters, the initial introduction of this type aircraft in the Fleet."[20] Two days later, in company with the rest of the Third Fleet, the *Essex*

[18] Col Frank G. Dailey ltr to Robert Sherrod, dtd 2Jun50, cited in Sherrod, *Marine Air Hist*, p. 343.

[19] BGen Frank G. Dailey ltr to Hd, HistBr, G–3 Div, HQMC, dtd 10Nov65, hereafter *Dailey ltr*.

[20] BGen William A. Millington ltr to Hd, HistBr, G–3 Div, HQMC, dtd 18Nov65.

steamed out of the anchorage bound for a series of strikes on Formosa and Luzon in the period 3–9 January 1945.[21] The weather during that week of operations was foul and solidly overcast for the greater portion of the time. At the end of their first days aboard the *Essex*, 9 of which were spent at sea, the two Marine squadrons had lost 7 pilots and 13 F4Us solely as a result of operational accidents during instrument flight conditions. One Marine aviator stated: "We just can't learn navigation and carrier operations in a week as well as the Navy does it in six months." [22]

On 10 January, Admiral Halsey's Third Fleet entered the South China Sea to log 3,800 miles in an 11-day series of strikes against targets on the coast of Indochina and on Hong Kong and Formosa.[23] Both Marine squadrons on the *Essex* participated in the TF 38 raids on Saigon, Hainan, Hong Kong, Swatow, and Formosa. Lieutenant Colonel Millington, the VMF–124 commander, became the air group commander of the *Essex* on 15 January, when Commander Otto Kinsman, the naval officer holding that position, was killed in action.[24]

After a last series of strikes on Formosa, the Pescadores, and Sakishima Gunto on 21 January, and following a photographic mission over Okinawa on the 22d, the Third Fleet retired from the South China Sea and set a course for Ulithi, arriving there on the 25th. At 0001, 27 January, the Third Fleet became the Fifth Fleet when Admiral Spruance assumed tactical command from Halsey.

In their first month of carrier operations, the two Marine squadrons claimed a total of 10 Japanese planes destroyed in the air and 16 on the ground. Marine pilots flew 658 sorties. Operational losses of the squadrons, 7 pilots and 15 aircraft, were considerably greater than the 1 pilot and 2 planes lost in combat.

By 10 February, TF 38—now TF 58—was ready to sortie against the enemy once more. The target this time was Tokyo, some 1,500 miles due north of Ulithi. Joining the Fifth Fleet were three other large carriers, each with two Marine fighter squadrons in its complement. On the *Bennington* were VMF–112 (Major Herman Hansen, Jr.) and –123 (Major Everett V. Alward); the *Wasp* had VMF–216 (Major George E. Dooley) and –217 (Major Jack R. Amend, Jr.); and VMF–221 (Major Edwin S. Roberts, Jr.) and –451 (Major Henry A. Ellis, Jr.) were on the *Bunker Hill*. Admiral Spruance's fleet now had a total of eight VMFs on four large carriers. Based on the lessons learned in the January operations, all of the Marine pilots "had received intensive navigational training at Ulithi" before boarding the carriers "and would get more en route to Japan 'in weather not previously considered suitable for CV operations.' " [25]

[21] For Third Fleet operations in this period, see Samuel E. Morison, *The Liberation of the Philippines—History of United States Naval Operations in World War II*, v. XIII (Boston: Little, Brown and Company, 1959), pp. 87–92, hereafter Morison, *Liberation of the Philippines*.

[22] Sherrod, *Marine Air Hist*, p. 333.

[23] Morison, *Liberation of the Philippines*, pp. 164–174.

[24] VMF–124 WarD, Jan45.

[25] Col Frank G. Dailey ltr to Robert Sherrod, dtd 2Jun50, *op. cit.*

After the task force had departed Ulithi, all hands learned that their target was to be Tokyo, and that these first carrier-plane raids on the enemy capital were to precede by three days the 19 February invasion of Iwo Jima by VAC troops. It was also announced that the Marine squadrons in TF 58 would furnish air support for the Iwo landing forces beginning on D-Day.

On 16 February, Vice Admiral Marc A. Mitscher's carriers launched their planes to hit the airfields and aircraft factories around Tokyo Bay. Lieutenant Colonel Millington led the first fighter strike from the deck of the *Essex* as Major David E. Marshall, skipper of VMF–213, took off with his squadron from the same carrier to lead the escort for torpedo and photo-reconnaissance planes headed for the Tokyo area. The other Marine squadrons were given equally important missions. Although the weather on the 16th was abominable, the fifth air sweep of the area launched that day by the *Essex* and the *Bunker Hill* found clear weather over their target and had the honor of being "the first Navy [and Marine] fighter planes to arrive over Tokyo." [26] Additional strikes were launched before and shortly after dawn on the 17th, but with the weather worsening rapidly and restricting further flight operations, Admiral Mitscher cancelled the remaining planned strikes, recovered all of his airborne planes, and laid a course for Iwo Jima.

When, on the morning of 19 February, TF 58 was approximately 100 miles away from Iwo, its planes began a series of prelanding strikes on the target. For 20 minutes, between H minus 55 and H minus 35, 120 fighters and bombers from the fast carriers hit the landing beaches and adjacent areas.[27] At 0642, Lieutenant Colonel Millington led a flight of two Marine and two Navy fighter squadrons—flying F4Us and F6Fs, respectively—on a mission to napalm, rocket, and strafe the flanks and high ground along the beaches.

> The attacks were delivered from a double-column approach with the divisions of planes breaking to port and starboard, dropping napalm on the first run, pulling out to seaward and repeating attacks with rockets and .50-caliber bullets until the time limit expired. The 48-plane flight then rendezvoused for an H minus 5 strafing attack along the landing beach. These attacks were delivered from north to south in steep dives, all planes pulling out sharply to the right to rejoin the tail element for repeated runs. The attack was moved inland gradually as the landing craft approached the beach so that the bullet-impact area remained 200 yards ahead of the troops. As the troops hit the beach, the bullet-impact area was shifted 500 yards inland to smother the fire from that area against the shore line. Because of naval gunfire in the same area, pull-outs were ordered at 600 feet [altitude]. The flight was ready to stand by for close-support missions but none was immediately assigned.[28]

Millington and Marshall had worked out the plan for this low-level attack with the commander of the Landing Force Air Support Control Unit, Colonel Vernon E. Megee, who later said that this was "one of the outstanding examples of

---

[26] Morison, *Victory in the Pacific*, p. 22.

[27] Bartley, *Iwo Jima*, p. 49.

[28] Sherrod, *Marine Air Hist*, p. 347.

effective precision beach strafing seen during the Pacific War." [29]

For several days following D-Day, Marine carrier pilots supported ground operations, and then TF 58 moved on to launch strikes at Chichi Jima. After one day at this target, the fast carriers set a northerly course for a high-speed run to Japan and another round of attacks on Tokyo. The first planes were launched on 25 February when the carriers were approximately 190 miles from the Japanese capital.[30] Again the weather was bad, even worse, as a matter of fact, than that which had been experienced during the attacks of 16–17 February. Most of the strikes were diverted to hit secondary targets or targets of opportunity. When flying conditions became impossible before noon, Admiral Mitscher cancelled other sweeps that had been scheduled for later that day. Because weather forecasts for the following day boded no better conditions, Mitscher decided to strike Nagoya and headed the task force in that direction. Early on the 26th, he realized that high winds and heavy seas would prohibit him from launching aircraft. The task force commander then cancelled the strikes, headed for a refueling area, refueled, and then steamed towards Okinawa for a series of attacks on that island on the 1st and 2d of March.

The weather over the Ryukyus was a considerable improvement over that experienced in the Home Islands, and all TF 58 pilots "accomplished the usual pattern of devastation, which now was almost routine." [31] Following the Okinawa strikes, the carriers returned to Ulithi, arriving there on 4 March, and refitted for an immediate return to sea. The destination of Mitscher's carriers was Kyushu, where TF 58 pilots were to hit the airfields in a series of strikes prior to the invasion of Oknawa. At the Ulithi fleet anchorage, VFM–124 and –213 were detached from the *Essex* on 10 March and were returned to the United States in the escort carrier *Long Island.* Three days later, the entire *Wasp* air group—including VMF–216 and –217—was replaced by an all-Navy group. Upon their detachment, the two Marine squadrons were transferred stateside by way of Ewa.

The ground crewmen of the four VMFs remained on the *Wasp* and *Essex* to service the Navy Corsairs, primarily because the Navy crewmen were largely unfamiliar with that type of plane. With the exception of those in the carrier-based VMFs, there were no other F4Us in TF 58 fighter squadrons up to that time. The *Essex* Marines remained with the carrier until early June; the Marines in the *Wasp* retired from the fighting much earlier when that ship was hit off Shikoku on 19 March and returned to Ulithi, and then steamed to Bremerton, Washington, via Pearl Harbor, for repairs.[32]

Although two of Admiral Mitscher's carriers lost their Marine squadrons before TF 58 steamed out of Ulithi on

[29] BGen Vernon E. Megee ltr to Dr. Jeter A. Isely, dtd 10Feb50, cited in Isely and Crowl, *Marines and Amphibious War*, p. 507.

[30] Morison, *Victory in the Pacific*, p. 57.

[31] Frederick C. Sherman, *Combat Command: The American Aircraft Carriers in the Pacific War* (New York: E. P. Dutton & Company, Inc., 1950), p. 342.

[32] USS *Wasp* (CV–18) WarDs, Mar, Apr45 (OAB, NHD).

14 March, the loss was made up when the *Franklin,* carrying VMF–214 (Major Stanley R. Bailey) and –452 (Major Charles P. Weiland), joined the Fifth Fleet. Following a refueling at sea on the 16th, Spruance's fleet began a high-speed run to its target, Kyushu.

The force was within 90 miles of the island when, just prior to dawn on the 18th, the first planes were launched against Kyushu airfields. The TF 58 pilots found few enemy planes on the ground at the target area, and none in the air, and so they bombed hangars and installations instead. Shortly after 0700, the missing Japanese aircraft appeared over the launching areas and began to attack the carriers. The *Yorktown, Enterprise,* and *Intrepid* were hit but not damaged badly enough to put them out of commission. The bomb that struck the *Enterprise* did not explode, and the *Intrepid* suffered only minor damage from a near miss when a Japanese plane crashed and exploded alongside. Seven crewmen were killed and 69 wounded in the *Kamikaze* attacks this day.[33]

On the next day, Mitscher sent his planes against Japanese shipping in the Inland Sea and in the harbors of Kobe and Kure. A short time after the carriers had launched their sweeps, sneak raids began to punish the force. In the case of nearly every carrier that was hit, the damage was caused by a single suicide plane that approached undetected and dove out of the clouds in an attempt to destroy the flattop below. At 0709, a suicider suddenly appeared over the *Wasp* and landed a bomb which penetrated the flight deck to the hangar deck, exploded a plane, and caused great damage and many casualties. Within 15 minutes after the attack, damage control parties had repaired vital facilities and put out the fires; by 0800, the *Wasp* was recovering her planes. A total of 101 men were killed or died of wounds, and 269 were wounded.[34] Despite her damage, the *Wasp* continued to operate with the fleet several more days before limping to Ulithi and on to the States for repair.

Just two minutes before the *Wasp* was attacked, the *Franklin* received two bombs from an enemy plane which had approached undetected. The "Big Ben," as she was called by her crewmen, was in the midst of launching her second strike of the morning, and her flight deck was studded with planes warming up and ready to take off. The aircraft were fully armed with bombs and rockets, and their fuel tanks loaded with highly flammable aviation gasoline. The first of the two bombs tore through the flight deck and exploded in the hangar deck, wrecking the forward ele-

[33] Morison, *Victory in the Pacific,* p. 94.

[34] *Ibid.,* p. 95. General Dailey commented on the carriers' Marine guard, "whose duties included manning the 40mm AA batteries, and at least on the *Bunker Hill* were given credit for several enemy aircraft shot down. They in turn had their casualties by bombings." He also noted that "it was SOP in the fleet that if an interception of an enemy aircraft was made but not completed before coming over the fleet, contact was to be broken and the ships AA fire would take over. At times, both Marine and Navy pilots would be so intent they would follow an enemy aircraft over the fleet regardless of the AA fire and if they didn't get a kill, follow right down until they (the enemy) splashed in the water." *Dailey ltr.*

310-224 O - 69 - 28

VMF(CVS)-511 CORSAIR *ready to be launched by catapult from the deck of USS* Block Island. *(USMC 43750)*

AN EXPLOSION *rips the critically damaged* Franklin *as crewmen run for safety. (USN 80-G-237900)*

vator; the second bomb hit the flight deck and immediately started fires that spread to the planes that were warming up. The bombs on the planes began exploding, and then the 11.75-inch rockets, "Tiny Tims," with which the aircraft were armed, began going off:

> Some screamed by to starboard, some to port, and some straight up the flight deck. The weird aspect of this weapon whooshing by so close is one of the most awful spectacles a human has ever been privileged to see. Some went straight up and some tumbled end over end. Each time one went off the fire-fighting crews forward would instinctively hit the deck.[35]

Three hours after being hit, the *Franklin* had lost all way and lay dead in the water. By noon, most of the fires had been extinguished or brought under control and all of the wounded had been evacuated to other ships standing close by. The *Pittsburg* passed a towline to the carrier and gradually began towing the critically wounded vessel out of the danger area. By 0300 on 20 March, the *Franklin* had begun to work up her own power and nine hours later she slipped her tow and headed for Ulithi and eventually New York.

In the flaming and exploding inferno following the bombing, *Franklin* lost 724 killed or missing and 265 wounded.[36] In these casualty figures, 65 of the dead were pilots and ground crewmen from the two Marine squadrons. Airborne at the time of the attack on the *Franklin*, VMF–214 and –452 pilots landed later either on the *Hancock* or the *Bennington*, from whose decks they continued attacks on Kyushu until 19 March, when the task force retired from the area. The two Marine squadrons were sent to Marine Corps Air Station, El Centro, California, where they remained until the war ended.

From 23 to 25 March, TF 58 began the last of the softening-up operations on Okinawa before the scheduled invasion. Together with the other squadrons of the *Bennington* air group, VMF–112 and –123 flew many sorties over the target area. On L-Day, only four Marine squadrons—VMF–112 and –123 in the *Bennington* and VMF–221 and –451 in the *Bunker Hill*—remained in the task force. All four squadrons together with the Navy pilots napalm bombed and strafed Hagushi beaches on 1 April, and then later in the day hit targets beyond the beachhead. Following the Okinawa ground support missions of the first few days of the operation, TF 58 planes, pilots, and ships were kept busy fending off the destructive *Kamikazes*. The battle against the Japanese suiciders was to keep the Fifth Fleet occupied for the rest of its stay in Okinawa waters. On 11 May, the *Bunker Hill* became the hapless target of a successful suicide attack, in which the carrier sustained such widespread damage that it was forced to limp to Bremerton for extensive repairs. After three months of almost continuous action, VMF–221 and –451 were out of the war. VMF–112 and –123, the last remaining Marine squadrons in TF 38 (the tactical designation changed again on 27 May, when Halsey replaced Spruance and the Fifth

---

[35] Report of the Executive Officer of the *Franklin* quoted in Morison, *Victory in the Pacific*, p. 96.

[36] *Ibid.*, p. 98.

once more became the Third Fleet),[37] operated from the carriers until 8 June, when, following strikes on Kyushu, the *Bennington* was detached from the force and headed for Leyte.

Even before the VMFs had begun operating from the decks of the large carriers with the fleet and the CVE program had gotten underway, other decisions affecting the future role of Marine Corps aviation were being made. Growing out of the deliberations of the Pearl Harbor conferees and the directives of Admiral King, the composition and strength of Marine aviation was to be adjusted.

On 2 November 1944, CominCh issued an order directing the decommissioning of four Marine medium bomber squadrons (VMBs). In reply, the Commandant pointed out that the Marine Corps had, at that time, 12 tactical VMBs and 4 in the replacement training program, and that CinCPac had indicated he needed 8 of these squadrons in the forward area. General Vandegrift further noted that, instead of the 11 agreed upon, 15 Marine squadrons had been recently decommissioned (11 in the 9th MAW and 4 in MarFAirWest)[38] in compliance with Admiral King's directive on 10 September. In view of these facts, the CMC recommended that no further Marine squadrons should be commissioned at that time.[39]

Admiral King's senior staff officers agreed with the Commandant's recommendations for several reasons. One was that it had become obvious that the decommissioning of the 15 squadrons had adversely affected the morale of Marine aviation personnel. King's Deputy CNO (Air) stated further that he did not believe it was the proper time to decommission four squadrons arbitrarily in view of the critical shortage of air support in the forward area and especially in the Philippines. He then recommended that no action be taken regarding the VMBs until future requirements for Marine and Army aircraft in the Pacific had been more firmly fixed, because the results of the Leyte campaign could be a determining factor.[40] The Deputy CominCh–CNO, Vice Admiral Richard S. Edwards, concurred in these recommendations and further recommended that the matter be studied before a final decision was made.[41] Admiral King agreed, directed that the Deputy CNO (Air) and the CMC appoint action officers to conduct the study, and ordered the decommissioning of the VMBs held in abeyance pending a report from these officers.[42]

A satisfactory solution to this problem, answering both the needs of Marine

[37] Regarding these changes in designations, Halsey commented, "Instead of the stagecoach system of keeping the drivers and changing the horses, we changed drivers and kept the horses. It was hard on the horses, but it was effective." Halsey and Bryan, *Halsey's Story*, p. 197.

[38] *FMF Air Status Rpts*, Oct–Nov44.

[39] CMC memo serial 008A30844 to CominCh, dtd 2Nov44 (OAB, NHD).

[40] Capt Wallace M. Beakley, USN, memo for Adm Cooke, dtd 4Nov44 (OAB, NHD).

[41] Deputy CominCh–CNO memo to Adm King, dtd 7Nov44 (OAB, NHD).

[42] CominCh memo FF1/A16–3, serial 003246, to Vice CNO and CMC, Subj: Decommissioning of 4 Marine Bombing Squadrons, dtd 8Nov44 (OAB, NHD).

aviation and improving the morale of Marine pilots, was found in the requirements of the CVE program. On 31 January 1945, VMB–621 and –622 of MAG–62, 9th MAW, were redesignated VMTBs and assigned to escort carriers. The next month, on 15 February, VMB–623 and –624 of the same organization were similarly redesignated and reassigned. At the same time, all four squadrons were transferred from the east to the west coast.[43]

As of 21 December 1944, Marine aviation was organized into 5 wings with 93 tactical squadrons, 29 replacement training squadrons, 3 operational training squadrons, and was assigned 2,342 aircraft as follows:

	CVE	Shore Based	Replacement Training	TOTAL
VMF VMFB	108	960	384	1452
VMTB	72	168	96	336
VMB		144	48	192
VMF (N)		72	24	96
VMR		120	45	165
VMD		36	12	48
MOTG–81			53	53
TOTAL	180	1500	662	2342[44]

	Proposed					
	CVE	Shore Based	Replacement Training	Fleet Training	TOTAL	Change
VMF VMFB	540	504	252	34	1330	−122
VMTB	360	72	36	20	488	+152
VMB		96			96	− 96
VMF (N)		72	30		102	+ 6
VMR		120	45		165	
VMD	12	24	12		48	
MOTG–81			65		65	+ 12
TOTAL	912	888	440	54	2294	— 48[45]

Reflecting the increased emphasis on the Marine CVE program, the Marine Corps proposed an aviation structure which reduced the number of land-based tactical squadrons from 93 to 52 and the replacement and operational training squadrons from 32 to 20. It was also proposed that the squadrons in the CVE program should be increased from 12 to 61, and also that they should be augmented by 2 fleet training squadrons. In effect, while the total Marine aircraft requirement would be reduced by 48 planes, the table above indicates that the fighters and torpedo bombers assigned to the CVE program would be increased from 180 to 912 planes. Al-

[43] *DivAvn OpDs,* Jan–Feb45; *FMF Air Status Rpts,* Dec44–Feb45.

[44] Dir, DivAvn, HQMC, memo Op–37–C–fgd for F–15 [AvnSec, Planes Div, CominCh], Subj: Proposed Program of Marine Corps Aviation, dtd 21Dec44 (OAB, NHD).

[45] *Ibid.*

though it was also proposed that the 9th Wing, a training unit, be decommissioned, and a Marine Air Training Command, East Coast, formed in its place, this change did not take place until after the war, in early 1946.

In approving the proposed changes, CominCh directed that, hereafter, the training program of Marine Corps squadrons was to emphasize close support of ground troops in amphibious operations. In addition, he stated that a sufficient number of squadrons were to be trained in carrier operations to permit an ultimate total of 16 CVE groups to be embarked simultaneously, and to furnish adequate spare groups."[46] February 1946 was the target date set for the completion of the carrier training program.

Concerning other aspects of the Marine aviation program, Admiral King stated that the remaining tactical squadrons were to be organized in two combat wings which would operate from shore bases; their primary mission was to be support of Marine ground troops and defense of bases to which they were assigned. He made it clear that the transition to the new program was not to interfere with the tactical employment of Marine squadrons in current and future operations. Finally, King stated that, when CinCPac so recommended, the 4th MAW was to be decommissioned and its squadrons gradually absorbed into the CVE program.[47]

Immediately after the formation of Marine Carrier Groups, AirFMFPac, its squadrons began training at the Marine Corps air stations at Mojave and Santa Barbara, California. In accordance with CominCh instructions, a *Commencement Bay*-class CVE was made available on the west coast to permit the pilots to train for carrier qualification. Four CVEs were to be in operation by 15 February 1945. Each ship was to have an air complement consisting of a VMF(CVS) with 18 Corsairs and Hellcats, and a VMTB (CVS) with 12 Avengers.

This aircraft strength figure was not adhered to, however, for MCVG–1 on the *Block Island* had "12 TBM, 10 F4U, 8 F6F night fighters and 2 F6F planes, planes, "because the *Block Island* was equipped with an SP (height finder) radar suitable for night intercepts, hence the day-night character of her assigned air group."[48] The MCVG–4 air complement on the *Cape Gloucester* consisted of 12 TBMs, 16 F4Us, and 2 F6F photo planes. By careful spotting of the aircraft on the carrier deck, it was found that two additional F4Us could be added to the strength of the group, and when it passed through Hawaii en route to the West Pacific, the *Cape Gloucester* took on board these two extra planes to give it a total of 32 operating aircraft.[49]

Before going on board the carriers, the Marine fliers underwent the same

[46] CominCh–CNO memo FF1/A1, serial 00130, to CMC, Subj: Marine Corps Aviation Program, Revision of, dtd 15Jan45 (OAB, NHD).

[47] *Ibid.* The 4th Wing was not disbanded until after the end of the war.

[48] BGen John F. Dobbin ltr to Hd, HistBr, G–3 Div, HQMC, dtd 3Nov65, hereafter *Dobbin ltr.*

[49] BGen Donald K. Yost ltr to Hd, HistBr, G–3 Div, HQMC, dtd 30Nov65, hereafter *Yost ltr.*

type of carrier training that Navy pilots experienced. The training of the Marines additionally emphasized the tactics and techniques to be employed during close support missions. Besides taking part in the regular squadron training program, the new carrier pilots had to learn or refresh their knowledge of the following subjects: communications and flight deck procedures; recognition, survival and first aid; map reading and navigation; ordnance and gunnery; and escape from submerged aircraft procedures effected from a training device called a "Dilbert Dunker." Subjects included in the flight training syllabus were air tactics, night flying, carrier landings, rocket firing, navigation, fixed and free gunnery, and bombing. The flight syllabus for VMTB pilots and crews also included radar search and torpedo drops.[50] Once the squadrons were assigned to the CVEs, as part of their shakedown preparations, the pilots had to make eight satisfactory carrier landings to become fully qualified. It was during this period, when the squadrons were training on the CVEs, that a considerable number of operational accidents occurred.

These were caused by a combination of pilot error and aircraft failure arising almost directly out of the design of the F4U–1s and F4U–1Ds, the Corsairs,[51] flown by Marine fighter squadrons and the flight deck characteristics of the carriers. In comparison with the much larger and faster CVs, the *Commencement Bay*-class escort carriers made a top speed of only 19 knots, and had flight decks that were only 75 feet wide and 553 feet long.

The length of both Corsair models was slightly more than 33 feet and their wing span was nearly 41 feet. Both F4U types were powered by 2,000-horsepower Pratt and Whitney radial engines. Driving a three-bladed propeller slightly more than 13 feet in diameter, these powerful Corsairs were the first naval aircraft operating in the war with a speed capability in excess of 400 miles per hour. In the opinion of one Marine ace who flew the plane in the Pacific: "The Corsair was a fine carrier plane, and most of us preferred it to the F6F. It was always called 'The Bent Wing Widowmaker.' "[52]

The cockpit of the Corsair lay well back in the fuselage, behind a long nose, which severely limited the vision of the pilot while the plane was in a landing attitude. The inverted, low gull wing located forward of and below the pilot restricted his vision during the critical carrier landing approach. The high torque characteristics of the engine, that is, the tendency of the plane to roll to the left and sometimes out of control if power was applied when the plane was travelling at slow speeds, gave the pilot a very small margin of error.[53] In land-based operations, hard right rudder and judicious application of full throttle when the plane was taking off, landing, or being waved off were required; aboard the small carriers, these aircraft-han-

[50] *Ibid.*

[51] The F4U–1 was assigned to Marine aviation in 1942 and the F4U–1D in 1944. HistBr, G–3 Div, HQMC, *Marine Corps Aircraft, 1913–1960—Marine Corps Historical Reference Series*, No. 20, reprinted 1964 (Washington, 1962).

[52] *Dobbin ltr.*

[53] Capt Harvey D. Bradshaw comments to HistBr, G–3 Div, HQMC, dtd 26Apr65.

dling techniques became even more critical.

Despite the aversion of the Navy to use of Corsairs on carriers, many changes had been made in the plane which made it suitable for such employment. These included:

> . . . raising the pilot's cabin to improve visibility (November 1942), improved aileron action (January 1943), larger bearings in the tail wheels (March 1943), installation of a spoiler on right wing to reduce violence of stalls when under acceleration and to provide new stall warning (November 1943), new oleo strut-filling procedure (May 1944).[54]

The relative inexperience of the new MCVG pilots in CVE landing operations and the inherent difficulty in flying the Corsair resulted in numerous other training and operational accidents. Although the average number of carrier landings required for pilot qualification was 8, a Corsair pilot had to make a minimum of 20–25 before he could attain a realistic proficiency level.[55]

Takeoffs from the carrier while at sea were the source of another major problem to the Corsair pilots. At best, the top speed of the CVE would provide only 19–19½ knots of headwind on a calm day. Under optimum takeoff conditions, the Corsairs required a minimum headwind of from 20 to 26 knots; a 30-knot headwind was ideal.[56] Unless optimum wind and speed conditions existed, a Corsair, heavily laden with bombs, rockets, armament, and fuel could not attain flying speed and would drop off the end of the flight deck into the sea ahead of the carrier. For this reason, the F4Us were launched by catapult in almost every case.

Lieutenant Colonel Royce W. Coln's MCVG–3 pilots on USS *Vella Gulf*, soon found:

> . . . that in practically any external load condition the risk was too unfavorable to try a fly away launch with the F4U. We therefore immediately adopted a SOP that all F4Us and F6Fs [launched] would be catapult shots rather than fly aways. TBMs which were usually spotted all the way aft and under 28–30 knots relative wind with a 2000 pound internal load could fly off with reasonable safety. We also found that with this system we could do a total launch of all aircraft in much less time.[57]

MCVG–1 Corsairs on the *Block Island* were "almost always catapulted," for the carrier "had two Cats. An H4 on the Port and an H2 on the Starboard bow. The H4 gave you a 4G slam and was the greatest feeling a pilot ever had, especially on a black night. You knew you were going all the way the minute you felt it." [58]

Sometimes the hydraulic-powered catapults failed to accelerate the aircraft to flying speed by the time the Corsair left the flight deck—a "cold-cat shot" in carrier pilot's parlance. In those cases, the pilot often was unable to keep the tail of his plane from dragging. If he could not gain the required flying speed, the plane would "mush" right into the water directly in the path of the sharp-prowed carrier before the

[54] Sherrod, *Marine Air Hist*, p. 330.

[55] Col Louis H. Steman comments to HistBr, G–3 Div, HQMC, dtd 26Apr65.

[56] *Ibid.*

[57] Col Royce W. Coln ltr to CMC dtd 16Nov65, hereafter *Coln ltr.*

[58] *Dobbin ltr.*

pilot could extricate himself from his Corsair. It was in these cases that Dilbert Dunker training proved its worth. More often than not, however, pilot and plane were lost. Despite the occurrence of these shake-down problems, the CVE program was soundly launched.[59]

The first of the Marine escort carriers commissioned was the USS *Block Island*, which embarked Lieutenant Colonel John F. Dobbin's MCVG-1 (VMF (CVS)-511, VMTB (CVS)-233, and CASD-1) at San Diego on 19 March and then headed for Pearl Harbor and duty with the fleet.[60]

On 29 April, the *Block Island* arrived at Ulithi and was ordered to support the Okinawa operation. From 10 May to 19 June, as a component of Task Unit 32.1.3 the carrier alternated between targets in the Sakishima Islands and on Okinawa itself. MCVG-1 planes helped reduce Shuri Castle with 2,000-pound bombs, "and did some close air support work with the Marine Divisions on Okinawa." ("Not as much as we would have liked," Dobbin noted.)[61]

In the six-week period of operations in the vicinity of Okinawa, the task unit sortied northward, where its planes conducted strikes against Kagoshima, and as an alternate target when it was weathered in, Amami O Shima. At no time did *Block Island* aircraft encounter opposition over their targets, nor was the carrier itself subject to *Kamikaze* attacks.

Following these operations, the CVE went to Leyte for replenishment. On 25 June, it went to sea again to participate in a three-day series of strikes in support of the landings at Balikpapan, Borneo. Immediately after Japan's surrender, in company with another Marine CVE, the *Gilbert Islands*, and five destroyer escorts, *Block Island* participated in a POW rescue operation. These ships steamed to Formosa before the island commander had formally capitulated, and took on board approximately 1,000 Allied POWs who had been imprisoned there.

The USS *Gilbert Islands*, the second Marine CVE commissioned, embarked Lieutenant Colonel William R. Campbell's MCVG-2 (VMF(CVS)-512, VMTB(CVS)-143, and CASD-2) on 6 March at San Diego, and left for the Pacific the following month.[62] On 25 May, the carrier arrived off Okinawa and flew its first CAP and close air support strikes. On 1 June, the *Gilbert Islands* joined the *Block Island* in TU 32.1.3, then neutralizing enemy installations in the Sakishima Gunto, and later participated in the Balikpapan preinvasion strikes.

Two more Marine CVEs arrived in the Pacific before the end of the war. The *Cape Gloucester*, with Lieutenant Colonel Donald K. Yost's MCVG-4

---

[59] Not all Marine aviators were enthused about being stationed aboard carriers, for as General Rogers recalled, he "despised them more than anything in the world. I was scared of them, actually. Well . . . I'd been flying so much longer than almost anyone else, and I would not trust myself to a [landing] signal officer. I couldn't believe that the signal officer was correct. I knew that I was correct." *Rogers interview.*

[60] USS *Block Island* (CVE-105) WarDs, Mar-Apr45 (OAB, NHD).

[61] *Dobbin ltr.*

[62] USS *Gilbert Islands* (CVE-107) WarDs, Mar-Apr45 (OAB, NHD).

(VMF(CVS)-351, VMTB(CVS)-132, and CASD-4) embarked, arrived at Okinawa on 4 July and was attached to Task Group 31.2 for duty.[63] After spending a few days covering minesweeping operations, the Marine CVE joined three other carriers, and steamed from Okinawa on 1 August to conduct antishipping operations in the East China Sea and to launch strikes against shipping in the Saddle and Parker Island groups near Shanghai at the mouth of Hangchow Bay.

The fighter complements on the Navy CVEs in the task group consisted primarily of FM-2s (the General Motors "Wildcat," a single engine fighter), which did not have the high-altitude performance characteristics of the Marine Corsair. The performance of the Corsair was improved by removal of its rocket rails and one pylon, permitting "VMF-351 pilots to bag four fast high-flying Japanese reconnaissance planes in addition to the one transport they caught on a course between Shanghai and the Japanese homeland." [64]

Following the surrender of Japan, the *Cape Gloucester* debarked its Marine aircraft group at Okinawa and proceeded to Nagasaki where it took on board and transported to Okinawa 260 liberated Australian POWs. MCVG-4 then reboarded the carrier which returned to Japanese waters, over which the Marine pilots provided an air cover for Fifth Fleet minesweeping and occupation forces en route to Sasebo. While flying a reconnaissance mission over Kyushu, the MCVG commander, Lieutenant Colonel Yost, flew his aircraft into high tension power lines strung across a valley which, because of rain and a low ceiling, he was following back to the coast and the carrier. The engine of his Corsair failed and he was forced to make a wheels-up landing at Omura airfield, "and he became a one-man premature 'invasion' force" [65] which preceded the occupation of Kyushu by approximately a week.

The fourth Marine CVE commissioned was the *Vella Gulf*, which had Lieutenant Colonel Royce W. Coln's MCVG-3 (VMF(CVS)-513, VMTB (CVS)-234, and CASD-3) on board. It sailed from San Diego on 17 June for Pearl Harbor, where it conducted further training. On 9 July, the carrier left for Saipan by way of Eniwetok and Guam. On 24 and 26 July, the Marine pilots flew strikes north of Guam against Pagan and Rota, two islands which Allied fliers had attacked many times before. The *Vella Gulf* then proceeded to Okinawa, where it arrived on 9 August, the day that the second atomic bomb was dropped. Following

[63] USS *Cape Gloucester* (CVE-109) WarD, Jul45 (OAB, NHD).

[64] *Yost ltr.*

[65] *Ibid.* Soon after he landed, a Japanese staff car drove up and an officer notified the Marine that the commanding general of the district wished to see him. The Japanese were very amicable and set up a guard around the plane upon request. At the Japanese headquarters, Lieutenant Colonel Yost was able to send a message to the American forces in Tokyo and to the Fleet notifying them of his whereabouts. Although the airfield was in poor shape a TBM from the *Cape Gloucester* was able to land and pick him up the next day. BGen Donald K. Yost comments to HistBr, G-3 Div, HQMC, dtd 14Mar67.

the surrender of Japan, the CVE was assigned to participate in the occupation.

The *Salerno Bay*, carrying MCVG–5 (VMF(CVS)–514, VMTB(CVS)–144, and CASD–5), and the *Puget Sound*, carrying MCVG–6 (VMF(CVS)–321, VMTB(CVS)–454, and CASD–6), had not yet arrived in the war zone when the conflict with Japan ended.[66] Thus, only four Marine CVEs saw any action in the Pacific. Although the Marine CVE project had been established to provide Marine fliers for close support of amphibious landings and ground operations, except for a few instances at Okinawa and Balikpapan, the carriers did not fulfill their intended functions.

The Marine CVE program was conceived and activated too late in the war to do more than just begin to prove itself. On the other hand, the *Block Island* and *Gilbert Islands* Marine pilots could probably have been employed to a greater extent in support of ground operations at Okinawa, for both CVEs were in the combat area for a long enough period.

As one of the carrier group commanders observed, the significant aspect of the carrier program was:

> The fact that, for the first time, Marine aviation would operate within its ideal 'conceptual' role. This being that Marine Air would perform pre-D-Day operations from carriers, then participate in the amphibious phase. . . .[67]

[66] In addition to those operating squadrons already assigned to carriers, a number of other squadrons, destined for carrier assignment were undergoing training in California at El Toro, Santa Barbara, and Mojave.

[67] *Coln ltr.*

Like many other projects that were born during the last stages of World War II, the concept underlying the program was soundly enough organized and firmly enough established, however, to become an important and integral facet of postwar Marine Corps amphibious warfare doctrine.

FINAL OPERATIONS [68]

Allied air and naval pressure on Japan continued unremitting following the fall of Okinawa and in the period that the ground and amphibious forces of both the Central Pacific and Southwest Pacific commands prepared for OLYMPIC. At the 29 June meeting of the JCS, when 1 November was confirmed as the date for the invasion of Kyushu,[69] the service chiefs also determined that the blockade from air bases not only on Okinawa and Iwo Jima but also in the Marianas and Philippines was to be intentified. They also agreed upon the following courses of action in the Pacific: defeat of enemy units in all of the Philippines; alloca-

[68] Unless otherwise noted, the material in this section is derived from: *CinCPac Ops*, Jun–Aug45; CinCPac WarDs, Jun–Aug 45 (OAB, NHD); *USAFMidPac G–5 Hist;* USSBS, *Japan's Struggle;* Butow, *Japan's Decision;* Herbert Feis, *Japan Subdued* (Princeton: Princeton University Press, 1961), hereafter Feis, *Japan Subdued;* Halsey and Bryan, *Halsey's Story;* King and Whitehill, *King's Naval Record;* Leahy, *I Was There;* Harry S. Truman, *Year of Decisions—Memoirs*, v. I (Garden City, N. Y.: Doubleday & Company, Inc., 1955), hereafter Truman, *Memoirs; War Reports.*

[69] On 30 June, CinCPac sent out a warning order for OLYMPIC to his entire command. CinCPac WarD, Jun45 (OAB, NHD).

tion of all of the forces necessary to guarantee the security of Western Pacific sea lanes prior to OLYMPIC; and acquisition of a sea route to Russian Pacific ports,[70] very likely a preparatory measure for the impending entry of Russia into the war with Japan.

During July, further steps were taken to revise and strengthen the preparations and forces for the final operation against Japan. On the 10th, the JCS ordered the China-based XX and XXI Bomber Commands deactivated. The personnel and planes of the former were transferred to Lieutenant General James H. Doolittle's Eighth Air Force, which had deployed from Europe to Okinawa. XXI Bomber Command squadrons were transferred to Lieutenant General Nathan F. Twining's [71] Twentieth Air Force, which was based in the Marianas. The Eighth and Twentieth together would comprise the United States Army Strategic Air Force in the Pacific (USASTAF), commanded by General Carl A. Spaatz. Strategic control of USASTAF would remain with the JCS in the same manner as it had controlled Twentieth Air Force, and similarly General Arnold would be its executive agent.

On 10 July, Nimitz turned over the Seventh Air Force, which had squadrons on Iwo, in the Marianas, and in the Ryukyus, to the commander of the Far East Air Forces (FEAF), General George C. Kenney, who had been air chief of the SWPA throughout the Pacific war. CinCPac also ordered Major General Louis E. Woods' Tactical Air Force on Okinawa to conduct operations in conjunction with the Eighth Air Force. Kenney's FEAF was expanded on 13 July, and was composed at this time of the Fifth Air Force, which was, for the most part, based in the Ryukyus, and the Thirteenth Air Force, which was based in the Philippines. For OLYMPIC, FEAF was to conduct tactical operations in support of the invasion, and USASTAF was to conduct the strategic bombing of the Home Islands.

Based on the JCS directive of 3 April to Nimitz and McArthur, on 19 July CinCPOA transferred to CinCAFPac the control of U. S.-held areas in the Ryukyus. In turn, Nimitz retained responsibility for the operations of naval units and installations in this area. On 26 July, General Stilwell was ordred to report to MacArthur with the Army forces under his command at 1200 on 31 July, at which time control of the Ryukyus passed from CinCPac-CinCPOA to CinCAFPac.[72]

While these administrative and command changes were taking place, aircraft from carrier task forces and land-based commands embarked upon an accelerated program of attacks designed to weaken Japan before the invasions of Kyushu and Honshu. From bases in the Marianas, B-29s averaged 1,200 sorties a week in July. These large bombers dropped 42,711 tons of explosives on 39 Japanese industrial centers during the month; a large percentage of the missions were mass incendiary raids.[73]

Okinawa airfields captured during the campaign, and others constructed later

[70] Leahy, *I Was There*, p. 385.

[71] General Twining is the brother of retired Marine General Merrill B. Twining.

[72] CinCPac WarD, Jul45 (OAB, NHD).

[73] *CinCPac Ops*, Jul45, p. 16.

on almost all of the suitable space on the island, were filled to overflowing with aircraft of all types by July. Bombers taking off from Okinawa to hit Japanese targets were often covered by Iwo Jima-based AAF fighters, which also flew fighter-bomber sweeps over the Empire. Kenney's fighters and medium bombers, and Marine F4Us and TBMs (operating with, but not under, the AAF) struck Japan day and night in July, hitting a wide assortment of vital targets on Kyushu, Shikoku, and Honshu in accordance with OLYMPIC preinvasion plans. In completion of the transfer of Okinawa forces from CinCPac to CinCAFPac, on 31 July the Seventh Air Force assumed operational control of the 2d MAW and ADC.

For the Navy, the final phase of the war against Japan opened at dawn on 1 July, when the Third Fleet, stated Halsey:

> . . . sortied from Leyte under a broad directive: we would attack the enemy's home islands, destroy the remnants of his navy, merchant marine, and air power, and cripple his factories and communications. Our planes would strike inland; our big guns would bombard coastal targets; together they would literally bring the war home to the average Japanese citizen.[74]

Halsey's Third Fleet accomplished what it set out to do. Strike day was 10 July, when the fast carrier task force arrived at launching positions and fighter sweeps were sent over Tokyo. Not a single enemy interceptor was in the air; two snooper planes, which investigated the American ships from beyond the ships' AA range, were quickly shot down by carrier aircraft circling overhead. For the rest of the month in company with TF 37, the fast carrier task force of the British Pacific Fleet which joined on 16 July, TF 38 ranged up and down the Pacific coast of the Japanese islands, maintaining a series of heavy air strikes and surface bombardments against selected targets. "The enemy's failure to hit us implied that he was hoarding his air power against an expected invasion, but most of us believed that he had little air power to hoard." [75]

As soon as possible after he had assumed office following the death of President Roosevelt, Truman concentrated his efforts on finding a way to end the war in the Pacific. As the American Commander in Chief, he was kept abreast by his military and civilian advisers of all developments in the war and briefed on plans proposed for future operations. He also faced the problem of getting to know the two other major Allied heads of state and establishing a rapport with them. Truman's heavy workload and the necessity of finding immediate solutions to pressing problems prevented him from leaving Washington for a Big Three meeting. Instead, as an interim measure, he sent Harry L. Hopkins and W. Averell Harriman, the Ambassador to the Soviet Union, to meet with Stalin and his advisers in May.[76] Basically, their assignment was to inform Stalin that "we wanted to carry

[74] Halsey and Bryan, *Halsey's Story*, p. 257.

[75] *Ibid.*, p. 260.

[76] Hopkins was sent because, as Roosevelt's trusted advisor, he had met Stalin earlier in the war, when he had made several trips to Moscow to carry out missions for the President.

out the Roosevelt policies." [77] Additionally, Harriman and Hopkins were to try to get Stalin to commit himself to Russia's early entry into the war against Japan and to obtain from him a firm date for that event. On 28 May, the two diplomats advised Truman that Stalin had set 8 August as the date he would declare war on Japan.

Stalin also told the Americans that, while he would remain a party to the policy of unconditional surrender, he believed that Japan would not surrender easily if the Allies insisted upon enforcing the provisions of this policy. Stalin concluded that if Japan sued for peace in hopes of obtaining terms that might possibly be less stringent than those implied in an unconditional surrender, the Allies should accept the offer and enforce their will upon the defeated enemy by occupying his homeland.

Truman stated later that he was:

> . . . reassured to learn from Hopkins that Stalin had confirmed the understanding reached at Yalta about Russia's entry into the war against Japan. Our military experts had estimated that an invasion of Japan would cost at least five hundred thousand American casualties even if the Japanese forces then in Asia were held on the Chinese mainland. Russian entry into the war was highly important to us.[78]

As the Allies drew closer to the heart of the Empire, Truman believed that this Russian action ". . . would mean the saving of hundreds of thousands of American casualties." [79] With further discussion of this matter as one of his more compelling reasons for attending a conference with Stalin and Churchill, Truman agreed to meet with them on 15 July at Potsdam, a suburb of Berlin.

On the day after the meeting had convened, Truman was informed that the first atomic bomb had been successfully exploded in a test on 16 July at Alamogordo, New Mexico. The next day, Secretary of War Henry L. Stimson flew to Potsdam to give the President the full details of the test. Truman recalled that:

> We were not ready to make use of this weapon against the Japanese, although we did not know as yet what effect the new weapon might have physically or psychologically, when used against the enemy. For that reason the military advised that we go ahead with the existing military plans for the invasion of the Japanese home islands.[80]

The atomic bomb project had been kept so secret that the JCS first learned of it as a group only after completion of the test. Marshall, however, had kept King abreast of the progress of the project.[81] On 24 July, Truman casually mentioned to Stalin that the United States had "a new weapon of unusual destructive force. The Russian Premier showed no special interest. All he said was that he was glad to hear of it and hoped that we would make 'good use of it against the Japanese.' " [82]

Despite the vast potential suspected of the new weapon—and all the possible implications inherent in its use—and the fact that OLYMPIC preparations and preinvasion operations were well

[77] Truman, *Memoirs*, p. 258.

[78] *Ibid.*, p. 265.

[79] *Ibid.*

[80] *Ibid.*, p. 415.

[81] King and Whitehill, *King's Naval Record*, p. 621.

[82] Truman, *Memoirs*, p. 416.

under way, Admirals Leahy and King and proponents of strategic bombing still held reservations about the need for invading Japan. They were even less in favor of dropping an atomic bomb on that country. They believed that Japan had already been defeated and was ready to surrender. King felt that the President gave his approval for the bomb to be dropped because the Chief Executive feared that too many American troops would be killed in an invasion. King agreed with this estimate, but he thought that:

> . . . had we been willing to wait, the effective naval blockade would, in the course of time, have starved the Japanese into submission through lack of oil, rice, medicines, and other essential materials. The Army, however, with its underestimation of sea power, had insisted upon a direct invasion and an occupational conquest of Japan.[83]

Faced with the prospect of either invading Japan or destroying that country with atomic bombs, Truman was presented with the suggestion that Japan might choose or even be induced to surrender and end the war sooner than expected. Late in May, Acting Secretary of State Joseph C. Grew, a veteran diplomat who had been American Ambassador to Japan for a 10-year period before the war, suggested that Truman issue a proclamation which called upon the Japanese to submit and guaranteed the continuation of the Emperor as head of state. The President favored the idea and instructed Grew to send his recommendation through regular channels for consideration by the JCS and the State-War-Navy Coordinating Committee. The latter was a group of senior State, War, and Navy Department officials who assisted their chiefs in handling politico-military matters. When Grew's recommendation was approved in principle, he further recommended that the President's message to the Japanese people be issued at once to coincide with the fall of Okinawa. The JCS demurred, for they wanted to wait until the United States was ready to follow up a Japanese refusal of the peace offer with an actual invasion. Truman then decided not to publish the proclamation until after the Potsdam Conference had begun in order to give his declaration greater weight by including Great Britain and China,[84] two of our cobelligerents in the war against Japan, as the joint issuing powers, and by issuing it from the scene of a conference of victorious Allies.

After he had become President, and when he first learned of the development of the atomic bomb, Truman determined that the bomb was to be employed strictly as a weapon of war against purely military targets. The final decision to drop the bomb was to be his alone. Before the Potsdam Conference, he had been given a list of suggested targets and the choice was finally narrowed to four cities in which were located important industrial and military complexes, according to intelligence reports. Hiroshima, Kokura, Niigata, and Nagasaki were chosen in that order, and on 24 July, General Spaatz was directed to send the B–29s of his USASTAF against one of the

[83] King and Whitehill, *King's Naval Record*, p. 621.

[84] China was neither a party to nor present at the Potsdam Conference.

targets on the first day after 3 August 1945 that weather conditions would permit visual bombing.

On 26 July, the United States, in company with the United Kingdom and the Republic of China, issued what has come to be known as the Potsdam Declaration.[85] This surrender ultimatum gave Japan the opportunity to end the war voluntarily or to face utter destruction. The terms offered the Japanese government included assurances that its people would not be enslaved by the victors. On the other hand, Japan's leaders were told that their country was to be disarmed, shorn of its conquests of 50-years' standing, and deprived of its war-making potential in all sectors. The authority and influence of the militarists were to be eliminated, and "until a new order is established and until there is conclusive proof that Japan's war-making power is destroyed, Japan shall be occupied by the Allies."[86] Further, the enemy was told that his countrymen would be permitted access to sources of raw materials and be given an opportunity to develop their own form of democratic self-government. Upon receipt of the Potsdam Declaration, Japan could choose to surrender unconditionally or face complete annihilation. Disregarding all reality by not accepting the terms of the declaration, Japan chose the suicidal course. On 28 July, Radio Tokyo broadcast Prime Minister Suzuki's statement that he believed:

> . . . the Joint Proclamation by the three countries is nothing but a rehash of the Cairo Declaration. As for the Government, it does not find any important value in it, and there is no other recourse but to ignore it entirely and resolutely fight for the successful conclusion of this war.[87]

With this tacit rejection, Truman decided that the fastest way of ending the war with a minimum of U. S. casualties was to drop the bomb.

The Potsdam Conference was recessed briefly from 26 to 28 July, while Prime Minister Churchill left for London to learn that, following the counting of the absentee ballots holding the vote of British servicemen, he had been defeated in the general elections and had been replaced by Clement Atlee. At the first evening session following the end of this recess, Stalin stated that on 13 July, Japan had approached the Soviet Union with a request that it mediate an end to the Pacific War, but that Russia would refuse to do so. Truman thanked him in the name of the signatories to the Potsdam Declaration, and restated their determination to hew to the ultimatum delivered in that document.

This was not the first time that Japan made peace overtures. In September 1944, the Swedish Minister in Tokyo had been approached by an unnamed high-level Japanese official, who said that, in order to obtain peace, Japan was prepared to surrender territories that were taken from Great Britain and would recognize all former British investments in East Asia. The Tokyo-based Swedish foreign officer passed word of this encounter to his seniors in Stockholm, ending his report with the comment: "Behind the man who gave me this message there stands one of the

[85] See Butow, *Japan's Decision to Surrender* pp. 243–244, for full text.

[86] *Ibid.*, Article 3 (b) of the Potsdam Declaration.

[87] Quoted in Feis, *Japan Subdued*, p. 97.

best known statesmen in Japan and there is no doubt that this attempt must be considered as a serious one."[88] Stockholm passed the information on to London, which had Lord Halifax, the British Ambassador to the United States, notify Secretary of State Cordell Hull of the conversation.

Hull concurred both with a British proposal to make no reply to the indirect Japanese approach and with London's suggestion to the Swedish Minister for Foreign Affairs that he "reply, if he so wishes, that the Swedish Government considered it useless to deliver such a message to His Majesty's Government in the United Kingdom."[89] Although nothing further came from this attempt of the Japanese another message was received in Stockholm from its Minister in Tokyo to the effect "that he had been advised that the Japanese Foreign Minister was himself preparing to approach the British Government."[90] Despite the fact that no additional material on this matter has appeared, it is safe to assume that the reaction of the Allies to this second approach would have been the same as it was to the first.

On 6 August, the first atomic bomb was dropped on Hiroshima. When the apparently heavy loss of life and widespread damage caused by this bombing did not impel the Japanese government to take any steps to end the war, General Spaatz was ordered to continue operations as planned. A second atomic bomb was dropped on Nagasaki on 9 August. During the interval between these atomic attacks, 550 B–29s and carrier aircraft struck at Japan in day and night raids, and on 8 August the Soviet Union delivered a declaration of war on Japan, effective the next day. On 10 August, Japan sued for peace on the basis of the terms enunciated in the Potsdam Declaration.

The next day, in reply to the Japanese suit, President Truman told the defeated power that a supreme commander would accept its surrender. In addition, Japan was told that the Emperor and the Japanese High Command would have to issue a cease fire to all Japanese armed forces before the Allies could accept its capitulation. Three days of frantic discussions ensued in Tokyo before the Mikado's government could agree on how best to accept Potsdam terms and what reply to give to Truman's note of the 11th.

Meanwhile, on 13–14 August, B–29s dropped 5½ million leaflets, printed in Japanese, which contained a text of the Japanese surrender offer, the American reply, and the terms of the Potsdam Declaration. Other leaflet drops were scheduled for 15 and 16 August.

Up to 13 August, the Japanese people were only vaguely aware of the fact that their country was losing the war, and had no idea that the government was suing for peace. Now that this was common knowledge, the Emperor's hand was strengthened and he could take final

[88] Paraphrase of a telegram from the Swedish Minister of Foreign Affairs to the British Foreign Office, dtd 24Sep44, quoted in Department of State, *Foreign Relations of the United States: Diplomatic Papers, 1944. The Near East, South Asia, and Africa. The Far East,* v. V (Washington: GPO, 1965), p. 1184.

[89] *Ibid.*

[90] Chief, Div of British Commonwealth Affairs, U. S. Department of State, memo to Adm William D. Leahy, CofS, U. S. Army and Navy, dtd 10Oct44, cited in *Ibid.*, p. 1186.

310-224 O - 69 - 29

action without worrying about extremists, who might have otherwise attempted to keep Japan in the war. Further strengthening the Emperor's position was the fact that Japan was to be permitted a conditional surrender, the condition being that the institution of the Emperor would be retained.

On 14 August, Hirohito asked the Swiss government to relay to the Allies a message stating that he had issued an Imperial Rescript that denoted Japanese acceptance of the provisions of the Potsdam Declaration. The message also stated that he was ordering his commanders to cease fire and to surrender their forces to, and to issue such orders as might be required by, representatives of the Supreme Commander for the Allied Powers, General MacArthur.

President Truman then notified the Japanese government that he regarded this message as a full acceptance of the Potsdam Declaration and ordered the Emperor to command all Japanese forces everywhere to cease fire immediately. The Japanese were ordered also to send envoys to Manila to discuss arrangements for the formal surrender with MacArthur and his aides. On 14 August, CinCPOA issued the following message to all of his forces in the Pacific Ocean Areas: "OFFENSIVE OPERATIONS AGAINST JAPANESE FORCES WILL CEASE AT ONCE X CONTINUE SEARCHES AND PATROLS X MAINTAIN DEFENSIVE AND INTERNAL SECURITY MEASURES AT HIGHEST LEVEL AND BEWARE OF TREACHERY."[91] At 1900 that same day, President Truman announced that a cease fire was in effect, and that the war was over.

[91] Quoted in Halsey and Bryan, *Halsey's Story*, p. 272.

THE SILENT GUNS[92]

Almost immediately after announcing the capitulation of Japan, President Truman issued a directive to General MacArthur, designating him Supreme Commander for the Allied Powers (SCAP) and giving him the power to accept the surrender of Japan for the governments of the United States, Republic of China, United Kingdom, and Union of Soviet Socialist Republics. Each of the heads of state of these governments was to designate a representative to be with MacArthur at the surrender ceremony and to sign the instrument of surrender for his country; Truman chose Admiral Nimitz as the American signatory. From the moment that the Japanese signed the surrender document, the authority of the Emperor and Japanese government to rule was subject to MacArthur, who, as SCAP, had supreme command over all Allied land, sea, and air forces that were to be allocated for employment on occupation duty in Japan.

On 20 August, when the Japanese emissaries arrived in Manila to review the instructions relating to the surrender, they received MacArthur's General Order No. 1, which had been prepared earlier in expectation of the end

[92] Unless otherwise noted, the material in this section is derived from: *CinCPac Ops*, Aug–Sep45; CinCPac WarDs, Aug–Sep45; *USAFMidPac G-5 Hist;* Butow, *Japan's Decision;* Feis, *Japan Subdued;* Halsey and Bryan, *Halsey's Story;* Truman, *Memoirs; War Reports*.

of the war. Under the terms of this document, Japanese commanders of forces in the Pacific islands south of Japan were to surrender to Nimitz or his representatives, and commanders of forces in Japan proper, the Philippines, and the southern section of Korea were to surrender to MacArthur or to his representatives. On 15 August, the JCS amplified General Order No. 1 by directing the occupation of the key areas of Japan, Korea, and the China coast. Under a system of priorities, the swift occupation of Japan was to be regarded as the supreme operation and would have first call on all available resources. Next in order was the early occupation of Seoul and acceptance of the surrender of Japanese forces in that area. Operations to be undertaken on the coast of China and on Formosa were to follow when forces and transport were available.

The immediate purpose of occupying the China coast by gaining control of key ports and communications centers was to extend such assistance to Chiang Kai-shek's Nationalist forces in China as was practicable without American involvement in a major land campaign. As the first of the Allies to go to war against Japan, the Chinese government was expected to accept the surrender of the Japanese on the mainland. The situation on Chinese soil, and especially in those great sections of the nation under Japanese control, was so confused, however, that it appeared impracticable, if not altogether impossible, for the Nationalist Government to fulfill its function as stated in MacArthur's General Order No. 1, *viz.*, to take the Japanese surrender. Although the Chinese Communists had fought the Japanese, they had been fighting the Nationalists as well, and were in fact, still trying to gain the upper hand in China when the war against Japan ended. To prevent large stores of Japanese arms and equipment from falling into the hands of the Communists, the Japanese forces in China were instructed to surrender only to Chiang Kai-shek or his representatives.

The conflict between the Communists and the Nationalists had been going on for many years before and slackened only slightly during World War II, when in the interests of national unity, both parties turned their attention to ousting the Japanese from the country. With the end of the war and the impending surrender of large Japanese forces, the Chinese civil war threatened to break out anew and on a larger scale than before, but this time with international implications which threatened the newly won peace. This, then, was what faced the Allies in China.

When Lieutenant General Albert C. Wedemeyer, commander of U.S. forces in China, apprised Washington of the explosive situation then existing in China and of the need to take vigorous action there to assist Chiang Kai-shek in re-establishing the authority of his government, he was directed to arrange for the movement of Chinese troops on American transport planes and ships into all areas in China and Formosa held by the Japanese in order to disarm and repatriate the defeated enemy. Previously prepared plans were then approved for sending in Marines and soldiers to help Nationalist forces secure key Chinese ports and communication centers.[93] At this time, while attention was focused on the situation in China briefly, MacArthur's headquarters in Manila prepared for the surrender ceremony in Tokyo Bay and prepared to implement plans approved earlier for the occupation of Japan.

At Manila, the Japanese delegation was informed that MacArthur would formally accept the surrender of Japan in Tokyo Bay on 31 August on board the USS *Missouri*. According to the preliminary plans, 150 AAF technicians were to land on 26 August at Atsugi Airdrome, 14 miles southwest of Tokyo, to prepare the way for a subsequent large-scale landing two days later by the 11th Airborne Division and advance headquarters of the Eighth Army, FEAF, and Army Forces, Pacific—a total of approximately 7,500 men in all.

Elements of the Third Fleet were to enter Sagami Bay on the 26th also, while Japanese harbor pilots were to maneuver other naval units directly into Tokyo Bay on the same day.

Two days later, the Fleet Landing Force (TG 31.3), comprised of the 4th Marines (Reinforced) and commanded by Brigadier General William T. Clement, ADC of the 6th Marine Division, was to go ashore on Miura Peninsula, 30 miles southwest of Tokyo, and take over the Yokosuka Naval Base.[94] On the same day, MacArthur was to land at Atsugi to discuss the conduct of the full occupation with members of the *IGHQ*. Further airborne and naval landings were to continue on the 29th and 30th, and on the 31st, as additional occupation forces landed, the surrender ceremony was to take place on the *Missouri*. Because a typhoon struck the Home Islands during the latter part of August, the entire schedule for the occupation was postponed two days, and the surrender ceremony was rescheduled for 2 September.

On 27 August, however, the transports carrying the Fleet Landing Force and its components had already arrived in Sagami Bay to find it congested with the warships making up the Fleet Flagship Group, which was waiting to enter Tokyo Bay for the surrender ceremonies. To relieve the congestion, *Missouri* and three destroyers steamed towards the channel leading to Tokyo Bay in order to pick up the Japanese pilots who would navigate the ships to their anchorages in the bay. After the Fleet Flagship Group had entered the bay, TG 30.2 (British Flagship Group), TF 35 (including TG 35.90 (Support Force), TF 37 (British Support Force), and TF 31 (Yokosuka Occupation Force) followed in that order. While an air umbrella of hundreds of planes from TF 38 carriers covered the task forces and groups slowly moving in Sagami Wan, many more land-based fighters and bombers from Okinawa and Iwo Jima patrolled the skies over Japan proper.

The 150 technicians from the Fifth Air Force landed at Atsugi on the 28th with their emergency communications and airfield engineering equipment and began operations preparatory to subsequent landings. On the 30th, the 11th Airborne Division and the various ad-

[93] See pt V below, "North China Marines," for a continuation and fuller exposition of this story.

[94] The occupation of Yokosuka by the 4th Marines will be treated in fuller detail in pt IV, "Occupation of Japan," below.

vance headquarters staffs arrived at Atsugi from Okinawa. Meanwhile, in conjunction with the arrival of the airborne division, an amphibious landing force comprising U. S. Marines and sailors, British sailors, and Royal Marines went ashore at Yokosuka and occupied the harbor forts off Miura Peninsula.

During the last day of the month, Fleet Landing Force troops consolidated their hold on the occupied naval base and prepared to send patrols down the peninsula to demilitarize outlying installations. By the close of 1 September, as the hour for the Japanese surrender approached, Allied troops had gained control of most of the strategic area along the shores of Tokyo Bay, excepting Tokyo itself.

At 0908, the instrument of surrender was signed on board the *Missouri*. Signing first for Japan on behalf of the Emperor and the Japanese Government was Foreign Minister Mamouri Shigemitsu. Next was General Yoshijiro Umezu, Chief of Staff of the Army, who signed for the *Imperial General Headquarters*. General MacArthur then signed as Supreme Commander for the Allied Powers. Flanking him were two officers who had been recently released from a prison camp near Mukden and invited by him to witness the surrender: Lieutenant General Jonathan M. Wainwright, the defender of Bataan and Corregidor, and Lieutenant General Arthur E. Percival, the British commander of Singapore at the time of its capture. Following his signing, MacArthur called upon the representatives of the Allied Powers to sign in the following order: the United States, the Republic of China, the United Kingdom, the Union of Soviet Socialist Republics, the Commonwealth of Australia, the Dominion of Canada, the Provisional Government of the French Republic, the United Kingdom of the Netherlands, and the Dominion of New Zealand. As these signatories fixed their names to the two documents—one for the Allies, and a duplicate for Japan—a mass flight of 450 aircraft from TF 38 "roared over the *Missouri* masthigh." [95] After all had signed, General MacArthur ended the ceremonies, saying :"Let us pray that peace be now restored to the world and that God will preseve it always. These proceedings are now closed." [96]

In addition to the principals and the signatories present at the signing many Allied flag and general officers who had participated in the war against Japan were witnesses. Ship's personnel from the *Missouri* and Marines from her detachment manned every possible vantage point. The Marine officers present were Lieutenant General Roy S. Geiger and his aide, Major Roy Owsley from FMFPac; Brigadier General William T. Clement, commander of the Fleet Landing Force; Brigadier General Joseph H. Fellows, from the staff of CinCPac–CinCPOA; and the officers of the Marine detachment of the *Missouri*, Captain John W. Kelley, and First Lieutenants Francis I. Fenton, Jr., Alfred E. W. Kelley, and Josiah W. Bill.[97]

Although the signing of the surrender document formally ended the war in the

[95] Halsey and Bryan, *Halsey's Story*, p. 283.

[96] Morison, *Victory in the Pacific*, p. 366.

[97] Records Group 24, Records of the Bureau of Naval Personnel, Log of USS *Missouri*, entry of 2Sep45 (National Archives).

AMONG THE FEW MARINES *present at the surrender ceremony on USS* Missouri *are (1) Lieutenant General Roy S. Geiger and (2) Brigadier General Joseph H. Fellows. (USA SC211066)*

CORSAIRS AND HELLCATS *fly in formation over Tokyo Bay during the surrender ceremonies. In the left foreground is the* Missouri. *(USN 80-G-421130)*

Pacific, it did not mean that the world would return to normalcy or that all Marines would once again take up peacetime types of duty. A drastic reduction in Marine Corps strength in the immediate postwar era did not necessarily mean a commensurate reduction in the number and types of responsible missions assigned the Corps, for the increased role of the United States in international affairs after the war had a direct bearing on what Marines were to do and where they were to do it. In addition to the assignment of part of the VAC to occupation duty in Japan and the deployment of IIIAC troops to China, some Marines participated in the surrender of Japanese-held islands and their later occupation, other Marines were assigned to reactivated peacetime garrisons in the Pacific Ocean Areas, and the majority was rotated back to the United States and released or discharged from active duty.

THE MARINE CORPS AT THE END OF THE WAR [98]

By V-J Day, the day that the surrender was signed, the Marine Corps had reached a peak strength of 485,833. Of this figure, 242,043 Marines were serving overseas. The major portion of the overseas figure—190,945—consisted of ground forces in six Marine divisions, a Fleet Marine Force headquarters, three amphibious force headquarters, and supporting service and tactical units. Total Marine Corps aviation strength on 31 August 1945 was 101,182; 61,098 Marines in this figure were serving overseas in four Marine aircraft wings, an Air, Fleet Marine Force headquarters, and supporting service and headquarters squadrons.

The major Marine ground commands in the Pacific at this time consisted of FMFPac at Oahu, IIIAC on Guam, and VAC on Maui. The Marine divisions were located as follows: 1st on Okinawa, 2d on Saipan, 3d on Guam, 4th on Maui, the 5th at sea en route to Japan, and the 6th, less the 4th RCT at Yokosuka, on Guam. Of the Marine aviation organizations, AirFMFPac was based at Ewa, the 1st MAW was at Mindanao, the 2d on Okinawa, the 3d at Ewa, and the 4th on Majuro. The groups and squadrons of these four wings were based either with the wing headquarters or on various islands throughout the Pacific. Attached to the 3d Wing was a Marine carrier group in four escort carriers that were under the operational control of Carrier Division 27.

Fleet Marine Force ground and training-replacement activities on the east coast of the United States in August 1945 consisted of the Marine Training Command and the 7th Separate Infantry Battalion at Camp Lejeune, the Marine Corps Base Depot at Norfolk, and the Training Battalion and the Field Artillery Training Battalion at the Marine Corps Schools, Quantico. On the west

[98] Unless otherwise noted, the material in this section is derived from: *CMC Rpt*, 1946; *FMFPac Hist; FMF Grd and Air Status Rpts*, Aug45; USMC HistDiv, Administrative History of the United States Marine Corps in the Post-war Period (1Sep45–1Oct46), dtd 19Aug47, hereafter *USMC Admin Hist;* CG, FMFPac, Administrative Activities of the Fleet Marine Force, Pacific, 1Sep45–1Oct46, encl A to CG, FMFPac ltr to CMC, ser 2863, dtd 24May47, hereafter *FMFPac Admin Hist*, 1945–1946.

coast, the Marine Training and Replacement Command, San Diego Area, was responsible for training and replacement functions at Camps Pendleton and Elliott.

Marine aviation operations in the United States were under the control of two commands. The 9th MAW, with headquarters at the Marine Corps Air Station, Cherry Point, N.C., was responsible for aviation training and replacement activities on the east coast. Marine Fleet Air, West Coast, with headquarters at the Marine Corps Air Station, Miramar, Calif., held similar responsibilities on the west coast.

Besides these Marine air and ground training and replacement commands in the United States were Marine Corps recruit depots at Parris Island, S. C., and San Diego, California, and numerous posts, stations, and independent guard detachments attached to various naval facilities. All of these, as well as the FMF organizations, were to face drastic revision as the Marine Corps began to revert to a peacetime status.

Four major personnel problem areas facing the Marine Corps, like the other Services at the beginning of the postwar period, concerned demobilization, disbandment of activities and suspension program; personnel procurement; and postwar, fiscal, and mobilization planning. The most immediate problem facing the Commandant in the period 1 September 1945 to 1 October 1946—when wartime and immediate postwar exigencies had eased—was the rapid demobilization of his Corps, for, in effect, all of the personnel problems of the Marine Corps related in one way or another to demobilization.

On 11 August 1945, the Commandant submitted to the Secretary of the Navy a general plan, commonly known later as the Point System, which governed the discharge and separation of enlisted Marines.[99] Approved on the 15th, this plan was intended to supplement, but not replace, existing Marine Corps policy and directives concerning discharges and releases. It also provided the most equitable means of establishing the priority in which Marines [100] were to be released by computing their service credits. Each Marine received 1 point for each month of service from 16 September 1940; 1 point for each month overseas or on duty afloat from 16 September 1940; 5 points for each decoration and for each campaign or engagement for which a battle star was awarded; and 12 points for each child under 18 years of age, but not more than 36 points for children. With 12 May 1945 as the cutoff date for computations, the critical score to be used when the plan first went into effect on 1 September was 85 points for male Marines and 25 points for Women Reservists. The original plan provided that the critical scores would be reduced to reflect changes in the missions, and therefore the personnel requirements, of the Marine Corps. The plan also provided that enlisted personnel with sufficient discharge points could remain on active duty for as long as they wanted to, and for the time that their services were required, without their having to sign an enlistment contract. Conversely, the Marine Corps found it necessary to

[99] *G-1 OpD*, Aug45; *USMC Admin Hist*, p. 2.

[100] The provisions of this plan did not apply to regular enlisted Marines serving under a four-year contract or the extension thereof.

retain on active duty, until their reliefs were procured and trained, certain key personnel who otherwise had amassed the required score for release and discharge.

Since the number of officers to be released was relatively smaller than the number of enlisted men involved, the Marine Corps gave individual consideration to the case of each officer. Because it had obtained its officers from various sources during the war, the Corps had to provide for either the demobilization, integration into the regular component, or return to enlisted ranks of each officer concerned.

On 10 October 1945, Marine Separation Centers were activated at the United States Naval Training Centers at Bainbridge, Maryland, and Great Lakes, Illinois. Initially, the Bainbridge center was set up to handle a maximum of 400 discharges a day, but by 19 November its quota had been exceeded, and two months later the center was processing 500 discharges a day.[101]

From time to time, as the current situation permitted, the Marine Corps demobilization plan was revised to increase the flow of discharges. In effect, the speed-up was accelerated by reducing the number of points required for separation. On 8 October, barely more than a month after the program had begun, the critical score was lowered to 60 points and all enlisted personnel with three or more dependent children under 18 years of age could request discharge. The point score was further reduced to 50 on 1 November and 45 on 1 February 1946. By 1 July 1946, the Marine Corps made it possible for inductees or reservists with 30 months of active duty to become eligible for discharge, regardless of the number of points each of them had acquired. The required discharge score for Women Reservists was comparably reduced each time that the score for male Marines was revised. Finally on 1 October, all reservists and selectees became eligible for discharge regardless of length of service time.[102]

By the end of June 1946, the Marine Corps demobilization program was entering its final stages and the strength of the Corps had been reduced to 155,592 Marines. This was a decrease of 68 percent from the September 1945 figure and 87 percent of the entire net decrease required to bring the Corps to the planned postwar limit of 108,200. The Fleet Marine Force, which had carried the offensive combat burden of the Corps during the war, was the hardest hit of all Marine activities during the demobilization. At the end of the fighting, FMFPac immediately took steps to begin reducing the strength of its forces commensurate with its commitments. On 1 October 1946, FMFPac was approximately 8 percent of its 1 September 1945 size, or to put it another way, there was a total of 21,343 Marines in air and ground units in the Pacific in late 1946.[103]

During the 13-month period from 1 September 1945 to 1 October 1946, FMFPac received 30,071 replacements. In turn, 102,115 Marines were returned to the United States from the Pacific

[101] *USMC Admin Hist*, p. 4.

[102] ALMAR 117, dtd 16Aug46.

[103] *FMF Air* and *Grd Status Rpts*, Aug45, Sep46.

and Far East. This unusually rapid rate of demobilization stripped FMFPac units of the majority of their experienced personnel—officer and enlisted—and caused a situation in which an insufficient number of trained regulars remained overseas to perform specialist duties properly. On-the-job training of remaining Marines and the arrival of replacement drafts containing some experienced personnel partially, but not sufficiently, alleviated the situation.

Nonetheless, FMFPac faced a particularly acute situation in this period because it was heavily committed with units carrying out either occupation, garrison, or repatriation duties in China and Japan, and on many of the Pacific islands, such as Truk, Guam, Kwajalein, and Eniwetok. During the immediate postwar months, many Marine units had been disbanded, some new ones activated on either a temporary or a permanent basis, and some garrison detachments formed and transferred to island and area commanders for operational control.[104] All units under FMFPac were reorganized to reflect currently effective tables of organization and prescribed personnel ceiling strengths. In the face of the various administrative and organizational changes occurring during this time, all units found it most difficult to perform their missions properly because of the excessive personnel turnover. Moreover, insufficient transportation to rotate home eligible Marines, who were scattered throughout the Pacific, created additional problems. Instances occurred when the return home of many of these Marines, whose early discharge was desired by the Marine Corps because of existing postwar plans, was delayed because troop transports were not immediately available.

The second major problem facing the Marine Corps in the postwar era was to convert a greatly expanded wartime organization into a competent peacetime instrument of national security. This changeover resulted ultimately in the consolidation or disbandment of many Marine activities, and a reduction in the mission of others to reflect their lessened size. From 1 September 1945 to 30 September 1946, 368 Marine organizations were disbanded and 104 activated. A majority of the latter, such as replacement or rotation drafts, had been activated on a temporary basis only, and some of the others actually were redesignated rather than activated. One of the most important aspects of the disbandment of activities and suspension program was its sensitivity to the progress of the de-

[104] As an example, appearing in FMF ground status reports for the first time in April 1946 are the following units: Marine Detachment (Provisional), Headquarters, Commander, Philippine Sea Frontier, which was attached to the FMFPac Representative, Marianas Area for operations; Marine Detachment (Provisional), Eniwetok, attached to Atoll Commander, Eniwetok, for operations; and Marine Detachment (Provisional), Samar, attached to FMFPac for operations. The last two detachments named were to revert to the Post-War Shore Establishment upon order. *FMF Grd Status Rpt*, Apr46. Appearing in the May status report was a provisional detachment on Bikini. By October, only the Kwajalein, Truk, and Eniwetok detachments remained, and the time left to them was limited. A more complete review of the organization and deployment of Marine security forces in the Pacific will be found in the next chapter, "Back to the Islands."

mobilization program. Because of this, close coordination in the conduct of both programs was essential, and Marine Corps personnel allowances had to be constantly revised in order to maintain a proper balance between Marine Corps missions and the number of Marines available to conduct those missions successfully.

An example of how one program affected the other may be seen in the close relationship of demobilization with the base roll-up program in FMFPac. That headquarters held the mission of closing down Marine supply installations and bases in the Pacific, and of disposing of millions of dollars of surplus property and goods therein. The early loss of large numbers of experienced supply, service, and clerical personnel from FMFPac logistics agencies imposed a particular hardship on those units which had the actual duty of closing out scattered bases and receiving from disbanding line organizations vast quantities of material which had to be stored, maintained, safeguarded, and finally disposed of.[105] Attesting to the enormity of the task is the fact that on 1 July 1945, the Marine Corps had on hand in the Pacific property valued in excess of $400 million at cost. In the following year, Marine Corps supply activities had disposed of some $207 million worth of items. At the end of the fiscal year, on 1 July 1946, the Marine Corps still had $68 million of surplus property to dispose of, but fewer Marines were available to do the job.[106]

[105] *FMFPac Admin Hist*, p. 5.

[106] *CMC Rpt*, 1946.

Personnel procurement was the third problem to confront the Marine Corps at the end of the war. Even with a massive separation and discharge program underway, the Corps had to return to a peacetime status almost immediately, and to reach its required manning level of 108,200 Marines. The officer procurement program in the postwar period featured the cessation of the mass officer candidate programs of the war years and the return to peacetime methods for the recruitment and training of regular personnel. The huge task of selecting 4,400 outstanding reserve and temporary officers for transfer to the regular establishment began after V-J Day, and was in its final stages by 30 September 1946.

One of the important sources for Marine Corps officers had been the Navy V–12 College Program, which provided a number of billets for Marines. At the end of the war, approximately 1,900 men remained on active duty in the Marine Corps portion of the V–12 program. The Corps, however, had no desire to bear the expense of educating officer candidates who would not be part of the peacetime establishment. Finally, after considerable study, the Marine Corps offered individuals who had completed seven or eight semesters of study an opportunity to accept reserve commissions and choose between immediate release to inactive duty or a brief tour of active service. Those who chose the latter might apply for a regular commission if they so desired. Undergraduates not eligible for a commission were permitted either to resign, transfer to general duty, or transfer to an NROTC unit. On 30 June 1946, the Marine Corps phase of the V–12 program was deactivated.[107]

Another source for officers along with the V–12 program was the wartime officer candidate course at Quantico. This, too, was allowed to lapse and so the principal postwar sources of permanent Marine officers were both the vast number of men who had been temporarily commissioned during the war and reservists mobilized at its beginning.

One important goal in the postwar period was to build up enlisted strength by recruiting as many enlisted regular Marines as possible and by reenlisting all of the regulars whose enlistment contracts had been or were about to be completed. On V-J Day, 72,843 Marines were serving on regular enlistment contracts; by 30 June 1946, however, 60 percent of these contracts were scheduled to expire. Since a postwar level of 100,000 male regular Marines had already been established, the Marine Corps found it necessary to initiate an intensive procurement program to recruit replacements for men scheduled for discharge and to acquire an additional number so that the postwar manning level could be achieved. By 1 October 1946, this goal had nearly been reached with a total of 95,000 regulars on active duty, and very few of them due for discharge until 1948. As a result of all of this sound planning, the Marine Corps personnel picture was consonant with postwar plans that had been developed.

The establishment of postwar, fiscal, and mobilization plans was the fourth major problem with which the Marine Corps had to contend. There were two distinctive phases in this planning—the recession phase and the postwar developmental phase. The first of these concerned the period September 1945–March 1946, when most of the measures adopted for expediency during the war by the Marine Corps expired. The second phase was a period in which the entire Marine Corps began to function in accordance with its established postwar roles and missions. The most important event of 1946 insofar as those objectives were concerned was the publication of the Navy Basic Post-War Plan No. 2.[108] This plan was to be used for planning purposes only, but Admiral Nimitz, who replaced Admiral King as CNO on 15 December 1945, indicated in his covering letter that, the Marine Corps would be fairly well established along the lines of the plan as it then stood. A note of the future was sounded in the final paragraph of CNO's covering letter, which read: "In all planning, it is essential that an effective, balanced, mobile fleet, including air components, have first priority. Economy in men, money, and materials is mandatory." [109]

General Vandegrift, Commandant of the Marine Corps since 1 January 1944, determined from this plan that the general task of the Marine Corps would be to perform the following functions:

> (a) To provide a balanced Fleet Marine Force, including its supporting air com-

[107] For a fuller story on the Marine Corps V–12 program, See Condit, Diamond, and Turnbladh, *Marine Corps Training*, chap 15, pp. 288–316.

[108] CNO ltr to distr list, dtd 21Mar46, Op–001/mac, (SC) A16–3/EN, Ser: 051P001, Subj: Basic Post-War Plan No. 2 (OAB, NHD).

[109] *Ibid.*

ponent, for service with the Fleet in the seizure or defense of advanced Naval Bases or for the conduct of such limited land operations as are essential to the prosecution of a Naval campaign.

(b) To continue the development of those aspects of amphibious operations which pertain to the tactics, technique, and equipment employed by landing forces.

(c) To provide detachments and organizations for service on armed vessels of the Navy.

(d) To provide security detachments for protection of Naval property at Naval stations and bases.[110]

To ensure that the Marine Corps would adequately perform these functions, it was determined that the strength of the Corps would be 108,200, or approximately 22 percent of the overall Navy postwar strength of 487,700. With this number, the Marine Corps was to maintain the Fleet Marine Force, ships' detachments, security forces for the naval establishment, Headquarters Marine Corps, and Marine supporting activities.

In his annual report for the fiscal year 1946 to the Secretary of the Navy, the Commandant expounded on the functioning of these four tasks as follows:

(a) *Fleet Marine Force:*

Experience in the war gives incontrovertible evidence that amphibious warfare is an essential adjunct of naval warfare. . . . In the war in the Pacific, the Fleet was able to play its historic role . . . only because of the existence of the Fleet Marine Force. . . .

The Fleet Marine Force, in conjunction with Headquarters Marine Corps and the Marine Corps Schools, will continue its role in the development of these aspects of amphibious operations which pertain to the tactics, techniques, and equipment employed by landing forces.

(b) *Detachments Afloat:*

On carriers, battleships, and cruisers, Marine detachments will provide a trained nucleus for the ship's landing force, gun crews as required, and local security for the vessels. On amphibious command ships, Marines will perform duty on staffs under the direction of amphibious force commanders, and communications duties as directed by the commanding officer of the vessel. Marine detachments on transports will perform transport quartermaster functions and provide local security as directed.

(c) *Security Forces:*

Marine Corps personnel will be assigned the task of providing necessary internal security for Naval Shore Establishments within and beyond the continental limits of the United States, and of providing external security in accordance with specifically assigned missions in such establishments outside the United States.

(d) *Supporting Activities:*

In order to maintain the Corps, it will be necessary to procure, equip, train, and administer Marine personnel in such a manner that assigned missions can be accomplished. Marines within supporting activities will therefore be serving at Logistic Establishments, Recruit Training Depots, Personnel Procurement offices, Headquarters establishments, training activities, and in non-available duty status.[111]

According to Basic Post-War Plan No. 2, the Fleet Marine Force was to consist of two Marine divisions and one Marine brigade, reinforced, and supporting naval units when and as re-

---

[110] *CMC Rpt*, 1946. Also see Col Thomas G. Roe, *et. al. A History of Marine Corps Roles and Missions: 1775–1962*—Marine Corps Historical Reference Series, No. 30 (Washington: HistBr, G–3 Div, HQMC, 1962) for a survey of these functions during the period indicated in the title.

[111] *CMC Rpt*, 1946.

quired. Normal locations would be: one division at Camp Lejeune and one at Guam, and the brigade on the west coast at Camp Pendleton. Marine aviation was to consist of two aircraft commands: one, AirFMFPac, with responsibility in the West Coast-Hawaii-Marianas area; and the other in overall command of six Marine carrier groups aboard as many escort carriers. The final components of the FMF were to be Headquarters, Fleet Marine Force, Pacific, and Headquarters, Fleet Marine Force, Atlantic (FMFLant), with Force Troops assigned to each command.

Ships' detachments were to consist of Marine detachments assigned to the larger combatant naval vessels, amphibious command ships, transports of the Atlantic and Pacific Fleets, and transports of the Naval Transportation Service. Small Marine aviation service detachments were to be assigned to the CVEs that had Marine carrier squadrons in their air complements. The security forces would consist of interior guards for naval establishments within and outside of the continental United States, and air warning and antiaircraft artillery units. Headquarters Marine Corps and supporting establishments were to be composed of the Marine headquarters at Washington, the recruit depots at Parris Island and San Diego, school activities, logistics activities, Marine barracks and camps located outside of naval establishments, Marine air stations in the continental United States, and Marine air stations in the Hawaiian area and the Marine Air Facility (in caretaker status) at St. Thomas, Virgin Islands.

These then, were the first postwar plans of the Marine Corps, and the forces with which they were to be accomplished. In the months immediately following V-J Day, Marine plans had frequently been revised to reflect new requirements, and the plans were changed even more drastically as each postwar year passed. Necessary steps were taken to reorganize the Marine Corps each time that the need arose. As the situation changed, the continuing reduction in the strength of the Corps overall had to be considered before new missions were assigned. As best it could, the Marine Corps built up its forces in the Pacific, the area where new peacetime responsibilities were waiting.

Return to the Islands

At war's end, Marine units not destined for an occupation assignment or deactivation went ashore to accept the surrender of Japanese forces on islands throughout the Pacific. After supervising the demilitarization and repatriation of the former enemy, the Marine units involved in these activities either were returned to the control of parent organizations, redesignated to reflect new duties as barracks or guard detachments, or deactivated.

Primarily, the story of the Marine Corps in the Pacific following the end of the war concerns Marine surrender and occupation duty, the activation of postwar garrison forces, and the many changes that the FMF underwent before the Marine Corps attained a peacetime stance.

SURRENDER AND OCCUPATION DUTY[1]

Long before Japan had indicated a willingness to sue for peace, staffs in Washington and at Pearl Harbor were specially constituted to work on plans for implementing the surrender of isolated Japanese garrisons on the many islands and occupied areas in the Pacific. The question facing these staff planners was whether the Japanese island and area commanders would follow the lead of their government and surrender or whether they would put up fanatic resistance that could continue long after V-J Day. On 14 August, the Emperor had issued an Imperial Rescript calling upon his commanders to surrender and cooperate with the victors. After they had received their Emperor's orders and were convinced of their authenticity, the outlying garrison commanders were more than willing to comply with them.

To provide for an orderly and systematic program of accepting the surrender of Japanese island garrisons and later occupying the islands, CinCPOA organized two task forces from the Marianas and Marshalls-Gilberts Area commands. These were: TF 94 (Commander, Marianas), established to operate the bases of that area and to occupy the East, Central, and Western Carolines (notably Truk, Yap, the Palaus, and the Bonins), and to evacuate enemy nationals from the Marianas; and TF 96 (Commander, Marshalls-Gilberts), organized for the same pur-

[1] Unless otherwise noted, the material in this section is derived from: CinCPac Rpt of Surrender and Occupation of Japan, dtd 11Feb46, hereafter *CinCPac Surrender and Occupation Rpt* (OAB, NHD); *FMF Grd Status Rpts;* Guam Island Comd WarDs, Aug45–May46; Peleliu Island Comd WarDs, Sep44–Feb46; Occupation For, Truk and Central Caroline Islands, WarDs, Dec46–Apr46, hereafter *Truk WarD,* with date; MarDet, Truk and Eastern Caroline Islands, AnnRpt of Activities, 1946.

poses as TF 94, but scheduled to operate in the Marshalls-Gilberts Area.[2]

Prior to V-J Day, American forces in the Marshalls and Gilberts had gained control of Eniwetok, Kwajalein, Majuro, and Tarawa Atolls. After Japan surrendered, Task Force 96 units had to set up occupation forces on such bypassed atolls and islands as Mille, Jaluit, and Wake, following the capitulation of the former enemy garrisons located there. In the Carolines, Ulithi, Peleliu, and Angaur had been in American hands since late 1944. Facing CinCPac elsewhere in the Pacific at the beginning of September 1945 was the surrender of Japanese forces in the Bonins, the rest of the Marianas and Palaus, and the formerly important strongholds of Yap, Truk, and Woleai, and their lesser satellite island garrisons. The occupation of some of these places, their demilitarization, and the evacuation of Japanese nationals was the responsibility of units of the Fleet Marine Force and small naval surface forces assigned to support them. As their tasks were accomplished, the Marine units either were gradually reduced in size until disbanded, withdrawn, or eventually replaced by small naval and Marine garrison forces. (See Map 25.)

The job of evacuating and repatriating Japanese military personnel and civilians was almost overwhelming. The enormity of the problem is best described by the numbers involved—100,000 military and 50,000 civilians. Aggravating the situation was the fact that most American and other shipping was employed to return home liberated Allied POWs and veterans scheduled for demobilization, and to carry back to the Far East replacement drafts and material required by the occupation forces in Japan and China. The solution to the shipping problem was eventually found in the employment of war-weary LSTs and demilitarized Japanese ships, none of which were suitable for other transportation requirements.

The vast majority of the island garrisons that were to surrender were in the area under the control of Vice Admiral George D. Murray, Commander, Marianas. To standardize the conduct of the surrender and occupation program, on 20 August he organized from within his force the Marianas Surrender Acceptance and Occupation Command (TG 94.3). Comprising this group were the following task units and their commanders: Truk Occupation Unit, Brigadier General Robert Blake; Bonins Occupation Unit, Colonel Presley M. Rixey; Palau Occupation Unit, Brigadier General Ford O. Rogers; Guam Evacuation Unit, Major General Henry L. Larsen; and three other units commanded by naval officers.[3]

A plan was devised standardizing the operations of these units and standardizing such specifics as the format of the

[2] On 5 January, Task Force 96 was dissolved as a separate command under CinCPOA. The subordinate commands formerly under TF 96 were absorbed by the Marianas command.

[3] ComMarianas OPln 4–45, dtd 20Aug45 (OAB, NHD).

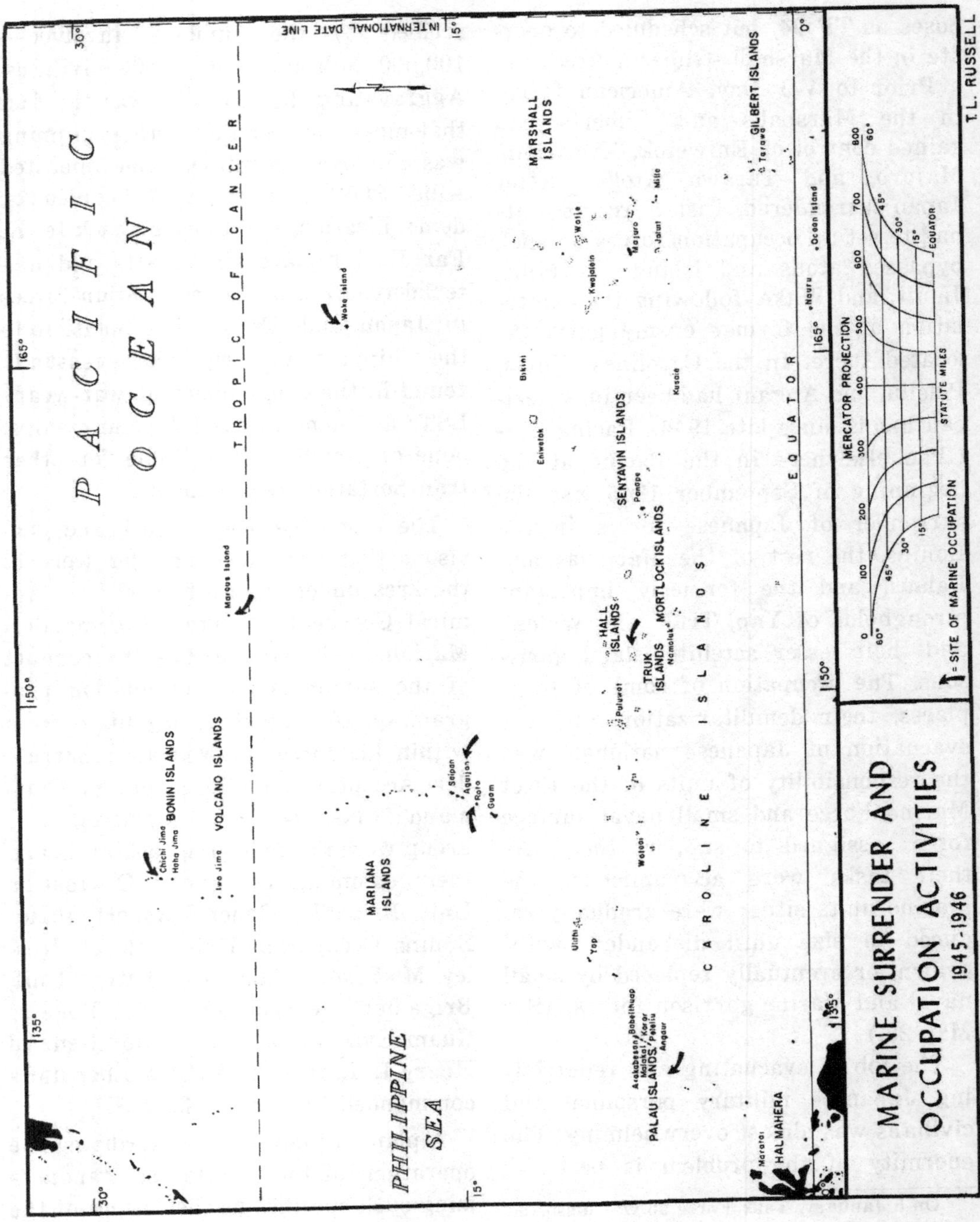
MARINE SURRENDER AND OCCUPATION ACTIVITIES
1945-1946
MAP 25
T.L. RUSSELL
= SITE OF MARINE OCCUPATION
MERCATOR PROJECTION
STATUTE MILES
PACIFIC OCEAN
PHILIPPINE SEA
TROPIC OF CANCER
INTERNATIONAL DATE LINE
EQUATOR
CAROLINE ISLANDS
BONIN ISLANDS
Chichi Jima
Haha Jima
VOLCANO ISLANDS
Iwo Jima
MARIANA ISLANDS
Saipan
Aguijan
Rota
Guam
Marcus Island
Wake Island
MARSHALL ISLANDS
Bikini
Eniwetok
Kwajalein
Wotje
Majuro
Jaluit
Mille
GILBERT ISLANDS
Tarawa
Ocean Island
Nauru
Kusaie
SENYAVIN ISLANDS
Ponape
MORTLOCK ISLANDS
Namoluk
HALL ISLANDS
TRUK ISLANDS
Puluwat
Woleai
Ulithi
Yap
PALAU ISLANDS
Arakabesan
Babelthuap
Malakal
Koror
Peleliu
Angaur
HALMAHERA
Morotai

surrender document [4] and the conduct of the surrender ceremony, the American flag raising, and the demilitarization of enemy forces, which were to be called thereafter "disarmed military personnel." In addition, the directive ordered each task unit commander to determine whether any Allied or American prisoners were present on the island being surrendered or if any had been kept there. The commanders also were to conduct spot interrogations and investigations to determine whether any war crimes or atrocities had been committed during the Japanese occupation.

According to the task force directive, relations between the victors and the vanquished were to be properly conducted at all times and all military courtesies were to be observed both before and after each surrender and flag-raising ceremony had been completed. Japanese interpretation of these orders often resulted in more than one enlisted Marine or sailor being saluted by all Japanese ranks.

The garrison on Mille Atoll, southwest of Majuro in the Marshalls, surrendered on 22 August—11 days before the Japanese government signed the surrender on the *Missouri*—and thus became the first enemy group in the Pacific Ocean Areas to do so. On the same day that Mille gave in, the garrison commander on Aka Shima in the Ryukyus surrendered to elements of the Tenth Army, which also accepted the surrender of the forces on Tokashiki Shima in the same island group the next day. On 29 August, the Japanese commander on Morotai, who controlled the garrisons of the entire Halmahera group, capitulated to the commander of the 93d Infantry Division.

Two days later, the garrison on Marcus Island, located between Wake and the Bonins, was the next major Japanese group to give up. Marcus figured prominently in Navy postwar plans. It was estimated that when the naval air base and terminal proposed for the island began operations, it would shorten the Honolulu-Tokyo air route by 1,049 miles. To administer the air base, CinCPac planned to establish a Marcus Island Command, but the numerous Japanese on the island had to be repatriated before the island could be developed. At the time of the surrender, there were 2,542 Japanese Army and Navy personnel and a number of Japanese civilians remaining—the majority suffering from all forms of tropical maladies and most of them severe malnutrition cases.

[4] Brigadier General Ford O. Rogers became Peleliu Island Commander on 7 August 1945, the day after the first atomic bomb was dropped on Japan. He recalled that his first conclusion after hearing of this bombing was that the war was all but over. Because he had not at that time received instructions from ComMarianas regarding how he was to conduct the surrender of Japanese forces in the Palaus, General Rogers directed his Chief of Staff, Colonel Charles L. Fike, and his Assistant Chief of Staff for Operations, Captain Melvin H. Bassett, USN, to prepare a surrender document to be used for that purpose. When the end of the war was announced, Rogers flew to Guam and submitted to Admiral Murray for approval the document his staff had prepared. "He took my surrender procedure word for word, and used it in all the other surrenders conducted under his command." *Rogers interview.* After ComMarianas had approved this document, Colonel Fike flew a spotter-type plane over Babelthuap, the headquarters of the Japanese Palaus command, and dropped a copy of the surrender terms to the enemy below. Fortunately, he was not shot at, although he had expected to receive ground fire. RAdm Melvin H. Basset, USN, comments to HistBr, G–3 Div, HQMC, dtd 27Oct66.

Not only were they unfit as laborers, their continued presence on the island presented a threat to the health of the occupation forces.

In order to evacuate the Japanese garrison as quickly and expeditiously as possible, the Marine 11th Military Police Company (Provisional), of the 5th Military Police Battalion, Island Command, Saipan, was sent to Marcus on 2 September. Arriving two days later, the unit remained on Marcus as the island guard until it was disbanded on 16 April 1946.[5] Two months earlier, on 18 February, the Heavy Antiaircraft Artillery Battery (Provisional) had been formed on the island. It was redesignated Marine Barracks, Marcus, on 10 April and passed to the administrative control of the Department of the Pacific (MarPac) that same day.[6] On 12 May, the Marine Barracks was disbanded, the same day that the Naval Air Base, Marcus, was deactivated.

The next Japanese area commander to yield was Lieutenant General Sadae Inoue, who surrendered on 2 September 1945 the entire Palau Group and all forces under his command, including those on Yap, to Brigadier General Ford O. Rogers, island commander on Peleliu.[7] The Peleliu Island Command had been organized on 16 July 1944, as the 3d Island Base Headquarters under the command of Brigadier General Harold D. Campbell; it was redesignated Island Command, Peleliu, on 16 November. Brigadier General Christian F. Schilt relieved General Campbell as Island Commander on 19 March 1945, and he was in turn relieved on 7 August by General Rogers.

Beginning 15 September, elements of the 111th Infantry occupied Koror, Malakal, and Arakabesan, and liberated 539 Indian and Javanese POWs, most of them suffering from marked malnutrition and beri-beri. By 5 October, all Japanese had been removed to Babelthuap, the largest island of the Palau group. A month later, the Indian and Javanese troops were repatriated. At the end of September, General Rogers sent a small force to reconnoiter islands and atolls in the vicinity of the Palaus to search for missing Allied personnel; none was found.[8]

[5] *FMFPac Admin Hist*, pt I, encl A, Schedule of Demobilization of FMFPac, 1Aug45–1Oct46, p. 1.

[6] See section, "Peacetime Garrison Forces," below for further information concerning this and other provisional and barracks detachments in 1946 and later.

[7] See also Garand and Strobridge, "Western Pacific Operations," pt III, "The Peleliu Campaign."

[8] General Rogers knew that the Japanese had taken one prisoner, the member of a B–24 crew which had been shot down. This man had jumped and crewmen of another plane in the raid had observed a Japanese boat picking up the American, who appeared alive and well. Rogers' first question to General Inoue after the surrender had been signed was: "You have one prisoner, where is he?" The Marine general was told that the prisoner had been killed by a Japanese officer, the commander of an antiaircraft battery, who had been given custody of the flyer in order to practice his English in conversations with him. The Japanese took the American "out to where our bombers had killed almost his entire antiaircraft command, began to brood over that, decided to punish him, and he shot him, buried him on the spot, and prayed over him." General Rogers ordered Inoue to produce the officer, but on the next day all he received was a picture of that individual laying on the ground with a hole in his head. He had committed suicide. *Rogers interview.*

On 26 October, less its 2d Battalion which had been disbanded, the 26th Marines arrived to relieve the 111th Infantry as the Peleliu garrison force. The following March, the Marine Detachment (Provisional), Peleliu, was activated and on the 15th, the 26th Marines was disbanded. The provisional force was redesignated Marine Barracks, Peleliu, on 15 April 1946, when administrative control of the unit passed to the Department of the Pacific.

On the same day that the Palaus surrendered, the commander of Japanese forces on Rota, located northeast of Guam, capitulated to Colonel Howard N. Stent, the representative of the Island Commander of Guam, Major General Henry L. Larsen. Rota was formally occupied on the 4th, and shortly thereafter all of the 2,651 disarmed Japanese military personnel, except for 5 patients, were transferred to POW stockades on Guam. Colonel Gale T. Cummings was appointed the temporary island commander of Rota, and Marine and Seabee forces under his command immediately began to repair the airstrip on the island, completing the task by the 1st of October.

The largest enemy force in the Central Pacific submitted on 2 September, when senior Japanese Army and Navy officers on Truk signed the instrument of surrender. Preparations for this act and the occupation of the former enemy territory were initiated on 30 August, when Brigadier General Leo D. Hermle, Deputy Island Commander, Guam, discussed with Vice Admiral Chuichi Hara, commander of the *Fourth Fleet*, and Lieutenant General Shunsaburo Mugikura, the *Thirty-first Army* commander, the steps to be taken for the surrender of all personnel and areas under their command.

Regarding his part in the presurrender discussions with the Japanese, General Hermle recalled:

> I carried out this mission under the orders of Vice Admiral Murray who furnished me a staff of about 12 officers, mostly technicians such as aviators, harbor defense [experts], engineers and two interpreters, etc. Contact with the Japanese on Truk was made from Guam via radio. We left Guam in the evening, one half the staff with me aboard a destroyer and the other half aboard a D.E. [destroyer escort]. We anchored off Dublon, Truk, the next morning.
>
> Japanese officials had been instructed, via radio, to approach the destroyer aboard a small boat displaying a white flag. An admiral and a general came aboard accompanied by a small staff. None of the Japanese would admit that they understood English, so all negotiations were conducted through the interpreters. I informed the Japanese that Vice Admiral Murray would take their formal surrender aboard his flagship and that they would receive further instructions concerning this matter by radio. They were informed that at the surrender they would be required to furnish lists of personnel, ships, planes, harbor defenses, etc. . . . At all times, they were very cooperative and the conference proceeded to a satisfactory conclusion in a few hours. During the conference, the captain of the destroyer gave them a light lunch for which they expressed great satisfaction.[9]

By signing the terms of surrender on 2 September on board USS *Portland*, Admiral Murray's flagship, General Mugikura committed the troops on the following islands under his control to

[9] LtGen Leo D. Hermle ltr to Hd, HistBr, HQMC, dtd 1Nov65.

lay down their arms and await American occupation: Truk, Wake, the Palaus, Mortlake (Nomoi), Ponape, Kusaie, Jaluit, Maloelap, Wotje, Puluwat, and Woleai, and Mille, Rota, and Pagan, which had already capitulated. In addition to these Army-controlled islands, the following bases under the control of the Navy were pledged to surrender at the same time by the signature of Admiral Hara: Namoluk, Nauru, and Ocean. When the military capitulated, Rear Admiral Aritaka Aihara, *IJN* (Retired), head of the Eastern Branch of the South Seas Government—a Japanese agency with headquarters on Truk—signed for the 9,000 civilians there and for those on the other islands within his jurisdiction.[10]

When an actual survey of the forces on Truk was made later, a total of 38,355 soldiers and sailors—including 3,345 Korean military personnel—was counted. In addition, a census of the civilians in the islands totalled 11,486, of which 1,338 were Japanese, 252 Korean, 9,082 native Caroline Islanders, 793 natives of Nauru, 8 Germans, 7 Spaniards, and 6 Chinese.

On the larger islands of Truk were such major Japanese military installations as bomber and fighter strips, seadromes, submarine and torpedo boat bases, ammunition magazines for weapons of all calibers and types, coast artillery defense installations, and other military facilities. All of these had to be demilitarized, dismantled, or destroyed. But first, the many sick Japanese on them had to be either treated and repatriated or evacuated.

Except for receipt of their regular share of American naval and air bombardments, some of the bypassed Japanese island garrisons did not fare too badly, especially if they had been based on one of the lush and fertile Pacific islands where they could raise their own food. There are some cases on record where Japanese commanders upon surrendering refused offers of food from the Americans because the garrison had a supply that was sufficient to maintain its members until they were embarked for return to the Home Islands.

Moen, one of the chief islands of Truk, was not a tropical paradise. When the occupation forces went ashore there, they found that bugs and worms had so ravaged the sweet potato crop, on which the Japanese garrison had so largely depended for subsistence, that all of the troops were suffering from malnutrition.

> Rank upon rank of 'living scarecrows lined up along the route of the inspecting party—men with ankles as thin as skinny wrists, with sunken cheeks, and with every rib showing sharply.'[11]

Three days before the surrender date, Brigadier General Robert Blake was designated Prospective Island Commander, Truk, a designation which was changed on 27 September to Prospective Commanding General, Occupation Forces, Truk and Central Carolines. The mission of his command was to occupy and develop Truk "as a fleet anchorage with facilities ashore limited to recreational purposes and for the support of

[10] *CinCPac Surrender and Occupation Rpt*, pp. 180–181.

[11] *Ibid.*, p. 181.

assigned aircraft and the servicing of transient aircraft."[12] With the aid of a small staff under the jurisdiction of the Island Commander, Guam, General Blake organized the unit that was to comprise the Truk Occupation Force. Because of the urgency of the Marine Corps demobilization program, the unit was formed slowly. Initial administrative duties were undertaken by the staff of the 2d Provisional Antiaircraft Artillery Group before it was disbanded. The headquarters of this group, however, provided the nucleus for the staff of the occupation force. The first detail of the new force to report in was the military government unit. A Base Headquarters Company (Provisional), was activated on 1 October 1945. It was formed according to the T/O of a Provisional Brigade Headquarters Company, and did not reach full strength until shortly before it departed for Truk.

In keeping with the future tasks of the force, elements of the 29th Naval Construction Battalion and Acorn 52[13] were assembled from bases all over the western Pacific. Both of these units also suffered from the loss of skilled and experienced artisans.

On 14 November, the 2d Battalion, 21st Marines, 3d Marine Division (then still on Guam), came under the control of the occupation force. The commander of 2/21, Lieutenant Colonel Lyman D. Spurlock, and a small detachment of Marines from the battalion had been sent to Truk in September to supervise the evacuation of Japanese personnel and Koreans, Okinawans, and Formosans, who had been members of Japanese-controlled forces. On 28 October, Spurlock was relieved by his executive officer, Major Robert J. Picardi, who remained in charge of the evacuation program until Lieutenant Colonel Spurlock returned to Truk on 25 November with the rest of 2/21 and the occupation force.

To that date, 6,696 Japanese civilian and military and Japanese-controlled forces, and their wives and children, had been repatriated; by December, this number had risen to 20,410, leaving 19,575 remaining in the islands. In January, 14,298 more evacuees left Truk, and in February, 1,426. At the end of April 1946, only 3,811 disarmed military personnel and their families remained, most of them working as laborers and assisting in the destruction of Japanese arms, fortifications, and munitions. The remarkable factor in the history of all of the former Japanese possessions in the Pacific that were surrendered to and occupied by American forces is the high degree of cooperation, docility, and lack of rancor on the part of the losers. There were few incidents of Japanese intransigence; those that did occur took place among the accused war criminals, who were usually more confused and contrite than sullen and unremitting.

One mission common to all Allied occupation and surrender groups was to investigate alleged Japanese war crimes and atrocities, and hunt down and imprison until their trial those accused of such acts.[14] By the time that war crimes

[12] G-3 WarD, encl (B) to Truk WarD, Dec45.

[13] An acorn was a naval unit designed to construct, operate, and maintain an advanced landplane and seaplane base and to provide facilities for operations.

[14] See below, this chapter, for Marine involvement in the conduct of war crimes trials.

tribunals had been convened, a considerable number of accused Japanese were being held in stockades on the various islands under Allied control throughout the Pacific. If evidence of an alleged crime was discovered after the accused had been repatriated to Japan, depositions were taken and presented to Allied tribunals convened in that country. During the first few months of the Truk occupation, General Blake's investigators had uncovered evidence sufficiently damning to warrant apprehension and detention of 42 individuals. Since no tribunals were held on Truk, the detainees were tried elsewhere, depending upon which of the Allied governments had paramount jurisdiction.[15]

Following the raising of the American flag on 25 November over the island group formerly held by the Japanese, General Blake's forces conducted a search of Truk and its neighboring islets for missing Allied personnel; none were found.

On 26 February 1946, the Base Headquarters Company, Occupation Forces, Truk and Central Caroline Islands, was redesignated the Marine Detachment (Provisional), Truk. Personnel to expand the new detachment to its authorized strength were low-point personnel transferred from 2/21. The next day, the battalion was detached from General Blake's command and returned to Guam, where it was disbanded on 5 March. On 15 April, the occupation forces command designation was changed to Commander, Truk and Central Caroline Islands. Exactly one month later, General Blake was relieved by a naval officer, who had the additional duties and title of Commander, Naval Air Base, Truk. In July 1946, the complement of the provisional Marine detachment was reduced from a strength of 256 to 42 men. On 12 October, administrative control of the detachment passed to MarPac.

Wake Island, which had been captured by the Japanese on 24 December 1941, was regained by the Americans on 4 September 1945, when Rear Admiral Shigematsu Sakaibara surrendered the forces under his command to the commander of the 4th MAW, Brigadier General Lawson H. M. Sanderson, who was the representative of the Commander, Marshalls-Gilberts Area, for the ceremony.[16]

Following his appearance on the *Levy* (DE–162) on 4 September to receive and sign the surrender documents, Admiral Sakaibara departed from the ship to make preparations ashore for the American flag to be raised that same afternoon. The first American again to set foot on Wake when General Sanderson's party arrived was Colonel Walter L. J. Bayler, famed as the "last man off Wake Island." [17] At 1330:

[15] *Truk WarDs*, Dec45, Jan–Apr46.

[16] This narrative of the Wake surrender and occupation is derived from: Comdr, Wake Island Surrender Acceptance Unit, ltr to CTG 96.14, dtd 7Sep45, Subj: Surrender Acceptance of Wake Atoll on 4Sep45, Narrative of (OAB, NHD), hereafter *Wake Island Surrender Rpt; CinCPac Surrender and Occupation Rpt*, pp. 186–187; ComMarianas ltr to Dir, Naval History, dtd 1Jan47, Subj: Narrative of the Marshalls Area Comd, 1Sep45–1Oct46, Anx V, Wake Island Comd Rpt for same period, hereafter *Marshalls Area Comd Hist* (OAB, NHD).

[17] See LtCol Walter L. J. Bayler, *Last Man Off Wake Island* (New York: Bobbs-Merrill, 1943).

TO THE COLORS *sounds as the American flag is raised over Wake Island for the first time since December 1941. (USMC 133686)*

> With the platoon at 'Present Arms,' with both American and Japanese saluting, the Colors were then hoisted and two-blocked while the notes of 'To the Colors' were sounded on the bugle. As the Colors reached the peak of the flag pole, the Levy commenced and completed firing a twenty-one gun salute.[18]

Although a Japanese garrison of 609 Army and 653 naval personnel had surrendered, this total was only a small fraction of the total number of Japanese that were isolated on the island from the time it was first bypassed by the Americans. Since that time, American bombs and shells had killed 600 of the enemy, 1,288 had died of malnutrition and disease, and 974 had been evacuated to the Home Islands as hospital cases. Of those remaining when the Americans arrived, 405 were ill—200 of these bedridden. Immediately upon occupying the island, American authorities sent food and medical supplies to succor the garrison.

In accordance with CinCPac plans, Wake was designated a Naval Air Facility on 4 September. Occupation forces arrived beginning the 7th, including a Marine detachment consisting of 2 officers and 54 enlisted from Engebi. These forces at once concentrated on repairing the airstrip, disposing of mines, destroying Japanese ammunition and bombs, and establishing a shore-based communication establishment. An inspection of existing air facilities disclosed that the east-west strip of the airfield was in good condition and capable of landing planes of any type or size. The seaplane lane formerly used by Pan American Airways flying clippers was re-marked during the first few weeks of the reoccupation, and mooring buoys for seaplanes were also placed during this period.

By the middle of September, all Japanese had been removed from Wake, the chief island of the atoll, to Peale and Wilkes Islands.[19] All Japanese, with the exception of Admiral Sakaibara and 16 commissioned and noncommissioned officers, were repatriated by 1 November 1945. The admiral and the others were temporarily detained before their transfer to Kwajalein for further investigation in their responsibility in the alleged execution of approximately 100 American civilian workers in October 1943.

Wake was officially commissioned as an Island Command and a Naval Air Base on 1 November, with a naval officer installed as the commander of both the island and the air base. On 14 January 1946, the Marine Detachment (Provisional), Wake, consisting of 5 officers and 110 enlisted was established. Less than a month later, on 10 February, the unit was redesignated Marine Detachment (Provisional), Eniwetok, and transferred there with orders to disband on conclusion of Operation CROSSROADS, the atomic bomb tests at Bikini Atoll. On 10 December, the detachment was disbanded.

Eight days after the provisional detachment left Wake, a Marine Heavy Antiaircraft Artillery Battery (Provisional), Wake, was activated. Like other provisional units formed in this period, the strength of the battery was 5 officers and 110 enlisted Marines. In

[18] *Wake Island Surrender Rpt*, p. 4.

[19] See Hough, Ludwig, and Shaw, *Pearl Harbor to Guadalcanal*, map on p. 97.

view of the decreasing importance of Wake in postwar plans, the battery was disbanded on 19 August 1946.

Another important Japanese capitulation occurred on 3 September 1945, when Lieutenant General Yosio Tachibana, senior commander of the Japanese forces in the Ogasawara Gunto (Bonin Islands) surrendered to Commodore John H. Magruder, Jr. aboard USS *Dunlap*, outside the harbor of Chichi Jima. Until the fall of Iwo Jima and his death, the former commander of the Bonins forces, Lieutenant General Tadamichi Kuribayashi had made his headquarters on that volcanic island. After Kuribayashi's death, the subcommander of the Bonins succeeded to command and moved the headquarters to Chichi Jima.

Approximately 140 nautical miles northeast of Iwo Jima, Chichi Jima was seriously considered by American planners as a potential target for an amphibious landing.[20] Chichi Jima was dropped in favor of Iwo Jima, because, although it had a good protected harbor, its terrain was too rugged to permit the rapid construction of airfields. Even more condemning were the results of photo-reconnaissance missions which showed Chichi Jima to have been more heavily fortified than Iwo.[21] Confirming this evaluation after the war was the report of the Bonin Occupation Forces Commander. Following some preliminary comments, Colonel Rixey wrote:

> This writer has seen Jap defenses from Tarawa to Iwo. Nothing previously seen can compare with coast and artillery defenses . . . surrounding Chichi harbor. Concrete emplacements, high in the mountains with steel door openings are too numerous to count. Artillery and machine gun fire which could have been placed on the airfield would have prevented *any* [force commander's italics] attempt at a landing there. With camouflage, as practiced by the Japs, in place, NGF spotters would have had a very difficult time locating these cleverly placed positions. . . . The location of many of the emplacements, which have to be seen to be appreciated indicate that the Jap plan was to permit an entrance into the harbor or onto the airfield, then to give us the 'works.' Most of these positions are inaccessible and many could not have been reached by NGF as they are situated on reverse slopes facing east.[22]

Survivors of the Japanese garrison on Chichi and Haha Jimas comprised 20,656 Army and Navy personnel and 2,285 civilian laborers who had been transported to and employed in the islands by the military. Additional Japanese garrison troops located on other islands near the Bonin group were evacuated by the U. S. Navy.

In mid-September 1945, at the same time that 2/21 was designated as the military element of the Truk Occupation Force, the 1st Battalion, 3d Marines, 3d Marine Division, was designated the military element of the Bonins Occupation Force. Immediately upon receipt of the orders detaching them from parent organizations, both battalions began reorganizing for the move and filling their ranks with volunteers, regulars, and low-point Marines.

On 10 October, the advance echelon of 1/3, consisting of Rixey's small staff

[20] See pt I, chap 1, *supra*.

[21] Morison, *Victory in the Pacific*, pp. 5–6.

[22] Bonins EvacRpt of Inspection 10–11Oct45, by Col Presley M. Rixey, encl (A) to ComMarianas ltr to CinCPac–CinCPOA, dtd 6Nov45, subj: Bonins EvacRpt (OAB, NHD).

and 20 military policemen, landed and met the Japanese liaison group headed by Major Yoshitaka Horie. When Colonel Rixey discovered that General Tachibana and Vice Admiral Kunizo Mori, the senior officer in tactical command at Chichi Jima, were not present in the group, he "sent for them to report to me at the dock, which they, of course, complied with." [23]

The Marines were the first American troops to set foot in the Bonins since Commodore Perry's expedition there in 1853.[24] Rixey's group had a primary mission of evacuating and repatriating the Japanese. A secondary task was to destroy the extensive Japanese defenses existing on the island. When the remainder of the battalion arrived on 13 December, it carried with it a large supply of explosives with which to accomplish this mission.

This main body had been designated the Bonins Occupation Force at Guam on 1 December. When it landed on Chichi Jima 12 days later, Colonel Rixey ordered the American flag raised over the former Japanese stronghold. After he had originally landed on 10 October, Colonel Rixey determined that the entire 1st Battalion, 3d Marines, would not be required to garrison the island, to supervise repatriation, and to demilitarize the defenses. He therefore recommended that the Occupation Forces be reduced in strength to 400 men only. He later found that even less troops could have been used because the Japanese were most cooperative and willing to please. It was not necessary to establish a manned boundary between the American and Japanese zones on the island; "A drawn line on a map was sufficient." [25] On 1 June 1946, after fulfilling its assigned mission, 1/3 was disbanded on Chichi Jima, and its Marines were transferred to other FMFPac units in the Pacific and the Far East.

During the several visits to the Bonins by American fast carrier task forces in 1944 and 1945 and the subsequent air and naval gunfire bombardments of those islands, one Marine and several Navy aviators were shot down and listed as missing in action. After Colonel Rixey had assumed his role as the commander of the Bonins, he instituted an investigation to determine the fate of these downed pilots. Soon, Rixey began hearing rumors and receiving anonymous reports concerning the inhumane and barbaric treatment American POWs had received at the hands of their Japanese captors.

Shortly after Colonel Rixey's arrival on Chichi Jima, a Japanese Coast Guard cutter entered the harbor. On board were Frederick Arthur Savory and his three uncles, all of whom were descendants of Nathaniel Savory, a Massachusetts whaler who had settled in the Bonin Islands in the 1830s. After the fall of Saipan, the Japanese had evacuated the American-Chamorro-Hawaiian family to the Home Islands. While in Japan, Fred Savory had heard rumors spread by soldiers repatriated from Chichi Jima regarding cannibalism on

---

[23] BGen Presley M. Rixey ltr to Hd, HistBr, HQMC, dtd 10Nov65, hereafter *Rixey ltr*.

[24] Aurthur and Cohlmia, *3d MarDiv Hist*, p. 334. It should be noted that Iwo Jima is in the Volcano Islands.

[25] *Rixey ltr*.

that island. He passed these stories on to Colonel Rixey.

The morbid story of the Chichi Jima garrison was related in full at the war-crimes trials held later at Guam. Two naval aviators—one captured in March 1944 and the other in August after they had parachuted from their disabled aircraft—were bayoneted to death at General Tachibana's orders following their interrogation. Five more American airmen, one a Marine, were executed after they, too, had been captured when they bailed out of their aircraft. Three were beheaded, one was bayoneted, and another beaten to death. It was upon the flesh of these five that certain members of the Japanese garrison fed. Testimony exonerated the majority of the Chichi Jima command from having been involved in this disgusting incident, and indicated that with the exception of the perpetrators of this foul deed, those who ate the flesh did not know what they were eating.

Reporting his reaction upon learning of the uncivilized actions of the guilty parties, Colonel Rixey wrote: "We were flabbergasted at first. We had expected beheadings, of course. But never cannibalism! What manner of men were these?" [26] The war crimes trials of 21 Chichi Jima officers and men were held on Guam during the fall of 1946, and entailed more than 1,000 pages of testimony and exhibits. Of the 21 accused, one officer who had no knowledge of the cannibalism was acquitted. The other 20 were found guilty and given various sentences ranging from death by hanging to life imprisonment and lesser penalties. One was hanged in June; General Tachibana and three of his other officers were executed at Guam on 24 September 1947.[27]

In quick profusion, the following former Japanese-garrisoned islands in the Central and Western Pacific islands were surrendered to CinCPac representatives in September 1945: Aguijan, Jaluit, Yap, Wotje, Maloelap, the Ryukyus, Kusaie, Ponape, Nauru, Lamotrek, Woleai; [28] and in October, Ocean. Some of these little-known islands with unfamiliar names were small and held nothing but a weather station manned by a few Japanese civilians and a slightly larger native population. Military garrisons of various sizes were on some of the larger islands, the size of the force determined by the strategic value that the Japanese had given the island. Regardless of the location or size of the former enemy garrison, the terms of the Potsdam Declaration bound the Allied Powers to permit, and by inference to assist, all Japanese military personnel to return to their homes after they and their organizations had been completely demilitarized. Because American shipping

[26] Cited in Sherrod, *Marine Air Hist*, p. 352.

[27] CO, MarBks, Guam, ltr to CMC, dtd 27May 48, Subj: Hist Narrative of Special War Crimes duties performed by personnel of the MarBks, Guam, hereafter, *MarBks, Guam, Hist Narrative.*

[28] Representing the Island Commander, Guam, Lieutenant Colonel Parker R. Colmer accepted the surrender of the Woleai Atoll Commander, Major General Kitamura, on 19 September. Two days earlier, a small Marine detachment under Lieutenant Colonel Colmer had landed on this tiny atoll in the Central Carolines to arrange for the demilitarization of the islet and the evacuation of its personnel. Island Comd, Guam, WarD, Sep45.

was fully committed to the support of MacArthur's occupation and surrender forces and in the return of U. S. servicemen to the States, until they could be repatriated, the disarmed enemy garrisons on the various atolls were supervised, but generally left undisturbed. The Japanese were allowed to fend for themselves from their own gardens until such time that Japanese shipping could be made available to transport them home.

Although American shipping was thoroughly involved in "Magic Carpet," the return home of combat veterans from the Pacific, and in operations in the Far East, the Japanese had 21 tankers, 101 transports, and 211 freighters still in operating condition after the war. General MacArthur said, however, that there was a more pressing need for these vessels to ship food and clothing to the Home Islands than to repatriate troops from outlying islands. "As of 7 October, according to *Domei* [the Japanese news agency], only 38,645 troops had been returned from overseas, including the continent of Asia, which meant that some 3,320,000 Japanese Army and 300,000 Navy personnel still remained outside the home islands." [29] CinCPac alleviated the situation somewhat in November, when it began to use amphibious vessels not suited for "Magic Carpet" in the repatriation of Japanese from the Marshalls-Gilberts and Marianas Areas. Liberty ships and LSTs in Philippines ports and at others in the Pacific were assigned to duty as Japanese repatriation vessels. By the end of December 1945, all Japanese military personnel had been evacuated from the Marshalls-Gilberts Area. By 10 January 1946, 73.9 percent of the Japanese nationals, military and civilian, on the islands in the Marianas had been evacuated to Japan. Not included in these groups were the Japanese who had been detained on either Guam or Kwajalein awaiting trial as war criminals or waiting to appear as witnesses at these trials. Nonetheless, before the middle of 1946, most disarmed military personnel had been returned to their homes in Japan, Korea, Okinawa, or Formosa, and thus many of the provisional Marine detachments that had been formed to supervise their repatriation could be deactivated. The Marine forces in the Pacific were ready then to phase into their postwar garrison programs.

[29] *CinCPac Surrender and Occupation Rpt*, p. 168.

PEACETIME GARRISON FORCES [30]

By 1 October 1946, of the 10 provisional Marine detachments and the military elements of occupation forces that

[30] Unless otherwise noted, the material in this section is derived from: *CMC Rpts*, 1946–1948; *USMC Admin Hist; G–1* and *G–3 OpDs*, 1Aug45–31Dec47; *FMF Air* and *Grd Status Rpts*, Aug45–Dec47; *FMFPac Admin Hist*, 1945–1947; CofS, MarPac, ltr to CG, MarPac, dtd 5Mar46, Subj: Hist of Hq, MarPac, During World War II, hereafter *MarPac WW II Hist;* CG, MarPac, ltr to CMC, dtd 20Mar47, Subj: Narrative of MarPac 1Sep45–1Oct46; CG, MarPac, ltr to CMC, dtd 15Apr47, Quarterly Summary of MarPac, 1Oct46–1Apr47; CG, MarPac ltr to CMC, dtd 10Jul47, same subject, period 1Apr–1Jul47, all hereafter *MarPac Hist*, with inclusive dates; MarGarForPac OpDs, for periods 1Sep45–1Oct46, 1Oct46–1Apr47, and 1Apr–1Jul47, all hereafter *MarGarForPac OpD*, with inclusive dates.

had been formed since the end of the war,[31] all except the ones on Wake, Kwajalein, and Eniwetok were disbanded, redesignated a Marine Barracks, or made a permanent Marine detachment.[32] Between April and July 1946, the following redesignations took place on the dates noted: 15 April, Marine Detachment (Provisional), Peleliu, became the Marine Barracks, Peleliu; 20 May, 8th Military Police Battalion (Provisional), became the Marine Barracks, Guam; 10 June, 5th Military Police Battalion (Provisional), became the Marine Barracks, Saipan; 20 June, Marine Detachment (Provisional), Samar, became the Marine Barracks, Samar; and 4 July, Marine Detachment (Provisional), Headquarters, Commander, Philippine Sea Frontier, became the Marine Detachment, Commander, Naval Forces, Philippines. The last-named organization was reorganized on 1 January 1947 as the Marine Barracks, Sangley Point, Philippine Islands.

The realignment and reduction of FMF units in the Far East also affected the organization of the garrison forces in the Pacific. For instance, in January 1946, CinCPac notified CNO that he anticipated a drastic reduction in the size and scope of fleet activities at Yokosuka for the remainder of the American occupation of Japan, and recommended that the 3d Battalion, 4th Marines—part of the Yokosuka Occupation Force—be redesignated as the 2d Separate Guard Battalion (Provisional), FMF, and remain there for interior guard duty. The change in title took place on 15 February 1946. Five months later, the size of the provisional battalion was reduced and on 15 June it became Marine Detachment, Fleet Activities, Yokosuka, and was placed under the administrative control of MarPac.

Based on the provisions of the postwar plans, all of these Marine units were given missions consisting of maintaining internal security and standing interior guard duty at the naval activities to which they had been assigned. In addition, the Marines on Truk, Marcus, and Peleliu islands manned antiaircraft artillery positions, while the garrisons on Guam, Saipan, and Truk had the important and difficult task of guarding and supervising nearly 7,000 Japanese war crimes prisoners and disarmed military personnel.

From 1 June 1946 until 5 May 1949, when the War Criminal Stockade was closed, the Guam garrison was responsible for the custody, discipline, feeding, clothing, and, in some cases, execution of Japanese war criminals confined in its custody. A total of 13 inmates was executed by hanging following their trial and conviction by the War Crimes Commission convened on Guam. Eleven executions were conducted by an Army Officer, a member of the Military Police Corps, who was an official hangman.

[31] See previous chapter, section entitled "The Marine Corps at the End of the War," p. 444*n*.

[32] On 10 December 1946, the provisional detachments on Wake and Eniwetok were disbanded. The Kwajalein unit was redesignated Marine Barracks, Kwajalein, and came under the administrative control of Marine Garrison Forces, Pacific. *FMFPac Admin Hist*, 1Oct46–1Apr47.

During the period he was attached to the Marine Barracks in order to carry out his duties, he trained two enlisted Marines "in the technique of execution by hanging." They executed the last two war criminals condemned to die after he left.[33]

On the date that the designations of the various garrisons changed, they were detached from FMFPac and placed under the administrative control of the Department of Pacific, or MarPac as it was both officially and familiarly abbreviated. Operational control of the Pacific garrisons was vested either with the naval activity or the senior naval command on the island to which the Marine unit was assigned.

On the effective date of attachment of the barracks and detachments, MarPac in turn placed them under the control of Marine Garrison Forces (MarGarFor), 14th Naval District, in Honolulu. MarGarFor had been formed on 13 December 1941 in order that all of the various Marine garrison forces in the 14th Naval District could be centrally administered.

Headquarters Marine Corps directed Brigadier General Harry K. Pickett to assume command as Commanding General, MarGarFor, 14th Naval District. Although General Pickett and subsequent commanders functioned as deputies of the commanding general of MarPac, they did not carry the title. At the end of the war, because of the increasing importance of the Honolulu-based command and to ensure an efficient administrative control of the widely separated Marine Corps posts in the Pacific, the MarPac commander recommended that the title of the MarGarFor commander be changed to Commanding General, Pacific Ocean Marine Garrison Force, and that he be assigned as the deputy commander of MarPac.[34] The Commandant approved the redesignation, and it became effective on 15 October 1946, when the new title, Commanding General, Marine Garrison Forces, Pacific, appeared.[35] The first officer assigned to this command was Brigadier General William A. Worton, who established his headquarters at the Marine Barracks, Pearl Harbor, T.H., and reported his assumption of command on the 15th to both CinCPac and MarPac.[36]

When the Marine Garrison Forces was first organized in 1941, there were only a few barracks and detachments under its command, and all of these were in the Hawaiian Islands and on Johnston and Palmyra Islands. At the end of the war, the number of subordinate units increased considerably; the mission assigned MarGarForPac at the time of its redesignation in 1946 made it responsible for all posts, detachments, offices, and other Marine Corps organizations in the Pacific Ocean Areas with the exception of FMF units, Marine Corps air stations, and shore-based air warning

[33] *MarBks, Guam, Hist Narrative.*

[34] *MarPac WW II Hist,* p. 4.

[35] CMC Spdltr ser MC–819560, dtd 27Sep46; MarGarForPac GenO 2–46.

[36] MajGen William A. Worton ltr to Hd, HistBr, HQMC, dtd 25Oct65.

units.[37] On 23 May 1947, MarGarForPac became an administrative command directly under the control of Headquarters Marine Corps. As before, CinCPac had operational control of the Marine posts and stations in the Pacific. The Marine Garrison Forces, Pacific, command was deactivated on 31 August 1948, and the Commanding General, FMFPac, assumed administrative control of all Marine security forces and supporting establishments in the Pacific formerly under the control of MarGarForPac.[38]

Although it was purely an administrative command throughout seven years of existence, MarGarForPac played an increasingly important role in supervising the constant change in the composition, designation, and number of Marine garrisons in the Pacific during the period 1945–1948. The assumption of control over these Marine outposts by FMFPac meant that the Marine Corps transition to peacetime status had been accomplished and security forces and supporting establishments in the Pacific were stabilized—for the time being, at any rate. Closely paralleling the steps leading to stabilization of these non-Fleet Marine Force organizations were the day-to-day changes that carried FMF units from a war to peacetime character.

STABILIZATION OF THE FMF[39]

By the beginning of 1947, the Marine Corps had adjusted to operating on a peacetime level with a complement drastically reduced from a peak strength; there were 92,222 Marines on active duty on 30 June 1947.[40] Because of major organizational changes and the constant turnover of personnel in activities where trained and experienced Marines were needed, the retention of key Marines was especially critical in Marine logistical, aviation, and recruit training units. The problem was all the more grave in face of the missions assigned to the Marine Corps for Fiscal Year 1947.[41] These missions were a combination of those previously assigned in accordance with the postwar program of the Corps and those foreshadowing what the National Security Act of 1947 and later amendments would assign.

---

[37] MarGarForPac consisted of the following in October 1946: H&S Co, MarGarForPac; MB, USNB, Pearl Harbor, T.H.; MB, NAD, Oahu, T.H.; MB, NAS, Ford Island, Pearl Harbor, T.H.; MB, NAS, Barber's Point, Oahu, T.H.; MB, NAF, Honolulu, Oahu, T.H.; MB, NAS, Kaneohe Bay, Oahu, T.H.; MB, NAS, Kahului, Maui, T.H.; MB, NOB, Midway Island; MB, NAS, Johnston Island; MB, Guam; MB, Saipan; MB, NOB, Okinawa; MB, NOB, Subic Bay, P.I.; MB, Peleliu, Palau Islands; MD, (Prov), USNBS, Marcus Island; MD, Truk; MD, Fleet Activities, Yokosuka; MD, Commander, Naval Forces, Philippines; MB, Kwajalein; MD, Fita-Fita Guard, Samoa. *Muster Rolls*, MarGarForPac, Oct46 (Diary Unit, Files Sec, Pers Dept, HQMC).

[38] CMC Spdltr ser MC–1034430, dtd 19Aug48, cited in *Ibid.*, Aug48.

[39] Unless otherwise noted, the material in this section is derived from: *CMC Rpts*, 1946–1949; *USMC Admin Hist; G–1, G–3*, and *DivAvn OpDs; FMF Air* and *Grd Status Rpts; FMFPac Admin Hist.*

[40] *CMC Rpt*, 1947.

[41] As other agencies of the Federal government do, the Marine Corps operates on a fiscal rather than a calendar year, or from 1 July of one year to 30 June of the following year. Fiscal year 1947 then would have been from 1 July 1946 to 30 June 1947.

In essence, the Marine Corps was:

> (a) To provide fleet marine forces of combined arms, together with supporting air components for service with the fleet in the seizure or defense of advance naval bases and for the conduct of such land operations as may be essential to the prosecution of naval campaigns.
>
> (b) To develop in coordination with other armed services, those phases of amphibious operations which pertain to the tactics, techniques and equipment employed by landing forces.
>
> (c) To provide detachments and organizations for service on armed vessels of the Navy.
>
> (d) To provide security detachments for the protection of naval property at naval stations and bases.
>
> (e) To be prepared in accordance with integrated joint mobilization plans for the expansion of the peacetime components to meet the needs of war.
>
> (f) In addition, to maintain such activities as necessary to insure the adequate administration, supply, training and technical directions of personnel and units engaged in accomplishment of the basic missions.[42]

To meet the challenges engendered by these missions and at the same time to reduce its size to reflect peacetime tasks, the Marine Corps reviewed and revamped the assignments given to its major components. In conjunction with Headquarters Marine Corps and the Marine Corps Schools, the Fleet Marine Force was to continue its role as stated in paragraph (b) quoted above. Marine detachments afloat on carriers, battleships, and cruisers were to provide a trained nucleus for the ships' landing forces, gun crews as required, and local security for the vessels. Marine detachments on transports would perform the functions of transport quartermasters and local security as directed. Marines in security forces assigned to naval shore establishments within and outside of the continental limits of the United States would provide necessary internal security; those detachments assigned to such establishments outside of the United States would provide external security in accordance with specifically assigned missions. In order to maintain the Marine Corps and to assist it in accomplishing its missions, Marine supporting activities, such as logistic establishments, recruit training depots, personnel procurement offices, headquarters establishments, and training activities, would procure, equip, train, and administer Marine personnel.[43]

During the last two war years, the FMF had been organized in accordance with the F-Series Tables of Organization, by which a Marine division consisted of a headquarters battalion, tank, engineer, and pioneer battalions, service troops, an artillery regiment, and three infantry regiments, totalling 843 officers and 15,548 enlisted, or 16,391 Marines overall. A reinforced regiment, or RCT, consisted of the infantry regiment itself, an artillery battalion, engineer, pioneer, transportation, ordnance, service and supply, medical, and tank companies, a reconnaissance platoon, and a band section, all totalling 4,585 Marines. Under the G-Series T/O (peace),[44] approved 4 September 1945, the Marine Corps reorganized to reflect its postwar size and conditions at that time. A Marine division consisted of the same components

[42] *CMC Rpt*, 1947.

[43] *CMC Rpt*, 1946.

[44] See App H for this T/O.

310-224 O - 69 - 31

as before, but its strength was increased to 962 officers and 17,182 enlisted Marines, or a total of 18,144. While the size of the division was increased, the number of Marine divisions scheduled for active service according to postwar plans had dropped to two. As soon as the G-Series T/O was authorized, all Marine organizations affected by it were reconstituted. Many units, because of their missions as well as their depleted states, could be reorganized at only 90 percent of their authorized strength. During the period February-April 1946, the 1st Marine Division in China and a few other units had dropped down to 80 percent of authorized strength. To rectify the situation, the Commandant ordered them to organize on the basis of the prevailing G-Series T/O, and beefed up their strengths slightly to reflect their current assignments.

In late 1945, two of the Marine divisions that had served so spectacularly in the Pacific during their tours in combat were disbanded. The 4th Marine Division, which had been formally activated on 16 August 1943 at Camp Pendleton, returned to the United States from Maui on 3 November 1945 and its units disbanded that month, again at Camp Pendleton. Following the earlier formation of some of its organic units, the 3d Division had been activated on 16 September 1942, also at Camp Pendleton like the 4th Division. A little more than three years later, on 28 December 1945, the 3d—less 1/3 in the Bonins and 2/21 on Truk—was disbanded on Guam. Replacement drafts consisting of low-point men boarded CVEs for the trip to China, where they were to join the 1st Marine Division; high-point Marines scheduled for discharge or reassignment were transferred to a transient center on Guam, where they awaited transportation to the States.

From September 1945 to June 1947, FMFPac was reduced in size until it approached the postwar form it would take. Many changes in the composition and designation of FMFPac units occurring in this period will be discussed later when the occupations of Japan and North China are considered. On 30 June 1947, FMFPac strength was 19,125 Marines in units at Camp Pendleton, the Hawaiian Islands, China, and Guam. In the months leading to this date, other major Marine commands, FMF and non-FMF, were formed.

On 22 January 1946, the Commandant directed the Commanding General, Marine Barracks, Quantico, to form a special infantry brigade to be prepared for expeditionary service and maintained in a state of readiness. The headquarters and two battalions of the 1st Special Marine Brigade was formed at Quantico, and the 3d Battalion was formed at Camp Lejeune. The following month, on 4 February, administrative and operational control of the brigade passed to the brigade commander, Brigadier General Oliver P. Smith. Four days later, General Smith was directed to maintain his command on two weeks readiness, and to report to the Commander in Chief, Atlantic Fleet (CinCLant), for planning purposes. Although organized along the basic lines of a FMF team, the brigade was not part of the Fleet Marine Force. In May 1946, it conducted the only major Marine Corps training mission undertaken during that fiscal year, a joint amphibious exercise in the Caribbean area. At the end of July, General Smith was directed to disband the 1st Battalion at Quantico on 10 August. The brigade headquarters and the 2d Battalion, which had joined the 3d Battalion at Camp Lejeune in March, were disbanded on 31 August.[45]

Another significant event occurring in 1946 was the activation on 16 December of Fleet Marine Force, Atlantic (FMFlant), under the operational control of CinCLant. Three days before its formation, the Commanding General, 2d Marine Division, at Camp Lejeune, was directed to activate FMFLant and to act as its commanding general in addition to his duties as head of the 2d Division. He was to assume command of the ground units comprising the force on the 16th, and to take over the aviation units on 2 January 1947.

In December 1946, Marine aviation commands in the Atlantic and Pacific areas were designated as subordinate units of the Fleet Marine Force, Atlantic and Pacific, respectively. In the process of this change, the 2d Marine Aircraft Wing at Cherry Point was redesignated Aircraft, Fleet Marine Force, Atlantic.

Marine aviation went through as many changes as the ground component following the end of hostilities, and was likewise concerned with reaching its postwar level as quickly as possible. These transitions required the rapid but controlled demobilization of personnel and deactivation of units and stations without the loss of a high state of combat readiness. The first phase of the program relating to the withdrawal of overseas units depended on future requirements of Marine aviation in the Pacific.

The deployments of the 1st Marine Aircraft Wing and its four groups to China and MAG–31 to Japan were the last World War II tactical operations of AirFMFPac. It then became possible for the aviation command to plan for the rotation of excess units and personnel to the States, and the redeployment began when transportation became available. The strength of AirFMFPac on 1 September 1945 totalled 43,819 Marines; 13 months later, this force had been reduced 90 percent to a total of 4,693. The loss of experienced personnel created as much of a problem in the air units as it had in the ground organizations. As demobilization progressed, approximately 80 percent of the replacements sent to AirFMFPac to fill in the gaps were inexperienced and untrained insofar as the technical requirements of that command were concerned. As a result, the squadrons and groups soon reached the point where they had insufficient numbers of experienced key personnel, not only for maintaining operational functions but also for training the inexperienced Marines. This situation was eased somewhat when the Commandant immediately ordered overseas a large number of fully qualified noncommissioned officers. As a result, AirFMFPac was able to continue essential operations with some degree of efficiency and, in effect, improve them as the new personnel were trained and became more fully qualified.

A second personnel problem engendered by demobilization was the

[45] The material concerning the formation and disbandment of the 1st Special Marine Brigade was derived from *CMC Rpts*, 1946 and 1947; *G–3 OpDs*, Jan–Jul46.

acute shortage of personnel to fill the jobs formerly held by aviation ground officers. During the war, most aviation ground billets were filled by non-flying officers specifically trained for those positions which could be filled by a ground officer. This practice permitted the full-time assignment of pilots to flying duties. Since the postwar aviation T/Os did not include provisions for the assignment of ground officers to the wings and their subordinate commands, those organizations had to set up extensive training programs to requalify pilots for the ground billets. Once they were trained in and fulfilling the functions of communications, ordnance, engineering, and air combat intelligence officers, the pilots continued regular flight duties in order to maintain proficiency in their primary assignment.

Owing to the constant transfer of pilots and skilled maintenance personnel, the operations of the fighter squadrons were reduced to a bare minimum. Even harder hit were the transport squadrons, whose pilots were released faster than those of the other tactical squadrons. When the workloads of the transport squadrons increased greatly because they were employed to carry supplies to the units in the Western Pacific and the Far East—and to transfer personnel scheduled for discharge back to the States—the demand for the services of transport pilots increased accordingly. To meet the demand, pilots qualified to fly other types of aircraft were transferred from overseas and Stateside tactical squadrons to the transport units for immediate retraining.

Concurrent with the release and transfer of aviation personnel were the deactivation and transfer of many major and subordinate aircraft organizations. On 31 December 1945, the 3d Marine Aircraft Wing was decommissioned at Ewa. The 2d MAW transferred from Okinawa to Cherry Point, where it was to be based, on 15 February 1946. The next month, on 13 March, the 4th MAW closed its headquarters on Guam and departed for the west coast, where it was to be based. The units attached to these Stateside-bound wings were transferred to the headquarters of other Marine organizations in the islands, and returned to the United States when transportation became available for their redeployment.

Completing the postwar roll-up of Marine aviation in the Pacific was the departure of MAG–31 from Yokosuka for the United States on 20 June. The 1st MAW remained in China with one transport and two observation squadrons and MAG–24, which was composed of three VMFs and one VMF(N). MAG–15, the only Marine aviation command remaining in the Hawaiian Islands, was based at Ewa with two VMRs; the group also had a fighter squadron based at Midway. With its peacetime functions stabilized by October 1946, AirFMFPac operations in the latter part of the year consisted mainly of training replacements and routing trans-Pacific supply and replacement personnel passenger flights.

In the 22 months following the signing of the Japanese surrender at Tokyo Bay, the strength of the Marine Corps was reduced from a peak of 485,837 on 1 September 1945 to a low of 92,222 on

30 June 1947. While this decrease represented a drop of 82 percent, the 1947 figure was considerably greater than 28,277, the size of the Corps on 1 July 1940. Although some problems occurred during the transition from war to peace, Marine units adjusted to the various situations with which they were confronted and continued to operate on a relatively high level of efficiency.

Made in the USA
Monee, IL
19 October 2025

32342071R00267